Breadth in Contemporary Topology

Proceedings of Symposia in
PURE MATHEMATICS

Volume 102

Breadth in Contemporary Topology

2017 Georgia International
Topology Conference
May 22–June 2, 2017
University of Georgia, Athens, Georgia

David T. Gay
Weiwei Wu
Editors

AMERICAN
MATHEMATICAL
SOCIETY
Providence, Rhode Island

2010 *Mathematics Subject Classification*. Primary 57-06, 20F36, 53D35, 53D12, 55P48, 57M25, 57M27, 57R17, 57R91.

The 2017 Georgia International Topology Conference was supported by a grant from the National Science Foundation (grant DMS-1719320). Any opinions, findings, and conclusions or recommendations expressed in this material are those of the authors and do not necessarilly reflect the views of the National Science Foundation.

Library of Congress Cataloging-in-Publication Data

Names: Georgia International Topology Conference (2017: University of Georgia) | Gay, David T., 1968– editor. | Wu, Weiwei, 1983– editor.

Title: Breadth in contemporary topology: 2017 Georgia International Topology Conference, May 22–June 2, 2017, University of Georgia, Athens, Georgia / David T. Gay, Weiwei Wu, editors.

Description: Providence, Rhode Island: American Mathematical Society, [2019] | Series: Proceedings of symposia in pure mathematics; volume 102 | Includes bibliographical references.

Identifiers: LCCN 2019001949 | ISBN 9781470442491 (alk. paper)

Subjects: LCSH: Topology–Congresses. | Homology theory–Congresses. | Manifolds (Mathematics)–Congresses. | AMS: Manifolds and cell complexes – Proceedings, conferences, collections, etc. msc | Group theory and generalizations – Special aspects of infinite or finite groups – Braid groups; Artin groups. msc | Differential geometry – Symplectic geometry, contact geometry – Global theory of symplectic and contact manifolds. msc | Differential geometry – Symplectic geometry, contact geometry – Lagrangian submanifolds; Maslov index. msc | Algebraic topology – Homotopy theory – Loop space machines, operads. msc | Manifolds and cell complexes – Low-dimensional topology – Knots and links in msc | Manifolds and cell complexes – Low-dimensional topology – Invariants of knots and 3-manifolds. msc | Manifolds and cell complexes – Differential topology – Symplectic and contact topology. msc | Manifolds and cell complexes – Differential topology – Equivariant algebraic topology of manifolds. msc

Classification: LCC QA611.A1 G456 2017 | DDC 514/.34–dc23

LC record available at https://lccn.loc.gov/2019001949

DOI: https://doi.org/10.1090/pspum/102

Color graphic policy. Any graphics created in color will be rendered in grayscale for the printed version unless color printing is authorized by the Publisher. In general, color graphics will appear in color in the online version.

♾ The paper used in this book is acid-free and falls within the guidelines
established to ensure permanence and durability.
Visit the AMS home page at `https://www.ams.org/`

10 9 8 7 6 5 4 3 2 1 24 23 22 21 20 19

Contents

Preface

Since 1961 the University of Georgia has organized and hosted an annual week-long topology conference and a major international topology conference every eight years. The tradition started in 1961 with a memorable four-week conference, the proceedings of which, *The Topology of 3-Manifolds and Related Topics*, were published by Prentice Hall. The 2017 Georgia International Topology Conference, continuing this long tradition, was held at the University of Georgia in Athens, Georgia, from May 22 to June 2, 2017. The conference had 41 speakers, ranging from the leading figures to talented young mathematicians, who covered a wide spectrum of the topics in topology and attracted a total of over 190 participants.

The main goals of our conference are to give wide exposure to new and important results and to encourage interaction among graduate students and researchers in different stages of their careers. Toward this end, we encouraged speakers to aim their talks at a broad audience of topologists, and we also scheduled a series of expository lectures that introduced graduate students to several of the themes of the conference.

The papers in this volume were written by both speakers and participants of the conference and cover topics ranging from symplectic topology to classical knot theory to the topology of 3- and 4-dimensional manifolds to geometric group theory. Several papers focus on open problems, especially the paper of Margalit on mapping class groups of surfaces. Other papers present new and insightful proofs of classical results, such as Pardon's paper on groups which can and cannot act on 2-manifolds. Still other papers lay out rich and technical foundational material or survey material presented in more depth elsewhere. Taken as a whole, this volume captures the spirit of the conference, both in terms of public lectures and informal conversations, and presents a sampling of some of the great new ideas generated in topology over the preceding eight years.

The conference was organized by David Gay, Jean Gutt, Will Kazez, Gordana Matic, Juanita Pinzón-Caicedo, Adam Saltz, Michael Usher, and Weiwei Wu. The Scientific Advisory Committee, consisting of Ian Agol, Yakov Eliashberg, David Gabai, Dusa McDuff, Tomasz Mrowka, Peter Ozsvath, Peter Teichner, and Karen Vogtmann, provided great assistance selecting speakers. The logistical support of the conference was due to the hard work of Julie McEver, Connie Poore, Gail Suggs, Laura Ackerley, and Christy McDonald. Finally, the organizers are grateful to the generous support of the National Science Foundation (grant DMS-1719320) and to the University of Georgia.

D. G.
W. W.

Proceedings of Symposia in Pure Mathematics
Volume **102**, 2019
https://doi.org/10.1090/pspum/102/01806

Structure of the flow and Yamada polynomials of cubic graphs

Ian Agol and Vyacheslav Krushkal

ABSTRACT. We establish a quadratic identity for the Yamada polynomial of ribbon cubic graphs in $\mathbb{R}^3$, extending the Tutte golden identity for planar cubic graphs. An application is given to the structure of the flow polynomial of cubic graphs at zero. The golden identity for the flow polynomial is conjectured to characterize planarity of cubic graphs, and we prove this conjecture for a certain infinite family of non-planar graphs.

Further, we establish exponential growth of the number of chromatic polynomials of planar triangulations, answering a question of D. Treumann and E. Zaslow. The structure underlying these results is the chromatic algebra, and more generally the SO(3) topological quantum field theory.

1. Introduction

Using the interplay between classical and quantum polynomials of graphs and ideas from topological quantum field theory (TQFT), we establish results on the structure of the Yamada and flow polynomials of cubic graphs. It has been known since the work of Birkhoff and Lewis in the 1940s [**3**] that the values $(3 \pm \sqrt{5})/2$ of the parameter play a special role in the theory of the chromatic polynomial χ_T of planar triangulations. In a series of papers [**18, 19**] in the 1960s, Tutte established further remarkable properties, including the golden identity. Formulated dually in terms of the flow polynomial F_G of planar cubic graphs G, it reads

$$F_G(\phi+2) = \phi^E \, F_G(\phi+1)^2, \tag{1.1}$$

where E is the number of edges of G and ϕ denotes the golden ratio $(1+\sqrt{5})/2$. The special role played by $\phi+1$, and more generally by the Beraha numbers [**2**] $B_n = 2 + 2\cos(2\pi/n)$, was conceptually explained in [**6**] where these results were placed in the context of SO(3) TQFT.

We show in Theorem 3.1 that Tutte's identity (1.1) admits an extension to the Yamada polynomial R_G of ribbon cubic graphs G in $\mathbb{R}^3$:

$$R_G(e^{\pi i/5}) = (-1)^{V-E} \, \phi^E \, R_G(e^{-2\pi i/5})^2, \tag{1.2}$$

2010 *Mathematics Subject Classification.* Primary 05C31; Secondary 05C10, 57M27, 57R56.

The first author was supported by NSF grant DMS-1406301, a Simons Investigator Award, and the Institute for Advanced Study (in particular the Simonyi Fund and the Fund for Mathematics).

The second author was supported in part by NSF grant DMS-1612159.

where V, E denote the number of vertices and edges of G, respectively. In fact, (1.2) is a common extension of (1.1) and of the identity for links, relating the 2-colored Jones polynomial at $e^{\pi i/10}$ and the square of the Jones polynomial at $e^{-\pi i/5}$ [**12**, Corollary 4.16], see section 3.

The Yamada polynomial [**23**] is a quantum invariant of ribbon graphs [**14**] in $\mathbb{R}^3$, corresponding to the adjoint representation of $U_q(\mathfrak{so}_3)$. Conceptually, as discussed in [**6**, Section 5] and also [**12**, Section 4.4], the reasons underlying the golden identity are the level-rank duality between the $\mathrm{SO}(3)_4$ and the $\mathrm{SO}(4)_3$ TQFTs, and the isomorphism $\mathfrak{so}(4) \cong \mathfrak{so}(3) \times \mathfrak{so}(3)$.

Concretely, the Yamada polynomial is defined by the contraction-deletion rule and the $SO(3)$ Kauffman skein relation, see section 2.1 for details. For *planar* graphs G, the Yamada polynomial coincides with a renormalization of the flow polynomial:

$$F_G(Q) = (-1)^{V-E} R_G(q), \text{ where } Q = q + 2 + q^{-1}.$$

For non-planar graphs (and for knotted embeddings of planar graphs) the Yamada polynomial carries a lot of information about the embedding of a ribbon graph in 3-space, and so in general the Yamada polynomial of a ribbon graph and the flow polynomial of the underlying abstract graph are quite different.

In contrast with (1.1), in [**1**] we formulated a conjecture that the Tutte golden identity for the flow polynomial characterizes planarity of cubic graphs:

CONJECTURE 1.1. *For any cubic bridgeless graph G,*

$$(-\phi)^E F_G(\bar{\phi}+2) \geq F_G(\bar{\phi}+1)^2, \tag{1.3}$$

where $\bar{\phi}$ denotes $(1-\sqrt{5})/2$. Moreover, G is planar if and only if (1.3) is an equality.

Similarly, numerical evidence points to the inequality

$$F_G(\phi+2) \leq \phi^E F_G(\phi+1)^2, \tag{1.4}$$

for any cubic graph G, with an equality if and only if G is planar. In section 5 we prove the conjecture for a family of near-planar graphs (which have a planar projection with a single crossing); Conjecture 1.1 in general remains open.

It is interesting to note that as a consequence of (1.1), there is a relation between the values of the flow polynomial of planar cubic graphs at 0 and 4:

$$F_G(0) \equiv 3^E F_G(4)^2 \pmod 5, \tag{1.5}$$

see Lemma 4.1. More generally, a congruence (mod 5) between the values $R_G(-1)$, $R_G(1)$ implies an extension of (1.5) to *all* cubic graphs,

$$F_G(0) \equiv 3^E R_G(1)^2 \pmod 5, \tag{1.6}$$

where the value $R_G(1)$ is also known as the Penrose number of G (defined up to a sign), see section 4. Using these relations, we give an application to the structure (mod 5) of the flow polynomial at zero for cubic graphs. This value is known [**4**] to count Eulerian equivalence classes of totally cyclic orientations.

THEOREM 1.2. *Let G be a cubic graph with V vertices. Then the value of the flow polynomial of G at zero satisfies the following.*

$$F_G(0) \equiv 0, 1, 4 \pmod 5 \ \text{ if } \ V/2 \ \text{ is even,} \tag{1.7}$$

$$F_G(0) \equiv 0, 2, 3 \pmod 5 \ \text{if} \ V/2 \ \text{is odd.} \tag{1.8}$$

Moreover, suppose G is a snark, that is a bridgeless cubic graph with chromatic index 4. *Then* $F_G(0) \equiv 0 \pmod 5$, *and therefore* $F_G(0)$ *is divisible by* 120.

The proofs of the results in this paper are given in the context of the chromatic algebra [**6**], and more generally the SO(3) TQFT. Using these methods, we also answer a question of D. Treumann and E. Zaslow [**15**] about the asymptotics of the number of chromatic polynomials of planar triangulations, motivated by their work on Legendrian surfaces [**16**].

THEOREM 1.3. *The number of chromatic polynomials of planar triangulations with n vertices grows exponentially in n.*

An exponential upper bound is well-known, and it can be deduced from Tutte's enumeration of planar triangulations [**17**]. We show that the chromatic algebra contains a free semigroup, yielding an exponential lower bound, see section 6. Conceptually, this may be viewed as an application of the Tits alternative for semigroups.

The key ingredients used throughout the paper – the chromatic algebra and the flow category – are recalled in section 2. The identity (1.2) for the Yamada polynomial is established in section 3. Section 4 discusses the structure of the flow polynomial at zero (mod 5) and gives a proof of Theorem 1.2. The proof of Conjecture 1.1 for a collection of near-planar graphs is given in section 5. Since the original Tutte golden identity (1.1) serves as the motivation for several results in this paper, for convenience of the reader we include its proof in section 7.

2. Graph polynomials, algebras and categories

This section summarizes the relevant background material and notation used in the paper.

2.1. The chromatic and flow polynomials. The flow polynomial $F_G(Q)$ of a graph G satisfies the contraction-deletion rule: given an edge e of G which is not a bridge,

$$F_G(Q) = F_{G/e}(Q) - F_{G\smallsetminus e}(Q). \tag{2.1}$$

If G contains a bridge, $F_G \equiv 0$. The flow polynomial of a graph consisting of a single vertex and n loops is defined to be $(Q-1)^n$, and the polynomial is multiplicative with respect to taking the disjoint union.

For planar graphs G, the flow polynomial is essentially the chromatic polynomial χ_{G^*} of the dual graph G^*:

$$F_G(Q) = Q^{-1}\chi_{G^*}(Q), \tag{2.2}$$

The flow polynomial was defined by Tutte [**20**], and in fact both the chromatic and flow polynomials are specializations of the 2-variable Tutte polynomial $T_G(x, y)$; (2.2) is a special case of the duality relation satisfied by the Tutte polynomial: $T_G(x, y) = T_{G^*}(y, x)$. Like the chromatic polynomial, the flow polynomial at positive integers admits a well-known combinatorial interpretation: for $n \in \mathbb{Z}_+$, $F_G(n)$ is the number of nowhere-zero flows with values in an abelian group of order n, cf. [**4**].

2.2. The Yamada polynomial [**23**] $R_G(q)$ is an invariant of spatial ribbon graphs G, i.e. ribbon graphs embedded in $\mathbb{R}^3$. A *ribbon* graph is an abstract graph G with an embedding into a surface S, so that the complement $S \smallsetminus G$ is a disjoint union of 2-cells. Given such an embedding, a neighborhood of G in S is a compact surface with boundary, which may be thought of as a choice of a 2-dimensional thickening of G. Such a thickening of vertices and edges may be also encoded using cyclic ordering of half-edges incident to each vertex.

The Yamada polynomial is defined using the contraction-deletion rule:

$$R_G(Q) = R_{G/e}(Q) + R_{G\smallsetminus e}(Q), \tag{2.3}$$

and a version of the $SO(3)$ Kauffman skein relations (cf. [**7**, Section 5]) applied to a planar projection of G:

$$R_{G \amalg \bigcirc} = (q + 1 + q^{-1})\, R_G, \tag{2.4}$$

(2.5) 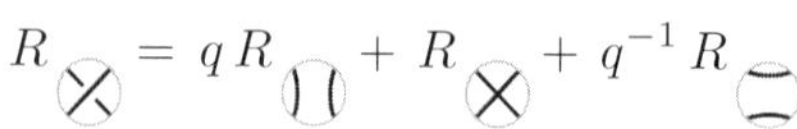

(2.6) 

As usual in skein-theoretic definitions, the graphs in each of these equations differ as shown, and are identical outside of the disk. If a planar diagram of G has a bridge, R_G is set to be zero. The Yamada polynomial is multiplicative under taking disjoint union, and $R_{G\vee H} = -R_G R_H$. R_G is an invariant of the isotopy class of an embedding of the ribbon graph G into $\mathbb{R}^3$. (In terminology of [**23**, Theorem 2], R_G is a *regular deformation* invariant of a planar diagram of G; in [**11**] this equivalence relation is called *rigid vertex isotopy*.)

Using just equations (2.4) – (2.6), one gets $(-1)^{V-E}$ times the SO(3) Kauffman polynomial of 4-regular ribbon graphs in $\mathbb{R}^3$, where V, E denote the number of vertices and edges of the graph. It is worth noting that usually (cf. [**11**], [**7**]) in the definition of the SO(3) Kauffman polynomial the term corresponding to the 4-valent vertex in equations (2.5), (2.6) has a minus sign. We follow the convention used in the literature on the Yamada polynomial, and as mentioned above the two polynomials are related by an overall sign $(-1)^{V-E}$.

Therefore the Yamada polynomial may be thought of as a renormalized version of the SO(3) Kauffman polynomial, extended to spatial ribbon graphs of arbitrary vertex degree using the contraction-deletion rule (2.3).

If G is *planar*, then there are no crossings to resolve, and the Yamada polynomial is determined by the contraction-deletion rule and its loop value (2.4); in this case $F_G(Q) = (-1)^{V-E} R_G(q)$, where $Q = q + 2 + q^{-1}$. In general, the Yamada polynomial carries a lot of information about the embedding of G into $\mathbb{R}^3$. For example, tying a knot into an edge of G results (up to a normalization) in multiplication of R_G by the SO(3) invariant of the knot. Therefore, in general the Yamada polynomial of a ribbon graph is quite different from the flow polynomial of the underlying abstract graph.

To describe the TQFT context for the results of this paper, next we give a brief summary of the relevant material on the Temperley-Lieb algebra, the chromatic algebra and their structure at roots of unity.

2.3. The Temperley-Lieb algebra. The Temperley-Lieb algebra, TL_n, is an algebra over $\mathbb{C}[d]$ consisting of linear combinations of 1-dimensional submanifolds, considered up to isotopy rel boundary, in a rectangle. Each submanifold meets both the top and the bottom of the rectangle in n points. By definition, deleting a simple closed curve has the effect of multiplying the element by d. Often d will be specialized to a complex number, and in this case the algebra will be denoted TL_n^d. The multiplication is given by vertical stacking of rectangles. The standard generators of TL_4 are shown in Figure 1.

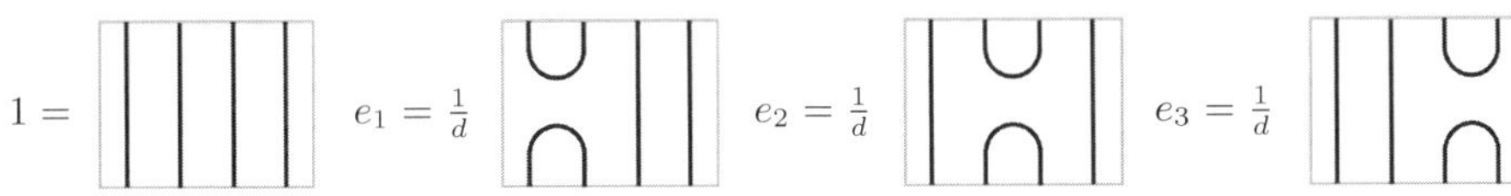

FIGURE 1. Generators of TL_4

The trace $\mathrm{tr}_d \colon \mathrm{TL}_n^d \longrightarrow \mathbb{C}$ is defined on rectangular pictures by connecting the top and bottom endpoints by disjoint arcs in the complement of the rectangle in the plane, and then evaluating $d^{\#\text{circles}}$. The Hermitian product is defined by $\langle a, b\rangle = \mathrm{tr}(a\,\bar{b})$, where the involution $^-$ reflects pictures in a horizontal line and replaces coefficients with their complex conjugates.

For special values of the parameter d, $d = 2\cos\left(\frac{\pi k}{n+1}\right)$, where $k, n+1$ are coprime, TL_n^d contains a non-trivial ideal, the *trace radical* consisting of the elements a such that $tr(ab) = 0$ for all $b \in \mathrm{TL}_n^d$. This ideal is generated by the *Jones-Wenzl projector* $P^{(n)}$ [**10**, **22**]; this fact will be interpreted as a local relation for the flow polynomial of planar graphs in section 2.5. At primitive roots of unity ($k = 1$) the Hermitian product descends to a positive definite inner product on the quotient of TL_n^d by the trace radical.

2.4. The chromatic algebra $\mathcal{C}_n$. The Temperley-Lieb algebra, discussed above, underlies the construction of SU(2) TQFT. Next we briefly summarize the definition and properties of the chromatic algebra, $\mathcal{C}_n$, corresponding to SO(3) TQFT; we refer the reader to [**6**, Section 2] for more details.

$\mathcal{C}_n$ is an algebra over $\mathbb{C}[Q]$, whose elements are formal linear combinations of *planar* cubic graphs (considered up to isotopy rel boundary) in a rectangle, modulo the local relations shown in Figure 2. The first relation is the analogue of the contraction-deletion rule for cubic graphs.

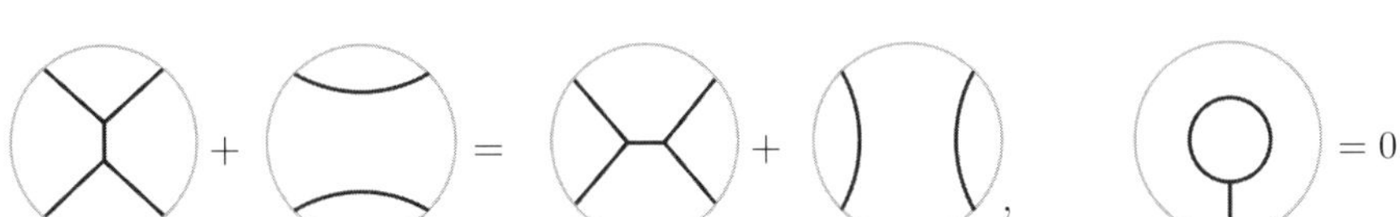

FIGURE 2. Relations defining the chromatic algebra.

The intersection of a graph with the boundary of the rectangle consists of $2n$ points: n points both at the top and the bottom, Figure 3. It is convenient to allow 2-valent vertices as well, and deleting a simple closed curve has the effect of multiplying the element by $Q-1$. When Q is specialized to a complex number, the algebra is denoted $\mathcal{C}_n^Q$.

FIGURE 3. Examples of graphs in $\mathcal{C}_3$.

The trace $tr\colon \mathcal{C}_n^Q \longrightarrow \mathbb{C}$ is defined by connecting the endpoints of G by disjoint arcs in the complement of the rectangle in the plane and evaluating the flow polynomial of the resulting graph at Q. (Or equivalently, the trace equals Q^{-1} times the chromatic polynomial of the dual graph.) The trace is well-defined since the local relations in Figure 2 are precisely the relations defining the flow polynomial of a planar cubic graph. The multiplication and the Hermitian product on $\mathcal{C}_n^Q$ are defined analogously to those in the Temperley-Lieb algebra.

There are two variations of the definition of the chromatic algebra, which are going to be useful. First, rather than using just cubic graphs modulo relations in Figure 2, $\mathcal{C}_n$ may also be defined using graphs with arbitrary vertex degrees, modulo the contraction-deletion rule, see [**6**, Section 5].

Second, instead of using planar graphs, one may consider ribbon graphs in the cylinder $D^2 \times [0,1]$, with n endpoints both at $D^2 \times 0$ and at $D^2 \times 1$, modulo the defining relations of the Yamada polynomial, (2.3) - (2.6). This is closely related to the definition of the SO(3) BMW algebra (cf. [**7**, Section 5]). Using the relations

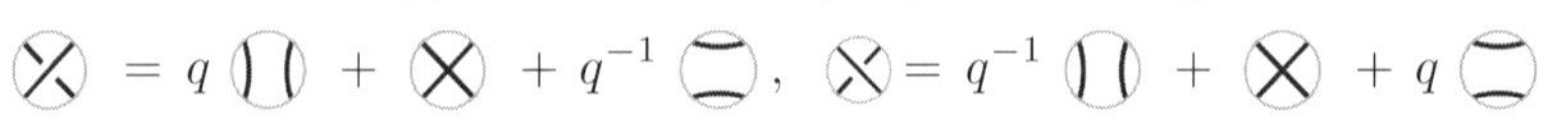

all crossings of a ribbon graph in the cylinder may be resolved to give an element of $\mathcal{C}_n$.

2.5. The map $\mathcal{C}_\mathbf{n} \longrightarrow \mathrm{TL}_{\mathbf{2n}}$. Consider the homomorphism $\Phi\colon \mathcal{C}_n^Q \longrightarrow \mathrm{TL}_{2n}^d$, where $Q=d^2$, replacing each edge of a graph with the second Jones-Wenzl projector, and resolving each vertex as shown in Figure 4. Moreover, for a trivalent graph G there is an overall factor $d^{V/2}$, where V is the number of vertices of G.

FIGURE 4. Definition of the homomorphism $\Phi\colon \mathcal{C}_n^Q \longrightarrow \mathrm{TL}_{2n}^d$, $Q=d^2$.

Φ induces a well-defined homomorphism of algebras $\mathcal{C}_n^Q \longrightarrow \mathrm{TL}_{2n}^d$, where $Q=d^2$, and moreover it is trace-preserving: the diagram

$$\begin{array}{ccc} \mathcal{C}_n^{d^2} & \xrightarrow{\ \Phi\ } & \mathrm{TL}_{2n}^d \\ \Big\downarrow{\scriptstyle tr} & & \Big\downarrow{\scriptstyle tr_d} \\ \mathbb{C} & \xrightarrow{\ =\ } & \mathbb{C} \end{array} \tag{2.7}$$

commutes [**6**, Lemmas 2.4, 2.5]. It follows that the pullback under Φ of the trace radical in TL^d is in the trace radical of $\mathcal{C}^{d^2}$. As mentioned in section 2.3, the trace radical in TL^d is non-trivial precisely for the values $d = 2\cos\left(\frac{\pi k}{n+1}\right)$. The elements of the trace radical in the chromatic algebra $\mathcal{C}^{d^2}$ are local relations (which hold in addition to the contraction-deletion rule) on graphs which preserve the flow polynomial, or equivalently the chromatic polynomial of the dual graph. When $d = \phi$, the trace radical of the Temperley-Lieb algebra is generated by the Jones-Wenzl projector $P^{(4)}$. In particular, the linear Tutte relation in $\mathcal{C}_4^Q$, where $Q = \phi + 1 = \phi^2$,

(2.8) 

may be seen as a consequence of the structure of TL^ϕ since it is mapped by Φ to $P^{(4)}$ [**6**, Section 2, p. 721]. Similarly, the linear relation

(2.9) 

holds in $\mathcal{C}_4^Q$, at the Galois conjugate value $Q = (3 - \sqrt{5})/2$.

2.6. The flow category. To study the flow polynomial of non-planar graphs, one may use abstract (not necessarily planar) graphs to define an algebra along the lines of the Temperley-Lieb and chromatic algebras, cf. [**21**]. Unlike the chromatic algebra case, here one does not have a map to the Temperley-Lieb algebra, and the values $Q = d^2$, $d = 2\cos\left(\frac{\pi}{n+1}\right)$ do not play a special role since these are artifacts of planarity. We will not use the algebra structure, the notion of a category is more suitable for our applications. We summarize the construction of the *flow category* below, following [**1**, Section 3.1].

The objects of the flow category are finite ordered sets $\overline{n} = \{1, \ldots, n\}$. Consider $\mathcal{G}_{m,n} = \{$finite graphs with $m + n$ marked univalent vertices$\}$, where the marked vertices are divided into two ordered subsets of m, respectively n vertices. The edges incident to the marked vertices are called boundary edges and the rest are *internal* edges.

The space of morphisms $\mathcal{F}^Q_{m,n}$ in the flow category between $\overline{m}$, $\overline{n}$ consists of formal $\mathbb{C}$-linear combinations of graphs $\mathcal{G}_{m,n}$, modulo the contraction-deletion relation which applies to internal edges, Figure 5. Graphs whose equivalence classes are elements of $\mathcal{F}^Q_{m,n}$ may be represented geometrically as in Figure 5. (It is important to note that unlike in sections 2.2, 2.4, over/under-crossings do not carry any information here since the figure represents the abstract graph structure and not a specific planar projection.) In addition, the loop value is set to be $Q - 1$, and graphs with a univalent vertex (other than the specified marked vertices) are set to be zero.

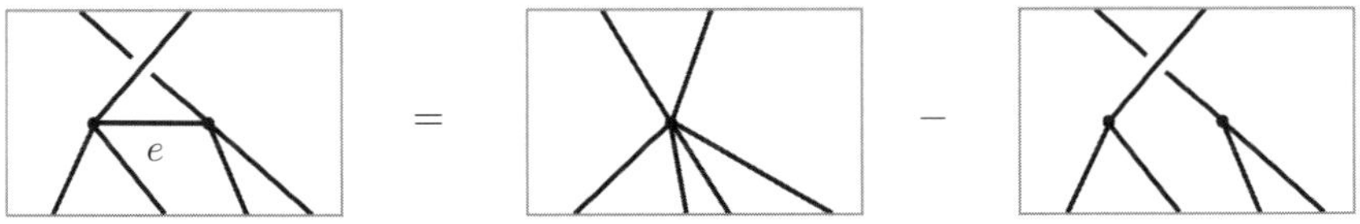

FIGURE 5. The contraction-deletion rule in $\mathcal{F}_{4,2}$.

A graph without marked vertices, and therefore no boundary edges, considered in $\mathcal{F}^Q_{0,0} \cong \mathbb{C}$, evaluates to its flow polynomial at Q. The pairing $\mathcal{F}^Q_{k,m} \times \mathcal{F}^Q_{m,n} \longrightarrow \mathcal{F}^Q_{k,n}$ is obtained by gluing along m boundary edges. For example, this pairing applied to two graphs $A \in \mathcal{G}_{0,m}$, $B \in \mathcal{G}_{m,0}$ gives $\langle A, B\rangle =$ the value of the flow polynomial $F_{A\cup B}(Q)$.

Given any graph representing an element of $\mathcal{F}^Q_{m,n}$, the contraction-deletion rule may be used to eliminate all internal edges. For example, the four graphs in Figure 6 form a basis of $\mathcal{F}^Q_{0,4}$. Three of these graphs, viewed relative to a fixed embedding of the marked vertices in the boundary, are planar, and one, denoted e_4, is non-planar.

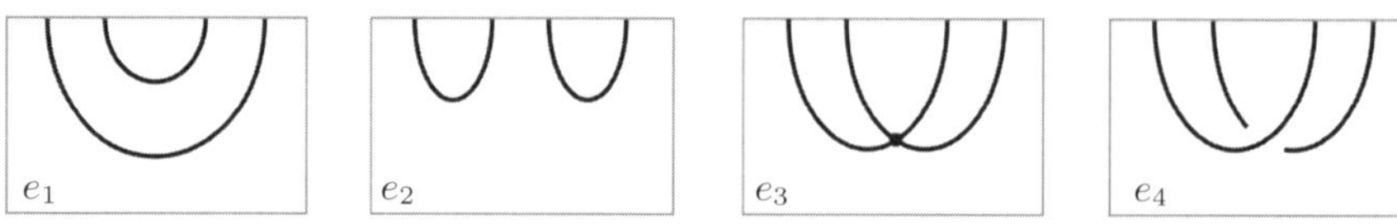

FIGURE 6. A basis of $\mathcal{F}^Q_{0,4}$.

It is convenient to introduce the notation $\mathcal{C}^Q_{m,n}$ for the planar analogue of $\mathcal{F}^Q_{m,n}$. That is, $\mathcal{C}^Q_{m,n}$ consists of formal $\mathbb{C}$-linear combinations of *planar* graphs with two ordered subsets of m, respectively n marked vertices, modulo the contraction-deletion relation which applies to internal edges.

2.7. Conventions and notation. Unless stated otherwise, in the following sections the flow polynomial of *planar* graphs and the Yamada polynomial of spatial ribbon graphs will be considered in the context of the chromatic algebra. On the other hand, the flow polynomial of abstract (non-planar) graphs does not fit in this context, and it will be studied in the setting of the flow category.

It is convenient to introduce a short-hand notation for the evaluation of graph polynomials. For example, when working with the Yamada polynomial of a graph G with a specified crossing , the notation $_x$ will stand for the evaluation of $R_G(x)$. Similarly, given two graphs G_1, G_2 in a disk with the same number of marked points on the boundary, the notation $\langle G_1, G_2\rangle_Q$ will stand for the relevant pairing. For example, in the context of the flow category, it will mean the value of the flow polynomial $F_{G_1\cup G_2}$ at Q, where the union of G_1, G_2 is taken along the marked points on the boundary.

3. The golden identity for the Yamada polynomial

The purpose of this section is to prove the extension (1.2) of the Tutte golden identity (1.1) to the Yamada polynomial of cubic ribbon graphs in $\mathbb{R}^3$. It is convenient to allow vertices of degree 2 in the statement of the theorem:

THEOREM 3.1. *Let G be a ribbon graph in $\mathbb{R}^3$, with vertices of degrees 2 and 3. Then*

$$R_G(e^{\pi i/5}) = (-1)^{V-E}\,\phi^{E'}\,R_G(e^{-2\pi i/5})^2, \tag{3.1}$$

where $E' = V_3 - \chi(G)$, V_3 is the number of trivalent vertices of G, $\chi(G)$ is the Euler characteristic, and $\phi = (1+\sqrt{5})/2$.

If there are no vertices of degree 2, then $E' = E$, and (3.1) is the same identity as (1.2). Moreover, since both $\chi(G)$ and $V_3(G)$ are topological invariants of G, introducing new 2-valent vertices (i.e. subdividing edges of G) does not affect (3.1).

In the special case where G has no trivalent vertices, (3.1) gives an identity for the SO(3) Kauffman polynomial of framed *links* in $\mathbb{R}^3$, where the factor $(-1)^{V-E}\,\phi^{E'}$ equals 1. Then the left-hand side of (3.1) may be interpreted as the 2-colored Jones polynomial of the link, and the right-hand side equals the square of the Jones polynomial. Therefore for links, (3.1) matches the identity stated in [**12**, Corollary 4.16].

PROOF OF THEOREM 3.1. Recall from section 2.2 that if G is *planar*, one has $F_G(Q) = (-1)^{V-E}\, R_G(q)$, where $Q = q+2+q^{-1}$. In particular, for planar graphs G,

$$R_G(e^{2\pi i/10}) = F_G(\phi+2), \;\; R_G(e^{-2\pi i/5}) = F_G(\phi+1).$$

Therefore in this case, (3.1) is reduced to the Tutte golden identity (1.1).[1] In the general case of a ribbon graph G in the statement of the theorem, the proof is by induction on the number c of crossings in a planar diagram of G. The base case corresponds to planar graphs, discussed above.

As in section 2.7, the notation  in the proof below will denote the evaluation of the Yamada polynomial at q.

By induction assume that graphs with fewer than c crossings satisfy (3.1). Consider a graph with c crossings. For brevity of notation denote $x := e^{2\pi i/10}$, $y := e^{-2\pi i/5}$. Combining the skein relation (2.5) with the contraction-deletion rule (2.3), one has

(3.2) 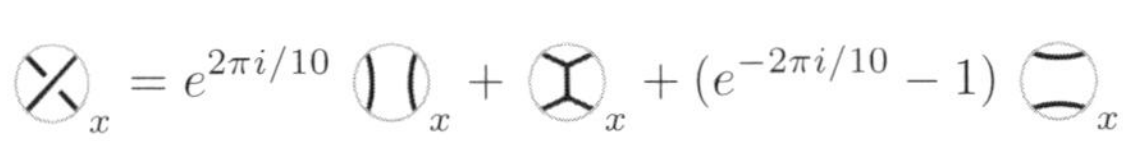

Since the three graphs on the right have $c-1$ crossings, (3.1) holds for them by the inductive assumption, thus

(3.3) 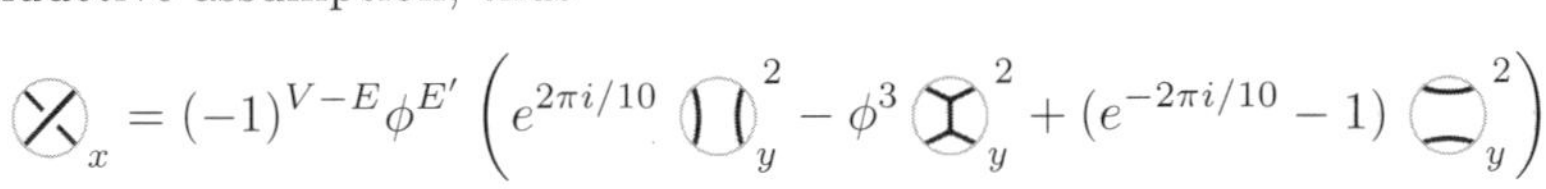

Because of the normalization sign $(-1)^{V-E}$ in the formula relating the flow and Yamada polynomials of planar graphs, the linear relation (2.8) at $q = e^{-2\pi i/5}$ (corresponding to $Q = \phi+1$) for the Yamada polynomial reads

(3.4) 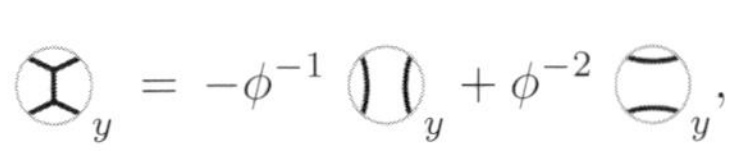

Applying this relation to  in (3.3) gives

(3.5) 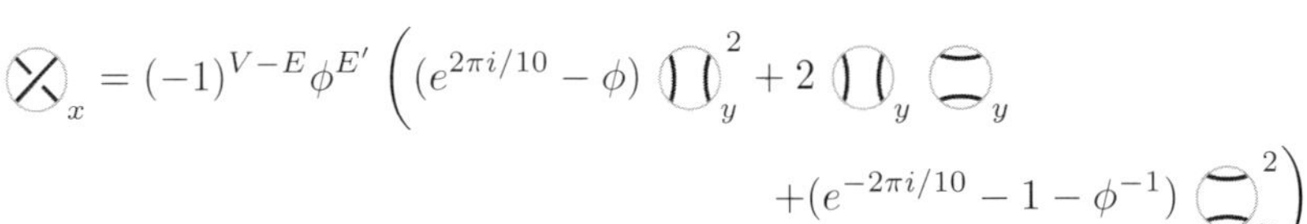

[1] Its proof is given in section 7.

To prove the inductive step, one needs to show that (3.5) $= (-1)^{V-E}\phi^{E'}$  ^2_y. Using the skein relation (2.5) and the contraction-deletion rule,

(3.6)

$(-1)^{V-E}\phi^{E'}$  $^2_y = (-1)^{V-E}\phi^{E'}\Big(e^{-2\pi i/5}$  $_y +$  $_y + (e^{2\pi i/5}-1)$  $_y\Big)^2.$

To complete the proof, replace  $_y$ in the formula above with (3.4), yielding

(3.7) $(-1)^{V-E}\phi^{E'}\Big((e^{-2\pi i/5}-\phi^{-1})$  $_y + (e^{2\pi i/5}-1+\phi^{-2})$  $_y\Big)^2$

$=(-1)^{V-E}\phi^{E'}\Big((y-\phi^{-1})^2$  $^2_y + 2(y-\phi^{-1})(y^{-1}-\phi^{-1})$  $_y$  $_y$

$+(y^{-1}-\phi^{-1})^2$  $^2_y\Big).$

Noting that $y+y^{-1}=\phi^{-1}$, one checks that the expression above matches (3.5):

coefficients of  $^2_y:\ x-\phi=-e^{-2\pi i/10}=e^{4\pi i/5}=(y-\phi^{-1})^2,$

coefficients of  $_y$  $_y:\ 2=2(y^{-1}-\phi^{-1})(y-\phi^{-1}),$

coefficients of  $^2_y:\ x^{-1}-1-\phi^{-1}=(y^{-1}-\phi^{-1})^2.$

□

4. Structure of the flow polynomial (mod 5)

In this section we establish the (mod 5) version of the golden identity, and prove a theorem stated in the introduction:

THEOREM 1.2. *Let G be a cubic graph with V vertices. Then the value of the flow polynomial of G at zero satisfies:*

(4.1) $$F_G(0)\equiv 0,1,4 \pmod 5 \text{ if } V/2 \text{ is even},$$

(4.2) $$F_G(0)\equiv 0,2,3 \pmod 5 \text{ if } V/2 \text{ is odd}.$$

Moreover, suppose G is a snark, that is a bridgeless cubic graph with chromatic index 4. Then $F_G(0)\equiv 0 \pmod 5$, and therefore $F_G(0)$ is divisible by 120.

REMARK 1. [**4**, Theorem 1.2] interpreted $|F_G(0)|$ as the number of Eulerian equivalence classes of totally cyclic orientations.

Proof of Theorem 1.2. We start by stating the (mod 5) version of the golden identity:

LEMMA 4.1. *For planar cubic graphs G,*

(4.3) $$F_G(0)\equiv 3^E\, F_G(4)^2 \pmod 5,$$

where E is the number of edges of G. More generally, let G be a ribbon cubic graph in $\mathbb{R}^3$. Then

(4.4) $$R_G(-1)\equiv (-1)^{V-E}\, 3^E\, R_G(1)^2 \pmod 5.$$

PROOF OF (4.3). The flow polynomial $F_G(Q)$ for $Q = (3-\sqrt{5})/2, (5-\sqrt{5})/2$ takes values in the ring $R = \mathbb{Z}[\frac{1+\sqrt{5}}{2}]$. Consider the ideal I generated by $\sqrt{5}$. Since $\phi\sqrt{5} = (5+\sqrt{5})/2 \in I$, it follows that $1+\phi \equiv 4 \pmod{\sqrt{5}}$, and $\phi \equiv 3 \pmod{\sqrt{5}}$. Plugging this into the golden identity (1.1), one gets $F_G(0) \equiv 3^E F_G(4)^2 \pmod{\sqrt{5}}$. Since $F_G(0)$ and $3^E F_G(4)^2$ are integers, the equivalence holds (mod 5). □

PROOF OF (4.4). The values of the Yamada polynomial $R_G(e^{\pi i/5}), R_G(e^{-2\pi i/5})$ are elements of $\mathbb{Z}[e^{\pi i/5}]$. Consider the ideal generated by $e^{\pi i/5}+1$. Modulo this ideal, $e^{\pi i/5}$ and $e^{-\pi i/5}$ are equivalent to -1, $e^{2\pi i/5}$ and $e^{-2\pi i/5}$ are equivalent to 1, and $\phi = e^{\pi i/5} + e^{-\pi i/5} \equiv 3$. □

REMARK 2. Another proof of Lemma 4.1 may be given by following (mod 5) the steps of the proofs of the golden identities (1.1), (3.1). For example, the (mod 5) version at $Q = 4$ of the Tutte linear relation (2.8) for the flow polynomial of planar graphs, 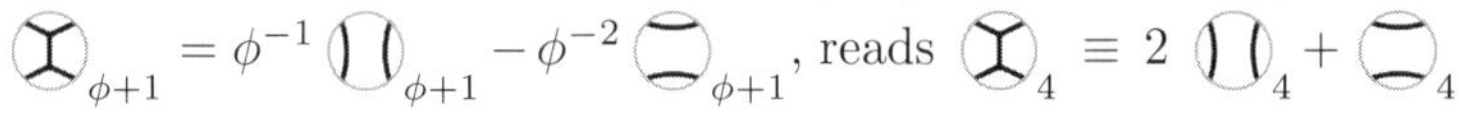 (mod 5).

Returning to the proof of Theorem 1.2, recall that for planar graphs G, $F_G(Q) = (-1)^{V-E} R_G(q)$, where $Q = q+2+q^{-1}$. In particular, for planar G, $R_G(-1) = (-1)^{V-E} F_G(0)$. The following identity holds for the flow polynomial of all (planar and non-planar) graphs:

(4.5) 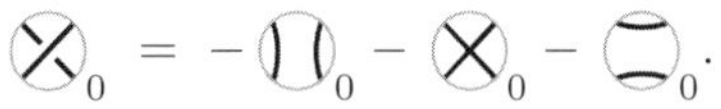

This identity is checked by pairing both sides of (4.5) with the four basis elements of the flow category $\mathcal{F}^Q_{0,4}$ in Figure 6. This relation coincides with the skein relation (2.5) for the Yamada polynomial, normalized by the factor $(-1)^{V-E}$, at $q=-1$. Thus $R_G(-1) = (-1)^{V-E} F_G(0)$ for *all* graphs G. Now it follows from (4.4) that for any abstract (planar or non-planar) cubic graph G,

$$F_G(0) \equiv (-1)^{V-E} 3^E R_G(1)^2 \pmod 5. \tag{4.6}$$

REMARK 4.1. Up to a sign, the value $R_G(1)$ of the Yamada polynomial of ribbon cubic graphs equals the *Penrose number* of G, introduced and studied in [**13**], [**9**, Section 2.3]. Indeed, the skein relation [**9**, Proposition 2], satisfied by the Penrose number, coincides with the version of the skein relation (2.5) for $(-1)^{V-E} R_G(1)$:

(4.7) 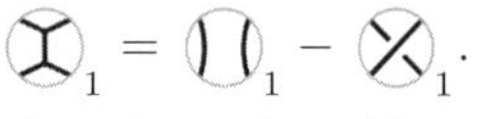

It is an invariant of ribbon graphs where the ribbon structure affects only its sign, so $|R_G(1)|$ is an invariant of abstract cubic graphs.

The congruence (4.6), the relation $3V = 2E$ for cubic graphs, and the fact that perfect squares are congruent to $0, 1$, or 4 (mod 5) conclude the proof of (4.1), (4.2).

To prove the last statement of the theorem, consider a planar diagram of a ribbon cubic graph G. In other words, the graph is immersed in the plane, with some edge crossings. The value $R_G(1)$ is independent of which strand is over/under in each crossing. Using the recursion relation (4.7) at $q=1$, one checks that $(-1)^{V-E} R_G(1)$ equals the signed count of 3-edge colorings of G, where the sign gets a (-1) factor for each pair of edges that cross with a different color. This is invariant of regular homotopy of the planar diagram, and it satisfies the skein relation: each 3-coloring of the graph on the left in (4.7) corresponds (with the same sign) to a 3-coloring of precisely one term on the right.

This implies that the Yamada polynomial of snarks at $q = 1$ is zero. (More generally, there is a well-known correspondence between 3-edge-colorings and 4-flows, cf. Proposition 6.4.5 (ii) in [**5**], so for any cubic graph $|R_G(1)| \leq F_G(4)$, with equality for planar graphs.) Now the congruence $F_G(0) \equiv 0 \pmod 5$ follows from (4.6). Finally, since the flow polynomial $F_G(Q)$ of a snark is divisible by $\prod_{k=1}^{4}(Q-k)$, $F_G(0)$ is also divisible by 4!. □

5. Golden inequality for non-planar cubic graphs

Unlike the golden identity for the Yamada polynomial, Theorem 3.1, which holds for any spatial cubic graph G, in [**1**] we stated a conjecture that the golden identity for the flow polynomial characterizes planarity:

CONJECTURE 5.1. *For any cubic bridgeless graph G,*

$$(-\phi)^E\, F_G((5-\sqrt{5})/2) \geq F_G((3-\sqrt{5})/2)^2, \tag{5.1}$$

Moreover, G is planar if and only if (5.1) *is an equality.*

In this section we develop methods to prove this conjecture for a certain family of non-planar cubic graphs. We will call a graph G *near-planar* if it admits a planar projection with a single crossing, for example $K_{3,3}$ in Figure 7. We will view such graphs as $G =$  $\cup\, \overline{G}$ where $\overline{G}$ is a planar graph in a disk (the complement in S^2 of the shaded disk in Figure 7) with 4 endpoints on the boundary circle.

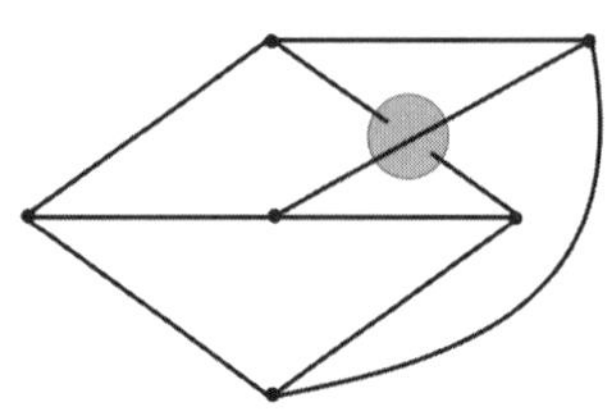

FIGURE 7

Considered as an element of the chromatic algebra $\mathcal{C}_2^Q$, $\overline{G}$ may be expressed as a linear combination of the three basis elements without internal edges. The coefficients depend on Q; denote by α, β, γ their values at $Q = (3-\sqrt{5})/2$, Figure 8. In the following lemma the graph $\overline{G}$ outside a disk will be fixed throughout the proof, and (as in section 2.7) the notation  $_Q$ will stand for the evaluation of the flow polynomial at Q.

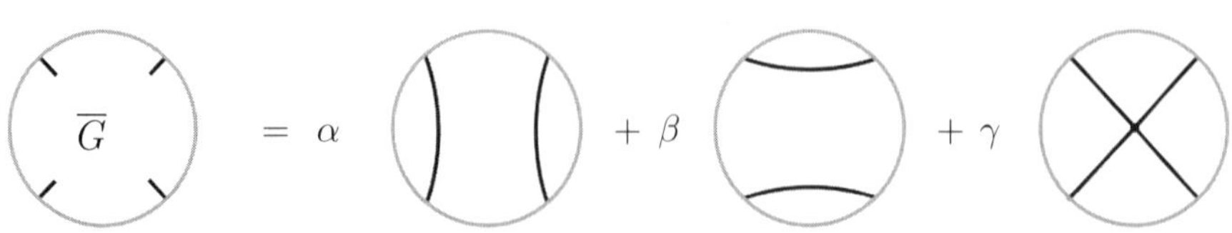

FIGURE 8. A planar cubic graph in $\mathcal{C}_2^Q$, $Q = (3-\sqrt{5})/2$, expressed as a linear combination of basis vectors.

LEMMA 5.2 (**Golden identity for the flow polynomial of near-planar graphs**). *Let* $G =$  $\cup \overline{G}$ *be a near-planar graph. Then the following identity holds for the evaluations of the flow polynomial of* G *and of the two planar graphs obtained from* G *by resolving the crossing:*

(5.2) 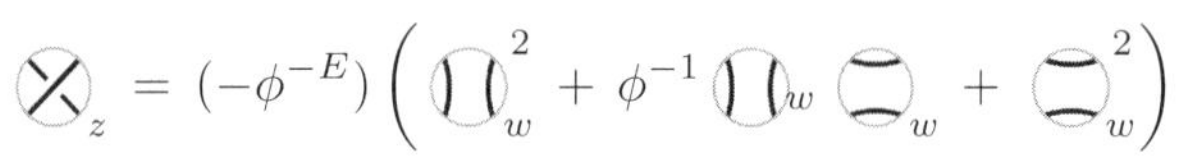

where $z = (5-\sqrt{5})/2, w = (3-\sqrt{5})/2$.

We remark that this lemma is motivated by D.W. Hall's version of the golden identity for constrained chromials [**8**]. The proof of Lemma 5.2 relies on the following statement.

PROPOSITION 5.3. *Fix* $Q \neq (3 \pm \sqrt{5})/2$, *and let* $G =$  $\cup \overline{G}$ *be a near-planar graph. Then*

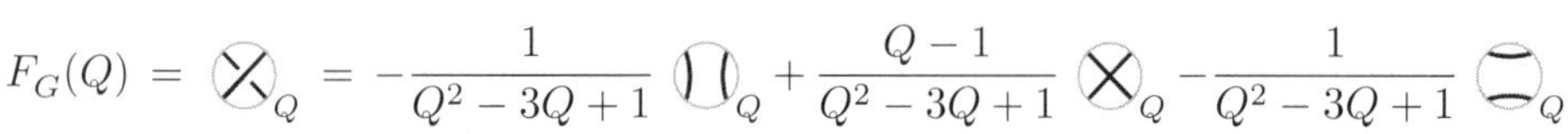

The proof of Proposition 5.3 amounts to checking that both sides have identical evaluations when paired up with the three basis elements ,  and  of the chromatic algebra $\mathcal{C}_2^Q$. (As discussed in section 2.5, the bilinear pairing $\langle ., . \rangle_Q$ on C_2^Q is non-degenerate precisely for $Q \neq (3 \pm \sqrt{5})/2$.)

PROOF OF LEMMA 5.2. Setting Q to equal $z = (5-\sqrt{5})/2$ in Proposition 5.3 and using the contraction-deletion rule, one has the following equality:

(5.3) 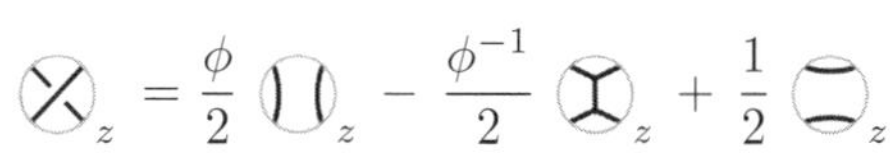

Equation (5.2) is obtained by applying the golden identity (1.1) to the three planar cubic graphs on the right in (5.3). □

It follows from Lemma 5.2 that Conjecture 5.1 for near-planar graphs is equivalent to the inequality

(5.4) 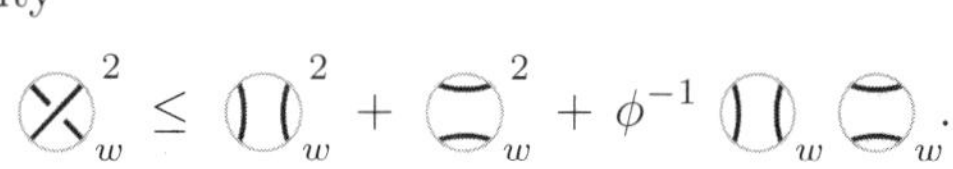

REMARK 5.1. It is interesting to note that the inequality (5.4) can be restated in terms of the Yamada polynomial:

(5.5) 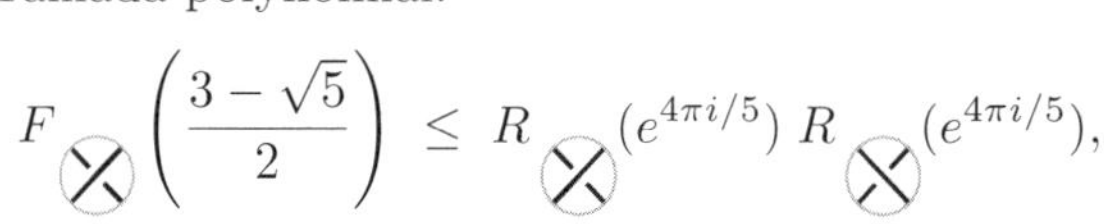

where the left-hand side is the evaluation of the flow polynomial of the abstract near-planar graph, and the right-hand side is the product of the Yamada polynomials of two spatial ribbon graphs corresponding to the two possible crossings. The proof of the equivalence of (5.4), (5.5) consists of applying the Yamada polynomial skein relations to the crossings, and simplifying using the linear relation (2.9).

We are now in a position to give a reformulation of conjecture 5.1 for near-planar graphs.

Lemma 5.4. *Let $\overline{G}$ be a cubic graph in a disk, with 4 marked points on the boundary. Considered in the chromatic algebra C_2^Q, $Q = (3-\sqrt{5})/2$, $\overline{G} = \alpha$*  $+$ β  $+\gamma$ *. Conjecture* 5.1 *for near-planar graphs is equivalent to the inequality*

$$(1+3\phi)\,\alpha\,\beta \;\leq\; \gamma\,(\alpha+\beta+\gamma). \tag{5.6}$$

Proof. The term being squared on the right-hand side of (5.1) equals:

$F_G(w) = \langle$  $, \overline{G}\rangle_w = \langle$  $, \alpha$  $+\beta$ 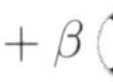 $+\gamma$  $\rangle_w$

$$= -\phi^{-1}(\alpha+\beta) + \phi^{-2}\gamma.$$

Analogous calculations for  and  in place of  yield

$$-\phi^{-1}\beta + \phi^{-2}(\alpha+\gamma), \quad -\phi^{-1}\alpha + \phi^{-2}(\beta+\gamma)$$

respectively. Equation (5.2) expresses the left-hand side in (5.1) in terms of  $_w$ and  $_w$. Multiplying out the resulting expressions, (5.1) is seen to be equivalent to (5.6). □

Next we establish the inequality (5.6) for an infinite family of near-planar graphs.

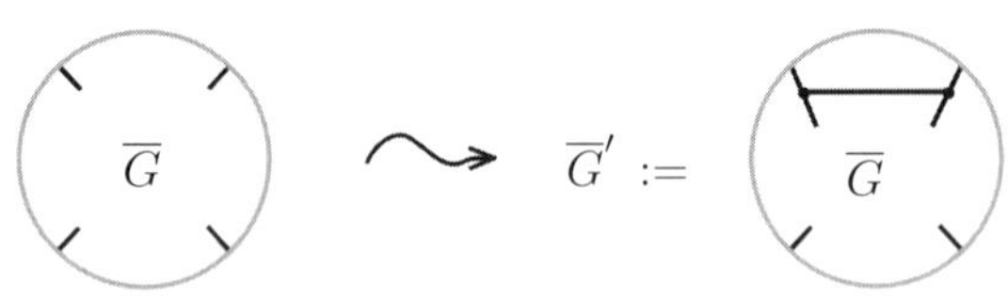

Figure 9. A modification of $\overline{G}$: addition of a *peripheral* edge (connecting two boundary edges of a cubic graph $\overline{G}$).

Lemma 5.5. *Conjecture* 5.1 *holds for the family of near-planar graphs $G =$*  $\cup\, \overline{G}$*, where $\overline{G}$ is inductively built by addition of peripheral edges.*

Examples of graphs, considered in this lemma, are shown in Figure 10. The graph on the left, capped with , is $K_{3,3}$. An analogous proof shows that the inequality (1.4) at the Galois conjugate values $\phi+1$, $\phi+2$ also holds for the same family of graphs.

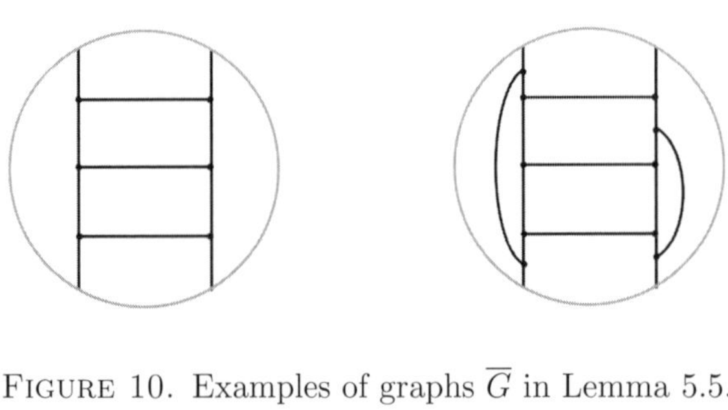

Figure 10. Examples of graphs $\overline{G}$ in Lemma 5.5.

PROOF OF LEMMA 5.5. It suffices to prove that the inequality (5.6) is preserved under addition of a peripheral edge. We will also show that whenever the coefficients α, β, γ are non-zero, they satisfy

$$\text{sign}(\alpha) = \text{sign}(\beta) = -\text{sign}(\gamma) = (-1)^{V/2}, \tag{5.7}$$

where V is the number of vertices of $\overline{G}$. The result of adding a peripheral edge to $\overline{G} = \alpha$  $+ \beta$  $+ \gamma$  is the graph $\overline{G}'$ shown in Figure 11. (There are 4 possible peripheral edges; by symmetry between α and β the proof below applies to each one.)

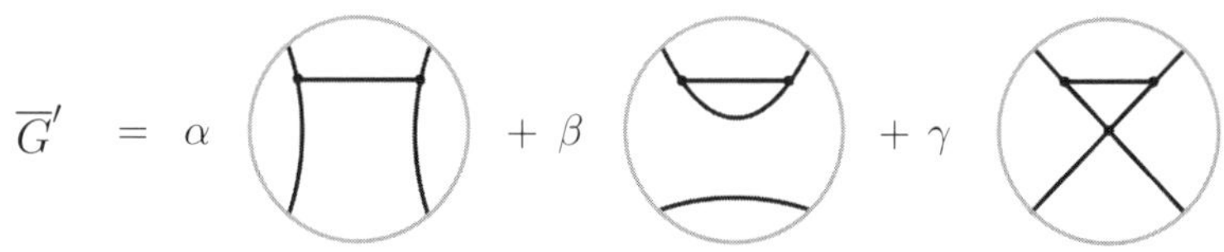

FIGURE 11. A peripheral edge added to $\overline{G}$.

Applications of the contraction-deletion rule give

$$\overline{G}' = -\alpha$$  $+ (-\phi\beta + \gamma)$  $+ (\alpha - \phi^2\gamma)$ .

Denote the coefficients of $\overline{G}'$ by α', β', γ':

$$\alpha' = -\alpha, \; \beta' = -\phi\beta + \gamma, \; \gamma' = \alpha - \phi^2\gamma. \tag{5.8}$$

This gives an inductive proof of the statement (5.7). A direct calculation shows that the desired inequality for the coefficients α', β', γ' of $\overline{G}'$,

$$(1 + 3\phi)\, \alpha' \beta' \;\leq\; \gamma' (\alpha' + \beta' + \gamma'), \tag{5.9}$$

is equivalent to

$$\phi^2\, \alpha\, \beta \;\leq\; \gamma\, (\alpha + \beta + \gamma). \tag{5.10}$$

Since sign(α)= sign(β), $\phi^2\, \alpha\, \beta \leq (1 + 3\phi)\, \alpha\, \beta$.Therefore the assumed inequality (5.6) for α, β, γ implies the inequality (5.9) for α', β', γ'.

Finally, consider the last statement in conjecture 5.1. Note that the inequality $\phi^2\, \alpha\, \beta \leq (1 + 3\phi)\, \alpha\, \beta$ in the previous paragraph is strict precisely when both α, β are non-zero. Suppose the inductive construction of the family of graphs in the statement of the lemma starts with , and consider the first time a horizontal peripheral edge is added. Before this step, the graph $\overline{G}$ is of the form shown on the left in Figure 12: there is a vertical line in the disk, intersecting the graph $\overline{G}$ in a single point. The vector space of graphs in a disk with three boundary points, modulo the defining local relations of the chromatic algebra, is 1-dimensional. Hence it is clear that $\overline{G} = \lambda$ 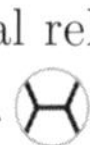 in $\mathcal{C}_2$, and the coefficient β (defined in Figure) 8 is zero. Then adding a horizontal edge gives a scalar multiple of the graph on the right in Figure 12, resulting in non-zero coefficients α, β. At this point, the graph $G =$  $\cup\, \overline{G}$ is still planar, since one of the crossing strands may be drawn within the rectangle in the graph on the right in the figure. Since both coefficients α, β are non-zero, the addition of any new peripheral edge after this step makes the inequality (5.9) strict. (And of course the graph becomes non-planar.) It follows from (5.8) that all

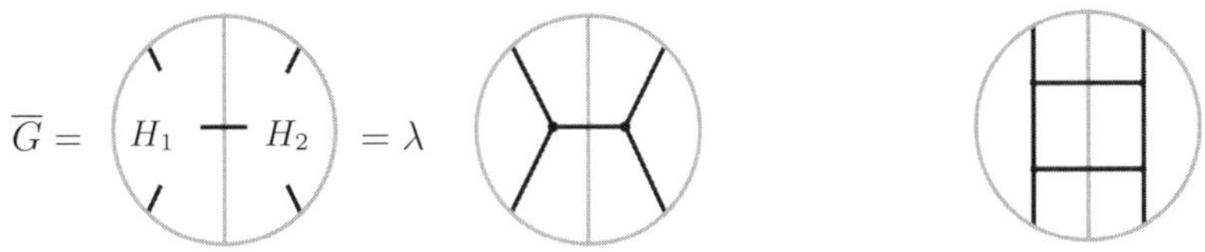

FIGURE 12

further additions of edges increase α, β in absolute value, so the difference of the two sides of the inequality (5.9) continues strictly increasing. Thus the inequality detects planarity in this family of graphs. □

6. Exponential growth of the number of chromatic polynomials

In this section we prove Theorem 1.3, working dually with the flow polynomial of planar cubic graphs. Consider $\mathcal{C}^Q_{1,3}$, the vector space consisting of $\mathbb{C}$-linear combinations of planar cubic graphs in a rectangle with one marked point at the bottom and three marked points at the top of the rectangle, modulo the relations in Figure 2. (This notion was also discussed at the end of section 2.6.) The loop value is $Q-1$, and for the remainder of this section we fix $Q = (3-\sqrt{5})/2$. As discussed in section 2.5, at this value of Q there is an additional local relation

(6.1) 

$\mathcal{C}^Q_{1,3}$ is a module over the chromatic algebra; the action is by vertical concatenation, matching three marked points on the boundary.

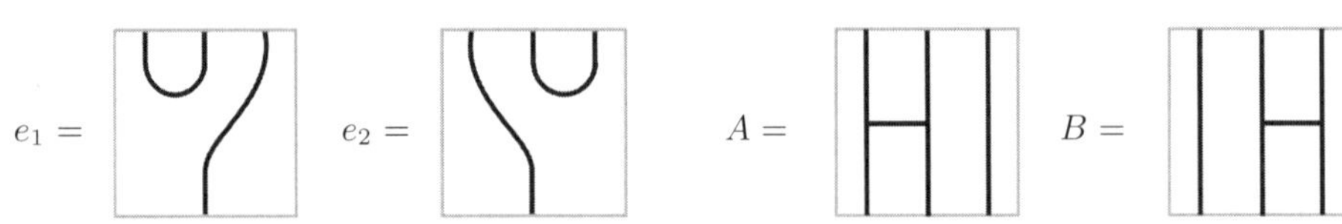

FIGURE 13. Elements $e_1, e_2 \in \mathcal{C}^Q_{1,3}$ and $A, B \in \mathcal{C}^Q_3$

For a generic Q, the vector space $\mathcal{C}^Q_{1,3}$ is 3-dimensional. At the specified value $Q = (3-\sqrt{5})/2$, the additional relation (6.1) reduces the dimension to two. Consider its basis e_1, e_2, shown on the left in Figure 13, and let A, B be elements of $\mathcal{C}_3$ shown on the right in the same figure. The action of A, B on e_1, e_2 is calculated using (6.1); the loop value is $Q - 1 = -\phi^{-1}$. The result is shown in Figures 14, 15.

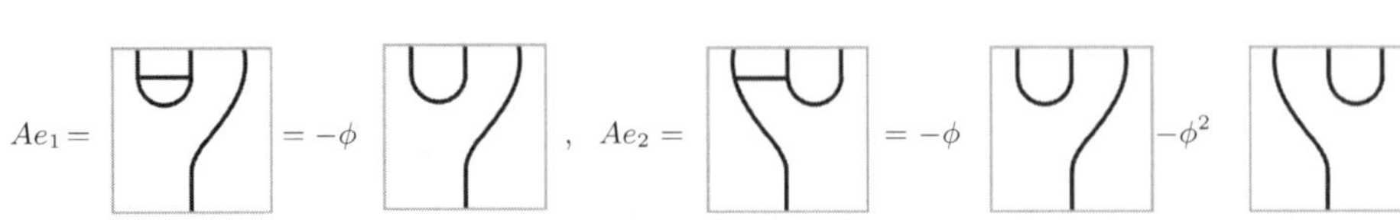

FIGURE 14. The action of A

FIGURE 15. The action of B

In other words, with respect to the chosen basis, A and B are represented by the matrices

$$A = -\phi \begin{bmatrix} 1 & 1 \\ 0 & \phi \end{bmatrix}, B = -\phi \begin{bmatrix} \phi & 0 \\ 1 & 1 \end{bmatrix}.$$

The squares of these matrices are given by

$$A^2 = \phi^2 \begin{bmatrix} 1 & \phi^2 \\ 0 & \phi^2 \end{bmatrix}, B^2 = \phi^2 \begin{bmatrix} \phi^2 & 0 \\ \phi^2 & 1 \end{bmatrix}.$$

It follows from the ping pong lemma for semigroups that A^2, B^2 generate a free semigroup in the chromatic algebra $\mathcal{C}_3^Q$. (Consider vectors v with positive components with respect to the basis e_1, e_2, and let $u = A^2v, w = B^2v$. Then the components of u, w satisfy $u_1 > u_2, w_1 < w_2$.) Therefore words $\{W\}$ of length n in A^2, B^2 represent 2^n distinct elements in $\mathcal{C}_3^Q$.

Given two elements $a, b \in \mathcal{C}_{1,3}^Q$, the product $a \cdot \bar{b}$ (where $\bar{b} \in \mathcal{C}_{3,1}^Q$ is obtained from b by reflection in a horizontal line) is an element of $\mathcal{C}_1^Q$. Here the product is given by vertical concatenation, matching three marked boundary points. $\mathcal{C}_1$ is 1-dimensional, and the trace $\mathcal{C}^Q \longrightarrow \mathbb{C}$ is an isomorphism. Denote $\langle a, b \rangle := \mathrm{tr}(a \cdot \bar{b})$. Note that the Gram matrix $\langle e_i, e_j \rangle_{i,j \in \{1,2\}}$ is non-degenerate at $Q = (3 - \sqrt{5})/2$. Therefore if two words W_1, W_2 in A^2, B^2 are not equal as 2×2 matrices, then $\langle W_1 e_i, e_j \rangle \neq \langle W_2 e_i, e_j \rangle$ for some $i, j \in \{1, 2\}$. Words in A^2, B^2 are represented geometrically by planar cubic graphs G in a rectangle, and (as discussed in section 2.4) $\langle We_i, e_j \rangle$ is the flow polynomial of the graph obtained by gluing e_i on the bottom of G, $\bar{e}_j$ on top, and then taking the trace, i.e. connecting the endpoints by an arc in the plane, Figure 16.

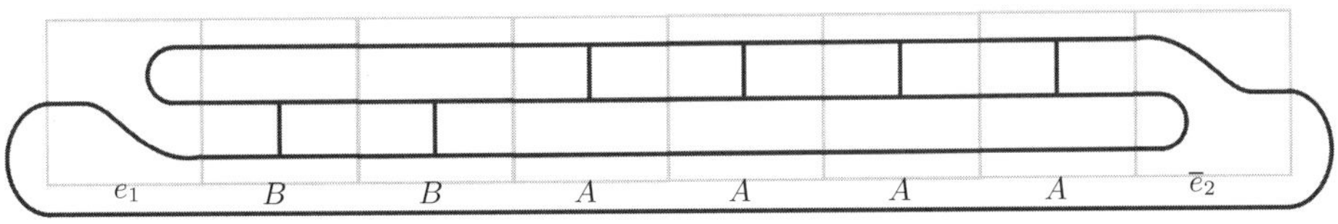

FIGURE 16. The flow polynomial of the pictured graph at $(3 - \sqrt{5})/2$ equals $\langle A^4 B^2 e_1, e_2 \rangle$. (The multiplication is represented horizontally, rather than vertically, to conserve space.)

Since words of length n in A^2, B^2 give 2^n distinct matrices, there are at least $2^{n/4}$ distinct values for one of the matrix entries. This translates into at least $2^{n/16}$ values $F_G((3 - \sqrt{5})/2)$ of the flow polynomial of planar trivalent graphs G with n vertices. □

Remark: After the paper was written, we discovered that one may also prove exponential growth at $Q = 4$, observing that $A^2 e_1 = 4e_1, A^2 e_2 = e_1 + e_2, B^2 e_1 = e_1 + e_2, B^2 e_2 = 4e_2$, and completing the argument as above.

7. Appendix: A proof of the golden identity for planar cubic graphs

This section gives a variation of the proof [**6**] of the Tutte golden identity [**19**] for the flow polynomial of planar cubic graphs:

$$F_G((5-\sqrt{5})/2) = (-\phi)^{-E}\, F_G((3-\sqrt{5})/2)^2, \tag{7.1}$$

We include a proof since this identity underlies several results in this paper. The version of the proof presented here has more of a computational flavor; it might be useful in numerical investigation of whether there are identities at other parameter values.

Denote $z := (5-\sqrt{5})/2, w := (3-\sqrt{5})/2$. Let $\overline{G}$ be a graph in the disk with 4 marked points on the boundary, and consider its coefficients with respect to the usual basis of C_2^Q (as in Figure 8),

$\overline{G} = \alpha_Q$  $_Q + \beta_Q$  $_Q + \gamma_Q$  $_Q$

for $Q = z, w$. Flips  $\leftrightarrow$  acts transitively on connected planar cubic graphs with a fixed number of vertices, and it suffices to prove that the golden identity for , , and  imply the golden identity for .

Consider the evaluations:

$\langle$  $, \overline{G}\rangle_Q = \alpha_Q(Q-1)^2 + \beta_Q(Q-1) + \gamma_Q(Q-1)^2,$

$\langle$  $, \overline{G}\rangle_Q = \alpha_Q(Q-1) + \beta_Q(Q-1)^2 + \gamma_Q(Q-1)^2,$

$\langle$  $, \overline{G}\rangle_Q = \alpha_Q(Q-1)(Q-2) + \gamma_Q(Q-1)(Q-2)^2,$

$\langle$  $, \overline{G}\rangle_Q = \beta_Q(Q-1)(Q-2) + \gamma_Q(Q-1)(Q-2)^2.$

The golden identity for $G = \overline{G}$, capped off with , , , , respectively, is equivalent to quadratic equations on the coefficients at z and w:

$$(A)\qquad \alpha_z\phi^{-4} + \beta_z\phi^{-2} + \gamma_z\phi^{-4} \;-\; (-\phi)^{-E}\left(\alpha_w\phi^{-2} - \beta_w\phi^{-1} + \gamma_w\phi^{-2}\right)^2 \;=\; 0,$$

$$(B)\qquad \alpha_z\phi^{-2} + \beta_z\phi^{-4} + \gamma_z\phi^{-4} \;-\; (-\phi)^{-E}\left(-\alpha_w\phi^{-1} + \beta_w\phi^{-2} + \gamma_w\phi^{-2}\right)^2 \;=\; 0,$$

$$(C)\qquad -\alpha_z\phi^{-3} + \gamma_z\phi^{-4} \;-\; (-\phi)^{-E-3}\left(\alpha_w - \gamma_w\phi\right)^2 \;=\; 0,$$

$$(D)\qquad -\beta_z\phi^{-3} + \gamma_z\phi^{-4} \;-\; (-\phi)^{-E-3}\left(\beta_w - \gamma_w\phi\right)^2 \;=\; 0.$$

The left-hand sides of these four equations satisfy the relation $A-B = C-D$, so the validity of any three of them implies the fourth. The proof of (7.1) is completed by comparing the loop values:  $_z$ $=$ 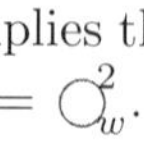 ^2_w. □

Acknowledgments

We would like to thank Gordon Royle for many discussions, and for sharing with us numerical data on graph polynomials. We also thank Kyle Miller and the referee for helpful comments.

References

[1] Ian Agol and Vyacheslav Krushkal, *Tutte relations, TQFT, and planarity of cubic graphs*, Illinois J. Math. **60** (2016), no. 1, 273–288. MR3665181

[2] S. Beraha, *Infinite non-trivial families of maps and chromials*, Thesis, Johns Hopkins University, 1975.

[3] G. D. Birkhoff and D. C. Lewis, *Chromatic polynomials*, Trans. Amer. Math. Soc. **60** (1946), 355–451, DOI 10.2307/1990348. MR0018401

[4] Beifang Chen and Richard P. Stanley, *Orientations, lattice polytopes, and group arrangements II: Modular and integral flow polynomials of graphs*, Graphs Combin. **28** (2012), no. 6, 751–779, DOI 10.1007/s00373-011-1080-8. MR2991172

[5] Reinhard Diestel, *Graph theory*, 3rd ed., Graduate Texts in Mathematics, vol. 173, Springer-Verlag, Berlin, 2005. MR2159259

[6] Paul Fendley and Vyacheslav Krushkal, *Tutte chromatic identities from the Temperley-Lieb algebra*, Geom. Topol. **13** (2009), no. 2, 709–741, DOI 10.2140/gt.2009.13.709. MR2469528

[7] Paul Fendley and Vyacheslav Krushkal, *Link invariants, the chromatic polynomial and the Potts model*, Adv. Theor. Math. Phys. **14** (2010), no. 2, 507–540. MR2721654

[8] Dick Wick Hall, *On golden identities for constrained chromials*, J. Combinatorial Theory Ser. B **11** (1971), 287–298. MR0289319

[9] François Jaeger, *On the Penrose number of cubic diagrams*, Discrete Math. **74** (1989), no. 1-2, 85–97, DOI 10.1016/0012-365X(89)90201-X. Graph colouring and variations. MR989125

[10] Vaughan F. R. Jones, *Subfactors and knots*, CBMS Regional Conference Series in Mathematics, vol. 80, Published for the Conference Board of the Mathematical Sciences, Washington, DC; by the American Mathematical Society, Providence, RI, 1991. MR1134131

[11] Louis H. Kauffman and Pierre Vogel, *Link polynomials and a graphical calculus*, J. Knot Theory Ramifications **1** (1992), no. 1, 59–104, DOI 10.1142/S0218216592000069. MR1155094

[12] Scott Morrison, Emily Peters, and Noah Snyder, *Knot polynomial identities and quantum group coincidences*, Quantum Topol. **2** (2011), no. 2, 101–156, DOI 10.4171/QT/16. MR2783128

[13] Roger Penrose, *Applications of negative dimensional tensors*, Combinatorial Mathematics and its Applications (Proc. Conf., Oxford, 1969), Academic Press, London, 1971, pp. 221–244. MR0281657

[14] N. Yu. Reshetikhin and V. G. Turaev, *Ribbon graphs and their invariants derived from quantum groups*, Comm. Math. Phys. **127** (1990), no. 1, 1–26. MR1036112

[15] D. Treumann, *How many chromatic polynomials of planar maps are there?*, https://mathoverflow.net/q/241734

[16] D. Treumann and E. Zaslow, *Cubic Planar Graphs and Legendrian Surface Theory*, arXiv:1609.04892. To appear in Adv. Theor. Math. Phys.

[17] W. T. Tutte, *A census of planar maps*, Canad. J. Math. **15** (1963), 249–271, DOI 10.4153/CJM-1963-029-x. MR0146823

[18] W. T. Tutte, *On chromatic polynomials and the golden ratio*, J. Combinatorial Theory **9** (1970), 289–296. MR0272676

[19] W. T. Tutte, *More about chromatic polynomials and the golden ratio*, Combinatorial Structures and their Applications (Proc. Calgary Internat. Conf., Calgary, Alta., 1969), Gordon and Breach, New York, 1970, pp. 439–453. MR0263698

[20] W. T. Tutte, *A contribution to the theory of chromatic polynomials*, Canadian J. Math. **6** (1954), 80–91, DOI 10.4153/cjm-1954-010-9. MR0061366

[21] K. Walker, Deletion-contraction relations, hard hexagons, and the shadow world, Talk at IPAM, 2007. Available at http://canyon23.net/math/talks/

[22] Hans Wenzl, *On sequences of projections*, C. R. Math. Rep. Acad. Sci. Canada **9** (1987), no. 1, 5–9. MR873400

[23] Shuji Yamada, *An invariant of spatial graphs*, J. Graph Theory **13** (1989), no. 5, 537–551, DOI 10.1002/jgt.3190130503. MR1016274

University of California, Berkeley, 970 Evans Hall #3840, Berkeley, California, 94720-3840

Email address: ianagol@berkeley.edu

Department of Mathematics, University of Virginia, Charlottesville, Virginia 22904-4137

Email address: krushkal@virginia.edu

Proceedings of Symposia in Pure Mathematics
Volume **102**, 2019
https://doi.org/10.1090/pspum/102/01805

Inequivalent Lefschetz fibrations on rational and ruled surfaces

R. İnanç Baykur

ABSTRACT. In this short note, we give an explicit construction of inequivalent Lefschetz pencils and fibrations of same genera on blow-ups of all rational and ruled surfaces. This complements our earlier results, concluding that *every* symplectic 4–manifold, after sufficiently many blow-ups, admits inequivalent Lefschetz pencils and fibrations, which cannot be obtained from one another even via any sequence of fibered Luttinger surgeries.

1. Introduction

A pair of Lefschetz pencils or fibrations[1] on a closed oriented smooth 4–manifold are said to be *equivalent* (or *isomorphic*) if there is an orientation-preserving self-diffeomorphism of the total space which commutes the two maps. Simon Donaldson's ground-breaking work in [**8**] provides arbitrarily large genera *pencils* on any symplectic 4–manifold. A fundamental problem then is to determine if there are relatively-minimal inequivalent pencils or fibrations of *the same genus and number of base points* on a given symplectic 4–manifold. For brevity, we refer to them as *inequivalent pencils* and *inequivalent fibrations*, where it should be understood that their genera and number of base points are already the same.

This note is an amendment to our results in [**4**], where we proved that any symplectic 4-manifold, which is not a rational or a ruled surface, after a number of blow-ups, admits arbitrarily many inequivalent Lefschetz pencils and fibrations. Moreover, they are not related via *fibered Luttinger surgeries* [**4**]. Here we will give similar examples of pencils and fibrations on *rational* and *ruled surfaces*. Recall that rational and ruled surfaces are birationally equivalent to the complex projective plane and S^2–bundles over Riemann surfaces, so they are diffeomorphic to blow-ups of $\mathbb{CP}^2$, $S^2 \times \Sigma_h$, $S^2 \widetilde{\times} \Sigma_h$, which are known to have unique symplectic structures up to symplectomorphisms and deformations.

2010 *Mathematics Subject Classification.* Primary 57R17; Secondary 57N13.
This work is supported by the NSF Grant DMS-1510395.

[1]We assume that a *pencil*, unlike a Lefschetz fibration, always has base points. We also assume that a Lefschetz pencil or fibration always has critical points, and each singular fiber contains only one critical point and no exceptional spheres (i.e. it is *relatively-minimal*).

Our examples in [**4**] were obtained from Donaldson pencils via *partial doubling sequences*, which involve blow-ups and degree doublings of pencils introduced in [**3**, **19**]. These pencils were distinguished by looking at the collection of exceptional spheres, which arise as *multisections* [**5**, **9**] of different degrees in respective pencils. Thus, blowing-up all the base points, we obtain inequivalent Lefschetz fibrations. However, the same strategy does not work for rational or ruled surfaces, since in this case, one is not guaranteed to have disjoint multisections representing a given collection of exceptional classes [**4**].

We will instead apply partial doublings to hand-picked pencils on rational and ruled surfaces in order to derive inequivalent pencils which have *different numbers of reducible fibers* versus irreducible ones. The point here is that "doublings are not created equal"; depending on which subcollections of base points we pick, we can arrive at pencils with topologically different singular fibers. Since the distinction comes from the number of reducible fibers, we get inequivalent fibrations after blowing-up all the base points. Our main result is:

THEOREM 1. *Any rational or ruled surface, after sufficiently many blow-ups, admits inequivalent relatively-minimal pencils and fibrations of the same genera and number of base points, which cannot be obtained from one another via any sequence of fibered Luttinger surgeries.*

Last part of the theorem comes easy for our examples: performing a fibered Luttinger surgery is equivalent to conjugating a subword of the corresponding monodromy factorization of a pencil by some Dehn twist; see [**1**, **4**]. This can be done whenever the twist curve is fixed, up to isotopy, by the mapping class given by the subword. Clearly, the number of reducible fibers is unchanged under this operation, so the examples with different numbers of reducible fibers we produce are also inequivalent up to fibered Luttinger surgeries.

There are many other examples of inequivalent Lefschetz pencils and fibrations. A pair of inequivalent *fibrations* on a blow-up of $T^2 \times \Sigma_2$, whose fibers have different divisibility in homology, was discussed by Ivan Smith in [**18**]. Several inequivalent fibrations on homotopy elliptic surfaces, distinguished by their monodromy groups, were discovered by Jongil Park and Ki-Heon Yun in [**16**, **17**]. As for inequivalent *pencils*, first examples were constructed by Kenta Hayano and the author on 4–manifolds homeomorphic to blow-ups of the K3 surface [**5**]. The construction in [**5**] uses pairs of lantern substitutions which modify pencils differently, but land on the same 4–manifold. Notably, Noriyuki Hamada recently constructed pairs of inequivalent pencils on irrational ruled surfaces $S^2 \tilde{\times} \Sigma_h$, which *only* differ by the configuration of their base points on the reducible fibers, so they do not yield inequivalent fibrations after blow-ups.

With our aforementioned result from [**4**], we arrive at the following conclusion, which demonstrates how diverse the Donaldson-Gompf correspondence between symplectic 4–manifolds and Lefschetz pencils and fibrations is:

COROLLARY 2. *Any closed symplectic 4–manifold, possibly after blow-ups, admits inequivalent relatively-minimal pencils and fibrations of the same genera and number of base points, which cannot be obtained from one another via any sequence of fibered Luttinger surgeries.*

2. Preliminaries

Here we will quickly review the basic notions and background results on Lefschetz pencils / fibrations, positive Dehn twist factorizations, fibered Luttinger surgeries and partial conjugations, and doublings of Lefschetz pencils. For further details on these topics, the reader can turn to [**10**] and [**2–4**].

Let X be a closed, oriented 4-manifold, and $B = \{b_j\}$, $C = \{p_i\}$ be finite, non-empty sets of points in X. A *Lefschetz pencil* (X, f) is given by a surjective map $f\colon X \setminus B \to S^2$, which is a submersion on $X \setminus C$, such that around each *base point* b_j and each *critical point* p_i, it conforms to local complex models $(z_1, z_2) \mapsto z_1/z_2$ and $(z_1, z_2) \mapsto z_1 z_2$, respectively. A *Lefschetz fibration* is defined similarly for $B = \emptyset$. After blowing-up every base point b_j of a pencil, we obtain a Lefschetz fibration with disjoint exceptional spheres S_j as its sections. We say (X, f) is a *genus–g* pencil / fibration, for g the genus of the regular fiber F. We will assume that every fiber contains at most one critical point, which can always be achieved after a small perturbation. A *singular fiber* has a nodal singularity at the critical point p_i, and is obtained by shrinking a simple loop a_i on F, called the *vanishing cycle*. We have a *reducible* fiber (with two connected components) if a_i is separating F, and *irreducible* if a_i is non-separating.

Let Σ_g^m denote a compact, oriented surface of genus g with m boundary components, where $\Sigma_g = \Sigma_g^0$. The *mapping class group* Γ_g^m is the group of orientation-preserving self-diffeomorphisms of Σ_g^m which restrict to identity along $\partial \Sigma_g^m$, modulo isotopies of the same type. Let $t_a \in \Gamma_g^m$ denote the positive (right-handed) Dehn twist along a simple loop a on Σ_g^m. A genus–g Lefschetz pencil (X, f) with m base points prescribes a *positive factorization* of the boundary multi-twist (or *monodromy factorization*), given by a relation

$$t_{\delta_1} \cdot \ldots \cdot t_{\delta_m} = t_{a_1} \cdot \ldots \cdot t_{a_\ell}$$

in Γ_g^m, where $\delta_1, \ldots, \delta_m$ are curves parallel to distinct boundary components of Σ_g^m (which we will also denote the boundary component by), a_i is a vanishing cycle corresponding to the critical point p_i of f, and $\ell = |C|$. Conversely, given a positive factorization as above, one can construct such a Lefschetz pencil (X, f). For $g \geq 2$, we indeed get a one-to-one correspondence between equivalence classes of Lefschetz pencils and positive factorizations up to *Hurwitz moves* (trading subwords $t_{a_i} t_{a_{i+1}}$ and $t_{t_{a_i}(a_{i+1})} t_{a_i}$) and *global conjugations* (replacing every t_{a_i} in the factorization with $t_{\phi(a_i)}$, for the same $\phi \in \Gamma_g^n$). All goes the same for Lefschetz *fibrations* for $m = 0$.

By the pioneering works of Donaldson and Gompf [**8, 10**], every symplectic 4-manifold admits a Lefschetz pencil whose fibers are symplectic, and conversely, the total space of a Lefschetz pencil or a fibration (recall $C \neq \emptyset$) can always be equipped with a symplectic form for which all regular fibers are symplectic.

As observed in [**2**], *Luttinger surgery* (where one removes a neighborhood of a Lagrangian torus in a symplectic 4–manifold and glues it back in differently so as to produce a new symplectic 4–manifold) can be performed in a way compatible with the pencil structure, when the restriction of the pencil map to the torus is a circle bundle over a loop on the base, away from the critical values [**1, 2, 4**]. This surgery is equivalent to modifying the corresponding positive factorization by replacing a subword $t_{a_i} \cdot \ldots \cdot t_{a_k}$ with a conjugate $t_{\phi(a_i)} \cdot \ldots \cdot t_{\phi(a_k)}$ for some $\phi = t_b^{\pm}$, which can be done if and only if the mapping class $t_{a_i} \cdot \ldots \cdot t_{a_k}$ fixes b, up to isotopy [**1**].

Another way to produce a new pencil from a given one, and on the same 4–manifold, is the *doubling* procedure, discussed for holomorphic pencils and Donaldson's pencils in [**19**], and for pencils obtained via branched coverings of $\mathbb{CP}^2$ in [**3**]. In general, by [**4**][Lemma 3.1], one can double *any* (topological) genus–g pencil (X, f) with m base points, provided $g \geq 2$ and $m \leq 2g-2$. The result is a genus–g' pencil (X, f') with m' base points, where $g' = 2g + m - 1$ and $m' = 4m$. If the original pencil is compatible with a given symplectic form, then the resulting one is also compatible with a deformation equivalent symplectic form.

Topologically, the doubling procedure can be interpreted roughly as follows: First break up the symplectic pencil (X, f) into two pieces; the regular neighborhood of a smooth fiber (the "concave piece"), which is a symplectic disk bundle of degree m over a genus–g surface, and its complement (the "convex piece"). Describe a standard higher genus pencil on the former piece, based on g and m —the model for which comes from holomorphic pencils. Then extend it over the convex piece, which now becomes a subpiece of the new pencil (X, f'). By the enterprise of [**3**], the monodromy of f' is explicitly determined by that of f [**3**][Theorem 4]. In particular, we observe the following: any *separating* Dehn twists in the positive factorization of f' comes, in a one-to-one fashion, from a separating Dehn twist in the positive factorization of f, which bounds a subsurface on F that does not contain a base point of f. (See Figure 1 for an example.)

3. Constructions

It is time to spell out our recipe in full detail. Take a positive factorization for a genus $g \geq 2$ pencil (X, f) with base locus B and regular fiber F, whose separating vanishing cycles are $a_1, \ldots, a_r$. Let $B' = \{b'_j\}$ and $B'' = \{b''_j\}$ be two subsets of the base locus B of f (where the base points correspond to boundary components in the positive factorization), so that $|B'| = |B''| \leq 2g - 2$. Set $K'_i, K''_i \in \{0, 1\}$ to be the number of components in $F \setminus a_i$ which do not intersect B' (resp. B'').

If $K' = \sum_{i=1}^{r'} K'_i$ is unequal to $K'' = \sum_{i=1}^{r''} K''_i$, we are game.

We blow-up all the base points of f in $B \setminus B'$ (resp. $B \setminus B''$) and double the resulting pencil on $\widetilde{X} = X \# |B \setminus B'| \overline{\mathbb{CP}}{}^2$ with $m = |B'|$ base points, which naturally come from the inclusion of X minus the blown-up base points into $\widetilde{X}$. In this way, we arrive at a new pencil $(\widetilde{X}, f')$ (resp. $(\widetilde{X}, f'')$). As we reviewed earlier, each one of the pencils f' and f'' on $\widetilde{X}$ has genus $2g + m - 1$ and $4m$ base points, whereas the number of reducible fibers they have are K' and K'', respectively. (Figure 1 illustrates 2 distinct partial doublings of sorts.) For each $0 \leq m_0 < m$, we get a pair of inequivalent pencils by blowing-up m_0 base points of both pencils, and for $m_0 = m$, we get a pair of inequivalent fibrations.

It remains to find the main ingredient for our recipe; a pencil (X, f) as prescribed above, on (blow-ups of) rational and ruled surfaces.

3.1. Inequivalent pencils and fibrations on rational surfaces. There is the following positive factorization of the boundary multi-twist in Γ_2^3, which was found by Sinem Onaran in [**15**]:

$$t_{\delta_1} t_{\delta_2} t_{\delta_3} \;=\; t_{a_3} t_{b_1} (t_{a_1} t_{a_2} t_{a_3} t_{b_1})^2 t_{a_3} t_{\psi(b_2)} t_{a_3} t_{b_2} t_{a_4} t_{a_5} t_{a_3} t_{b_2} t_{a'_4} t_{a_6}\,.$$

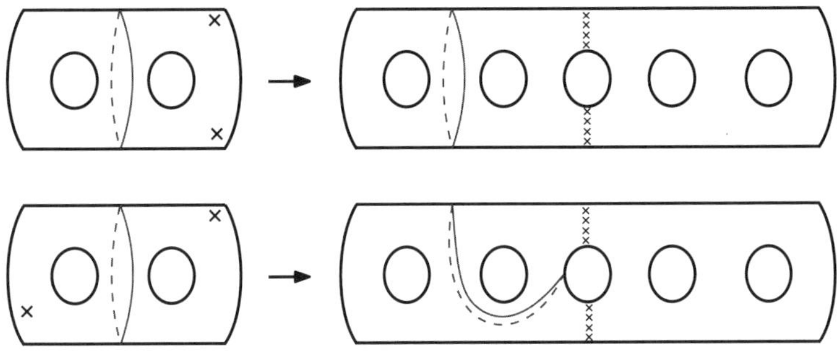

FIGURE 1. The lifts of a separating vanishing cycle (in blue) under doubling of a genus–2 pencil with 2 different pairs of base points, marked with $\times$ in the figure.

Here the twist curves a_i, b_j and a'_4 are as in Figure 2, and $\psi = t_{a_5}^{-1} t_{a_4}^{-1}$. (See [**15**][Figure 6]. This is a lift of a well-known positive factorization of identity in Γ_2 [**6**], which corresponds to a hyperelliptic Lefschetz fibration on $\mathbb{CP}^2 \# 13\overline{\mathbb{CP}}^2$.)

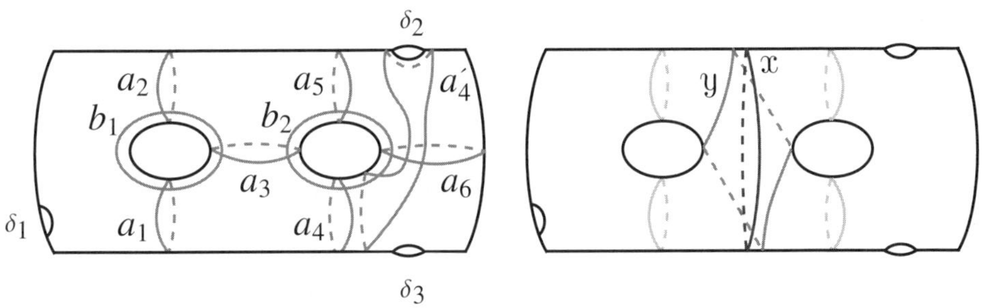

FIGURE 2. Twist curves a_i, b_j and a'_4 on the surface Σ_2^3. The boundary components are $\delta_1, \delta_2, \delta_3$. The new curves x, y are given in the second copy of Σ_2^3, where the lantern sphere sits in the middle. The separating curve x is in blue.

We rewrite the above expression as

$$
\begin{aligned}
& t_{\delta_1} t_{\delta_2} t_{\delta_3} \\
& \quad = t_{a_3} t_{b_1} t_{a_1} t_{a_2} t_{a_3} t_{b_1} t_{a_1} t_{a_2} t_{a_4} t_{a_5} \psi (t_{a_3} t_{b_1} t_{a_3} t_{\psi(b_2)} t_{a_3} t_{b_2}) \psi^{-1} t_{a_3} t_{b_2} t_{a'_4} t_{a_6} \\
& \quad = t_{a_3} t_{b_1} t_{a_1} t_{a_2} t_{a_3} t_{b_1} \underline{t_{a_1} t_{a_2} t_{a_4} t_{a_5}} t_{\psi(a_3)} t_{\psi(b_1)} t_{\psi(a_3)} t_{\psi^2(b_2)} t_{\psi(a_3)} t_{\psi(b_2)} t_{a_3} t_{b_2} t_{a'_4} t_{a_6}
\end{aligned}
$$

and then apply a lantern substitution along the underlined subword to obtain a new positive factorization:

$$
\begin{aligned}
& t_{\delta_1} t_{\delta_2} t_{\delta_3} \\
& \quad = t_{a_3} t_{b_1} t_{a_1} t_{a_2} t_{a_3} t_{b_1} \underline{t_{a_3} t_x t_y} t_{\psi(a_3)} t_{\psi(b_1)} t_{\psi(a_3)} t_{\psi^2(b_2)} t_{\psi(a_3)} t_{\psi(b_2)} t_{a_3} t_{b_2} t_{a'_4} t_{a_6} ,
\end{aligned}
$$

which involves new twist curves x, y given in Figure 2. Note that x is the only separating twist curve here. Importantly, the "lantern sphere" we utilized did not contain the base points. The final positive factorization we obtained above describes a genus–2 Lefschetz pencil (X, f) with 3 base points.

The Seiberg-Witten adjunction inequality implies that in any closed symplectic 4–manifold, except for rational or ruled surfaces, we have $2g(F)-2 = -\mathrm{e}(F) \geq [F]^2$ for any connected symplectic surface F [**13, 20**]. However, (X, f), equipped with the Gompf-Thurston symplectic form, has a symplectic fiber F of genus $g(F) = 2$ and self-intersection $[F]^2 = 3$ (the number of base points), violating the inequality. So we conclude that X is a rational or ruled surface. We can then determine its exact diffeomorphism type through a simple calculation of its algebraic topological invariants. The first integral homology group $H_1(X)$, which is isomorphic to the quotient of $H_1(F)$ by the subgroup normally generated by homology classes of vanishing cycles, is easily seen to be trivial: the vanishing cycles a_1, b_1, a_3 and b_2 already kill the whole group. Since the Euler characteristic of X is given by $\mathrm{e}(X) = 4 - 4g + |C| - |B|$, where C and B are the critical locus and the base locus, we calculate $\mathrm{e}(X) = 12$. It follows that $X \cong \mathbb{CP}^2 \# 9\overline{\mathbb{CP}}^2$.

Now let b_j be the base point of (X, f) corresponding to the boundary components δ_j in the factorization, for $j = 1, 2, 3$. For $B' = \{b_2, b_3\}$ and $B'' = \{b_1, b_2\}$, our recipe generates a pair of inequivalent genus–5 pencils f' and f'' on the rational ruled surface $\widetilde{X} \cong \mathbb{CP}^2 \# 11\,\overline{\mathbb{CP}}^2 \cong S^2 \times S^2 \# \, 10\,\overline{\mathbb{CP}}^2$ with 8 base points, where f' has 1 reducible fiber but f'' has none. (As illustrated in Figure 1.)

3.2. Inequivalent pencils and fibrations on irrational ruled surfaces. We next consider the following positive factorization of the boundary multi-twist in Γ^3_{2h}, for each $h \geq 1$, due to Noriyuki Hamada [**11**]:

$$t_{\delta_1} t_{\delta_2} t_{\delta_3} \;=\; t_{B_{0,1}} t_{B_{1,1}} \cdots t_{B_{h,1}} t_{C_1} t_{B_{0,2}} t_{B_{1,2}} \cdots t_{B_{h,2}} t_{C_2}\,,$$

where the curves $B_{i,k}, C_k$ for $i = 0, 1, \ldots, h$ and $k = 1, 2$ are as in Figure 3. (This is the positive factorization "W_{IIA} for even genus" in [**11**][Figure 23] with only 3 boundary components we picked and relabeled here. It is a lift of a well-known positive factorization of identity in Γ_2 [**7, 12, 14**], which corresponds to a hyperelliptic Lefschetz fibration on $S^2 \times \Sigma_h \# 4\overline{\mathbb{CP}}^2$.) Corresponding to this positive factorization is a genus–$2h$ pencil (X, f) with 3 base points, where $X \cong S^2 \times \Sigma_h \# \,\overline{\mathbb{CP}}^2$. Note that f has 2 separating vanishing cycles.

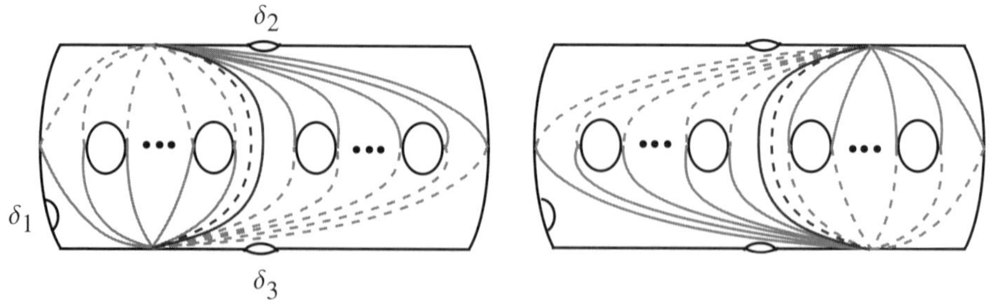

FIGURE 3. From left to right: twist curves $B_{0,1}, B_{1,1}, \ldots, B_{h,1}, C_1$, and $B_{0,2}, B_{1,2}, \ldots, B_{h,2}, C_2$ on (identical copies of) the surface Σ^3_{2h}. The boundary components are $\delta_1, \delta_2, \delta_3$. The separating curves C_1 and C_2 are given in blue.

Once again, let b_j be the base point of (X, f) corresponding to the boundary component δ_j in the factorization, for $j = 1, 2, 3$. We see that for $B' = \{b_2, b_3\}$ and

$B'' = \{b_1, b_2\}$, our recipe generates a pair of inequivalent genus-$(4h+1)$ pencils f' and f'' on the irrational ruled surface $\widetilde{X} \cong S^2 \times \Sigma_h \# 2\overline{\mathbb{CP}}^2 \cong S^2 \widetilde{\times} \Sigma_h \# 2\overline{\mathbb{CP}}^2$ with 8 base points, where f' has 2 reducible fibers but f'' has 1.

3.3. Generalizations. We end with a few remarks on our recipe for constructing inequivalent pencils and fibrations in general:

1. From each pair of inequivalent pencils f' and f'' we obtained, we can easily derive infinitely many more pairs of inequivalent pencils and fibrations of arbitrarily high genera. This is achieved by taking further doubles of f' and f''; the base points of the new pencil will lie only on one component of each reducible fiber in f' or f'' (see the monodromy in [**3**][Theorem 4]), so the number of reducible fibers will remain different.

2. One can produce more than just a pair of inequivalent pencils and fibrations, whenever there is an input pencil (X, f) with several subcollections of base points $B', B'', \ldots, B^{(n)}$ of equal rank, which have different numbers $K', K'', \ldots, K^{(n)}$ of reducible fiber components they miss. For instance, a further lift of Hamada's positive factorization to Γ^4_{2h} in [**11**] can be used to generate yet another pencil f''' with no reducible fibers.

3. Our recipe can be improved in several ways. For example, we only compared the *number* of reducible fibers, yet one can as well compare the *topological types* of reducible fibers (determined by the genera of the subsurfaces they bound). It seems plausible that one can produce *arbitrarily many* inequivalent fibrations and pencils by employing doubling *sequences* as in [**4**] for pencils on blow-ups of the same manifold, which contain different topological types of reducible fibers.

Acknowledgments

These examples of inequivalent pencils and fibrations came to life during my visit to Seoul, Korea in 2015, and were first announced during my lectures at the highly stimulating *2015 Seoul National University Topology Winter School.*

References

[1] Denis Auroux, *Mapping class group factorizations and symplectic 4-manifolds: some open problems*, Problems on mapping class groups and related topics, Proc. Sympos. Pure Math., vol. 74, Amer. Math. Soc., Providence, RI, 2006, pp. 123–132, DOI 10.1090/pspum/074/2264537. MR2264537

[2] D. Auroux, S. K. Donaldson, and L. Katzarkov, *Luttinger surgery along Lagrangian tori and non-isotopy for singular symplectic plane curves*, Math. Ann. **326** (2003), no. 1, 185–203, DOI 10.1007/s00208-003-0418-9. MR1981618

[3] Denis Auroux and Ludmil Katzarkov, *A degree doubling formula for braid monodromies and Lefschetz pencils*, Pure Appl. Math. Q. **4** (2008), no. 2, Special Issue: In honor of Fedor Bogomolov., 237–318, DOI 10.4310/PAMQ.2008.v4.n2.a2. MR2400878

[4] R. İnanç Baykur, *Inequivalent Lefschetz fibrations and surgery equivalence of symplectic 4-manifolds*, J. Symplectic Geom. **14** (2016), no. 3, 671–686, DOI 10.4310/JSG.2016.v14.n3.a2. MR3548482

[5] R. İnanç Baykur and Kenta Hayano, *Multisections of Lefschetz fibrations and topology of symplectic 4-manifolds*, Geom. Topol. **20** (2016), no. 4, 2335–2395, DOI 10.2140/gt.2016.20.2335. MR3548468

[6] Joan S. Birman and Hugh M. Hilden, *On the mapping class groups of closed surfaces as covering spaces*, Advances in the theory of Riemann surfaces (Proc. Conf., Stony Brook, N.Y., 1969), Princeton Univ. Press, Princeton, N.J., 1971, pp. 81–115. Ann. of Math. Studies, No. 66. MR0292082

[7] Carlos Alberto Cadavid, *On a remarkable set of words in the mapping class group*, ProQuest LLC, Ann Arbor, MI, 1998. Thesis (Ph.D.)–The University of Texas at Austin. MR2699379

[8] S. K. Donaldson, *Lefschetz pencils on symplectic manifolds*, J. Differential Geom. **53** (1999), no. 2, 205–236. MR1802722

[9] Simon Donaldson and Ivan Smith, *Lefschetz pencils and the canonical class for symplectic four-manifolds*, Topology **42** (2003), no. 4, 743–785, DOI 10.1016/S0040-9383(02)00024-1. MR1958528

[10] Robert E. Gompf and András I. Stipsicz, *4-manifolds and Kirby calculus*, Graduate Studies in Mathematics, vol. 20, American Mathematical Society, Providence, RI, 1999. MR1707327

[11] N. Hamada, *Sections of the Matsumoto-Cadavid-Korkmaz Lefschetz fibration,* preprint; https://arxiv.org/abs/1610.08458.

[12] Mustafa Korkmaz, *Noncomplex smooth 4-manifolds with Lefschetz fibrations*, Internat. Math. Res. Notices **3** (2001), 115–128, DOI 10.1155/S107379280100006X. MR1810689

[13] T. J. Li and A. Liu, *Symplectic structure on ruled surfaces and a generalized adjunction formula*, Math. Res. Lett. **2** (1995), no. 4, 453–471, DOI 10.4310/MRL.1995.v2.n4.a6. MR1355707

[14] Yukio Matsumoto, *Lefschetz fibrations of genus two—a topological approach*, Topology and Teichmüller spaces (Katinkulta, 1995), World Sci. Publ., River Edge, NJ, 1996, pp. 123–148. MR1659687

[15] Sinem Çelik Onaran, *On sections of genus two Lefschetz fibrations*, Pacific J. Math. **248** (2010), no. 1, 203–216, DOI 10.2140/pjm.2010.248.203. MR2734172

[16] Jongil Park and Ki-Heon Yun, *Nonisomorphic Lefschetz fibrations on knot surgery 4-manifolds*, Math. Ann. **345** (2009), no. 3, 581–597, DOI 10.1007/s00208-009-0366-0. MR2534109

[17] Jongil Park and Ki-Heon Yun, *Lefschetz fibrations on knot surgery 4-manifolds via Stallings twist*, Michigan Math. J. **66** (2017), no. 3, 481–498, DOI 10.1307/mmj/1497513628. MR3695347

[18] I. Smith, *Symplectic geometry of Lefschetz fibrations,* Ph.D. thesis, University of Oxford (1998).

[19] Ivan Smith, *Lefschetz pencils and divisors in moduli space*, Geom. Topol. **5** (2001), 579–608, DOI 10.2140/gt.2001.5.579. MR1833754

[20] Clifford Henry Taubes, *The Seiberg-Witten and Gromov invariants*, Math. Res. Lett. **2** (1995), no. 2, 221–238, DOI 10.4310/MRL.1995.v2.n2.a10. MR1324704

Department of Mathematics and Statistics, University of Massachusetts, Amherst, Massachusetts 01003-9305

Email address: `baykur@math.umass.edu`

Proceedings of Symposia in Pure Mathematics
Volume **102**, 2019
https://doi.org/10.1090/pspum/102/01807

Virtual and welded periods of classical knots

Hans U. Boden and Andrew J. Nicas

ABSTRACT. We show that any virtual or welded period of a classical knot can be realized as a classical period. A direct consequence is that a classical knot admits only finitely many virtual or welded periods.

In [**Fox62**], Fox gave an overview of the Smith Conjecture, which was unresolved at the time, concerning periodic transformations of the 3-sphere preserving a simple closed curve. Question 7 of [**Fox62**] deals with the periodic symmetries of a knot, which he observed come in eight different varieties. Of special interest are the two cases corresponding to knots admitting a "free period" or a "cyclic period", respectively. A *symmetry* of a (tame) knot $K \subset S^3$ is a diffeomorphism $\varphi\colon S^3 \to S^3$ of finite order p (the *period* of φ) such that $\varphi(K) = K$. A knot K is said to be:

- *freely periodic* if there is a fixed point free, orientation preserving symmetry of K,
- *(cyclically) periodic* if there is an orientation preserving symmetry of K whose fixed point set is non-empty and disjoint from K.

In the second case, the solution of the Smith Conjecture [**Mor84**] implies that the fixed point set is an unknotted circle and that the symmetry is conjugate in the diffeomorphism group of S^3 to a rotation about an axis in S^3. A fundamental result of E. Flapan asserts that a nontrivial classical knot admits only finitely many periods (see [**Fla85**], where she also shows that any knot, other than a torus knot, admits only finitely many free periods).

Virtual knots and links were defined by Kauffman in [**Kau99**] as equivalence classes of virtual knot diagrams by the extended Reidemeister moves. Very briefly, a virtual knot diagram is a 4-valent graph in the plane with two types of crossings, classical and virtual crossings. Classical crossings are depicted with over and under-crossings as usual, and virtual crossings are circled. A diagram without virtual crossings is called classical because it represents a classical knot or link. In [**GPV00**], it is shown that for classical knots, equivalence through virtual knot

---

2010 *Mathematics Subject Classification.* Primary: 57M25, Secondary: 57M27.

Key words and phrases. Virtual and welded knots, periodic knots, group automorphisms, Nielsen realization.

The authors were supported by grants from the Natural Sciences and Engineering Research Council of Canada.

diagrams implies equivalence as classical knots, and thus virtual knot theory is an extension of classical knot theory.

Allowing the first forbidden move (see [**Kau99**]) leads to the coarser notion of *welded knots*. Satoh showed that welded knots represent ribbon knotted tori in S^4 [**Sat00**]. Two classical knots that are equivalent as welded knots are equivalent as classical knots, and so classical knot theory embeds into the theory of welded knots. In this paper, we are concerned with symmetries of classical knots that arise from periodic virtual knot diagrams, which were first introduced by S. Y. Lee in [**Lee12**] and which are extended here to include welded knots.

DEFINITION 1. Let p be a positive integer. A virtual knot diagram that misses the origin is said to be *p-periodic* if it is invariant under a rotation about the origin in the plane by an angle of $2\pi/p$. A virtual or welded knot K is said to be *p-periodic* if it admits a p-periodic virtual knot diagram, and in that case, we say that p is a *virtual period* or *welded period* for the virtual or welded knot K.

S. Y. Lee asked whether classical knots or links can admit "exotic" virtual periods, cf. [**Lee12**, Question 4], and we rephrase his question to include welded knots as follows.

QUESTION 2. Can a classical knot admit virtual or welded periods which are different from its classical periods?

The following theorem answers this question and is our main result.

THEOREM 3. *If a classical knot admits a p-periodic virtual or welded knot diagram, then it admits a p-periodic classical knot diagram. The sets of virtual and welded periods of a given classical knot are the same as its set of classical (cyclic) periods. In particular, a nontrivial classical knot admits only finitely many virtual and welded periods.*

Let $N(K)$ be a regular neighborhood of a knot $K \subset S^3$. The *exterior* of K is the compact manifold with torus boundary: $E_K = S^3 \smallsetminus \mathrm{Int}(N(K))$. Let $\mathrm{Mcg}(E_K)$ denote the *mapping class group of E_K*, that is, the group of diffeomorphisms of E_K modulo the subgroup of diffeomorphisms isotopic to the identity. Let $G_K = \pi_1(E_K)$ denote knot group of K.

The proof of Theorem 3 will be presented in 5. We sketch the argument in the special case where K is a hyperbolic knot, that is, $\mathrm{Int}(E_K)$ admits a complete hyperbolic metric of finite volume. Let $\mathrm{Out}(G_K)$ be the group of outer automorphisms of G_K (see 3) and let $\mathrm{Isom}(E_K)$ be the group of isometries of the complete hyperbolic metric of finite volume.

THEOREM 4 (see Theorem 10.5.3 [**Kaw96**]). *If K is a hyperbolic knot, then there are isomorphisms* $\mathrm{Mcg}(E_K) \cong \mathrm{Out}(G_K) \cong \mathrm{Isom}(E_K)$.

In Theorem 4, the isomorphism $\mathrm{Mcg}(E_K) \cong \mathrm{Out}(G_K)$ is a special case of [**Wal68**, Corollary 7.5], and the isomorphism $\mathrm{Out}(G_K) \cong \mathrm{Isom}(E_K)$ makes use of Mostow's rigidity theorem [**Mos68**], as extended to complete, finite volume hyperbolic manifolds by Prasad [**Pra73**].

Assume that K admits a p-periodic virtual knot diagram. Such a diagram gives rise to an automorphism $\varphi\colon G_K \to G_K$ which, by Proposition 5, must have order p. The fundamental group of any complete, finite volume hyperbolic manifold is torsion free and has trivial center and so G_K, when K is a hyperbolic knot, has these properties. This observation allows us to conclude that the image of φ in $\mathrm{Out}(G_K)$ also has order p, and Theorem 4 applies to show that φ corresponds to an isometry $\varphi_* \in \mathrm{Isom}(E_K)$ of order p. There is a flat torus T obtained as a cross section of the cusp associated to the hyperbolic structure on $\mathrm{Int}(E_K)$ and a compact submanifold $E'_K \subset \mathrm{Int}(E_K)$, with $T = \partial E'_K$, that is also an exterior for K and is invariant under φ_*. By [**Luo92**, Theorem 2], $(\varphi_*)|_{E'_K}$ extends to an order p diffeomorphism $\widehat{\varphi}_*\colon S^3 \to S^3$ preserving K. (Alternatively, one can conclude this from the main theorem of [**GL89**] as follows. The diffeomorphism $(\varphi_*)|_{E'_K}\colon E'_K \to E'_K$ maps a meridian of K to another meridian for K, and any p-periodic diffeomorphism defined on the boundary torus ∂V of a solid torus V which sends a meridian to another meridian can be extended to a p-periodic diffeomorphism of V.) Furthermore, Proposition 9 can be used to show $\widehat{\varphi}_*$ is orientation preserving. While this argument produces an orientation preserving symmetry of K of order p, it does not immediately exclude the possibility that $\widehat{\varphi}_*$ could be a free symmetry. That exclusion is accomplished using our Theorem 15 as in the proof of Theorem 3 given in 5.

1. Periodic virtual knots and Wirtinger presentations

Let K be a virtual knot, and let G_K denote its knot group. Given a virtual knot diagram for K, we obtain a Wirtinger presentation for G_K, and the isomorphism class of G_K and its peripheral structure P_K are invariants of the underlying welded equivalence class of K [**Kim00**]. If K is classical, then $G_K = \pi_1(E_K)$, the fundamental group of the exterior of the knot $E_K = S^3 \smallsetminus \mathrm{Int}(N(K))$, and P_K is the conjugacy class of the image of the fundamental group of ∂E_K in G_K.

Suppose K is a classical or virtual knot and D is a p-periodic virtual knot diagram representing K. The periodic transformation of the diagram induces an automorphism $\varphi\colon G_K \to G_K$ as follows. Recall that virtual knots can be represented as knots in thickened surfaces $\Sigma \times I$. From this point of view, the knot group is the fundamental group of the complement of K in the cone space $C(\Sigma \times I) = \Sigma \times I/\!\sim$ obtained by identifying $(x, 1) \sim (y, 1)$ for all $x, y \in \Sigma$. Notice that the periodic transformation of the diagram defines a homeomorphism of $C(\Sigma \times I)$ fixing the cone point, which we take to be the base point in the fundamental group.

Using symmetry, it follows that D must have np real crossings and also np arcs, for some $n > 0$, which are paths from one real undercrossing to the next undercrossing, ignoring virtual crossings. We denote the quotient knot by K_*, it is represented by the diagram D_* obtained by taking the quotient of D by the $\mathbb{Z}/p$ action. Notice that D_* has n real crossings.

We will use the symmetry of the diagram to produce a particularly useful Wirtinger presentation for the knot group G_K, following the approach given in [**BZH14**, Ch. 14D], cf. [**BNW18**, §3.1]. Let

$$c_{1,0}, \dots, c_{n,0}, c_{1,1}, \dots, c_{n,1}, \dots, c_{1,p-1}, \dots, c_{n,p-1}$$

denote the undercrossings of D in the order in which they are encountered as one travels around the knot, and let $\varepsilon_{i,j}$ denote the writhe of $c_{i,j}$ for $1 \leqslant i \leqslant n$ and $0 \leqslant j \leqslant p-1$. Notice that by symmetry, $\varepsilon_{i,j} = \varepsilon_{i,j+1}$ for $0 \leqslant j \leqslant p-1$, so we will write ε_i instead of $\varepsilon_{i,j}$. We choose the generator φ of $\mathbb{Z}/p$ that acts on the diagram D by sending $c_{i,j}$ to $c_{i,j+1}$, where $1 \leqslant i \leqslant n$ and $0 \leqslant j \leqslant p-1$ and $j+1$ is taken mod p (i.e., if $j = p-1$, then $j+1$ is taken to equal 0).

Using this setup, we can label the arcs of the diagram D

$$a_{1,0}, \dots, a_{n,0}, a_{1,1}, \dots, a_{n,1}, \dots, a_{1,p-1}, \dots, a_{n,p-1}$$

so that the arc $a_{i,j}$ starts at the crossing $c_{i-1,j}$ and ends at the crossing $c_{i,j}$ for $i = 2, \dots, n$. (In case $i = 1$, the arc $a_{1,j}$ starts at $c_{n,j-1}$ and ends at $c_{1,j}$.) The automorphism acts on the arcs by sending $a_{i,j}$ to $a_{i,j+1}$ for $1 \leqslant i \leqslant n$ and $0 \leqslant j \leqslant p-1$ with j taken mod p as above.

For example, Figure 1 depicts a 3-periodic diagram of a virtual knot with crossings $c_{i,j}$ and arcs $a_{i,j}$ labelled according to our conventions. Under clockwise rotation of $2\pi/3$ about the center, the automorphism φ acts by sending the crossing $c_{i,j}$ to $c_{i,j+1}$ and the arc $a_{i,j}$ to $a_{i,j+1}$ for $i = 1, 2$ and $j = 0, 1, 2$ with $j+1$ taken mod 3.

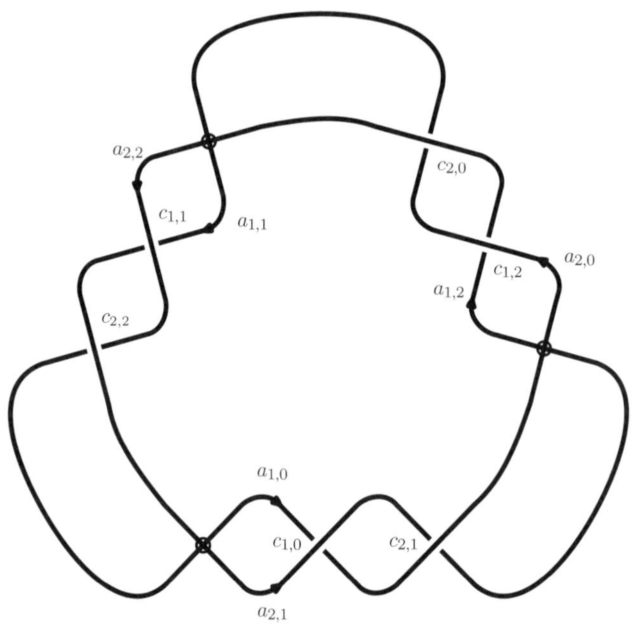

FIGURE 1. A 3-periodic virtual knot

In general, for a periodic diagram, the Wirtinger relation of the crossing $c_{i,j}$ is given by

$$r_{i,j} = (a_{k,\ell})^{-\varepsilon_i}\, a_{i,j}\, (a_{k,\ell})^{\varepsilon_i}\, (a_{i+1,j})^{-1}$$

for $i = 1, \dots, n-1$ and $j = 0, \dots, p-1$. When $i = n$, we get the relation $r_{n,j} = (a_{k,\ell})^{-\varepsilon_n}\, a_{n,j}\, (a_{k,\ell})^{\varepsilon_n} (a_{1,j})^{-1}$.

The resulting Wirtinger presentation of the p-periodic virtual knot K is then

$$G_K = \langle a_{i,j} \mid r_{i,j} \rangle, \tag{1}$$

where $1 \leqslant i \leqslant n$, $0 \leqslant j \leqslant p-1$. This presentation admits a $\mathbb{Z}/p$ symmetry, and the Wirtinger presentation for the knot group of the quotient knot K_* is obtained by adding the relations $a_{i,0} = a_{i,1} = \cdots = a_{i,p-1}$ for $1 \leqslant i \leqslant n$, which gives the presentation for the quotient group

$$G_{K_*} = \langle a_1, \dots, a_n \mid r_1, \dots, r_n \rangle,$$

where a_i refers to the equivalence class $\{a_{i,0}, \ldots, a_{i,p-1}\}$ of generators and r_i is the relation $(a_k)^{-\varepsilon_i} a_i (a_k)^{\varepsilon_i} (a_{i+1})^{-1}$ with $i+1$ taken mod n.

Let $\pi\colon G_K \to G_{K_*}$ denote the quotient map.

For an abstract group G, we use $\mathrm{Aut}(G)$ to denote its group of automorphisms.

PROPOSITION 5. *Suppose K is a virtual knot with meridian-longitude pair $\mu, \lambda \in G_K$ such that λ is nontrivial. If K admits a p-periodic virtual knot diagram D, then the corresponding automorphism $\varphi \in \mathrm{Aut}(G_K)$ has order p and preserves the peripheral system P_K. In fact, $\varphi(\mu) = g\mu g^{-1}$ and $\varphi(\lambda) = g\lambda g^{-1}$ for some $g \in G_K$.*

PROOF. Clearly $\varphi^p = \mathrm{id}$, so it suffices to show that $\varphi^k \neq \mathrm{id}$ for any $1 \leqslant k < p$. We first claim that $\varphi \neq \mathrm{id}$. We show this by contradiction. Assuming that $\varphi = \mathrm{id}$, it follows that the generators and the relations for G_K in the presentation (1) satisfy $a_{i,j} = a_{i,j+1}$ and $r_{i,j} = r_{i,j+1}$ for $1 \leqslant i \leqslant n$. These observations reveal that if $\varphi = \mathrm{id}$, then the quotient map $\pi\colon G_K \to G_{K_*}$ is an isomorphism of groups.

On the other hand, choose a meridian μ and longitude λ for K as follows. Let $\mu = a_{1,0} \in G_K$, and starting on the arc $a_{1,0}$ and going once around the knot, let ω be the word obtained by recording the overcrossing arc $(a_{k,\ell})^{\varepsilon_i}$ according to its writhe ε_i at each undercrossing $c_{i,j}$. We then set $\lambda = \omega\, (a_{1,0})^{-m}$, where m denotes the exponent sum of ω. Notice that the homology classes of $\mu, \lambda \in G_K$ can be represented by curves in $T^2 \subset \partial(\Sigma \times I \smallsetminus \mathrm{Int}(N(K)))$ with $[\mu] \cdot [\lambda] = 1$. We can choose a compatible meridian and longitude $\mu^*, \lambda^* \in G_{K_*}$ for K_* such that $\pi(\mu) = \mu^*$ and $\pi(\lambda) = (\lambda^*)^p$.

However, since π is an isomorphism, there is an element $\zeta \in G_K$ with $\pi(\zeta) = \lambda^*$ and $\zeta^p = \lambda$. Thus λ admits a p-th root. In that case, $[\mu]\cdot[\lambda] = p[\mu]\cdot[\zeta]$ is a multiple of p, which contradicts the fact that $[\mu] \cdot [\lambda] = 1$.

Now suppose $\varphi \in \mathrm{Aut}(G_K)$ has order k for some $1 < k < p$. If k is relatively prime to p, then writing $1 = \ell k + mp$, it follows that $\varphi = \mathrm{id}$, which is a contradiction. The only other possibility is when $p = \ell k$ for some ℓ. In this case, notice that K is ℓ-periodic, and set $\varphi' = \varphi^k$. Arguing as before shows that $\varphi' \neq \mathrm{id}$, and it follows that φ cannot have order k. □

The hypothesis in Proposition 5 that the longitude is nontrivial is essential. There are nontrivial virtual knots K with $G_K \cong \mathbb{Z}$. In that case, the longitude is trivial and the automorphism arising from a periodic diagram is also trivial, see Figure 2 for a specific example.

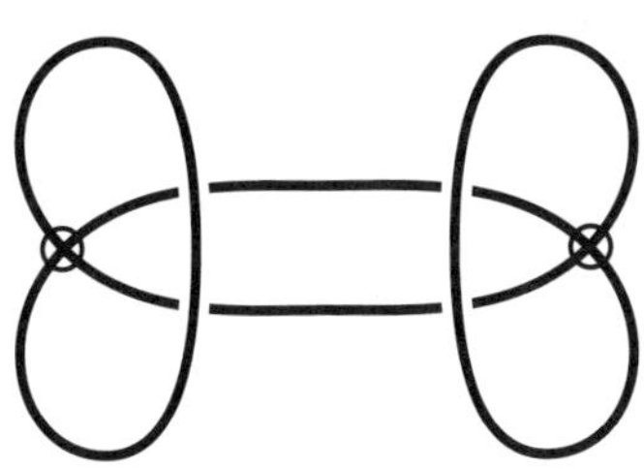

FIGURE 2. The Kishino knot (pictured) is 2-periodic and has infinite cyclic knot group.

QUESTION 6. Does there exist a periodic virtual or welded (non-classical) knot with "nontrivial" knot group, that is $G_K \not\cong \mathbb{Z}$, and trivial longitude for which the induced automorphism $\varphi\colon G_K \to G_K$ is the identity?

S.-G. Kim has shown that any classical knot group is the group of a (non-classical) virtual knot with trivial longitude, see [**Kim00**, Corollary 7].

QUESTION 7. Does there exist a periodic virtual or welded knot whose knot group is isomorphic to that of a classical knot such that the induced automorphism is not conjugate to any automorphism arising from a symmetry of the classical knot?

Note that any automorphism of the knot group of a prime classical knot preserves its peripheral structure, (see [**Tsa88**, Corollary 4.1]), and so the knot group in Question 7 should be that of a composite classical knot.

2. The group of outer automorphisms

Given a group G, an *inner automorphism* is an element of $\mathrm{Aut}(G)$ of the form $c_g\colon G \to G$, where $g \in G$ and $c_g(x) = gxg^{-1}$ for all $x \in G$ (conjugation by g). The subset of all inner automorphisms, $\mathrm{Inn}(G)$, is a normal subgroup of $\mathrm{Aut}(G)$ and the group of *outer automorphisms* of G is defined as the quotient group $\mathrm{Out}(G) = \mathrm{Aut}(G)/\mathrm{Inn}(G)$. There is an exact sequence

$$1 \longrightarrow Z(G) \xrightarrow{\ i\ } G \xrightarrow{\ C\ } \mathrm{Inn}(G) \longrightarrow 1,$$

where $Z(G) = \{x \in G \mid gx = xg \text{ for all } g \in G\}$ is the *center* of G, $i\colon Z(G) \hookrightarrow G$ is inclusion and $C(g) = c_g$ for $g \in G$. Hence $g \in Z(G)$ if and only if $c_g = \mathrm{id}_G$.

Let $Q\colon \mathrm{Aut}(G) \to \mathrm{Out}(G)$ denote the quotient homomorphism. The following proposition will be used in the proof of Theorem 12.

PROPOSITION 8. *Let G be a torsion free group with trivial center. If $\varphi \in \mathrm{Aut}(G)$ has finite order $n \geqslant 1$, then $Q(\varphi) \in \mathrm{Out}(G)$ also has order n.*

PROOF. Let m be the order of $Q(\varphi)$. Note that m divides n and $\varphi^m = c_g$ for some $g \in G$. Let $s = n/m$. Then $\mathrm{id}_G = \varphi^n = \varphi^{ms} = (c_g)^s = c_{g^s}$. Since the center of G is trivial it follows that $g^s = 1$ and so $g = 1$ because G is torsion free. Hence $\varphi^m = 1$ and so $m = n$. □

The following "relative" versions of automorphism and outer automorphism groups are relevant in knot theoretic applications. Let P be the conjugacy class of some subgroup H of G, that is, $P = \{gHg^{-1} \mid g \in G\}$. Let $\mathrm{Aut}(G,P)$ be the subgroup of $\mathrm{Aut}(G)$ consisting of all automorphisms that preserve P. Observe that $\mathrm{Inn}(G)$ is a normal subgroup of $\mathrm{Aut}(G,P)$. Let $\mathrm{Out}(G,P) = \mathrm{Aut}(G,P)/\mathrm{Inn}(G)$. Note that $\mathrm{Out}(G,P)$ is a subgroup of $\mathrm{Out}(G)$.

3. Automorphisms of knot groups

Given a knot K in S^3, the inclusion $\partial E_K \hookrightarrow E_K$ together with a choice of basepoint induces a homomorphism on fundamental groups that is injective if K is not the unknot. The conjugacy class of the image of this homomorphism, which we denote by P_K, is the *peripheral system* associated to K. It is known that the pair (G_K, P_K) is a complete knot invariant, that is, two (classical) knots K and K' are equivalent if and only if the pairs (G_K, P_K) and $(G_{K'}, P_{K'})$ are isomorphic.

A homeomorphism $F\colon E_K \to E_K$ induces an isomorphism of relative homology groups $F_*\colon H_3(E_K, \partial E_K) \to H_3(E_K, \partial E_K)$. Since E_K is a compact, connected orientable 3-manifold, the homology group $H_3(E_K, \partial E_K)$ is infinite cyclic and so F_* is either the identity, in which case we say F_* is *orientation preserving*, or -1 times the identity, in which case we say F_* is *orientation reversing*. By Alexander Duality, E_K is a homology circle and thus $H_2(E_K) = H_3(E_K) = 0$. The homology long exact sequence of the pair $(E_K, \partial E_K)$, together with its naturality, yield a commutative square

$$\begin{array}{ccc} H_3(E_K, \partial E_K) & \xrightarrow{F_*} & H_3(E_K, \partial E_K) \\ {\scriptstyle\delta}\downarrow & & \downarrow{\scriptstyle\delta} \\ H_2(\partial E_K) & \xrightarrow{(F|_{\partial E_K})_*} & H_2(\partial E_K), \end{array}$$

where the connecting homomorphism $\delta\colon H_3(E_K, \partial E_K) \to H_2(\partial E_K)$ is an isomorphism. It follows that F_* is the identity (respectively, -1 times the identity) if and only if $(F|_{\partial E_K})_*$ is the identity (respectively, -1 times the identity), that is, F is orientation preserving (respectively, reversing) if and only if $F|_{\partial E_K}$ is orientation preserving (respectively, reversing).

Assume that K is not the unknot. We assign to an automorphism $\varphi \in \mathrm{Aut}(G_K, P_K)$ the status "orientation preserving", respectively, "orientation reversing", as follows. Let $H \in P_K$. There exists $g \in G_K$ such that $\varphi(H) = g^{-1}Hg$. The composition in group homology

$$H_2(H) \xrightarrow{(\varphi|_H)_*} H_2(g^{-1}Hg) \xrightarrow{(c_g)_*} H_2(H) \tag{2}$$

is an isomorphism. This isomorphism is independent of the element g satisfying $\varphi(H) = g^{-1}Hg$ because if $g_1^{-1}Hg_1 = g_2^{-1}Hg_2$, then $u = g_1g_2^{-1} \in N_{G_K}(H)$, the normalizer of H in G_K, and hence $u \in H$ since $N_{G_K}(H) = H$ by [**Hei81**, Corollary, p. 148] (this is also a consequence of [**Sim76**, Theorem 2]). Observe that $g_1 = ug_2$ and so $c_{g_1} = c_{ug_2} = c_u c_{g_2}$. Since H is abelian and $u \in H$, $c_u = \mathrm{id}_H$ and so $c_{g_1} = c_{g_2}$.

Note that $H_2(H) \cong H_2(\partial E_K)$ is infinite cyclic. Hence the isomorphism (2) is either the identity, in which case we say φ is *orientation preserving*, or -1 times the identity, in which case we say φ is *orientation reversing*. It is straightforward to show that our definition of orientation preserving or reversing for φ is independent of the choice of $H \in P_K$. The set of all orientation preserving automorphisms is a subgroup, denoted $\mathrm{Aut}^+(G_K, P_K)$, of $\mathrm{Aut}(G_K, P_K)$ of index at most 2. Observe that any inner automorphism $\varphi = c_g$ is orientation preserving because $c_{g^{-1}}\varphi = c_{g^{-1}}c_g = \mathrm{id}_H$, and so orientation preserving and orientation reversing are well-defined notions for outer automorphisms. We denote the subgroup of orientation preserving outer automorphisms by $\mathrm{Out}^+(G_K, P_K)$.

The geometrically and algebraically defined versions of orientation preserving (or reversing) are related as follows. Let $x_0 \in \partial E_K$ be a basepoint and let

$$H = \mathrm{image}\,(i_*\colon \pi_1(\partial E_K, x_0) \to \pi_1(E_K, x_0)) \subset G_K = \pi_1(E_K, x_0).$$

If $F\colon E_K \to E_K$ is a homeomorphism and σ is a path in E_K from x_0 to $F(x_0)$, then the composition

$$\pi_1(E_K, x_0) \xrightarrow{F_*} \pi_1(E_K, F(x_0)) \xrightarrow{\sigma_\#^{-1}} \pi_1(E_K, x_0)$$

is an automorphism $\varphi \in \operatorname{Aut}(G_K, P_K)$, where P_K is the conjugacy class of H. Note that if the path σ lies in ∂E_K, then $\varphi(H) = H$. There is a commutative diagram

$$\begin{array}{ccccc} \pi_1(\partial E_K, x_0) & \xrightarrow{(F|_{\partial E_K})_*} & \pi_1(\partial E_K, F(x_0)) & \xrightarrow{\sigma_\#^{-1}} & \pi_1(\partial E_K, x_0) \\ \downarrow h & & \downarrow h & & \downarrow h \\ H_1(\partial E_K) & \xrightarrow{(F|_{\partial E_K})_*} & H_1(\partial E_K) & = & H_1(\partial E_K), \end{array} \tag{3}$$

where h denotes the Hurewicz homomorphism, which is an isomorphism in the above diagram because $\pi_1(\partial E_K, x_0)$ is abelian. Since ∂E_K is a 2-torus, the Pontryagin product gives an isomorphism

$$\Lambda^2 H_1(\partial E_K) \longrightarrow H_2(\partial E_K), \tag{4}$$

where $\Lambda^2 H_1(\partial E_K)$ is the second exterior power of $H_1(\partial E_K)$. It follows from (2), (3) and (4) that φ is orientation preserving (respectively, reversing) if and only if F is orientation preserving (respectively, reversing); moreover, F is orientation preserving if and only if $\det\left((F|_{\partial E_K})_*\colon H_1(\partial E_K) \to H_1(\partial E_K)\right) = 1$, equivalently, φ is orientation preserving if and only if $\det\left(c_g \circ \varphi|_H \colon H \to H\right) = 1$.

PROPOSITION 9. *Let K be a classical knot that admits a p-periodic virtual knot diagram D and let $\varphi \in \operatorname{Aut}(G_K, P_K)$ be the automorphism associated to D. Then φ is orientation preserving.*

PROOF. Let H be the peripheral subgroup of G_K generated by the meridian-longitude pair $\mu, \lambda \in G_K$ as in Proposition 5. Also by Proposition 5, $\varphi(\mu) = g^{-1}\mu g$ and $\varphi(\lambda) = g^{-1}\lambda g$ for some $g \in G_K$. Hence $H \xrightarrow{\varphi|_H} g^{-1}Hg \xrightarrow{c_g} H$ is the identity. It follows that the induced homomorphism $(c_g \circ \varphi|_H)_*\colon H_2(H) \to H_2(H)$ is also the identity and so φ is orientation preserving. □

4. Lifting periodic homeomorphisms

We give a criterion for a periodic homeomorphism of a space to lift to a periodic homeomorphism, of the same period, of its universal cover.

Convention. Let α and β be paths, parametrized by the unit interval, in a space Z such that $\alpha(1) = \beta(0)$. We denote path multiplication by $\alpha \cdot \beta$. Moreover, to avoid an excess of notation, for paths α, α' we write $\alpha = \alpha'$ to mean α is homotopic to α' keeping endpoints fixed. The isomorphism of fundamental groups induced by a path α in Z is denoted by $\alpha_\#\colon \pi_1(Z, \alpha(0)) \to \pi_1(Z, \alpha(1))$.

Let X be a path connected, locally path connected, semi-locally simply connected space and let $p\colon Y \to X$ be a universal covering projection. Let $x_0 \in X$ be a basepoint and $y_0 \in Y$ a basepoint covering x_0, that is, $p(y_0) = x_0$.

PROPOSITION 10. *Let $F\colon X \to X$ be an n-periodic homeomorphism where $n > 1$. Let $\sigma\colon [0,1] \to X$ be a path from x_0 to $F(x_0)$. Let $\widetilde{\sigma}\colon [0,1] \to Y$ be the unique lift of σ such that $\widetilde{\sigma}(0) = y_0$ and let $\widetilde{F}\colon Y \to Y$ be the unique lift of F such that $\widetilde{F}(y_0) = \widetilde{\sigma}(1)$. Then $\widetilde{F}$ is an n-periodic homeomorphism if and only if the element $[\tau] \in \pi_1(X, x_0)$ represented by the loop $\tau = \sigma \cdot (F \circ \sigma) \cdot (F^2 \circ \sigma) \cdots (F^{n-1} \circ \sigma)$ is the identity element.*

PROOF. Since $F^n = \mathrm{id}_X$, $(\widetilde{F})^n$ is a lift of id_X, that is, a covering transformation. Furthermore, $(\widetilde{F})^n(y_0) = \widetilde{\tau}(1)$ where $\widetilde{\tau} = \widetilde{\sigma} \cdot (\widetilde{F} \circ \widetilde{\sigma}) \cdots ((\widetilde{F})^{n-1} \circ \widetilde{\sigma})$. Note that $\widetilde{\tau}$ is the unique lift of τ such that $\widetilde{\tau}(0) = y_0$. Hence $(\widetilde{F})^n$ is the covering transformation corresponding to the element $[\tau] \in \pi_1(X, x_0)$. It follows that $(\widetilde{F})^n = \mathrm{id}_Y$ if and only if $[\tau]$ is the identity element. □

For $p\colon Y \to X$ as above we have the following sufficient condition for lifting an n-periodic homeomorphism of X to an n-periodic homeomorphism of Y.

THEOREM 11. *Let $F\colon X \to X$ be an n-periodic homeomorphism where $n > 1$. Assume that the center of $\pi_1(X, x_0)$ is trivial and that F is homotopic to a map $f\colon X \to X$ such that $f(x_0) = x_0$ and that $f_*\colon \pi_1(X, x_0) \to \pi_1(X, x_0)$ has order m, with m dividing n, in* $\mathrm{Aut}(\pi_1(X, x_0))$. *Then F lifts to an n-periodic homeomorphism $\widetilde{F}\colon Y \to Y$.*

PROOF. Let $H\colon f \simeq F$ be a homotopy from f to F. Let $\sigma(t) = H(x_0, t)$ for $t \in [0, 1]$. Then σ is a path from x_0 to $F(x_0)$ and there is a commutative diagram

$$\begin{array}{ccc} \pi_1(X, x_0) & \xrightarrow{F_*} & \pi_1(X, F(x_0)) \\ \| & & \uparrow{\scriptstyle \sigma_\#} \\ \pi_1(X, x_0) & \xrightarrow{f_*} & \pi_1(X, x_0). \end{array}$$

Let $\tau^{(1)} = \sigma$ and for $j = 2, \ldots, n$, let $\tau^{(j)}$ be the path from x_0 to $F^j(x_0)$ given by

$$\tau^{(j)} = \sigma \cdot (F \circ \sigma) \cdots (F^{j-1} \circ \sigma).$$

For $j = 1, \ldots, n$, we have commutative diagrams

$$\begin{array}{ccc} \pi_1(X, F^j(x_0)) & \xrightarrow{F_*} & \pi_1(X, F^{j+1}(x_0)) \\ {\scriptstyle \tau^{(j)}_\#}\uparrow & & \uparrow{\scriptstyle (F\circ\tau^{(j)})_\#} \\ \pi_1(X, x_0) & \xrightarrow{F_*} & \pi_1(X, F(x_0)), \end{array}$$

and so

$$F_*\tau^{(j)}_\# = (F \circ \tau^{(j)})_\# F_* = (F \circ \tau^{(j)})_\# \sigma_\# f_* = (\sigma \cdot (F \circ \tau^{(j)}))_\# f_* = \tau^{(j+1)}_\# f_*.$$

We have $F_* = \sigma_\# f_* = \tau^{(1)}_\# f_*$ and so $F_*^2 = F_*\tau^{(1)}_\# f_* = \tau^{(2)}_\# f_*^2$. Proceeding inductively, we obtain $F_*^n = \tau^{(n)}_\# f_*^n$. By hypothesis, $F^n = \mathrm{id}_X$ and f_*^n is the identity. Hence $\tau^{(n)}_\#$ is the identity isomorphism. Since $\tau = \tau^{(n)}$ is a closed loop, $\tau_\#$ is conjugation by $[\tau]$ and hence $[\tau] \in \pi_1(X, x_0)$ is the identity element because, by hypothesis, $\pi_1(X, x_0)$ has a trivial center. The conclusion of the theorem follows from Proposition 10. □

We recall some background from [**CR77**, §3] that will be needed. Let $\mathcal{E}(X)$ be the set of self-homotopy equivalences of X. Give it the compact-open topology and, in order to avoid pathologies, also assume in the subsequent discussion that X is a locally compact, separable metric ANR, for example, a metrizable topological manifold. Let $\mathcal{E}(X, x_0)$ be the subspace of $\mathcal{E}(X)$ consisting of basepoint preserving maps. Both $\mathcal{E}(X)$ and $\mathcal{E}(X, x_0)$ are H-spaces (with composition of maps as multiplication). The map $\Theta\colon \mathcal{E}(X, x_0) \to \mathrm{Aut}(\pi_1(X, x_0))$ given by $\Theta(f) = f_*$ induces a group homomorphism

$$\Theta_*\colon \pi_0(\mathcal{E}(X, x_0), \mathrm{id}_X) \to \mathrm{Aut}(\pi_1(X, x_0)). \tag{5}$$

Assume that X is path connected. Given $h \in \mathcal{E}(X)$ and a path σ in X from x_0 to $h(x_0)$ the composition

$$\pi_1(X, x_0) \xrightarrow{h_*} \pi_1(X, h(x_0)) \xrightarrow{\sigma_\#^{-1}} \pi_1(X, x_0)$$

yields an automorphism of $\pi_1(X, x_0)$. Another choice of path from x_0 to $h(x_0)$ gives an automorphism that agrees with $\sigma_\#^{-1} h_*$ up to an inner automorphism yielding a well-defined map $\Psi\colon \mathcal{E}(X) \to \mathrm{Out}(\pi_1(X, x_0))$ that induces a group homomorphism

$$\Psi_*\colon \pi_0(\mathcal{E}(X), \mathrm{id}_X) \to \mathrm{Out}(\pi_1(X, x_0)). \tag{6}$$

Let $i\colon \mathcal{E}(X, x_0) \hookrightarrow \mathcal{E}(X)$ be the inclusion map. The diagram

$$\begin{array}{ccc} \pi_0(\mathcal{E}(X, x_0), \mathrm{id}_X) & \xrightarrow{\Theta_*} & \mathrm{Aut}(\pi_1(X, x_0)) \\ {\scriptstyle \pi_0(i)}\downarrow & & \downarrow{\scriptstyle Q} \\ \pi_0(\mathcal{E}(X), \mathrm{id}_X) & \xrightarrow{\Psi_*} & \mathrm{Out}(\pi_1(X, x_0)) \end{array} \tag{7}$$

is commutative.

Let $\mathrm{Homeo}(X) \subset \mathcal{E}(X)$ be the subspace consisting of self-homeomorphisms of X. Note that if X is locally connected and locally compact, then $\mathrm{Homeo}(X)$ with the compact-open topology is a topological group. The composition

$$\mathrm{Homeo}(X) \hookrightarrow \mathcal{E}(X) \xrightarrow{\Psi} \mathrm{Out}(\pi_1(X, x_0)) \tag{8}$$

is a group homomorphism which we also denote by Ψ to avoid notation proliferation.

Recall that a space X is *aspherical* if it has a contractible universal covering space. If X is aspherical, then the homomorphisms Θ_* and Ψ_* in (5) and (6), respectively, are both isomorphisms. For any group G, let G_{fin} denote the subset of G consisting of elements of finite order (G_{fin} need not be a subgroup).

THEOREM 12. *Let X be an aspherical space with universal cover Y. Let $F\colon X \to X$ be an n-periodic homeomorphism where $n > 1$. Assume that $\pi_1(X, x_0)$ is torsion free and its center is trivial. If $\Psi(F) \in \mathrm{image}\,(Q\colon \mathrm{Aut}(\pi_1(X, x_0))_{fin} \to \mathrm{Out}(\pi_1(X, x_0))\,)$, then F lifts to an n-periodic homeomorphism $\widetilde{F}\colon Y \to Y$.*

PROOF. Since $\Psi\colon \mathrm{Homeo}(X) \to \mathrm{Out}(\pi_1(X, x_0))$ is a group homomorphism, $\Psi(F)$ has order m dividing n. By hypothesis, there exists $\varphi \in \mathrm{Aut}(\pi_1(X, x_0))$ of finite order such that $\Psi(F) = Q(\varphi)$. By Proposition 8, φ also has order m. Since Θ_* and Ψ_* in diagram (7) are isomorphisms (because X is aspherical) there exists $f \in \mathcal{E}(X, x_0)$ such that F is homotopic to f and $\Theta(f) = \varphi$. The conclusion of the Theorem follows from Theorem 11. □

REMARK 13. If an aspherical space is homotopy equivalent to a finite dimensional CW complex, then its fundamental group is torsion free.

We make use of the following theorem of P. A. Smith.

THEOREM 14 (P. A. Smith). *A periodic homeomorphism of $\mathbb{R}^3$ has a fixed point.*

PROOF. Let $F\colon \mathbb{R}^3 \to \mathbb{R}^3$ be a periodic homeomorphism. Regarding S^3 as the one-point compactification of $\mathbb{R}^3$, the map F extends to a periodic homeomorphism $\widehat{F}\colon S^3 \to S^3$. Suppose F has no fixed points. Then the point at infinity is the only fixed point of $\widehat{F}$. By [**Smi39**, Theorem 4], the fixed point set of a periodic homeomorphism of S^3 (other than the identity) is either empty or is homeomorphic to one of S^0, S^1 or S^2. None of these possible fixed point sets is a singleton, a contradiction. Hence F must have a fixed point. □

THEOREM 15. *Let $K \subset S^3$ be a knot that is not a torus knot. Let $F\colon E_K \to E_K$ be a periodic homeomorphism. If*

$$\Psi(F) \in \text{image}\,(Q\colon \operatorname{Aut}(G_K, P_K)_{\mathit{fin}} \to \operatorname{Out}(G_K, P_K)\,),$$

then F has a fixed point in the interior of E_K.

PROOF. Since K is not a torus knot, the center of G_K is trivial, [**BZ66**] (also see [**Sim76**, Theorem 4]). Let M be the interior of E_K. Note that the inclusion $M \hookrightarrow E_K$ is a homotopy equivalence. Choosing a basepoint $x_0 \in M$, we have $\pi_1(M, x_0) \cong G_K$. By [**Wal68**, Theorem 8.1], the universal covering space of M is homeomorphic to 3-dimensional Euclidean space, $\mathbb{R}^3$. Hence M is aspherical and moreover G_K is torsion free (see Remark 13). Let n be the period of F. Observe that F restricts to an n-periodic homeomorphism of M, which we write as $F'\colon M \to M$, and $\Psi(F) = \Psi(F')$. By Theorem 12, F' lifts to an n-periodic homeomorphism $\widetilde{F'}\colon \mathbb{R}^3 \to \mathbb{R}^3$. By Theorem 14, $\widetilde{F'}$ has a fixed point $y \in \mathbb{R}^3$. The image of y under the covering projection $\mathbb{R}^3 \to M$ is a fixed point of F' and hence also of F. □

Note that if $F\colon E_K \to E_K$ is an n-periodic homeomorphism with a fixed point $x \in E_K$ and σ is a path from a basepoint $x_0 \in E_K$ to x, then the composite

$$\pi_1(E_K, x_0) \xrightarrow{\sigma_\#} \pi_1(E_K, x) \xrightarrow{F_*} \pi_1(E_K, x) \xrightarrow{\sigma_\#^{-1}} \pi_1(E_K, x_0)$$

is an automorphism $\varphi \in \operatorname{Aut}(G_K, P_K)$ of order m dividing n; furthermore, if K is not a torus knot, then $m = n$ by the main theorem of [**FY83**].

COROLLARY 16. *Let $K \subset S^3$ be a knot that is not a torus knot. Let $F\colon E_K \to E_K$ be a nontrivial periodic homeomorphism. Then F is a free symmetry if and only if*

$$\Psi(F) \notin \text{image}\,(Q\colon \operatorname{Aut}(G_K, P_K)_{\mathit{fin}} \to \operatorname{Out}(G_K, P_K)_{\mathit{fin}}\,).$$

For example, the hyperbolic knot 9_{48} is freely periodic with order $p = 3$ (see [**Har81**] and [**Ril82**]), and so the associated element in $\operatorname{Out}(G_K)$ has order 3 but is not in the image of $\operatorname{Aut}(G_K)_{\mathit{fin}}$.

5. Proof of Theorem 3

The proof of Theorem 3 divides into two cases, the case K is a torus knot and the case K is not a torus knot.

CASE 1. **Torus knots**. Let K be the (r, s)-torus knot. The knot group of K admits the well-known presentation:

$$G_K = \langle a, b \mid a^r = b^s \rangle.$$

In [**Sch24**], Schreier determined structure of the automorphism group of G_K showing that

$$\mathrm{Aut}(G_K) = \langle I, J, H \mid J^r = J^s = H^2 = (HI)^2 = (HJ)^2 = I \rangle,$$

where I is conjugation by $a \in G_K$ and J is conjugation by $b \in G_K$ and H is the order two automorphism given on generators by $H(a) = a^{-1}$ and $H(b) = b^{-1}$.

The subgroup of $\mathrm{Aut}(G_K)$ generated by I and J has index two and is isomorphic to the free product $\mathbb{Z}/r * \mathbb{Z}/s$, so we get a short exact sequence:

$$1 \to \mathbb{Z}/r * \mathbb{Z}/s \to \mathrm{Aut}(G_K) \to \mathbb{Z}/2 \to 1.$$

Assume that $\varphi \in \mathrm{Aut}(G_K)$ is an automorphism induced from a p-periodic welded diagram of K. Proposition 5 implies that φ has order p. By Proposition 9, φ is orientation preserving and thus φ lies in the kernel $\mathrm{Aut}(G_K) \to \mathbb{Z}/2$. Thus φ is an element in $\mathbb{Z}/r * \mathbb{Z}/s$, and we apply the Kurosch Subgroup Theorem [**Kur34**] to conclude that p must divide r or s. Note that a suitably symmetric picture of an (r, s)-torus knot drawn on a torus makes it evident that such a knot indeed has cyclic periods of order p dividing r or s. This establishes the conclusion of Theorem 3 in the case K is a torus knot.

CASE 2. **Non-Torus knots**. Assume that K is not a torus knot. Since K is not a torus knot, its knot group G_K is torsion free with trivial center (see the proof of Theorem 15). Thus, if $\varphi \in \mathrm{Aut}^+(G_K, P_K)$ has order p, then so does its image $Q(\varphi) \in \mathrm{Out}^+(G_K, P_K)$ by Proposition 8.

For a compact 3-manifold M, let $\mathrm{Diff}(M)$, $\mathrm{Homeo}_{PL}(M)$ and $\mathrm{Homeo}(M)$ be, respectively, the groups of diffeomorphisms, PL-homeomorphisms and homeomorphisms of M. It is a consequence of the solution of the Smale Conjecture by Hatcher [**Hat83**] that the inclusions

$$\mathrm{Diff}(M) \hookrightarrow \mathrm{Homeo}_{PL}(M) \hookrightarrow \mathrm{Homeo}(M)$$

are homotopy equivalences, and hence induce isomorphisms

$$\mathrm{Mcg}(M) = \pi_0(\mathrm{Diff}(M), \mathrm{id}_M) \cong \pi_0(\mathrm{Homeo}_{PL}(M), \mathrm{id}_M) \cong \pi_0(\mathrm{Homeo}(M), \mathrm{id}_M).$$

As in (8), we denote the composite

$$\mathrm{Homeo}_{PL}(E_K) \hookrightarrow \mathcal{E}(E_K) \xrightarrow{\Psi} \mathrm{Out}(G_K)$$

by Ψ. Note that for any nontrivial knot K in S^3, the knot exterior E_K is a Haken manifold. Waldhausen's result [**Wal68**, Corollary 7.5], which applies to Haken manifolds, asserts that Ψ as above induces an isomorphism

$$\pi_0(\mathrm{Homeo}_{PL}(E_K), \mathrm{id}_{E_K}) \cong \mathrm{Out}(G_K, P_K).$$

Thus if $Q(\varphi)$ has order p, then there is a unique $\varphi_* \in \mathrm{Mcg}(E_K)$ (or, equivalently, in $\pi_0(\mathrm{Homeo}_{PL}(E_K), \mathrm{id}_{E_K})$) of order p such that $\Psi_*(\varphi_*) = Q(\varphi)$.

Assume that K admits a p-periodic virtual or welded diagram D. As explained in 1, D gives rise to an element $\varphi \in \mathrm{Aut}^+(G_K, P_K)$ which by Proposition 5 has order p. As observed above, $Q(\varphi) \in \mathrm{Out}^+(G_K, P_K)$ also has order p. Zimmermann's "Nielsen Realization Theorem", [**Zim82**, Satz 0.1] together with its Addendum, [**Zim82**, p. 358] (where the case of a 3-manifold with torus boundary components is discussed, also see [**HT87**]) asserts that φ_* can be realized as a PL homeomorphism $F_\varphi \colon E_K \to E_K$, which again has order p. Furthermore, F_φ is orientation preserving because φ is orientation preserving (see the discussion in 3).

By [**Luo92**, Theorem 2], F_φ extends to a p-periodic homeomorphism of S^3 that preserves K. This shows that the classical knot K admits a symmetry of order p, either as a cyclic period or as a free period. However, since $Q(\varphi) = \Psi(F_\varphi)$ is the image of an element in $\mathrm{Aut}(G_K, P_K)$ of finite order, Theorem 15 asserts that F_φ must have a fixed point in the interior of E_K. In particular, F_φ is not a free symmetry, and so K is a p-periodic classical knot. This completes the proof of Theorem 3. □

In [**Hil84**], Hillman showed that a nontrivial classical link admits only finitely many periods, extending Flapan's result for classical knots [**Fla85**]. We do not know whether the same conclusion is true for the virtual and welded periods of classical links with more than one component. Theorem 3, if extended to links, would imply that the answer is no, that is, any virtual or welded period of a link is also a classical period. Note that an alternative approach would be needed to deal with split links, since the exterior of a split link is not Haken and so Waldhausen's result [**Wal68**, Corollary 7.5] no longer applies. Extending Theorem 3 to links remains an interesting and open problem.

Acknowledgments

We would like to thank Micah Chrisman, Robin Gaudreau, Matthias Nagel, and an anonymous referee for their helpful comments.

References

[BNW18] Hans U. Boden, Andrew J. Nicas, and Lindsay White, *Alexander invariants of periodic virtual knots*, Dissert. Math. **530** (2018), 1–59.

[BZ66] Gerhard Burde and Heiner Zieschang, *Eine Kennzeichnung der Torusknoten* (German), Math. Ann. **167** (1966), 169–176, DOI 10.1007/BF01362170. MR0210113

[BZH14] Gerhard Burde, Heiner Zieschang, and Michael Heusener, *Knots*, Third, fully revised and extended edition, De Gruyter Studies in Mathematics, vol. 5, De Gruyter, Berlin, 2014. MR3156509

[CR77] P. E. Conner and Frank Raymond, *Deforming homotopy equivalences to homeomorphisms in aspherical manifolds*, Bull. Amer. Math. Soc. **83** (1977), no. 1, 36–85, DOI 10.1090/S0002-9904-1977-14179-7. MR0467777

[Fla85] Erica Flapan, *Infinitely periodic knots*, Canad. J. Math. **37** (1985), no. 1, 17–28, DOI 10.4153/CJM-1985-002-4. MR777036

[Fox62] R. H. Fox, *Knots and periodic transformations*, Topology of 3-manifolds and related topics (Proc. The Univ. of Georgia Institute, 1961), Prentice-Hall, Englewood Cliffs, N.J., 1962, pp. 177–182. MR0140101

[FY83] Michael Freedman and Shing Tung Yau, *Homotopically trivial symmetries of Haken manifolds are toral*, Topology **22** (1983), no. 2, 179–189, DOI 10.1016/0040-9383(83)90030-7. MR683759

[GL89] C. McA. Gordon and J. Luecke, *Knots are determined by their complements*, J. Amer. Math. Soc. **2** (1989), no. 2, 371–415, DOI 10.2307/1990979. MR965210

[GPV00] Mikhail Goussarov, Michael Polyak, and Oleg Viro, *Finite-type invariants of classical and virtual knots*, Topology **39** (2000), no. 5, 1045–1068, DOI 10.1016/S0040-9383(99)00054-3. MR1763963

[Har81] Richard Hartley, *Knots with free period*, Canad. J. Math. **33** (1981), no. 1, 91–102, DOI 10.4153/CJM-1981-009-7. MR608857

[Hat83] Allen E. Hatcher, *A proof of the Smale conjecture*, $\mathrm{Diff}(S^3) \simeq \mathrm{O}(4)$, Ann. of Math. (2) **117** (1983), no. 3, 553–607, DOI 10.2307/2007035. MR701256

[Hei81] W. Heil, *Normalizers of incompressible surfaces in 3-manifolds* (English, with Serbo-Croatian summary), Glas. Mat. Ser. III **16(36)** (1981), no. 1, 145–150. MR634302

[HT87] Wolfgang Heil and Jeffrey L. Tollefson, *On Nielsen's theorem for 3-manifolds*, Yokohama Math. J. **35** (1987), no. 1-2, 1–20. MR928368

[Hil84] Jonathan A. Hillman, *Links with infinitely many semifree periods are trivial*, Arch. Math. (Basel) **42** (1984), no. 6, 568–572, DOI 10.1007/BF01194056. MR756900

[Kau99] Louis H. Kauffman, *Virtual knot theory*, European J. Combin. **20** (1999), no. 7, 663–690, DOI 10.1006/eujc.1999.0314. MR1721925

[Kaw96] Akio Kawauchi, *A survey of knot theory*, Birkhäuser Verlag, Basel, 1996. Translated and revised from the 1990 Japanese original by the author. MR1417494

[Kim00] Se-Goo Kim, *Virtual knot groups and their peripheral structure*, J. Knot Theory Ramifications **9** (2000), no. 6, 797–812, DOI 10.1142/S0218216500000451. MR1775387

[Kur34] Alexander Kurosch, *Die Untergruppen der freien Produkte von beliebigen Gruppen* (German), Math. Ann. **109** (1934), no. 1, 647–660, DOI 10.1007/BF01449159. MR1512914

[Lee12] Sang Youl Lee, *Genera and periodicity of virtual knots and links*, J. Knot Theory Ramifications **21** (2012), no. 4, 1250037, 15, DOI 10.1142/S0218216511009881. MR2890463

[Luo92] Feng Luo, *Actions of finite groups on knot complements*, Pacific J. Math. **154** (1992), no. 2, 317–329. MR1159514

[Mor84] John W. Morgan, *The Smith conjecture*, The Smith conjecture (New York, 1979), Pure Appl. Math., vol. 112, Academic Press, Orlando, FL, 1984, pp. 3–6, DOI 10.1016/S0079-8169(08)61632-3. MR758460

[Mos68] G. D. Mostow, *Quasi-conformal mappings in n-space and the rigidity of hyperbolic space forms*, Inst. Hautes Études Sci. Publ. Math. **34** (1968), 53–104. MR0236383

[Pra73] Gopal Prasad, *Strong rigidity of* $\mathbf{Q}$*-rank* 1 *lattices*, Invent. Math. **21** (1973), 255–286, DOI 10.1007/BF01418789. MR0385005

[Ril82] R. Riley, *Seven excellent knots*, Low-dimensional topology (Bangor, 1979), London Math. Soc. Lecture Note Ser., vol. 48, Cambridge Univ. Press, Cambridge-New York, 1982, pp. 81–151. MR662430

[Sat00] Shin Satoh, *Virtual knot presentation of ribbon torus-knots*, J. Knot Theory Ramifications **9** (2000), no. 4, 531–542, DOI 10.1142/S0218216500000293. MR1758871

[Sch24] Otto Schreier, *Über die gruppen* $A^aB^b = 1$ (German), Abh. Math. Sem. Univ. Hamburg **3** (1924), no. 1, 167–169, DOI 10.1007/BF02954621. MR3069424

[Sim76] Jonathan Simon, *Roots and centralizers of peripheral elements in knot groups*, Math. Ann. **222** (1976), no. 3, 205–209, DOI 10.1007/BF01362577. MR0418079

[Smi39] P. A. Smith, *Transformations of finite period. II*, Ann. of Math. (2) **40** (1939), 690–711, DOI 10.2307/1968950. MR0000177

[Tsa88] Chichen M. Tsau, *Isomorphisms and peripheral structure of knot groups*, Math. Ann. **282** (1988), no. 2, 343–348. MR963021

[Wal68] Friedhelm Waldhausen, *On irreducible* 3*-manifolds which are sufficiently large*, Ann. of Math. (2) **87** (1968), 56–88, DOI 10.2307/1970594. MR0224099

[Zim82] Bruno Zimmermann, *Das Nielsensche Realisierungsproblem für hinreichend große 3-Mannigfaltigkeiten* (German), Math. Z. **180** (1982), no. 3, 349–359, DOI 10.1007/BF01214175. MR664520

Department of Mathematics & Statistics, McMaster University, Hamilton, Ontario, Canada L8S 4K1

URL: math.mcmaster.ca/~boden

Email address: boden@mcmaster.ca

Department of Mathematics & Statistics, McMaster University, Hamilton, Ontario, Canada L8S 4K1

Email address: nicas@mcmaster.ca

Proceedings of Symposia in Pure Mathematics
Volume **102**, 2019
https://doi.org/10.1090/pspum/102/01813

Transverse universal links

Roger Casals and John B. Etnyre

ABSTRACT. We show that there exists a transverse link in the standard contact structure on the 3–sphere such that all contact 3–manifolds are contact branched covers over this transverse link.

1. Introduction

In 1982, William Thurston showed that there exists a six component link in the 3–sphere such that any closed smooth oriented 3–manifold is the branched cover over the 3–sphere with branch locus this link [**21**]. He called such a link universal. Later H. Hilden, M. Lozano, and J.M. Montesinos showed that there exists a universal knot in the 3–sphere [**11**, **12**]. Since then there have been several articles showing certain knots and links are universal or not universal. For example the figure eight knot, Whitehead link, and Borromean rings are all universal [**13**, **14**].

In 2002, Emmanuel Giroux showed that any contact 3–manifold is a 3–fold simple branched cover of the 3–sphere with the standard contact structure (S^3, ξ_{std}) and branch locus a transverse link [**8**]. This is a contact strengthening of the Hilden-Montesinos Theorem [**10**, **17**] for smooth 3–manifolds. These contact constructions are useful for constructing open books for some contact 3–manifolds and embedding contact 3–manifolds in the standard contact structure on the 5–sphere [**5**].

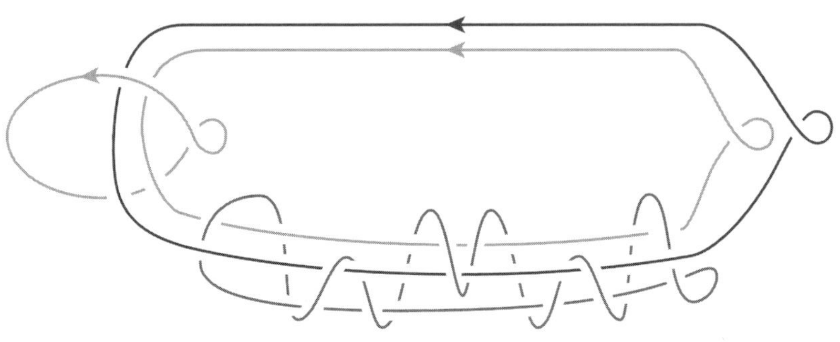

FIGURE 1. The universal transverse link in Theorem 1.1.

2010 *Mathematics Subject Classification.* Primary 53D10; Secondary 53D15, 57R17.

The first author is supported by the NSF grants DMS-1608018 and, subsequently, DMS-1841913, and a BBVA Research Fellowship.
The second author was partially supported by NSF grant DMS-1608684.

In view of the work of W. Thurston and [**11, 12**], it is a natural question in low-dimensional contact topology to ask whether there exists a universal transverse link in (S^3, ξ_{std}). That is, a transverse link $L \subseteq (S^3, \xi_{std})$ such that *any* contact 3–manifold is the contact branched cover of (S^3, ξ_{std}) along L. This question was first considered by M. Casey in [**3**] where she showed that no transverse knot in the knot type of the figure eight can be (contact) universal and that many covers of the Whitehead link and Borromean rings yield only overtwisted contact structures, raising doubts as to whether or not there exist universal transverse links. The main result in the present work is that universal transverse links exist:

THEOREM 1.1. *Any contact 3-manifold (M, ξ) can be realized as a contact branched cover of (S^3, ξ_{std}) branched along the transverse link L shown in Figure* 1.

The proof presented here directly adapts the argument for the existence of topologically universal links in [**11**] to the contact case. It is interesting to note that similar constructions of universal links that lead to universal knots cannot be easily adapted to the contact setting, leaving open the natural question

(a) Does there exist a universal transverse knot $K \subseteq (S^3, \xi_{std})$?

The Figure-8 knot is a smooth universal knot, but it cannot be a contact universal knot after the results from M. Casey's thesis [**3**]. This leads to the following problem:

(b) Find sufficient conditions for a smooth universal knot or link to admit transverse representatives which are contact universal.

In line with the first question, we would also like to know the answer to the following question:

(c) Let (S^3, ξ_{ot}) be an overtwisted contact structure. Does there exist a transverse knot $K \subseteq (S^3, \xi_{ot})$ such that the contact branched cover of (S^3, ξ_{ot}) along K is contactomorphic to (S^3, ξ_{std}) ?

It is possible to produce branched covers of overtwisted structures on S^3 which are tight *cf.* [**3**], yet it is apparent that such K in the question — should it exist — cannot be a smooth unknot.

Finally, a natural continuation of Theorem 1.1, in line with [**16, 20**], would be to answer the following question:

(d) Does there exist a symplectic surface $S \subseteq (D^4, \lambda_{std})$ such that any Weinstein 4-fold (W, λ) is a branched cover of (D^4, λ_{std}) along S ?

2. Background

In this section we review the necessary definitions on transverse knots and branched covers which we will use in our proof of Theorem 1.1. We also show, in the last subsection, that many standard modifications of the branch locus for branched covering maps that are known in the topological setting also hold in the contact geometric setting as well.

2.1. Transverse knots. For more details on transverse knots the reader is referred to [**4**]. Here we briefly review the aspects that will be relevant for our main result.

We will consider our knots in $(\mathbb{R}^3, \xi_{std}) \subset (S^3, \xi_{std})$ where on $\mathbb{R}^3$ we have the standard contact structure

$$\xi_{std} = \ker(dz - y\,dx),$$

and (x, y, z) are Cartesian coordinates on $\mathbb{R}^3$. A knot K is *transverse* if K is transverse to ξ_{std} at every point of K. Since ξ_{std} is co-oriented by the contact form and $\mathbb{R}^3$ is oriented, K will have an orientation induced on it so that K and ξ_{std} intersect positively.

We will study transverse knots via their *front projection*, that is, via the image of K under the projection $\pi : \mathbb{R}^3 \to \mathbb{R}^2 : (x, y, z) \mapsto (x, z)$. Notice that being transverse to ξ_{std} implies that the y–coordinate of K satisfies

$$y < \frac{dz}{dx},$$

where we think of the page as the xz-plane so that the positive y-axis points into the page. It is easy to see that any diagram for an oriented knot in the xz-plane determines a transverse knot up to isotopy through transverse knots as long as no portion of the knot is as shown in Figure 2. Moreover, Type II and III Reidemeister

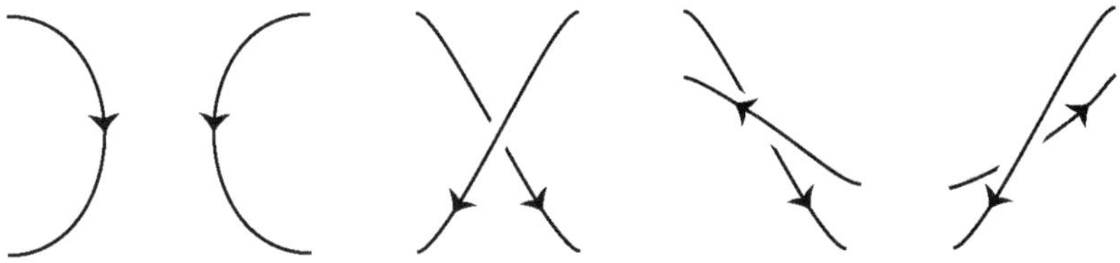

FIGURE 2. Forbidden portions of the front diagram of a transverse knot.

moves are allowed in front diagrams as long as no portion of the move contains a forbidden diagram from Figure 2. More specifically, there can be no vertical tangencies with the oriented tangent vector pointing down and there can be no crossings where the over strand has greater slope and is oriented right to left, while the under crossing is oriented left to right.

In addition to the *front projections*, there is a different presentation of transverse knots that will be useful for us as well. Consider $S^3 = \{(z_1, z_2) \in \mathbb{C}^2 : |z_1|^2 + |z_2|^2 = 1\}$ as the unit sphere in $\mathbb{C}^2$ with the standard contact structure $\xi_{std} = TS^3 \cap i(TS^3)$ given as the set of complex tangencies to S^3. The link $H = \{z_1 = 0\} \cup \{z_2 = 0\}$ is a Hopf link with each component being an unknot with self-linking number -1. It is easy to check that the contact structure on the complement $S^3 - H = T^2 \times (0, 1)$ is given by

$$\xi_{std} = \ker\left(\cos\left(\frac{\pi}{2}t\right)\,d\phi + \sin\left(\frac{\pi}{2}t\right)\,d\theta\right),$$

where (ϕ, θ) are angular coordinates on $T^2 = S^1 \times S^1$ and t is the coordinate on $(0, 1)$. We can now consider transverse knots in $T^2 \times (0, 1)$ and their *front projection* will be obtained by projecting out the t-coordinate. Once again a transverse knots can be recovered from its front projection. The only real difference with the situation described above is that the tangent vector to the projection cannot have both ϕ and θ component negative.

A closed transverse braid in $\mathbb{R}^3$ can always be assumed to lie in the thickened torus $T^2 \times (0,1)$ and such braids are in one-to-one correspondence curves in $T^2 \times (0,1)$ whose oriented tangents always have positive θ-coordinate [**1**].

2.2. Contact branched covers. A *branched covering map* is a smooth map $p : M \to Y$ between smooth 3–manifolds such that there is a link $L \subseteq Y$, called the *branch locus*, such that p restricted to $M - p^{-1}(L)$ is a covering map from $M - p^{-1}(L)$ to $Y - L$ and each component of $p^{-1}(L)$ has a neighborhood $N = S^1 \times D^2$ such that $p(N) = S^1 \times D^2$ and in these coordinates p is of the form $(\phi, z) \mapsto (n\phi, z^m)$ for some integers n and m where ϕ is the angular coordinate on S^1 and z is the complex coordinate on D^2 thought of as the unit disk in $\mathbb{C}$. We say the component has *order* m and we say p is *ramified* at the component if $m > 1$. An branched covering map of degree k is called *simple* if the pre-image of any point has size either k or $k-1$.

Given a contact submanifold, it is quite easy to construct contact structures on branched covering spaces from a contact structure on the base space.

THEOREM 2.1 (Geiges 1997, [**6**]; Öztürk and Niederkrüger 2007, [**19**]). *Let $p : M \to Y$ be a branched covering map with branch locus L. Given a contact structure ξ on Y with contact from α such that L is transverse to ξ, then $p^*\alpha$ may be deformed by an arbitrarily small amount near $p^{-1}(L)$ to give a contact form defining a unique, up to contact isotopy, contact structure ξ_L on M.*

The starting point for our main result is Giroux's proof that every contact 3–manifold can be constructed as a branched cover over the standard contact structure on S^3.

THEOREM 2.2 (Giroux 2002, [**8**]). *Given a contact structure ξ on a 3–manifold M, there is a 3–fold simple branched covering map $p : M \to S^3$ such that ξ is the contact structure induced on M by p, the standard contact structure on S^3, and some transverse realization of the branch locus of p.*

For each contact 3–manifold (M, ξ), Theorem 2.2 asserts the existence of a transverse link $L(M, \xi) \subseteq (S^3, \xi_{std})$, depending on (M, ξ), such that (M, ξ) is the contact branched cover of (S^3, ξ_{std}) along $L(M, \xi)$. Our main result Theorem 1.1 shows that the transverse link $L(M, \xi)$ can be chosen to be independent of (M, ξ), thus encoding all the complexity of the contact structure on (M, ξ) in the coverings of $S^3 \setminus L$.

Let us briefly recall the argument for Theorem 2.2. E. Giroux first shows that any contact 3–manifold is supported by an open book decomposition, whose binding can be assumed to be connected. Building on work of J.S. Birman [**2**] and D.L. Goldsmith [**9**], J.M. Montesinos and H.R. Morton showed in [**18**] that any open book decomposition on a smooth 3–manifold M with connected binding is pulled back from the open book decomposition of S^3 with disk page by a 3–fold branched cover map $p : M \to S^3$ whose branch locus is a braid. Since a braid is naturally a transverse knot the branched cover can be taken to be a contact branched cover and it is easy to see that the induced contact structure on the covering space is supported by the original open book, thus concluding the statement.

We end this subsection with two simple observations. The first follows immediately from the definition of contact branched cover.

LEMMA 2.3. *Suppose* $p : (M', \xi') \to (M, \xi)$ *is a contact branched covering map with branch set* $B \subset M$. *Let* L *be a link in* M *disjoint from* B *and* $L' = p^{-1}(L)$. *If* $p' : (M'', \xi'') \to (M', \xi')$ *is a contact branched covering map with branch set* L', *then* $p' \circ p : (M'', \xi'') \to (M, \xi)$ *is a contact branched covering map with branch set* $B \cup L$. □

The second fact is also standard in contact topology.

LEMMA 2.4. *The contact manifold obtained from any branched cover of* (S^3, ξ_{std}) *branched along a transverse unknot with self-linking number* -1 *is contactomorphic to* (S^3, ξ_{std}).

PROOF. It is clear that smoothly the branched cover of S^3 branched over the unknot is diffeomorphic to S^3. Moreover if U is the transverse unknot in the standard contact structure on S^3 with self-linking number -1 then $S^3 - U = S^1 \times \mathbb{R}^2$ with the contact structure $\xi = \ker(d\phi + r^2\, d\theta)$, where ϕ is the angular coordinate on S^1 and (r, θ) are polar coordinates on $\mathbb{R}^2$. Now any finite covering space of $(S^1 \times \mathbb{R}^2, \xi)$ is contactomorphic to $(S^1 \times \mathbb{R}^2, \xi)$, from which the result follows. □

The hypothesis on the self-linking number in Lemma 2.4 is meaningful, since branched covers along stabilized transverse knots are overtwisted.

2.3. Monodromy representations. A covering map $p : M \to Y$, for M connected, is determined by its monodromy representation. Let $x_0 \in Y$ be a fixed base point and label the points in $p^{-1}(x_0) = \{x_1, \ldots, x_n\}$. Then, given a loop γ in Y based at x_0 we can define the element σ_γ of the symmetric group S_n by $\sigma_\gamma(i)$ being the index on $\widetilde{\gamma}(1)$ where $\widetilde{\gamma}$ is the lift of γ based at x_i. This defines a transitive representation of the fundamental group

$$r : \pi_1(Y) \to S_n = \mathrm{Aut}(p^{-1}(x_0)).$$

Moreover, given such a representation the subgroup $G = \{g \in \pi_1(Y) : r(g)(1) = 1\}$ gives a covering space M that corresponds to the representation.

Let $p : M \to Y$ be a branched covering map with branch locus $L \subseteq Y$. By definition, the restriction $p|_{(M - p^{-1}(L))} : M - p^{-1}(L) \to Y - L$ is a covering space and, by the above discussion, this restriction is determined by a representation $\pi_1(Y - L) \to S_n$, with $n = |p^{-1}(x_0)|$ the cardinality of the fiber over any $x_0 \in Y - L$. In turn, such a representation determines the initial branched cover map $p : M \to Y$. Indeed, the representation of the fundamental group determines a cover of $Y - L$ and, in addition, any covering of $Y - L$ can be uniquely extended to a branched covering map of Y by using the local model

$$L \times \mathbb{C} \to L \times \mathbb{C}, \quad (\theta, z) \mapsto (m\theta, z^n),$$

near the branch set, where the target $L \times \mathbb{C}$ is a sufficiently small neighborhood of $L \subseteq Y$ and $\mathbb{C}$ provides normal coordinates for this inclusion. The domain $L \times \mathbb{C}$ models part of a sufficiently small neighborhood of $p^{-1}(L) \subseteq M$. See [**5**, Section 2.6] and [**6**] for further details.

In the case that Y is smoothly S^3, the fundamental group $\pi_1(Y - L)$ is well-known to be generated by meridians to L. In particular, given a diagram for L we have the Wirtinger presentation for $\pi_1(Y - L)$ with generators x_i corresponding to the strands in the diagram and relations coming from the crossings: $x_i = x_k^{-1} x_j x_k$, respectively $x_i = x_k x_j x_k^{-1}$, at a right handed, respectively left handed, crossing, where strand k goes over incoming strand j and outgoing strand i. Thus a branched covering map over S^3 with branch set L is determined by labeling a diagram of L with elements of S_n that satisfy the required relations at the crossings.

It is easy to see whether a branched covering map is simple using the monodromy representation: one just needs the image of the representation to consist only of transpositions, as higher length permutations correspond to more than two points coming together.

We now discuss some modifications one can make to the branch locus of a branched covering map without affecting the induced contact covering space.

LEMMA 2.5. *Let L be a transverse link in a contact 3–manifold (Y,ξ) and $p\colon M \to Y$ be a branched covering map with branch locus L inducing the contact structure ξ' on M. If part of a diagram for L is as shown on one side of a row in Figure 3 then replacing that portion of L with the other diagram shown in that row will result in a new contact branched covering of Y that still yields the same contact manifold (M,ξ').*

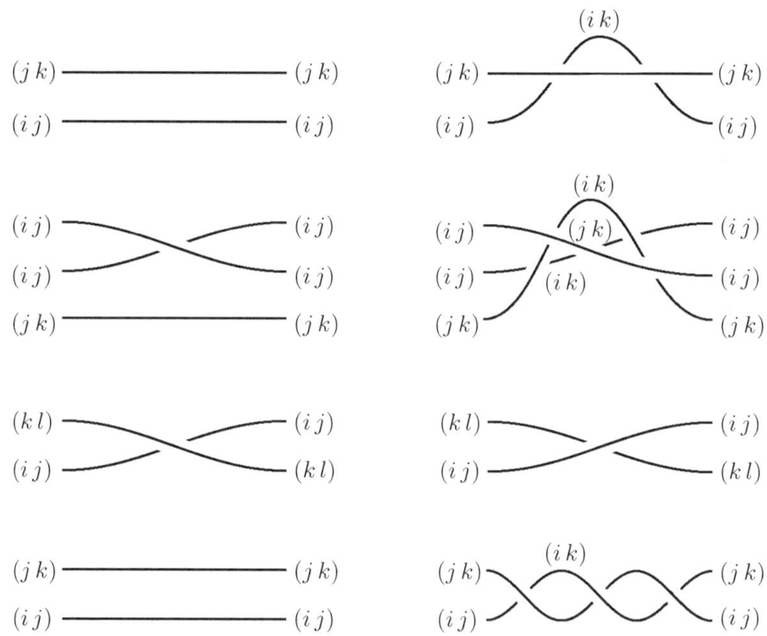

FIGURE 3. Replacing a portion of the branch locus of a simple cover with one of the figures in a row with the other does not change the manifold or contact structure described by the contact branched cover. All strands are oriented from left to right.

PROOF. The changes to the diagram depicted in the first two rows of Figure 3 correspond to contact isotopies of the underlying branch locus. This induces an isotopy of the branch cover map, thus does not change the branched cover or the contact structure.

The change in the fourth row was proven to leave the contact branched cover unchanged in [**3**] and can easily be seen by observing that the branched cover of the ball containing either branched loci is simply a ball and the contact structure on it is tight. See [**3**] for details.

That the change in the third row does not affect the contact branched cover follows exactly as for the fourth row. That is, the pre-image by the covering map of a ball containing the diagram on either side is a union of balls and the contact structure on each ball is tight. □

There is a second simple modification to the transverse branched locus which preserves the contact isotopy type of the branched cover, it adds a standard transverse unknot to the branch locus at the cost of increasing the degree of the branched cover:

LEMMA 2.6. *Let L be a transverse link in a contact 3–manifold (Y, ξ) and $p\colon M \to Y$ be an n–fold branched covering map with branch locus L inducing the contact structure ξ' on M. Let L' be the link consisting of L together with a maximal self-linking unknot U that is contained in a ball that is disjoint from L.*

The monodromy data for L' given by using the monodromy data on L and labelling the meridian of U by $(n, n+1)$ defines an $(n+1)$–fold branch covering map from M to Y that induces the same contact structure ξ' on M.

PROOF. We begin with a simple observation. Let $K = K_0 \cup K_1$ be a link with monodromy data for an n–fold branched covering map, such that there is a sphere S that separates K_0 and K_1. We can assume that S is a convex sphere and cut S^3 into two contact 3–balls B_0 and B_1 along S so that B_i contains K_i. Then cap off B_i by a tight contact 3–ball C_i to get a tight contact 3-sphere S^3_i that contains K_i. Let $p_i : M_i \to S^3_i$ be the n–fold branched covering map determined by K_i and its monodromy data. Note that the pre-image $p_i^{-1}(C_i)$ consists of n distinct 3–balls. Moreover if $p : Y \to S^3$ is the branched covering map corresponding to K then $Y \setminus p^{-1}(S)$ is contactomorphic to $(Y_0 \setminus p_0^{-1}(C_0)) \cup (Y_1 \setminus p_1^{-1}(C_1))$. In consequence, the contact manifold Y is the contact n–fold, possibly internal, connect sum of Y_0 and Y_1.

Now given the link $L' = L \cup U$ in the lemma notice that $Y_0 = Y \cup S^3$ since Y_0 will be an $(n+1)$–fold cover of S^3 branched along L, where the monodromy data for L only has labels between 1 and n, so the cover is disconnected and the $(n+1)$–sheet is disjoint from the rest. Similarly Y_1 is the union of n copies of 3–spheres by Lemma 2.4. It then follows that the $(n+1)$–fold contact connected sum of Y_0 and Y_1 results in a manifold contactomorphic to Y. □

3. Transverse universal links

In this section we prove Theorem 1.1 by showing the link L in Figure 1 is a universal transverse link $L \subseteq (S^3, \xi_{std})$. The proof will follow from two following lemmas.

LEMMA 3.1. *Let L' be the link shown in Figure 4. There is a branched cover*

$$p : (S^3, \xi_{std}) \to (S^3, \xi_{std}),$$

branched along the Hopf link discussed at the end of Section 2.1 such that the link $L_{m,n}$ shown in Figure 5 is a sub-link of $p^{-1}(L')$.

LEMMA 3.2. *Any contact 3-manifold (Y, ξ) can be realized as a contact branched cover of (S^3, ξ_{std}) branched along the transverse link $L_{n,m}$ shown in Figure 5, for some n and m.*

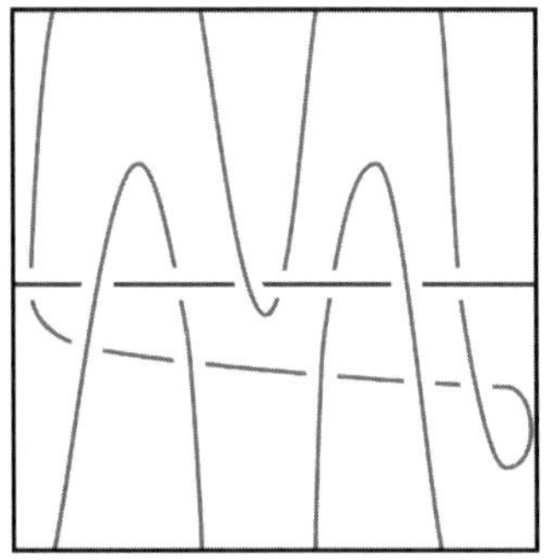

FIGURE 4. The front projection of the link L' depicted in the contact manifold $(T^2 \times (0,1), \xi_{\rm st}) \cong (S^3 - H, \xi_{\rm st}) \subseteq (S^3, \xi_{\rm st})$.

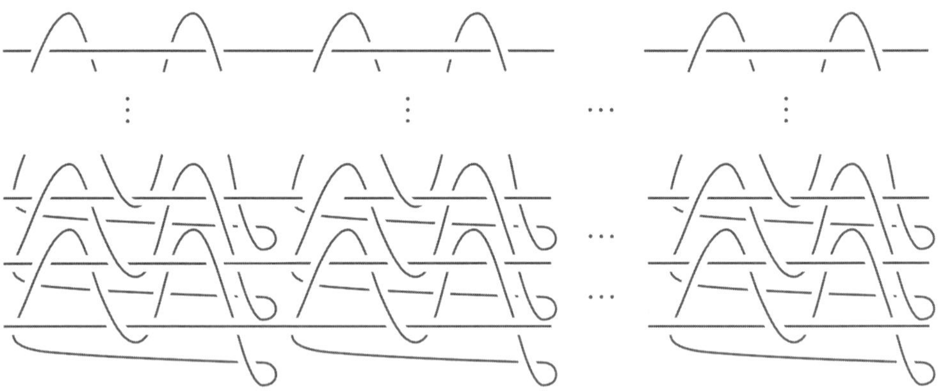

FIGURE 5. The transverse link $L_{m,n}$ in Lemma 3.1. There are n horizontal lines that are closed into a trivial braid and there are m vertical columns of clasping unknots. Lemma 3.2 states that for any contact manifold (Y, ξ) there is a choice of $n, m \in \mathbb{N}$ such that (Y, ξ) is the contact branched cover of (S^3, ξ_{std}) over this transverse link.

Lemmas 3.1 and 3.2, which will be momentarily proven, suffice to conclude our main result.

PROOF OF THEOREM 1.1. Let (M, ξ) be any given contact 3–manifold, we can apply Lemma 3.2 to construct a branched covering map $p' : (M, \xi) \to (S^3, \xi_{std})$ with branched set $L_{n,m}$. By Lemma 3.1 there is also a branched covering map $p : (S^3, \xi_{std}) \to (S^3, \xi_{std})$ with branched set the green Hopf link H depicted in Figure 1 such that $L_{n,m}$ is a sub-link of $p^{-1}(L')$, where $L' \subseteq T^2 \times (0,1) \simeq S^3 \setminus H$ is the link presented in Figure 4. Then Lemma 2.3 implies that the composition $p' \circ p : (Y, \xi) \to (S^3, \xi_{std})$ is a contact branched covering map with branch set the transverse link $L' \subseteq (T^2 \times (0,1), \xi_{\rm st}) \cong (S^3 - H, \xi_{\rm st})$ union the transverse Hopf link $H \subseteq (S^3, \xi_{\rm st})$.

In order to conclude Theorem 1.1 it suffices to identify the link in Figure 4 with the red and blue sub-link in Figure 1. This will be done in two steps. The first step is to show that the link in $(S^3, \xi_{\rm st})$ shown in Figure 4 is the same as the link in $(\mathbb{R}^3, \xi_{rot} = \ker\{dz + xdy - ydx\})$ shown in Figure 6, where $(\mathbb{R}^3, \xi_{rot})$ is contactomorphic to the complement of a point in (S^3, ξ_{st}) disjoint from the link. Then in the second step we explain why the link in Figure 6 is the link referred to in Theorem 1.1, depicted in Figure 1, by identifying the contact structure $(\mathbb{R}^3, \xi_{std} = \ker\{dz - ydx\})$ with $(\mathbb{R}^3, \xi_{rot})$.

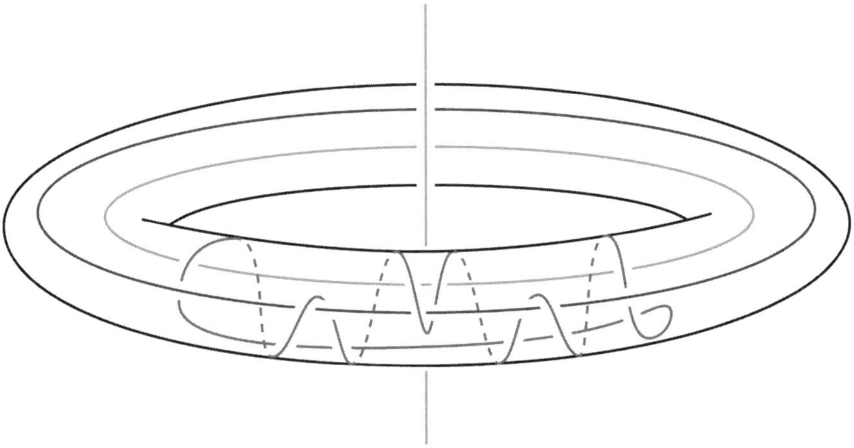

FIGURE 6. The link L' in ξ_{rot}

For the first step, the stereographic coordinate map from $S^3 - \{(0,0,1)\}$ to $\mathbb{R}^3$ pushes forward the contact structure ξ_{st} to the contact structure on $\mathbb{R}^3$ given by the kernel of the contact form $\alpha_1 = r^2\, d\theta + zr\, dr + \frac{1}{2}\left(1 + z^2 - r^2\right) dz$, where (z, r, θ) are cylindrical coordinates on $\mathbb{R}^3$. See the proof of [**7**, Proposition 2.1.8] for the detailed computation. The link in Figure 4 is mapped under the stereographic projection to the link in Figure 6. Given that the contact structure is $(\mathbb{R}^3, \ker\alpha_1)$, we need to apply a diffeomorphism to arrive at $(\mathbb{R}^3, \xi_{rot})$. It is shown in [**7**, Page 57] that the diffeomorphism

$$f : \mathbb{R}^3 \longrightarrow \mathbb{R}^3, \quad f(r, \theta, z) = \left(r, \theta - z, \frac{1}{2}z\left(1 + \frac{1}{3}z^2 + r^2\right)\right)$$

is a contactomorphism which pull-backs the contact structure ξ_{rot} to $\ker\alpha_1$. Since the components of the link in Figure 6, with the exception of the vertical green line, can be assumed to have r-coordinate arbitrarily close to 1 and z-coordinate arbitrarily close to 0, we see that the transverse link in Figure 6 represents the transverse link shown in Figure 4.

For the second step, we now need to identify the contact structures $(\mathbb{R}^3, \xi_{rot})$ and $(\mathbb{R}^3, \xi_{std})$. The contactomorphism from the first to the second is explicitly given by the diffeomorphism $\varphi(x, y, z) = (x, 2y, xy + z)$.

It is a simple exercise, see for instance [**15**, Figure 6], to show that the horizontal green and the blue curve in Figure 6 map to the large green and blue curve in Figure 1. Similarly the vertical green curve, perturbed not to go through infinity, in Figure 6 maps to the other green curve in Figure 1.

We must now see that the red curve in one figure maps to the red curve in the other. For any given $\varepsilon \in \mathbb{R}^+$, the region containing the red component can be confined inside the region $|x| \leq \varepsilon$, and then we see that the contactomorphism

$\varphi : (\mathbb{R}^3, \xi_{rot}) \to (\mathbb{R}^3, \xi_{std})$ sends the red component of the transverse link in Figure 6 to the red component in the transverse link in Figure 1, as desired. □

Let us now prove the two Lemmas 3.1 and 3.2.

PROOF OF LEMMA 3.1. Let $p_{m,n} : (T^2 \times (0,1)) \to (T^2 \times (0,1))$ be the covering map that unwinds the θ-circle m times and the ϕ-circle n times, where we are using the coordinates from Section 2.1. This covering map smoothly extends to a branched covering map $p : S^3 \to S^3$. Moreover, since $(p_{m,n})^*(\xi_{std})$ is contactomorphic to ξ_{std}, it is clear the contact structure induced on S^3 by this branched covering map is the standard contact structure. By construction, the link $L_{m,n}$ in Figure 5 is a sub-link of the pre-image $p^{-1}(L')$ of L' via this explicit branched cover. □

PROOF OF LEMMA 3.2. Given a contact manifold (M, ξ), we apply Theorem 2.2 to construct a 3–fold simple branched cover from M onto (S^3, ξ_{std}) with branch locus a transverse braid B, such that ξ is induced from the covering. Note that since the covering map is simple, the labels on the strands of B are all transpositions. We will now modify the transverse braid B without changing the contact isotopy type of the branched cover (M, ξ) until the B has the desired form depicted in Figure 5.

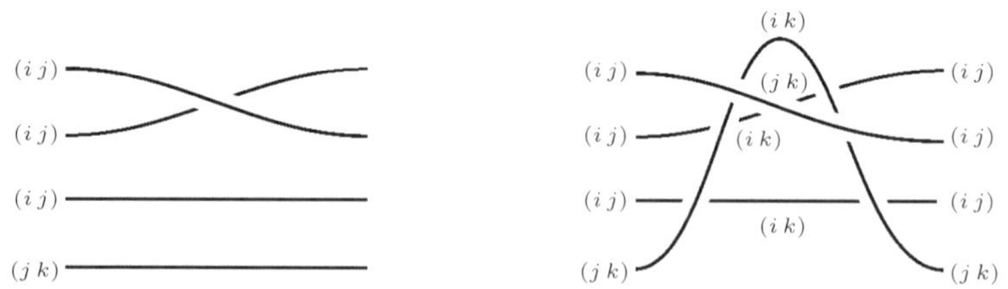

FIGURE 7. Arranging all crossings to have multiple labelings on the strands.

STEP 1. *Change B so that at each crossing the strands have three distinct labels.* The main idea in this step, described in detail below, is contained in Figure 7. Thinking of B as the closure of a braid we assume B has n strands. We will think of a diagram for B as n horizontal strands the xz-plane with twists at distinct x values, i.e. presented using standard generators of the braid group. Each of the n different strands at any x value must be labelled with at least two different transpositions in the symmetric group. If there is a single x value where there is just one then all strands in the braid will be labelled by this transposition and the covering is not a 3–folding covering, but just a cyclic 2–fold covering.

Now, let us suppose c is a crossing in the transverse braid B. The strands at c will either be labelled with three distinct transpositions or just one. In the former case, there is nothing to do, whereas in the latter case some other strand of the braid with the same x-value as the crossing must be labelled with a different transposition. Taking the strand with a different permutation that is closest to the crossing we can push it past the crossing as shown in Figure 7. This is done by a sequence of moves shown in the first two rows of Figure 3 and results in a braid with distinct labels at each crossing.

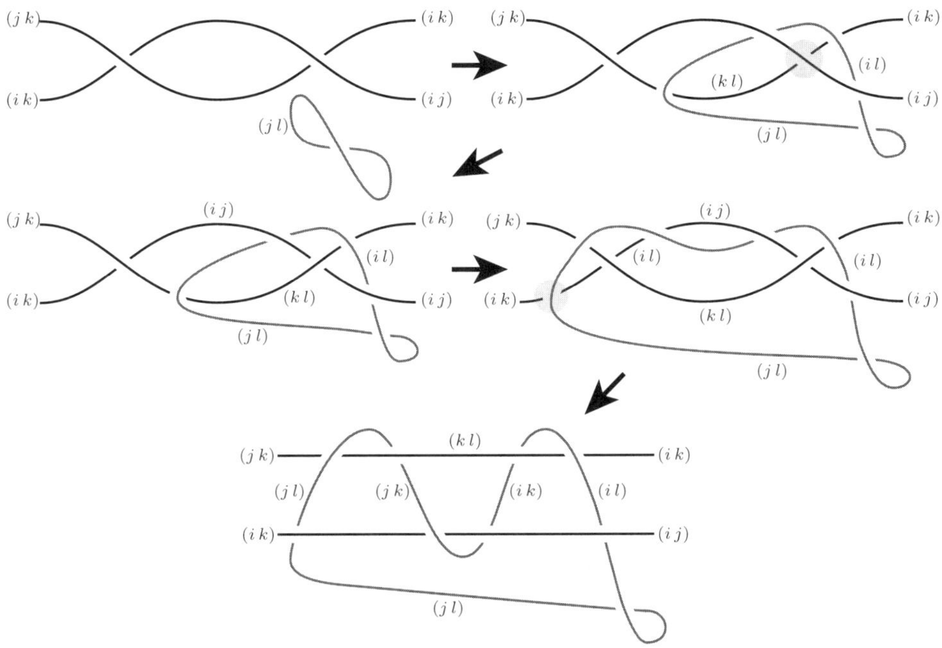

FIGURE 8. Adding an unknotted component to the branch locus.

STEP 2. *Change B so that it has only positive crossings and they occur in pairs.* We can apply the move shown in row four of Figure 3 to each negative crossing in order to change it into a pair of two positive crossings. Note that we also need to use the move in row one to remove adjacent positive and negative crossings. For each unpaired positive crossing we can similarly add three more positive crossing and change it into two paired positive crossings. For convenience we can also isotope the diagram so that each pair of positive crossings lies in a distinct vertical strip of the projection.

STEP 3. *Change B so that it is the trivial braid with linking unknots as in Figure* 5. For each pair of positive crossings we can use Lemma 2.6 to introduce an unknot with label $(j\,l)$, where when considering this crossing we already have an $(l-1)$–fold branched cover and j is chosen as shown in Figure 8. Now Figure 8 shows a sequence of isotopies of the unknot and applications of the move in row three of Figure 3 which manages to remove the pair of crossings at the expense of inserting a linked unknot. We will call such an unknot a *clasping unknot.*

If the braid after Step 2 had k pairs of crossings, then after Step 3, there are k clasping unknots and each one in a distinct vertical strip of the projection. Then adding $(n-2)$ new clasping unknots labelled with the identity permutation to the strands above and below each existing clasping unknot yields the link in Figure 5. Lemma 3.2 is now proven once we note that adding a knot to the branch locus whose strands are labelled with the identity permutation does not change the contact branched covering. □

Acknowledgments

We thank the referee for helpful comments that improved the paper. The first author also thanks Adán Medrano Martín del Campo for useful conversations.

References

[1] Daniel Bennequin, *Entrelacements et équations de Pfaff* (French), Third Schnepfenried geometry conference, Vol. 1 (Schnepfenried, 1982), Astérisque, vol. 107, Soc. Math. France, Paris, 1983, pp. 87–161. MR753131

[2] Joan S. Birman, *A representation theorem for fibered knots and their monodromy maps*, Topology of low-dimensional manifolds (Proc. Second Sussex Conf., Chelwood Gate, 1977), Lecture Notes in Math., vol. 722, Springer, Berlin, 1979, pp. 1–8. MR547448

[3] Meredith Casey. *Branched covers of contact manifolds.* PhD thesis, Georgia Institute of Technology, 2013.

[4] John B. Etnyre, *Legendrian and transversal knots*, Handbook of knot theory, Elsevier B. V., Amsterdam, 2005, pp. 105–185, DOI 10.1016/B978-044451452-3/50004-6. MR2179261

[5] John B. Etnyre and Ryo Furukawa, *Braided embeddings of contact 3-manifolds in the standard contact 5-sphere*, J. Topol. **10** (2017), no. 2, 412–446, DOI 10.1112/topo.12014. MR3653317

[6] Hansjörg Geiges, *Constructions of contact manifolds*, Math. Proc. Cambridge Philos. Soc. **121** (1997), no. 3, 455–464, DOI 10.1017/S0305004196001260. MR1434654

[7] Hansjörg Geiges, *An introduction to contact topology*, Cambridge Studies in Advanced Mathematics, vol. 109, Cambridge University Press, Cambridge, 2008. MR2397738

[8] Emmanuel Giroux, *Géométrie de contact: de la dimension trois vers les dimensions supérieures* (French, with French summary), Proceedings of the International Congress of Mathematicians, Vol. II (Beijing, 2002), Higher Ed. Press, Beijing, 2002, pp. 405–414. MR1957051

[9] Deborah L. Goldsmith, *Symmetric fibered links*, Knots, groups, and 3-manifolds (Papers dedicated to the memory of R. H. Fox), Princeton Univ. Press, Princeton, N.J., 1975, pp. 3–23. Ann. of Math. Studies, No. 84. MR0380766

[10] Hugh M. Hilden, *Every closed orientable 3-manifold is a 3-fold branched covering space of S^3*, Bull. Amer. Math. Soc. **80** (1974), 1243–1244, DOI 10.1090/S0002-9904-1974-13699-2. MR0350719

[11] Hugh M. Hilden, M. T. Lozano, and José María Montesinos, *Universal knots*, Bull. Amer. Math. Soc. (N.S.) **8** (1983), no. 3, 449–450, DOI 10.1090/S0273-0979-1983-15114-5. MR693959

[12] Hugh M. Hilden, María Teresa Lozano, and José María Montesino, *Universal knots*, Knot theory and manifolds (Vancouver, B.C., 1983), Lecture Notes in Math., vol. 1144, Springer, Berlin, 1985, pp. 25–59, DOI 10.1007/BFb0075011. MR823281
are universal.

[13] Hugh M. Hilden, María Teresa Lozano, and José María Montesinos, *The Whitehead link, the Borromean rings and the knot 9_{46} are universal*, Collect. Math. **34** (1983), no. 1, 19–28. MR747855

[14] Hugh M. Hilden, María Teresa Lozano, and José María Montesinos, *On knots that are universal*, Topology **24** (1985), no. 4, 499–504, DOI 10.1016/0040-9383(85)90019-9. MR816529

[15] Tirasan Khandhawit and Lenhard Ng, *A family of transversely nonsimple knots*, Algebr. Geom. Topol. **10** (2010), no. 1, 293–314, DOI 10.2140/agt.2010.10.293. MR2602837

[16] Andrea Loi and Riccardo Piergallini, *Compact Stein surfaces with boundary as branched covers of B^4*, Invent. Math. **143** (2001), no. 2, 325–348, DOI 10.1007/s002220000106. MR1835390

[17] José M. Montesinos, *Three-manifolds as 3-fold branched covers of S^3*, Quart. J. Math. Oxford Ser. (2) **27** (1976), no. 105, 85–94, DOI 10.1093/qmath/27.1.85. MR0394630

[18] José M. Montesinos-Amilibia and H. R. Morton, *Fibred links from closed braids*, Proc. London Math. Soc. (3) **62** (1991), no. 1, 167–201, DOI 10.1112/plms/s3-62.1.167. MR1078219

[19] Ferit Öztürk and Klaus Niederkrüger, *Brieskorn manifolds as contact branched covers of spheres*, Period. Math. Hungar. **54** (2007), no. 1, 85–97, DOI 10.1007/s-10998-007-1085-3. MR2310369

[20] R. Piergallini and D. Zuddas, *A universal ribbon surface in* B^4, Proc. London Math. Soc. (3) **90** (2005), no. 3, 763–782, DOI 10.1112/S0024611504015072. MR2137830
[21] William P. Thurston. Universal links. (Preprint).

Massachusetts Institute of Technology, Department of Mathematics, 77 Massachusetts Avenue Cambridge, Massachusetts 02139
Email address: `casals@mit.edu`

School of Mathematics, Georgia Institute of Technology, 686 Cherry Street, Atlanta, Georgia 30332-0160
Email address: `etnyre@math.gatech.edu`

Proceedings of Symposia in Pure Mathematics
Volume **102**, 2019
https://doi.org/10.1090/pspum/102/01812

Groups and polytopes

Stefan Friedl, Wolfgang Lück, and Stephan Tillmann

ABSTRACT. In a series of papers the authors associated to an L^2-acyclic group Γ an invariant $\mathcal{P}(\Gamma)$ that is a formal difference of polytopes in the vector space $H_1(\Gamma;\mathbb{R})$. This invariant is in particular defined for most 3-manifold groups, for most 2-generator 1-relator groups and for all free-by-cyclic groups. In most of the above cases the invariant can be viewed as an actual polytope.

In this survey paper we will recall the definition of the polytope invariant $\mathcal{P}(\Gamma)$ and we state some of the main properties. We conclude with a list of open problems.

1. Introduction

1.1. The Grothendieck group of polytopes. A *polytope in a finite dimensional real vector space* V is defined as the convex hull of a finite non-empty subset of V. Given a polytope $\mathcal{P}$ we denote by

$$\overline{\mathcal{P}} \; := \; \{-x \,|\, x \in \mathcal{P}\}$$

the mirror image of $\mathcal{P}$ in the origin.[1] We say that two polytopes $\mathcal{P}$ and $\mathcal{Q}$ are *translation-equivalent* if there exists a vector $v \in V$ with $v + \mathcal{P} = \mathcal{Q}$. We denote by $\mathfrak{P}(V)$ the set of all translation-equivalence classes of polytopes in V.

The *Minkowski sum* of two polytopes $\mathcal{P}$ and $\mathcal{Q}$ in V is defined as the polytope

$$\mathcal{P} + \mathcal{Q} \; := \; \{p + q \,|\, p \in \mathcal{P} \text{ and } q \in \mathcal{Q}\}.$$

This turns $\mathfrak{P}(V)$ into an abelian monoid, where the identity element 0 is represented by any polytope consisting of a single point.

It is straightforward to show, see e.g. [**Sc93**, Lemma 3.1.8], that $\mathfrak{P}(V)$ has the cancellation property, i.e. for any $\mathcal{P}, \mathcal{Q}, \mathcal{R} \in \mathfrak{P}(V)$ with $\mathcal{P} + \mathcal{Q} = \mathcal{P} + \mathcal{R}$ we have $\mathcal{Q} = \mathcal{R}$. We denote by $\mathfrak{G}(V)$ the set of all equivalence classes of pairs

2010 *Mathematics Subject Classification.* Primary 20J05; Secondary 20F65, 22E40, 57R19 .
Key words and phrases. Finitely presented group, L^2-acyclic groups, polytopes, BNS invariant.

The first author is supported by the SFB 1085 "Higher invariants" at the University of Regensburg, funded by the Deutsche Forschungsgemeinschaft (DFG)..

The second author is financially supported by the Leibniz-Preis granted by the DFG and the ERC Advanced Grant "KL2MG-interactions" (no. 662400) granted by the European Research Council.

The third author is partially supported under the Australian Research Council's Discovery funding scheme (project number DP140100158).

[1]In the literature the mirror image of $\mathcal{P}$ in the origin is often denoted by $-\mathcal{P}$. In our paper $-\mathcal{P}$ will have a very different meaning.

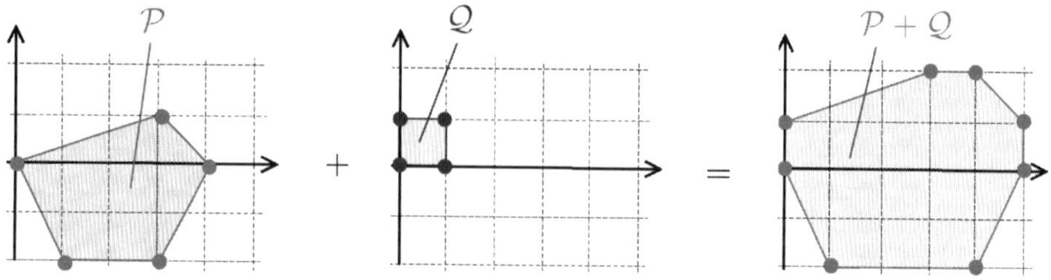

FIGURE 1

$(\mathcal{P},\mathcal{Q}) \in \mathfrak{P}(V)^2$, where we say that $(\mathcal{P},\mathcal{Q}) \sim (\mathcal{P}',\mathcal{Q}')$ if $\mathcal{P}+\mathcal{Q}' = \mathcal{P}'+\mathcal{Q}$. Note that $\mathfrak{G}(V)$ is an abelian group, and since $\mathfrak{P}(V)$ has the cancellation property it follows that the map $\mathfrak{P}(V) \to \mathfrak{G}(V)$ given by $\mathcal{P} \mapsto (\mathcal{P}, 0)$ is a monomorphism. We will use this monomorphism to identify $\mathfrak{P}(V)$ with its image in $\mathfrak{G}(V)$. Given $\mathcal{P}$ and $\mathcal{Q} \in \mathfrak{P}(V)$ we may write $\mathcal{P} - \mathcal{Q} = (\mathcal{P},\mathcal{Q})$. We refer to $\mathfrak{G}(V)$ as the *Grothendieck group of polytopes*.

Now let Γ be a finitely generated group. We write $\mathfrak{P}(\Gamma) = \mathfrak{P}(H_1(\Gamma;\mathbb{R}))$ and we say a polytope in $H_1(\Gamma;\mathbb{R})$ is *integral* if all the vertices lie in the image of $H_1(\Gamma;\mathbb{Z}) \to H_1(\Gamma;\mathbb{R})$. Similarly we write $\mathfrak{G}(\Gamma) = \mathfrak{G}(H_1(\Gamma;\mathbb{R}))$ and we say an element in $\mathfrak{G}(\Gamma)$ is *integral* if it can be represented by the difference of two integral polytopes.[2] A group homomorphism $\varphi\colon \Gamma \to \Pi$ induces natural homomorphisms $\varphi_*\colon \mathfrak{P}(\Gamma) \to \mathfrak{P}(\Pi)$ and $\varphi_*\colon \mathfrak{G}(\Gamma) \to \mathfrak{G}(\Pi)$.

1.2. L^2-acyclic groups and the Atiyah Conjecture. We say that a group Γ is of *type F* if it admits a finite model for $K(\Gamma, 1)$. (A group of type F is well-known to be torsion-free, see e.g. [**Br82**, Corollary VIII.2.5] for a proof.) We say Γ is *L^2-acyclic* if all its L^2-Betti numbers $b_n^{(2)}(\Gamma)$ vanish. The following gives examples of L^2-acyclic groups of type F.

(1) Fundamental groups of admissible 3-manifolds (here a 3-manifold is *admissible* if it is connected, orientable, irreducible, its boundary is empty or a disjoint union of tori, and its fundamental group is infinite).
(2) Free-by-cyclic groups, i.e. groups that can be written as a semidirect product $F \rtimes \mathbb{Z}$ with F a free group.[3]
(3) Torsion-free groups with two generators and one non-trivial relator.

We refer to [**LoL95**, Theorem 0.1], [**Lü02**, Section 4.2] and [**DL07**] for proofs that these groups are indeed L^2-acyclic of type F.

In this paper we are mostly interested in groups for which the Whitehead group is trivial. It is conjectured that the Whitehead group is trivial for any torsion-free group. This conjecture has now been proven for large classes of groups. For example the Whitehead group is known to be trivial for all torsion-free hyperbolic groups [**BLR08**], all virtually solvable [**We15**] and the aforementioned three types of groups, see e.g. [**AFW15**, (C.36)] and [**Wa78**, p. 249 and p. 250] for details.

[2]The invariants of a group π we will define later on will lie in $\mathfrak{G}(H_1(\pi;\mathbb{R}))$. In fact they will lie in a subgroup given by polytopes with integral vertices. Therefore it makes sense to study the polytope group of integral polytopes, which has also been studied in its own right by Funke [**Fu16**].

[3]Strictly speaking these groups should be called "free-by-infinite cyclic." But we follow the common convention to just say free-by-cyclic.

We refer to [**KLR16**, Theorem 2] for a detailed list of all groups for which it is currently known that the Whitehead group is trivial.

Furthermore, for the most part we want to restrict ourselves to groups that satisfy the Atiyah Conjecture. A torsion-free group Γ satisfies the Atiyah Conjecture if given any $(m \times n)$-matrix A over $\mathbb{Z}[\Gamma]$ the L^2-dimension of the kernel of the map $r_A\colon l^2(\Gamma)^m \to l^2(\Gamma)^n$ defined by $v \mapsto v \cdot A$ is a natural number. The class of torsion-free groups that are known to satisfy the Atiyah Conjecture is considerably smaller than the class of torsion-free groups for which it is known that the Whitehead group is trivial. For us it is mostly of interest that the Atiyah Conjecture is known for the following classes of groups:

(1) groups that are residually torsion-free elementary amenable,
(2) fundamental groups of most admissible 3-manifolds
(3) free-by-cyclic groups.

We refer to [**AFW15**, (H.21)] and [**Li93**, Theorem 1.5] for proofs. We also refer to [**LiL18**, Theorem 2.3] for a comprehensive summary of what is known about the Atiyah Conjecture.

In the following we refer to a group with trivial Whitehead group and which satisfies the Atiyah Conjecture as a Wh-AC-*group*. It is an open question whether all torsion-free groups are Wh-AC-groups.

1.3. The polytope invariant of L^2-acyclic groups of type F. In Section 2 we will use Reidemeister torsion over an appropriate skewfield to associate to an L^2-acyclic Wh-AC-group Γ of type F an integral element $\mathcal{P}(\Gamma)$ of $\mathfrak{G}(\Gamma)$. We refer to $\mathcal{P}(\Gamma)$ as the *polytope invariant of* Γ. In Section 3.1 we will see that there exists an L^2-acyclic Wh-AC-group Γ of type F such that neither $\mathcal{P}(\Gamma)$ nor $-\mathcal{P}(\Gamma)$ is represented by an actual polytope.

The following theorem summarizes some of the structural properties of this invariant. We refer to Section 2.4 for details.

THEOREM 1.1.

(1) *Suppose A, B and C are L^2-acyclic* Wh-AC-*groups of type F. Let $\varphi\colon C \to A$ and $\phi\colon C \to B$ be monomorphisms. We consider the corresponding amalgamated product $A *_C B$ and denote by a, b and c the monomorphisms from A, B and C into $A *_C B$. Then $A *_C B$ is also L^2-acyclic of type F. Furthermore, if $A *_C B$ is a* Wh-AC-*group, then*
$$\mathcal{P}(A *_C B) \;=\; a_*(\mathcal{P}(A)) + b_*(\mathcal{P}(B)) - c_*(\mathcal{P}(C)).$$
(2) *Let*
$$1 \to K \xrightarrow{i_*} G \to B \to 1$$
be an exact sequence of groups of type F. Suppose G is a Wh-AC-*group and K is L^2-acyclic. Then*
$$\mathcal{P}(G) \;=\; i_*(\mathcal{P}(K)) \cdot \chi(B).$$
(3) *Let M be an n-dimensional compact orientable manifold such that the boundary is either empty or a union of $(n-1)$-dimensional tori. If $\Gamma := \pi_1(M)$ is an L^2-acyclic*[4] Wh-AC-*group of type F, then*
$$\mathcal{P}(\Gamma) \;=\; (-1)^{n+1} \cdot \overline{\mathcal{P}(\Gamma)} \in \mathfrak{G}(\Gamma).$$

[4]The Singer Conjecture [**Sin77**][**Lü02**, p. 421] predicts that M has zero L^2-Betti numbers if and only if $\chi(M) = 0$.

The polytope invariant is in general rather difficult to calculate. The following theorem summarizes what is known for special classes of groups. In the interest of readability we keep the language somewhat informal. We refer to Section 3 for more carefully formulated statements.

THEOREM 1.2.

(1) *If $N \ncong S^1 \times D^2$ is an admissible 3-manifold that is not a closed graph manifold, then $\mathcal{P}(N)$ equals the dual of the Thurston norm ball. In particular $\mathcal{P}(N)$ is an integral polytope and it determines and is determined by the Thurston norm.*
(2) *If Γ is a group that admits a presentation $\pi = \langle x, y | r \rangle$ such that r is non-empty, reduced and cyclically reduced, then $\mathcal{P}(\Gamma)$ can be easily read off the word r.*
(3) *If N is a closed admissible 3-manifold that is not a graph manifold and if N admits a CW-structure with one 0-cell, two 1-cells, two 2-cells and one 3-cell, then $\mathcal{P}(N)$ can be determined immediately from the corresponding chain complex of the universal cover.*

It follows in particular from Theorem 1.2 that for 3-manifolds to which (2) or (3) applies one can easily obtain the Thurston norm from the fundamental group or the chain complex.

Next we turn to the question of what information the polytope invariant contains. The following theorem gives a partial answer to that question. It is again formulated in a slightly informal way, the precise statements will be given later in Sections 4 and 5.

THEOREM 1.3.

(1) *If $N \neq S^1 \times D^2$ is an admissible 3-manifold that is not a closed graph manifold, then we can mark some of the vertices of the polytope $\mathcal{P}(\pi_1(N))$ such that a class $\phi \in H^1(\pi_1(N); \mathbb{R})$ pairs maximally with a marked vertex if and only if it represents an element in the BNS-invariant* [**BNS87**] *$\Sigma(\pi_1(N))$.*
(2) *Let Γ be a group with $b_1(\Gamma) = 2$ that admits a presentation $\pi = \langle x, y | r \rangle$ such that r is non-empty, reduced and cyclically reduced. Then the following hold:*
    (a) *The "thickness" of $\mathcal{P}(\Gamma)$ is determined and determines the minimal complexity of "HNN-splittings" of Γ along groups.*
    (b) *We can mark some of the vertices of the polytope $\mathcal{P}(\Gamma)$ such that a class $\phi \in H^1(\Gamma; \mathbb{R})$ pairs maximally with a marked vertex if and only if it represents an element in the BNS-invariant $\Sigma(\Gamma)$.*
(3) [Funke–Kielak] *If Γ is an descending HNN-extension of a free group on two generators, then the geometry of $\mathcal{P}(\Gamma)$ is closely related to $\Sigma(\Gamma)$.*

Without too much effort statement (2) of Theorem 1.3 can also be formulated for groups $\pi = \langle x, y | r \rangle$ with $b_1(\pi) = 1$. To keep the paper short we leave it to the reader to make the fairly straightforward adjustments.

Most of the results in this paper are already explicit or at least implicit in our previous papers [**FT15**, **FST17**, **FL16**, **FL17**]. Only Theorem 1.2 (3) is a completely new statement.

The paper is organized as follows. In Section 2 we will use Reidemeister torsion to introduce the polytope invariant $\mathcal{P}(\Gamma)$ for any L^2-acyclic Wh-AC-group

Γ of type F. Furthermore we prove several statements regarding polytopes and Reidemeister torsions that will in particular imply Theorem 1.1. In Section 3 we give a more detailed discussion of the various statements of Theorem 1.2. In Sections 4 and 5 we will explain in more detail the statement and the references for Theorem 1.3. We conclude this paper with a long list of questions in Section 6.

2. Definition of the polytope invariant of groups

2.1. Ore localizations. Let R be a ring. (Throughout this paper a ring is understood to be associative and unital but not necessarily commutative.) We say R has the *Ore property* if given any two non-zero $x, y \in R$ there exist non-zero $r, s \in R$ such that $xr = ys$. If R has the Ore property, then one can define the classical ring of fractions which is a skew field. We refer to [**Lü02**, Section 8.2.1] and [**Ste75**] for more details on Ore localizations.

2.2. Review of division and rational closure. Let R be a subring of a ring S. The *division closure* $\mathcal{D}(R \subseteq S) \subseteq S$ is the smallest subring of S which contains R and is division closed, i.e., any element $x \in \mathcal{D}(R \subseteq S)$ which is invertible in S is already invertible in $\mathcal{D}(R \subseteq S)$. The *rational closure* $\mathcal{R}(R \subseteq S) \subseteq S$ is the smallest subring of S which contains R and is rationally closed, i.e., for any natural number n and matrix $A \in M_{n,n}(\mathcal{R}(R \subseteq S))$, if A is invertible over S, then A is already invertible over $\mathcal{R}(R \subseteq S)$. The division closure and the rational closure always exist. Obviously $R \subseteq \mathcal{D}(R \subseteq S) \subseteq \mathcal{R}(R \subseteq S) \subseteq S$.

Consider a group Γ. Let $\mathcal{N}(\Gamma)$ be the group von Neumann algebra which can be identified with the algebra $\mathcal{B}(L^2(\Gamma), L^2(\Gamma))^{\Gamma}$ of bounded left Γ-equivariant operators $L^2(\Gamma) \to L^2(\Gamma)$. Denote by $\mathcal{U}(\Gamma)$ the algebra of operators that are affiliated to the group von Neumann algebra, see [**Lü02**, Section 8] for details. This is the same as the Ore localization of $\mathcal{N}(\Gamma)$ with respect to the multiplicatively closed subset of non-zero divisors in $\mathcal{N}(\Gamma)$, see [**Lü02**, Theorem 8.22 (1)]. By the right regular representation we can embed $\mathbb{C}\Gamma$ and hence also $\mathbb{Z}\Gamma$ as a subring in $\mathcal{N}(\Gamma)$. We will denote by $\mathcal{R}(\Gamma)$ and $\mathcal{D}(\Gamma)$ the division and the rational closure of $\mathbb{Z}\Gamma$ in $\mathcal{U}(\Gamma)$. Summarizing we get a commutative diagram of inclusions of rings

$$\begin{array}{ccc} \mathbb{Z}\Gamma & \longrightarrow & \mathcal{N}(\Gamma) \\ \downarrow & & \\ \mathcal{D}(\Gamma) & & \downarrow \\ \downarrow & & \\ \mathcal{R}(\Gamma) & \longrightarrow & \mathcal{U}(\Gamma). \end{array}$$

We will use these inclusions to identify each ring with its monomorphic images.

The following lemma is well-known to the experts, full references are given in [**FL17**, Lemma 1.21].

LEMMA 2.1. *Let C_* be a finite chain complex of free left-$\mathbb{Z}\Gamma$-modules. Then the following assertions are equivalent:*

(1) *The L^2-Betti numbers of $\mathcal{N}(\Gamma) \otimes_{\mathbb{Z}[\Gamma]} C_*$ are all zero. (We refer to [**Lü02**] for the definition of the L^2-Betti numbers of a chain complex of $\mathcal{N}(\Gamma)$-modules.)*

(2) *The $\mathcal{R}(\Gamma)$-chain complex $\mathcal{R}(\Gamma) \otimes_{\mathbb{Z}\Gamma} C_*$ is contractible.*

LEMMA 2.2. *Let Γ be a group. If Γ is torsion-free and if Γ satisfies the Atiyah Conjecture, then the rational closure $\mathcal{R}(\Gamma)$ agrees with the division closure $\mathcal{D}(\Gamma)$ of $\mathbb{Z}\Gamma \subseteq \mathcal{U}(\Gamma)$. Furthermore, $\mathcal{D}(\Gamma)$ is a skew field.*

This lemma is proved in [**Lü02**, Lemma 10.39] for the Atiyah Conjecture over $\mathbb{C}$ with the division closure of $\mathbb{C}[\Gamma]$ instead of the division closure of $\mathbb{Z}[\Gamma]$, but the proof verbatim implies Lemma 2.2.

2.3. The polytope homomorphism. Throughout this section let Γ be a torsion-free group that satisfies the Atiyah Conjecture. By Lemma 2.2 $\mathcal{D}(\Gamma)$ is a skew field. Our goal is to associate to any element in the multiplicative group of units $\mathcal{D}(\Gamma)^\times = \mathcal{D}(\Gamma) \setminus \{0\}$ an integral element of $\mathfrak{G}(\Gamma)$. We outline the main steps in the construction and refer to [**FL16**], where this map was first introduced, for more details.

Write $H := H_1(\Gamma;\mathbb{Z})/\text{torsion}$ and view H as a multiplicative group. Let $\text{pr}\colon \Gamma \to H$ be the canonical projection and K be the kernel of pr. Choose a map of sets $s\colon H \to \Gamma$ with $\text{pr} \circ s = \text{id}_H$. We denote by $\mathbb{Z}[K] *_s H$ the ring, often referred to as the crossed product of $\mathbb{Z}[K]$ and H, whose underlying abelian group is given by finite formal sums $\sum_{h \in H} a_h h$ with each $a_h \in \mathbb{Z}[K]$ and for which multiplication is extended from the multiplication on $\mathbb{Z}[K]$ by the rule $g \cdot h = s(g) \cdot s(h) \cdot s((gh)^{-1}) \cdot gh$ for $g, h \in H$ and by the rule $h \cdot k = (hks(h)^{-1}) \cdot h$ for $h \in H$ and $k \in K$. It is straightforward to see that $\mathbb{Z}[K] *_s H$ is a ring and that

$$\begin{array}{ccc} \mathbb{Z}[\Gamma] & \to & \mathbb{Z}[K] *_s H \\ \sum\limits_{g \in \Gamma} a_g g & \mapsto & \sum\limits_{g \in \Gamma} a_g g s(\text{pr}(g))^{-1} \cdot \text{pr}(g) \end{array}$$

is a ring isomorphism, which we will sometimes use to identify these two rings. Given a non-zero element $\sum_{h \in H} a_h h \in \mathbb{Z}[K] *_s H$ we consider the integral polytope

$$\mathcal{P}\Big(\sum_{h \in H} a_h h\Big) = (\text{convex hull of } h \text{ with } a_h \neq 0) \in \mathcal{P}(H) = \mathcal{P}(\Gamma).$$

Defining $\mathcal{D}(K) *_s H$ analogously, given any non-zero element $f \in \mathcal{D}(K) *_s H$ we obtain the corresponding polytope $\mathcal{P}(f)$. Since $\mathcal{D}(K)$ is a skew field, in particular a domain, we obtain from elementary arguments that for any non-zero $f, g \in \mathcal{D}(K) *_s H$ we have

$$\mathcal{P}(f \cdot g) = \mathcal{P}(f) + \mathcal{P}(g). \tag{1}$$

We denote by $T = (\mathcal{D}(K) *_s H) \setminus \{0\}$ the set of all non-zero elements of the domain $\mathcal{D}(K) *_s H$. We can form the Ore localization $T^{-1}(\mathcal{D}(K) *_s H)$—a proof of this fact is for example given in [**DLMSY03**, Theorem 6.4] or [**Lü02**, Example 8.16]. There exists a canonical isomorphism

$$T^{-1}(\mathcal{D}(K) *_s H) \xrightarrow{\cong} \mathcal{D}(\Gamma),$$

which we will use to identify these rings, see e.g. [**Lü02**, Lemma 10.69]. (Again the lemma in [**Lü02**] is stated for the division closure of $\mathbb{C}[G]$ but it also holds with the same proof for the division closure of $\mathbb{Z}[G]$.) Now let $r = fg^{-1} \in \mathcal{D}(\Gamma) = T^{-1}(\mathcal{D}(K) *_s H)$ be non-zero. We define

$$\mathcal{P}(r) := \mathcal{P}(f) - \mathcal{P}(g) \in \mathfrak{G}(\Gamma).$$

It follows from (1) that this is well-defined and that this map defines a group homomorphism $\mathcal{P}\colon \mathcal{D}(\Gamma)^\times \to \mathfrak{G}(\Gamma)$. Since the target is abelian this descends to a group homomorphism

$$\mathcal{P}\colon \mathcal{D}(\Gamma)^\times/[\mathcal{D}(\Gamma)^\times,\mathcal{D}(\Gamma)^\times] \to \mathfrak{G}(\Gamma).$$

One easily checks that this homomorphism is independent of the choice of s.

2.4. The polytope invariant of an L^2-acyclic group of type F. Given a ring R we denote by $K_1(R)$ the usual K_1-group, as defined in [**Sil81**, **Ro94**]. If $\mathbb{K}$ is a skew field, then the Dieudonné determinant, see [**Sil81**, Corollary 4.3 on page 133] and [**Ro94**], gives rise to an isomorphism

$$\det\colon K_1(\mathbb{K}) \xrightarrow{\cong} \mathbb{K}^\times_{\mathrm{ab}} := \mathbb{K}^\times/[\mathbb{K}^\times,\mathbb{K}^\times].$$

In the following, given a torsion-free group Γ that satisfies the Atiyah Conjecture we will use this canonical isomorphism to identify $K_1(\mathcal{D}(\Gamma))$ with $\mathcal{D}(\Gamma)^\times_{\mathrm{ab}}$.

DEFINITION.

(1) An *L^2-acyclic pair* (X,φ) consists of a finite connected CW-complex and a homomorphism $\varphi\colon \pi_1(X)\to\Gamma$ such that $b_i^{(2)}(X,\varphi)=0$ for all i. Here $b_i^{(2)}(X,\varphi)$ denotes the L^2-Betti numbers of the covering space of X corresponding to φ, viewed as a Γ-CW-complex.

(2) Suppose that $(X,\varphi\colon \pi_1(X)\to\Gamma)$ *is an L^2-acyclic pair* and suppose that Γ is torsion-free and that it satisfies the Atiyah Conjecture. We denote by $\widetilde{X}$ the universal cover of X. By picking orientations of the cells of X and by picking lifts of the cells of X to $\widetilde{X}$ we can view $C_*(\widetilde{X})$ as a chain complex of based free $\mathbb{Z}[\pi_1(X)]$-left modules. Similarly we can view $\mathcal{D}(\Gamma)\otimes_{\mathbb{Z}[\Gamma]}C_*(\widetilde{X})$ as a chain complex of based free $\mathcal{D}(\Gamma)$-left modules. We had assumed that (X,φ) is L^2-acyclic. Together with Lemmas 2.1 and 2.2 this implies that the chain complex $\mathcal{D}(\Gamma)\otimes_{\mathbb{Z}[\Gamma]}C_*(\widetilde{X})$ is acyclic. We denote by $\tau(X,\varphi)\in K_1(\mathcal{D}(\Gamma))/\pm\Gamma=\mathcal{D}(\Gamma)^\times_{\mathrm{ab}}/\pm\Gamma$ the corresponding torsion as defined in [**Mi66**]. Furthermore we define the *polytope invariant of* (X,φ) as

$$\mathcal{P}(X,\varphi) := -\mathcal{P}(\tau(X,\varphi)) \in \mathfrak{G}(\Gamma).$$

The minus sign in the definition of $\mathcal{P}(X,\varphi)$ might initially be a little surprising. This choice of sign ensures that in many situations of interest, the polytope invariant $\mathcal{P}(X,\varphi)$ lies in $\mathfrak{P}(\Gamma)$, i.e. it indeed can be represented by a polytope.

The following proposition summarizes some of the key properties of the polytope invariant of an L^2-acyclic pair. The proof of the proposition follows from standard properties of Reidemeister torsion and the fact that the map $\mathcal{P}\colon K_1(\mathcal{D}(\Gamma))\to\mathfrak{G}(\Gamma)$ is a homomorphism. We refer to [**FL17**, Theorem 2.5] for details.

PROPOSITION 2.3. *Let $(X,\varphi\colon\pi_1(X)\to\Gamma)$ be an L^2-acyclic pair and suppose that Γ is a torsion-free group that satisfies the Atiyah Conjecture.*

(1) (Induction) *If $\delta\colon\Gamma\to\Pi$ is a monomorphism to a torsion-free group Π that satisfies the Atiyah Conjecture, then $\mathcal{P}(X,\delta\circ\varphi)=\delta_*(\mathcal{P}(X,\varphi))$, where $\delta_*\colon\mathfrak{G}(\Gamma)\to\mathfrak{G}(\Pi)$ is the induced map.*

(2) (Simple homotopy invariance) *If $f\colon W\to X$ is a simple homotopy equivalence of CW-complexes, i.e. if* $\mathrm{Wh}(f)=0\in\mathrm{Wh}(\pi_1(X))$, *then $\mathcal{P}(W,\varphi\circ f_*)=\mathcal{P}(X,\varphi)$.*

(3) (Homeomorphism invariance) *If $f\colon W \to X$ is a homeomorphism, then we have $\mathcal{P}(W, \varphi \circ f_*) = \mathcal{P}(X, \varphi)$.*

Now we can make the following two definitions.

(1) We say that a group Γ is *L^2-acyclic of type F* if it admits a finite $K(\Gamma, 1)$ and if all its L^2-Betti numbers vanish. If Γ is a Wh-AC-group, then we define

$$\mathcal{P}(\Gamma) \;:=\; \mathcal{P}(X, \mathrm{id})$$

where X is any finite $K(\Gamma, 1)$. It follows from Proposition 2.3 (2) that this definition does not depend on the choice of X. We refer to $\mathcal{P}(\Gamma)$ as the *polytope invariant* of Γ.

(2) Let M be a compact manifold and let $\varphi\colon \pi_1(M) \to \Gamma$ be a homomorphism to a group Γ that satisfies the Atiyah Conjecture. As above we say that (M, φ) is L^2-acyclic if $b_n^{(2)}(M, \varphi) = 0$ for all $n \in \mathbb{N}_0$. Now suppose that (M, φ) is L^2-acyclic. We pick a CW-structure X for M and we define $\mathcal{P}(M, \varphi) := \mathcal{P}(X, \varphi)$. It follows from Proposition 2.3 (3) that this definition does not depend on the choice of X. Sometimes we write $\mathcal{P}(M) := \mathcal{P}(M, \mathrm{id})$.

The following proposition collects a few more structural properties of the polytope invariant. These are again a consequence of [**FL17**, Theorem 2.5].

PROPOSITION 2.4.

(1) *Let $X = A \cup_C B$ be a decomposition of a finite CW-complex X into two connected CW-complexes A and B such that $C := A \cap B$ is also connected. Let $\varphi\colon \pi_1(X) \to \Gamma$ be a homomorphism to a torsion-free group Γ that satisfies the Atiyah Conjecture. We denote by $a\colon A \to X$, $b\colon B \to X$ and $c\colon C \to X$ the inclusion maps. If $(A, \varphi \circ a_*)$, $(B, \varphi \circ b_*)$ and $(C, \varphi \circ c_*)$ are L^2-acyclic, then (X, φ) is also L^2-acyclic and*

$$\mathcal{P}(X, \varphi) \;=\; a_*(\mathcal{P}(A, \varphi \circ a_*)) + b_*(\mathcal{P}(B, \varphi \circ b_*)) - c_*(\mathcal{P}(C, \varphi \circ c_*)).$$

(2) *Let*

$$1 \to F \xrightarrow{i} E \to B \to 1$$

be a fibration of finite CW-complexes. Let $\varphi\colon \pi_1(E) \to \Gamma$ be a homomorphism to a torsion-free group Γ that satisfies the Atiyah Conjecture. If $(F, \varphi \circ i_*)$ is L^2-acyclic, then

$$\mathcal{P}(E, \varphi) \;=\; \chi(B) \cdot i_*(\mathcal{P}(F, \varphi \circ i_*)).$$

(3) *Let M be an n-dimensional compact orientable manifold such that the boundary is either empty or a union of $(n-1)$-dimensional tori. Let $\varphi\colon \pi_1(M) \to \Gamma$ be a homomorphism to a group Γ that satisfies the Atiyah Conjecture. If (M, φ) is L^2-acyclic, then*

$$\mathcal{P}(M, \varphi) \;=\; (-1)^{n+1} \cdot \overline{\mathcal{P}(M, \varphi)} \in \mathfrak{G}(\Gamma).$$

It is clear that Theorem 1.1 is an immediate consequence of Propositions 2.3 and 2.4 and the definitions.

3. Examples

3.1. Elementary examples. We first consider the infinite cyclic group $\langle t\rangle$. In this case we take $X = K(\langle t\rangle, 1) = S^1$. The cellular chain complex of the universal cover of $\mathbb{R}$ is then isomorphic to

$$0 \to \mathbb{Z}[t^{\pm 1}] \xrightarrow{\cdot(1-t)} \mathbb{Z}[t^{\pm 1}] \to 0.$$

It follows that $\tau(X, \mathrm{id}) = (1-t)^{-1}$ and therefore $\mathcal{P}(\langle t\rangle) = -[0,1] \in \mathfrak{G}(\langle t\rangle) = \mathfrak{G}(\mathbb{R})$. In this case $\mathcal{P}(\langle t\rangle)$ is therefore not represented by a polytope.

Now consider $A = \langle s\rangle \times F_3$, where F_3 denotes, as usual, the free group on three generators and we also consider $B = \Gamma = \langle t\rangle \times (F_3 \times F_5)$. (Note that both groups satisfy the Atiyah Conjecture, this follows for example from the fact that both groups are residually torsion-free nilpotent [**DK92**, pp. 387 and 389] and [**LiL18**, Theorem 2.3].)

It follows from the above calculation and from Theorem 1.1 (2) that $\mathcal{P}(A)$ is represented by an interval of length $-\chi(F_3) = 2$. Similar to the above we see that $\mathcal{P}(B)$ is represented by minus an interval of length $\chi(F_3) \cdot \chi(F_5) = 8$. Now let $C = \mathbb{Z}$ and pick a monomorphism $\varphi\colon C \to A$ that factors through a monomorphism $C \to F_3$ and similarly pick a monomorphism $\psi\colon C \to B$ that factors through a monomorphism $C \to F_3 \times F_5$. We denote by $\Gamma = A *_C B$ the corresponding amalgamated product. (Strictly speaking we do not know of a proof that Γ satisfies the Atiyah Conjecture, the following discussion implicitly assumes that this is the case so that we can define $\mathcal{P}(\Gamma)$.) It is a consequence of Theorem 1.1 (1) that $\mathcal{P}(\Gamma)$ is a difference of two intervals and it is straightforward to see that $\mathcal{P}(\Gamma)$ is neither represented by a polytope nor by minus a polytope.

3.2. Groups with two generators and one relator. In general it is very hard to compute the polytope invariant, the main difficulty lies in determining the Dieudonné determinant of a square matrix. There is only one situation in which the calculation of the Dieudonné determinant is straightforward, namely when the matrix is a 1×1-matrix. As we will see, this observation makes it straightforward to determine the polytope invariant for groups with two generators and one relator.

We say that a presentation $\pi = \langle x, y | r\rangle$ is *nice* if it satisfies the following conditions:

(1) r is a non-empty, reduced and cyclically reduced word, and
(2) $b_1(\Gamma_\pi) = 2$, where Γ_π denotes the group defined by the presentation π.

Following [**FT15**] we will associate to a nice presentation $\pi = \langle x, y|r\rangle$ a polytope $\mathcal{S}(\pi)$. The definition is illustrated in Figure 2. First we identify $H_1(\Gamma_\pi;\mathbb{Z})$ with $\mathbb{Z}^2$ such that x corresponds to $(1,0)$ and y corresponds to $(0,1)$. Then the relator r determines a discrete walk on the integer lattice in $H_1(\Gamma_\pi;\mathbb{R}) = \mathbb{Z}^2$ and the polytope $\mathcal{S}(\pi)$ is obtained from the convex hull of the trace of this walk in the following way:

(1) Start at the origin and walk across $\mathbb{Z}^2$ reading the word r from the left.
(2) Take the convex hull $\mathcal{C}$ of the set of all lattice points reached by the walk.
(3) An elementary argument, using the fact that r is reduced and cyclically reduced, shows that one can take the Minkowski difference with the square $\mathcal{Q}$ of length one, i.e. there exists a polytope $\mathcal{S}(\pi)$ with $\mathcal{S}(\pi) + \mathcal{Q} = \mathcal{C}$.

Figure 2 illustrates the construction of the polytope for the nice presentation

$$\pi = \langle x, y \,|\, yx^4yx^{-1}y^{-1}x^2y^{-1}x^{-2}y^2xy^{-1}xy^{-1}x^{-1}y^{-2}x^{-3}y^2x^{-1} \rangle.$$

For the final Minkowski difference see also Figure 1.

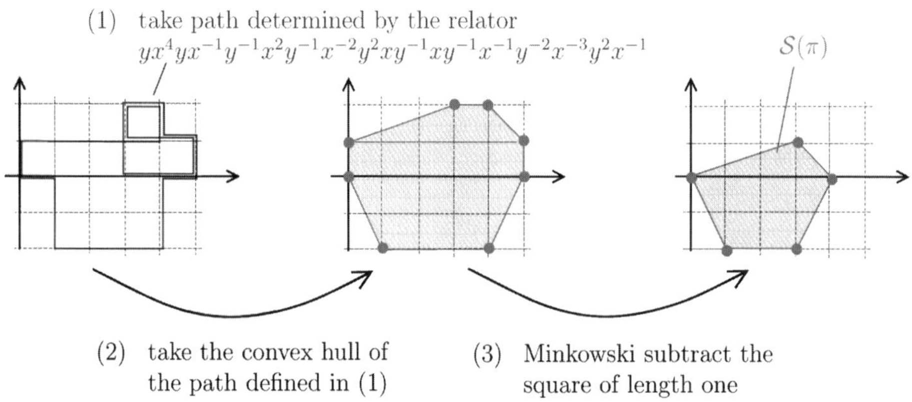

FIGURE 2. The polytope $\mathcal{S}(\pi)$ for a presentation π

We now view $\mathcal{S}(\pi)$ as an element in $\mathfrak{P}(\mathbb{Z}^2) = \mathfrak{P}(\Gamma_\pi)$, where we identify $\mathbb{Z}^2$ with $H_1(\Gamma_\pi;\mathbb{Z})$ by sending $(0,1)$ to x and $(0,1)$ to y. A priori $\mathcal{S}(\pi) \in \mathfrak{P}(\Gamma_\pi)$ is an invariant of the presentation and not of the underlying group Γ_π.

The following theorem gives another strong indication that the polytope is an invariant of the group.

THEOREM 3.1. *Let $\pi = \langle x, y \,|\, r\rangle$ be a nice presentation. If Γ_π is torsion-free*[5] *and if Γ_π is a* Wh-AC*-group, then Γ_π is L^2-acyclic of type F and*

$$\mathcal{S}(\pi) \;=\; \mathcal{P}(\Gamma_\pi) \in \mathfrak{P}(\Gamma_\pi).$$

REMARK. Recall that we had mentioned in the introduction that if Γ_π is residually elementary amenable and torsion-free, it satisfies the Atiyah Conjecture. It follows from Theorem 3.1 that in this case the polytope invariant $\mathcal{P}(\Gamma_\pi)$ is an invariant of π. (A proof of this fact without use of the Atiyah Conjecture is given in [**FT15**, Theorem 1.3].) In [**Wi14**, Conjecture 1.9] Wise (see also [**BK16**, p. 2]) conjectured that most hyperbolic groups $\langle x, y|r\rangle$ act properly and cocompactly on a CAT(0) cube complex. By Agol's Theorem [**Ag13**] a proof of this conjecture would imply that such groups are virtually special. By [**FST17**, Theorem 3.8] together with [**AFW15**, (H.26)] this would imply that any word hyperbolic torsion-free group with a nice presentation $\langle x, y|r\rangle$ is residually a group that is elementary-amenable and torsion-free. Summarizing, a proof of Wise's Conjecture would show that $\mathcal{S}(\pi)$ is an invariant of the underlying group for torsion-free hyperbolic groups with a nice $(2,1)$-presentation.

PROOF OF THEOREM 3.1. By [**LS77**, Proposition III.11.1] the 2-complex X corresponding to the given presentation is aspherical. We denote the universal

[5]The group Γ_π is torsion-free if and only if the word r is not a proper power of another word, see e.g. [**LS77**, Proposition II.5.17] for details.

cover of X by $\widetilde{X}$ and we write $\Gamma = \Gamma_\pi$. As usual we view $\mathbb{Z}[\Gamma]$ as a subring of $\mathcal{D}(\Gamma)$. The chain complex $\mathcal{D}(\Gamma) \otimes_{\mathbb{Z}[\Gamma]} C_*(\widetilde{X})$ is isomorphic to

$$0 \to \mathcal{D}(\Gamma) \xrightarrow{\begin{pmatrix} r_x & r_y \end{pmatrix}} \mathcal{D}(\Gamma)^2 \xrightarrow{\begin{pmatrix} x-1 \\ y-1 \end{pmatrix}} \mathcal{D}(\Gamma) \to 0$$

where $r_x := \frac{\partial r}{\partial x}$ and $r_y := \frac{\partial r}{\partial y}$ denote the Fox derivatives [**Fo53**] of the word r with respect to the generators x and y. By [**DL07**] and Lemmas 2.1 and 2.2 this chain complex is acyclic. Since $y - 1 \neq 0 \in \mathcal{D}(\Gamma)$ it follows from standard arguments, see e.g. [**Tu01**, Theorem 2.2] or [**DFL16**, Lemma 3.1] that the torsion of the chain complex equals $r_x \cdot (y-1)^{-1} \in \mathcal{D}(\Gamma)$. So we obtain

$$\mathcal{P}(\Gamma) \;=\; \mathcal{P}\big(\tau\big(\mathcal{D}(\Gamma) \otimes_{\mathbb{Z}[\Gamma]} C_*(\widetilde{X})\big)\big) \;=\; \mathcal{P}(r_x \cdot (y-1)^{-1}) \;=\; \mathcal{P}(r_x) - \mathcal{P}(y-1).$$

But by [**FT15**, Proposition 3.5] the polytope $\mathcal{S}(\pi)$ agrees with $\mathcal{P}(r_x) - \mathcal{P}(y-1)$ in $\mathfrak{P}(\Gamma_\pi)$. □

In [**FT15**, Section 8] we also deal with presentations of the form $\langle x, y | r \rangle$, where r is non-empty, reduced and cyclically reduced, but where $b_1(\Gamma_\pi) = 1$. In this case we can also 'naively' define a polytope (i.e. an interval) $\mathcal{S}(\pi)$ in $H_1(\Gamma_\pi; \mathbb{R}) = \mathbb{R}$ and the obvious analogue of Theorem 3.1 holds.

3.3. 3-manifolds. Let N be an admissible 3–manifold. For each $\phi \in H^1(N;\mathbb{Z})$ there is a properly embedded oriented surface Σ, such that $[\Sigma] \in H_2(N, \partial N; \mathbb{Z})$ is the Poincaré dual to ϕ. Letting $\chi_-(\Sigma) = \sum_{i=1}^k \max\{-\chi(\Sigma_i), 0\}$, where $\Sigma_1, \dots, \Sigma_k$ are the connected components of Σ, the *Thurston norm* of $\phi \in H^1(N;\mathbb{Z})$ is defined as

$$x_N(\phi) := \min\big\{\chi_-(\Sigma) \,|\, \Sigma \text{ is a properly embedded surface with } \mathrm{PD}([\Sigma]) = \phi\big\}.$$

Thurston [**Th86**] showed that x_N is a seminorm on $H^1(N;\mathbb{Z})$ and elementary arguments show that x_N extends to a seminorm x_N on $H^1(N;\mathbb{R})$. We denote by

$$\mathcal{T}(N) := \{v \in H_1(N;\mathbb{R}) \,|\, \phi(v) \leq 1 \text{ for all } \phi \in H^1(N;\mathbb{R}) \text{ with } x_N(\phi) \leq 1\}$$

the dual of the unit norm ball of x_N. Thurston [**Th86**] showed that $\mathcal{T}(N)$ is a polytope with integral vertices.

The following theorem is [**FL17**, Theorem 3.35]. The proof relies on the recent work of Agol [**Ag08, Ag13**], Przytycki-Wise [**PW18**] and Wise [**Wi09, Wi12a, Wi12b**]. (See also [**FK14, CF17, GM17**] for alternative proofs.)

THEOREM 3.2. *For any admissible 3-manifold $N \neq S^1 \times D^2$ that is not a closed graph manifold we have*

$$\mathcal{T}(N) \;=\; 2 \cdot \mathcal{P}(\pi_1(N)) \;\in\; \mathfrak{P}(\pi_1(N)).$$

The combination of Theorems 3.1 and 3.2 gives us the following corollary. This corollary was first proved in [**FST17**] using a different approach.

COROLLARY 3.3. *Let N be an admissible 3-manifold such that $\pi_1(N)$ admits a nice presentation $\pi = \langle x, y | r \rangle$. Then*

$$\mathcal{T}(N) \;=\; 2 \cdot \mathcal{S}(\pi).$$

3.4. Closed 3-manifolds with a two-generator fundamental group. The following theorem says that for "small" closed 3-manifolds one can easily obtain the Thurston norm from the chain complex of the universal cover. This theorem can be viewed as a version of Corollary 3.3 for closed 3-manifolds. It applies, in particular, to all closed admissible 3–manifolds of Heegaard genus equal to two. It is known through work of Kobayashi [**Ko88**] and Hempel [**He01**] in combination with Perelman's solution of the geometrisation conjecture that if the splitting distance of a genus two Heegaard splitting of a 3-manifold is larger than two, then the 3-manifold is hyperbolic (whence admissible). Maher [**Ma10**] used this fact to show that *most* 3–manifolds of Heegaard genus two are hyperbolic.

THEOREM 3.4. *Let N be a closed admissible 3-manifold that is not a graph manifold. We write $\pi = \pi_1(N)$. Suppose N admits a CW-structure with one 0-cell, two 1-cells, two 2-cells and one 3-cell. The corresponding cellular chain complex of the universal cover $\widetilde{N}$ is of the form*

$$0 \to \mathbb{Z}[\pi] \xrightarrow{\begin{pmatrix} c_1 & c_2 \end{pmatrix}} \mathbb{Z}[\pi]^2 \xrightarrow{\begin{pmatrix} b_{11} & b_{12} \\ b_{21} & b_{22} \end{pmatrix}} \mathbb{Z}[\pi]^2 \xrightarrow{\begin{pmatrix} a_1 \\ a_2 \end{pmatrix}} \mathbb{Z}[\pi] \to 0.$$

Then there exist $i, j \in \{1, 2\}$ with $c_i \neq 0$ and $a_j \neq 0$, and

$$\mathcal{T}(N) \;=\; \mathcal{P}(b_{3-i,3-j}) - \mathcal{P}(c_i) - \mathcal{P}(a_j).$$

PROOF. As we pointed out earlier, the chain complex $\mathcal{D}(\pi) \otimes_{\mathbb{Z}[\pi]} C_*(\widetilde{N})$ is acyclic. It thus follows that there exist $i, j \in \{1, 2\}$ with $c_i \neq 0$ and $a_j \neq 0$. Since $\mathcal{D}(\pi)$ is a skew field one can show, similar to [**Tu01**, Theorem 2.2], that

$$\tau(N) \;=\; c_i^{-1} \cdot b_{3-i,3-j} \cdot a_j^{-1}.$$

This implies that $\mathcal{P}(\pi_1(N)) = \mathcal{P}(b_{3-i,3-j}) - \mathcal{P}(c_i) - \mathcal{P}(a_j)$. The theorem now follows from Theorem 3.2. $\square$

4. Marked polytopes

A *marked polytope* is a pair $\mathcal{M} = (\mathcal{P}, \mathcal{V})$, where $\mathcal{P}$ is a polytope and $\mathcal{V}$ is a (possibly empty) subset of the set of vertices of $\mathcal{P}$. We refer to the vertices in $\mathcal{V}$ as the *marked vertices*. If $\mathcal{M}$ and $\mathcal{N}$ are two marked polytopes, then we define the (*marked*) *Minkowski sum $\mathcal{M} + \mathcal{N}$ of $\mathcal{M}$ and $\mathcal{N}$* as the Minkowski sum of the corresponding polytopes, where the marked vertices of the Minkowski sum are precisely those that are the sum of a marked vertex of $\mathcal{M}$ and a marked vertex of $\mathcal{N}$. An example is given in Figure 3, where the marked vertices are indicated by a dot.

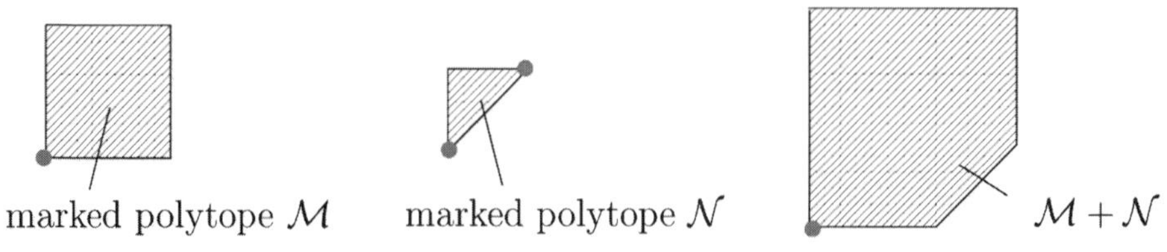

FIGURE 3. Example of the Minkowski sum of two markedpolytopes.

Marked polytopes appear naturally in many contexts:

(1) Let N be a 3-manifold. A class $\phi \in H^1(N;\mathbb{R})$ is called *fibered* if it can be represented by a non-degenerate closed 1-form. By [**Ti70**] an integral class $\phi \in H^1(N;\mathbb{Z}) = \operatorname{Hom}(\pi_1(N),\mathbb{Z})$ is fibered if and only if there exists a locally trivial fiber bundle $p\colon N \to S^1$ such that $p_* = \phi\colon \pi_1(N) \to \pi_1(S^1) = \mathbb{Z}$. Thurston [**Th86**] showed that we can turn $\mathcal{T}(N)$ into a marked polytope $\mathcal{M}(N)$ which has the property that $\phi \in H^1(N;\mathbb{R}) = \operatorname{Hom}(H_1(N;\mathbb{R}),\mathbb{R})$ is fibered if and only if it pairs maximally with a marked vertex. This means that there exists a marked vertex v of $\mathcal{M}(N)$, such that $\phi(v) > \phi(w)$ for any vertex $w \neq v$ in the underlying polytope $\mathcal{T}(N)$.

(2) Let Γ be a torsion-free group. We will now see that we can associate a marked polytope to a non-zero element in $\mathbb{Z}[\Gamma]$. (In our subsequent discussion of this assignment we will use the notation introduced in Section 2.3.) We recall that we have an identification $\mathbb{Z}[\Gamma] = \mathbb{Z}[K]*_s H$, where $H = H_1(\Gamma;\mathbb{Z})/\text{torsion}$ and $K = \operatorname{Ker}(\Gamma \to H)$. Given a non-zero element $f = \sum_{h\in H} a_h h$ we consider the marked polytope $\mathcal{M}(f)$ which is given by the polytope

$$\mathcal{P}\Big(\sum_{h\in H} a_h h\Big) = (\text{convex hull of } h \text{ with } a_h \neq 0)$$

where we mark a vertex h of $\mathcal{P}(f)$ if $a_h = \pm k$ for some $k \in K$. If $\mathbb{Z}[\Gamma]$ is a domain, then for any non-zero $f, g \in \mathbb{Z}[\Gamma]$ we have $\mathcal{M}(f \cdot g) = \mathcal{M}(f) + \mathcal{M}(g)$. (Note that if Γ satisfies the Atiyah Conjecture, then $\mathbb{Z}[\Gamma]$ is a domain, see e.g. [**Lü02**, Lemma 10.15].)

(3) Let $\pi = \langle x, y | r\rangle$ be a nice presentation such that Γ_π is torsion-free. We showed in [**FT15**, Proposition 3.5] that there exists a unique marked polytope $\mathcal{M}(\pi)$ with $\mathcal{M}(\pi) + \mathcal{M}(y-1) = \mathcal{M}(r_x)$ where $r_x = \frac{\partial r}{\partial x}$ is as before the Fox derivative.

The set of marked polytopes in a vector space forms a monoid under Minkowski sum. As illustrated in Figure 4 this monoid does *not* have the cancellation property. This implies that the monoid of marked polytopes does *not* inject into the corresponding Grothendieck group. Therefore we can not associate a meaningful notion of a difference of marked polytopes to an L^2-acyclic group π of type F at this point of time.

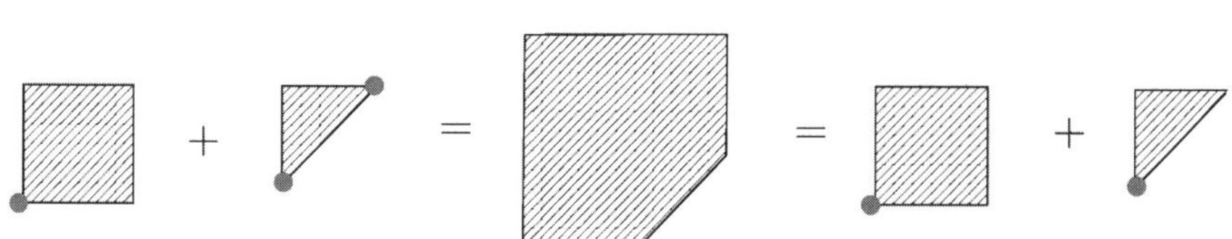

FIGURE 4. The Minkowski sum of marked polytopes does not have the cancellation property.

5. The polytope invariant and intrinsic properties of the group

In Section 2.4 we associated to an L^2-acyclic Wh-AC-group Γ of type F its polytope invariant $\mathcal{P}(\Gamma) \in \mathfrak{G}(\Gamma)$. In this section we want to determine information encoded in this invariant.

5.1. The thickness of the polytope invariant and splittings of the group. Given a polytope $\mathcal{P}$ in a real vector space V and given $\phi \in \operatorname{Hom}(V, \mathbb{R})$ we refer to

$$\operatorname{th}_\phi(\mathcal{P}) \;=\; \max\{\phi(x) - \phi(y) \mid x, y \in \mathcal{P}\} \;\in\; \mathbb{R}_{\geq 0}$$

as the *thickness of* $\mathcal{P}$ *in the* ϕ*-direction.* Since thickness is translation invariant we can also define $\operatorname{th}_\phi(\mathcal{P}) \in \mathbb{R}_{\geq 0}$ for any $\mathcal{P} \in \mathfrak{P}(V)$ and since thickness is additive under Minkowski sum we can also define $\operatorname{th}_\phi(\mathcal{P}) \in \mathbb{R}$ for any $\mathcal{P} \in \mathfrak{G}(V)$.

Given a polytope $\mathcal{P}$ in V we refer to

$$\mathcal{P}^{\mathrm{sym}} := \{\tfrac{1}{2}(p - q) \mid p, q \in \mathcal{P}\}$$

as the *symmetrization of* $\mathcal{P}$. This definition extends to a symmetrization map on $\mathfrak{P}(V)$ and $\mathfrak{G}(V)$. It is clear that the thickness of $\mathcal{P}$ depends only on the symmetrization of $\mathcal{P}$.

The following theorem is now a straightforward consequence of Theorem 3.2.

THEOREM 5.1. *For any admissible 3-manifold $N \neq S^1 \times D^2$ that is not a closed graph manifold and any $\phi \in H^1(N; \mathbb{R})$ we have*

$$\operatorname{th}_\phi(\mathcal{P}(\pi_1(N))) \;=\; x_N(\phi).$$

Ideally one would like to generalize the statement of Theorem 5.1 to larger classes of groups. The first problem that arises is that there is no satisfactory purely group theoretic definition of the Thurston norm. We refer to [**FSW15**, **FSW16**] for several ideas and approaches.

In an attempt to generalize Theorem 5.1 we will work with the notion of a splitting of a group. Let Γ be a finitely presented group and let $\phi\colon \Gamma \to \mathbb{Z}$ be an epimorphism. Let B be a finitely generated group. A *splitting of* (Γ, ϕ) *over* B is an isomorphism

$$f\colon \Gamma \xrightarrow{\cong} \langle A, t \mid \mu(B) = tBt^{-1}\rangle$$

such that the following hold:

(1) A is finitely generated,
(2) B is a subgroup of A and $\mu\colon B \to A$ is a monomorphism,
(3) $(\phi \circ f^{-1})(a) = 0$ for $a \in A$ and $(\phi \circ f^{-1})(t) = 1$.

It is well-known, see e.g. [**BS78**] or [**Str84**, Theorem B*], that any such pair (Γ, ϕ) admits a splitting over a finitely generated group. We define the *splitting complexity of* (Γ, ϕ) as

$$c(\Gamma, \phi) \;:=\; \min\{\operatorname{rank}(B) \mid (\Gamma, \phi) \text{ splits over } B\},$$

where $\operatorname{rank}(B)$ is defined as the minimal number of generators of B.

THEOREM 5.2. *Let Γ be a torsion-free group that admits a nice presentation $\langle x, y | r\rangle$. Furthermore suppose that Γ satisfies the Atiyah Conjecture. Then for any epimorphism $\phi\colon \Gamma \to \mathbb{Z}$ we have*

$$c(\Gamma, \phi) - 1 \;=\; \operatorname{th}_\phi(\mathcal{P}(\Gamma)).$$

REMARK. In [**FSW15**] it was shown that the statement of the theorem also holds if π is the fundamental group of a knot complement. But the equality of Theorem 5.2 does not hold for the fundamental groups of all 3-manifolds. For example, if N is the 3-torus, then $\pi_1(N) = \mathbb{Z}^3$ and it follows from Theorem 1.1 and the discussion in Section 3.1 that $\mathcal{P}(\pi_1(N))$ is a point. In particular the thickness for each ϕ is zero. On the other hand it is straightforward to see that for any epimorphism $\phi\colon \pi_1(N) \to \mathbb{Z}$ we have $c(\pi_1(N), \phi) = 2$.

PROOF. The inequality $c(\Gamma, \phi) - 1 \le \mathrm{th}_\phi(\mathcal{P}(\Gamma))$ was shown in [**FT15**, Proposition 7.3]. As was suggested to us by Nathan Dunfield, the proof of this inequality follows from a careful reading of the first step in the proof of the Freiheitssatz.

The proof of the reverse inequality follows along the same lines as the proof [**FT15**, Proposition 7.6] but one needs to replace the Ore localizations of the elementary-amenable and torsion-free groups by $\mathcal{D}(\Gamma)$. More precisely, it follows easily from the combination of Theorem 3.6(4) and Lemmas 4.3, 6.12 and 6.13 of [**FL16**] together with a slight variation on Theorem 4.1 of [**FL16**]. We leave the details to the reader. $\square$

5.2. The polytope invariant and its relation to the BNS-invariant. Let Γ be a finitely generated group. The Bieri–Neumann–Strebel [**BNS87**] invariant $\Sigma(\Gamma)$ of Γ is by definition a subset of $S(\Gamma) := (\mathrm{Hom}(\Gamma, \mathbb{R}) \setminus \{0\})/\mathbb{R}_{>0}$. We refer to [**BNS87**] for the precise definition, but in order to give a flavour of the invariant we recall three properties:

(1) An epimorphism $\phi \in \mathrm{Hom}(\Gamma, \mathbb{Z})$ represents an element in $\Sigma(\Gamma)$ if and only if it corresponds to an *ascending HNN-extension*. To be precise: if and only if there exists an isomorphism
$$f\colon \Gamma \ \to\ \langle A, t \mid A = t^{-1}\varphi(A)t\rangle$$
where A is a finitely generated group and $\varphi\colon A \to A$ is a monomorphism, such that ϕ corresponds under f to the epimorphism given by $t \mapsto 1$ and $a \mapsto 0$ for all $a \in A$.[6]

(2) A non-trivial homomorphism $\phi \in \mathrm{Hom}(\Gamma, \mathbb{Z})$ has the property that ϕ and $-\phi$ represent elements in $\Sigma(\Gamma)$ if and only if $\mathrm{Ker}(\phi)$ is finitely generated.

(3) $\Sigma(\Gamma)$ is an open subset of $S(\Gamma)$.

The first two properties follow from [**BNS87**, Proposition 4.3] (see also [**Br87**, Corollary 3.2]) and the third one is [**BNS87**, Theorem A].

It is shown in [**BNS87**, Theorem E] that given a 3-manifold N the fibered classes of $H^1(N; \mathbb{R}) = \mathrm{Hom}(\pi_1(N), \mathbb{R})$ correspond precisely to the classes that lie in $\Sigma(\pi_1(N))$. The following is now a reformulation of the statement of the first example in Section 4.

THEOREM 5.3. *Let $N \neq S^1 \times D^2$ be an admissible 3-manifold that is not a closed graph manifold. Then the polytope $\mathcal{P}(N)$ admits a marking with the property that for any non-trivial $\phi \in \mathrm{Hom}(\pi_1(N), \mathbb{R})$ we have*
$$[\phi] \in \Sigma(\pi_1(N)) \quad \Longleftrightarrow \quad \phi \textit{ pairs maximally with a marked vertex of } \mathcal{P}(N).$$

For groups with a nice presentation $\langle x, y | r\rangle$ we obtain a similar theorem.

[6] Some authors, e.g. [**FK16**], call this a *descending* HNN-extension. We follow the convention used in [**BS78, BNS87**].

THEOREM 5.4. *Let $\pi = \langle x, y \,|\, r\rangle$ be a nice presentation. A non-trivial class $\phi \in H^1(\Gamma_\pi;\mathbb{R})$ represents an element in $\Sigma(\Gamma_\pi)$ if and only if ϕ pairs maximally with a marked vertex of $\mathcal{M}(\pi)$.*

The theorem is stated as [**FT15**, Theorem 1.1]. In that paper the theorem is proved using the generalised Novikov rings of Sikorav [**Sik87**]. It can also be viewed as a reformulation of Brown's algorithm [**Br87**, Theorem 4.3].

Finally a relationship between the polytope invariant and the BNS-invariant was also found for a different class of groups by Funke–Kielak [**FK16**]. In order to state their theorem we need to introduce a few more definitions.

(1) Let $\varphi\colon \pi \to \pi$ be a monomorphism of a group π. We denote by
$$\pi*_\varphi \;:=\; \langle \pi, t \,|\, t^{-1}gt = \varphi(g), g \in \pi\rangle$$
the corresponding descending HNN-extension.[7] We refer to the epimorphism $\pi*_\varphi \to \mathbb{Z}$ given by $g \mapsto 0$ for all $g \in \pi$ and $t \to 1$ as the *canonical epimorphism.*

(2) Let $\mathcal{P}$ be a polytope in a vector space V and let $\phi \in \operatorname{Hom}(V,\mathbb{R})$. Following [**FK16**] we refer to
$$F_\phi(\mathcal{P}) \;=\; \{v \in \mathcal{P} \,|\, \phi(v) \le \phi(w) \text{ for all } w \in \mathcal{P}\}$$
as the *ϕ-minimal face of $\mathcal{P}$*.

(3) Let V be a vector space and let $\mathcal{S} = [\mathcal{P}] - [\mathcal{Q}] \in \mathfrak{G}(V)$. Following [**FK16**] we say that $\phi, \psi \in \operatorname{Hom}(V,\mathbb{R})$ are *$\mathcal{S}$-equivalent* if $F_\phi(\mathcal{P}) = F_\psi(\mathcal{P})$ and $F_\phi(\mathcal{Q}) = F_\psi(\mathcal{Q})$. Note that this definition is independent of the choice of $\mathcal{P}$ and $\mathcal{Q}$.

The following is the main result of [**FK16**].

THEOREM 5.5. *Let $\varphi\colon F_2 \to F_2$ be a monomorphism of the free group on two generators. Let $\phi\colon \Gamma := F_2*_\varphi \to \mathbb{R}$ be a homomorphism such that the class $[\phi] \in S(\Gamma)$ is not represented by the canonical epimorphism. Then there exists an open neighborhood U of $[\phi]$ in $S(\Gamma)$ such that for every non-trivial $\psi\colon \Gamma \to \mathbb{R}$ with $[\psi] \in U$ that is $\mathcal{P}(\Gamma)$-equivalent to ϕ we have*
$$[-\phi] \in \Sigma(\Gamma) \quad\Longleftrightarrow\quad [-\psi] \in \Sigma(\Gamma).$$

6. Questions

We conclude this survey with a long list of questions and conjectures.

CONJECTURE 6.1. *If Γ is a free-by-cyclic group, then $\mathcal{P}(\Gamma) \in \mathfrak{G}(\Gamma)$ can be represented by a polytope.*

QUESTION 6.2. Let $\Gamma \neq \mathbb{Z}$ be an L^2-acyclic Wh-AC-group of type F that admits a 2-dimensional $K(\Gamma,1)$. Can $\mathcal{P}(\Gamma) \in \mathfrak{G}(\Gamma)$ be represented by a polytope?

An affirmative answer to Question 6.2 also proves Conjecture 6.1.

Before we state our next question we recall that the Thurston norm of an admissible hyperbolic 3-manifold N is a norm. (This is a direct consequence of the fact that hyperbolic admissible 3-manifolds are atoroidal, i.e. any embedded torus is boundary parallel, see e.g. [**BP92**, Proposition D.3.2.8].) This implies that if N

[7] In [**FK16**] we authors refer to such an HNN-extension as an "ascending HNN-extension", whereas here we stick with the convention established above and that follows [**BS78, BNS87**].

is an admissible hyperbolic 3-manifold with $b_1(N) \geq 1$, then $\mathcal{P}(N) = \mathcal{P}(\pi_1(N))$ is not a point.

QUESTION 6.3. Let Γ be an L^2-acyclic Wh-AC-group of type F that admits a 2-dimensional $K(\Gamma, 1)$. Furthermore suppose that Γ is a non-elementary hyperbolic group and that $b_1(\Gamma) \geq 1$. Does it follow that $\mathcal{P}(\Gamma) \neq 0 \in \mathfrak{G}(\Gamma)$?

CONJECTURE 6.4. *Let Γ be an L^2-acyclic* Wh-AC-*group of type F. If Γ is amenable and if Γ is not virtually $\mathbb{Z}$, then $\mathcal{P}(\Gamma) = 0 \in \mathfrak{G}(\Gamma)$.*

Recently the previous conjecture was proved by Funke [**Fu17**, Corollary 5.4] for amenable groups that contain an non-abelian elementary amenable subgroup as a normal subgroup. In fact [**Fu17**, Theorem 5.3] suggests that the polytope invariant should be trivial for every L^2-acyclic Wh-AC-group of type F that contains an amenable non-abelian *normal* subgroup.

QUESTION 6.5. Let Γ be an L^2-acyclic Wh-AC-group of type F. Is the BNS-invariant related to $\mathcal{P}(\Gamma)$? Does an analogue of Theorem 5.5 hold?

In general it might be too optimistic to expect a positive answer. It seems more likely that the question can be answered in the affirmative if Γ has a 2-dimensional $K(\Gamma, 1)$.

It is known that the BNS invariants of metabelian and 3–manifold groups are polyhedral. An affirmative (even partial) answer to Question 6.5 may establish polyhedrality of BNS invariants for new families of groups, as well as whether the polyhedra are rational or not. We can therefore also ask the following question.

CONJECTURE 6.6. *Let Γ be a hyperbolic L^2-acyclic group of type F with a 2-dimensional $K(\Gamma, 1)$. Then there exists an integral marked polytope $\mathcal{M} \subset H_1(\Gamma; \mathbb{R})$ such that a non-trivial $\phi \in H^1(\Gamma; \mathbb{R})$ represents an element in $\Sigma(\Gamma)$ if and only if ϕ pairs maximally with a marked vertex of $\mathcal{M}$.*

The combination of Theorem 3.6(4) and Lemmas 4.3, 6.12 and 6.13 of [**FL16**] shows that if Γ is an L^2-acyclic Wh-AC-group of type F, then for any epimorphism $\phi\colon \Gamma \to \mathbb{Z}$ we have

$$\sum_{n \in \mathbb{N}_0} (-1)^{n+1} \cdot b_n^{(2)}(\mathrm{Ker}(\phi\colon \Gamma \to \mathbb{Z})) \;=\; \mathrm{th}_\phi(\mathcal{P}(\Gamma)).$$

CONJECTURE 6.7. *Let Γ be an L^2-acyclic* Wh-AC-*group of type F and let $n \in \mathbb{N}_0$. Then there exists a polytope $\mathcal{P}$ such that*

$$b_n^{(2)}(\mathrm{Ker}(\phi\colon \Gamma \to \mathbb{Z})) \;=\; \mathrm{th}_\phi(\mathcal{P})$$

for any epimorphism $\phi\colon \Gamma \to \mathbb{Z}$.

In Section 5.1 we saw that for 3-manifold groups and many two-generator one-relator groups the thickness of the polytope invariant is related to the complexity of splittings of the underlying group.

QUESTION 6.8. Let Γ be an L^2-acyclic Wh-AC-group of type F. What information does the thickness of the polytope invariant contain?

As we mentioned before, the thickness of a polytope is an invariant of the symmetrized polytope.

QUESTION 6.9. What information does the polytope invariant contain, that cannot be obtained from the symmetrized polytope invariant?

One partial answer is given by the discussion of the BNS-invariant. This invariant is in general not symmetric and if the polytope invariant is related to the BNS-invariant it cannot be symmetric in general. Nonetheless, there are many groups with empty BNS-invariant and non-symmetric polytope invariant.

Finally we want to discuss which elements of $\mathfrak{G}(\mathbb{Z}^n)$ can be realized as the polytope invariant of a manifold. Here we say that $\mathcal{P} \in \mathfrak{G}(\mathbb{Z}^n)$ *can be realized by a d-dimensional manifold* if there exists a pair (N, φ) where N is a closed L^2-acyclic d-dimensional manifold N such that $\pi_1(N)$ is a Wh-AC-group and where $\varphi\colon \mathbb{Z}^n \to H_1(N;\mathbb{Z})/\text{torsion}$ is an isomorphism, such that $\varphi_*(\mathcal{P}) = \mathcal{P}(N) \in \mathfrak{G}(\pi_1(N))$.

From our above results we know that not all $\mathcal{P} \in \mathfrak{G}(\mathbb{Z}^n)$ can be realized by manifolds. More precisely, in Proposition 2.4 we showed that polytopes realized by closed orientable manifolds have a symmetry. Furthermore in Theorem 3.2 we showed that the polytope invariant of an admissible 3-manifold $N \neq S^1 \times D^2$ that is not a closed graph manifold can be represented by a polytope. So we can only hope to realize elements of $\mathfrak{P}(\mathbb{Z}^n) \subset \mathfrak{G}(\mathbb{Z}^n)$ by 3-dimensional manifolds.

We have the following realization result.

LEMMA 6.10. *Let $n \in \mathbb{N}$ and let $\mathcal{P} \in \mathfrak{G}(\mathbb{Z}^n)$. For any $d \geq 7$ we can realize $\mathcal{P} + (-1)^{d+1}\overline{\mathcal{P}}$ by a d-dimensional manifold.*

PROOF. Let F be a free group on $n-1$ generators. We set $\Gamma = F \times \mathbb{Z}$. Then $K(\Gamma, 1)$ is 2-dimensional and it is L^2-acyclic by [**Lü02**, Theorem 1.35 (4)]. Furthermore Γ is a Wh-AC-group by [**KLR16**, Theorem 2] and [**LiL18**, Theorem 2.3]. We pick an isomorphism $\varphi\colon \mathbb{Z}^n \to H_1(\Gamma;\mathbb{Z})$ and we pick a set-theoretic section $s\colon H_1(\Gamma;\mathbb{Z}) \to \Gamma$ of the projection map $\Gamma \to H_1(\Gamma;\mathbb{Z})$.

Now let $\mathcal{P} \in \mathfrak{G}(\mathbb{Z}^n)$. We can find $a, b \in \mathbb{Z}[\mathbb{Z}^n]$ with $\mathcal{P} = \mathcal{P}(a) - \mathcal{P}(b)$. We write

$$\omega := s(\varphi(a)) \cdot s(\varphi(b))^{-1} \in K_1(\mathcal{D}(\Gamma)) = \mathcal{D}(\Gamma)^\times_{ab}.$$

It follows immediately from the definitions that $\mathcal{P}(\omega) = \varphi(\mathcal{P}) \in \mathfrak{G}(\Gamma)$. Now let $d \geq 7$. It follows easily from a slight generalization of [**FL17**, Lemma 2.9] that there exists a closed orientable d-dimensional manifold N with $\pi_1(N) = \Gamma$ such that

$$\tau(N, \mathrm{id}_{\pi_1 N}) = \omega \cdot \overline{\omega}^{(-1)^{d+1}} \in K_1(\mathcal{D}(\Gamma)) = \mathcal{D}(\Gamma)^\times_{ab}/\pm\Gamma.$$

It follows immediately that $\mathcal{P}(N) = \varphi(\mathcal{P}) + (-1)^{d+1}\overline{\varphi(\mathcal{P})} \in \mathfrak{G}(\Gamma)$. □

QUESTION 6.11.

(1) Let $\mathcal{P} \in \mathfrak{P}(\mathbb{Z}^n)$. Does there exist a closed orientable admissible 3-manifold that realizes $\mathcal{P} + \overline{\mathcal{P}}$?
(2) Let $\mathcal{P} \in \mathfrak{G}(\mathbb{Z}^n)$ and let $d \in \{4, 5, 6\}$. Does there exist an L^2-acyclic d-dimensional closed orientable manifold that realizes $\mathcal{P} + (-1)^{d+1}\overline{\mathcal{P}}$?

The first question seems to be very hard since by Theorem 3.2 it is a reformulation of the question of which Thurston norm balls are realized by aspherical 3-manifolds. This question has been open since the 1970's. The second question might be more accessible, especially in dimensions 5 and 6. It is conceivable that in dimension 4 the answer depends on whether one studies topological or smooth manifolds.

Acknowledgments

We wish to thank the many people who shared their ideas and insight with us as we developed the subject. We are especially grateful to Nathan Dunfield, Florian Funke, Dawid Kielak and Kevin Schreve for their contributions. We have also received many helpful comments from an anonymous referee.

References

[Ag08] Ian Agol, *Criteria for virtual fibering*, J. Topol. **1** (2008), no. 2, 269–284, DOI 10.1112/jtopol/jtn003. MR2399130

[Ag13] Ian Agol, *The virtual Haken conjecture*, Doc. Math. **18** (2013), 1045–1087. With an appendix by Agol, Daniel Groves, and Jason Manning. MR3104553

[AFW15] Matthias Aschenbrenner, Stefan Friedl, and Henry Wilton, *3-manifold groups*, EMS Series of Lectures in Mathematics, European Mathematical Society (EMS), Zürich, 2015. MR3444187

[BLR08] Arthur Bartels, Wolfgang Lück, and Holger Reich, *Equivariant covers for hyperbolic groups*, Geom. Topol. **12** (2008), no. 3, 1799–1882, DOI 10.2140/gt.2008.12.1799. MR2421141

[BP92] Riccardo Benedetti and Carlo Petronio, *Lectures on hyperbolic geometry*, Universitext, Springer-Verlag, Berlin, 1992. MR1219310

[BNS87] Robert Bieri, Walter D. Neumann, and Ralph Strebel, *A geometric invariant of discrete groups*, Invent. Math. **90** (1987), no. 3, 451–477, DOI 10.1007/BF01389175. MR914846

[BS78] Robert Bieri and Ralph Strebel, *Almost finitely presented soluble groups*, Comment. Math. Helv. **53** (1978), no. 2, 258–278, DOI 10.1007/BF02566077. MR0498863

[Br82] Kenneth S. Brown, *Cohomology of groups*, Graduate Texts in Mathematics, vol. 87, Springer-Verlag, New York-Berlin, 1982. MR672956

[Br87] Kenneth S. Brown, *Trees, valuations, and the Bieri-Neumann-Strebel invariant*, Invent. Math. **90** (1987), no. 3, 479–504, DOI 10.1007/BF01389176. MR914847

[BK16] J. O. Button and R. P. Kropholler, *Nonhyperbolic free-by-cyclic and one-relator groups*, New York J. Math. **22** (2016), 755–774. MR3548122

[CF17] D. Cooper and D. Futer, *Ubiquitous quasi-Fuchsian surfaces in cusped hyperbolic 3-manifolds*, Preprint (2017), arXiv:1705.02890. To appear in Geomety & Topology.

[DL07] Warren Dicks and Peter A. Linnell, *L^2-Betti numbers of one-relator groups*, Math. Ann. **337** (2007), no. 4, 855–874, DOI 10.1007/s00208-006-0058-y. MR2285740

[DLMSY03] Józef Dodziuk, Peter Linnell, Varghese Mathai, Thomas Schick, and Stuart Yates, *Approximating L^2-invariants and the Atiyah conjecture*, Comm. Pure Appl. Math. **56** (2003), no. 7, 839–873, DOI 10.1002/cpa.10076. Dedicated to the memory of Jürgen K. Moser. MR1990479

[DFL16] Jérôme Dubois, Stefan Friedl, and Wolfgang Lück, *The L^2-Alexander torsion of 3-manifolds*, J. Topol. **9** (2016), no. 3, 889–926, DOI 10.1112/jtopol/jtw013. MR3551842

[DK92] G. Duchamp and D. Krob, *The lower central series of the free partially commutative group*, Semigroup Forum **45** (1992), no. 3, 385–394, DOI 10.1007/BF03025778. MR1179860

[Fo53] Ralph H. Fox, *Free differential calculus. I. Derivation in the free group ring*, Ann. of Math. (2) **57** (1953), 547–560, DOI 10.2307/1969736. MR0053938

[Fr07] Stefan Friedl, *Reidemeister torsion, the Thurston norm and Harvey's invariants*, Pacific J. Math. **230** (2007), no. 2, 271–296, DOI 10.2140/pjm.2007.230.271. MR2309160

[FK14] Stefan Friedl and Takahiro Kitayama, *The virtual fibering theorem for 3-manifolds*, Enseign. Math. **60** (2014), no. 1-2, 79–107, DOI 10.4171/LEM/60-1/2-5. MR3262436

[FL16] S. Friedl and W. Lück, *L^2-Euler characteristics and the Thurston norm*, Preprint (2016), arXiv:1609.07805. To appear in Commentarii Math. Helv.

[FL17] Stefan Friedl and Wolfgang Lück, *Universal L^2-torsion, polytopes and applications to 3-manifolds*, Proc. Lond. Math. Soc. (3) **114** (2017), no. 6, 1114–1151, DOI 10.1112/plms.12035. MR3661347

[FST17] Stefan Friedl, Kevin Schreve, and Stephan Tillmann, *Thurston norm via Fox calculus*, Geom. Topol. **21** (2017), no. 6, 3759–3784, DOI 10.2140/gt.2017.21.3759. MR3693575

[FSW15] Stefan Friedl, Daniel S. Silver, and Susan G. Williams, *Splittings of knot groups*, Math. Ann. **362** (2015), no. 1-2, 401–424, DOI 10.1007/s00208-014-1123-6. MR3343883

[FSW16] Stefan Friedl, Daniel Silver, and Susan Williams, *The Turaev and Thurston norms*, Pacific J. Math. **284** (2016), no. 2, 365–382, DOI 10.2140/pjm.2016.284.365. MR3544305

[FT15] S. Friedl and S. Tillmann, *Two-generator one-relator groups and marked polytopes*, Preprint (2015), arXiv:1501.03489. To appear in Annales de l'Institut Fourier

[Fu16] F. Funke, *The integral polytope group*, Preprint (2016), arXiv:1605.01217.

[Fu17] F. Funke, *The L^2-torsion polytope of amenable groups*, Preprint (2017), arXiv:1704.07164.

[FK16] Florian Funke and Dawid Kielak, *Alexander and Thurston norms, and the Bieri-Neumann-Strebel invariants for free-by-cyclic groups*, Geom. Topol. **22** (2018), no. 5, 2647–2696, DOI 10.2140/gt.2018.22.2647. MR3811767

[GM17] D. Groves and J. Manning, *Quasiconvexity and Dehn filling*, Preprint (2017), arXiv:1708.07968.

[He01] John Hempel, *3-manifolds as viewed from the curve complex*, Topology **40** (2001), no. 3, 631–657, DOI 10.1016/S0040-9383(00)00033-1. MR1838999

[KLR16] Holger Kammeyer, Wolfgang Lück, and Henrik Rüping, *The Farrell-Jones conjecture for arbitrary lattices in virtually connected Lie groups*, Geom. Topol. **20** (2016), no. 3, 1275–1287, DOI 10.2140/gt.2016.20.1275. MR3523058

[Ko88] Tsuyoshi Kobayashi, *Casson-Gordon's rectangle condition of Heegaard diagrams and incompressible tori in 3-manifolds*, Osaka J. Math. **25** (1988), no. 3, 553–573. MR969018

[Li93] Peter A. Linnell, *Division rings and group von Neumann algebras*, Forum Math. **5** (1993), no. 6, 561–576, DOI 10.1515/form.1993.5.561. MR1242889

[LiL18] Wolfgang Lück and Peter Linnell, *Localization, Whitehead groups and the Atiyah conjecture*, Ann. K-Theory **3** (2018), no. 1, 33–53, DOI 10.2140/akt.2018.3.33. MR3695363

[LoL95] John Lott and Wolfgang Lück, *L^2-topological invariants of 3-manifolds*, Invent. Math. **120** (1995), no. 1, 15–60, DOI 10.1007/BF01241121. MR1323981

[Lü02] Wolfgang Lück, *L^2-invariants: theory and applications to geometry and K-theory*, Ergebnisse der Mathematik und ihrer Grenzgebiete. 3. Folge. A Series of Modern Surveys in Mathematics [Results in Mathematics and Related Areas. 3rd Series. A Series of Modern Surveys in Mathematics], vol. 44, Springer-Verlag, Berlin, 2002. MR1926649

[LS77] Roger C. Lyndon and Paul E. Schupp, *Combinatorial group theory*, Springer-Verlag, Berlin-New York, 1977. Ergebnisse der Mathematik und ihrer Grenzgebiete, Band 89. MR0577064

[Ma10] Joseph Maher, *Random Heegaard splittings*, J. Topol. **3** (2010), no. 4, 997–1025, DOI 10.1112/jtopol/jtq031. MR2746344

[Mi66] J. Milnor, *Whitehead torsion*, Bull. Amer. Math. Soc. **72** (1966), 358–426, DOI 10.1090/S0002-9904-1966-11484-2. MR0196736

[PW18] Piotr Przytycki and Daniel T. Wise, *Mixed 3-manifolds are virtually special*, J. Amer. Math. Soc. **31** (2018), no. 2, 319–347, DOI 10.1090/jams/886. MR3758147

[Ro94] Jonathan Rosenberg, *Algebraic K-theory and its applications*, Graduate Texts in Mathematics, vol. 147, Springer-Verlag, New York, 1994. MR1282290

[Sc93] Rolf Schneider, *Convex bodies: the Brunn-Minkowski theory*, Encyclopedia of Mathematics and its Applications, vol. 44, Cambridge University Press, Cambridge, 1993. MR1216521

[Sik87] J.-C. Sikorav, *Homologie de Novikov associée à une classe de cohomologie réelle de degré un*, Thèse Orsay, 1987.

[Sil81] John R. Silvester, *Introduction to algebraic K-theory*, Chapman & Hall, London-New York, 1981. Chapman and Hall Mathematics Series. MR629979

[Sin77] I. M. Singer, *Some remarks on operator theory and index theory*, K-theory and operator algebras (Proc. Conf., Univ. Georgia, Athens, Ga., 1975), Springer, Berlin, 1977, pp. 128–138. Lecture Notes in Math., Vol. 575. MR0467848

[Ste75] Bo Stenström, *Rings of quotients*, Springer-Verlag, New York-Heidelberg, 1975. Die Grundlehren der Mathematischen Wissenschaften, Band 217; An introduction to methods of ring theory. MR0389953

[Str84] Ralph Strebel, *Finitely presented soluble groups*, Group theory, Academic Press, London, 1984, pp. 257–314. MR780572

[Th86] William P. Thurston, *A norm for the homology of* 3*-manifolds*, Mem. Amer. Math. Soc. **59** (1986), no. 339, i–vi and 99–130. MR823443

[Ti70] D. Tischler, *On fibering certain foliated manifolds over* S^1, Topology **9** (1970), 153–154, DOI 10.1016/0040-9383(70)90037-6. MR0256413

[Tu01] Vladimir Turaev, *Introduction to combinatorial torsions*, Lectures in Mathematics ETH Zürich, Birkhäuser Verlag, Basel, 2001. Notes taken by Felix Schlenk. MR1809561

[Wa78] Friedhelm Waldhausen, *Algebraic K-theory of generalized free products. I, II*, Ann. of Math. (2) **108** (1978), no. 1, 135–204, DOI 10.2307/1971165. MR0498807

[We15] Christian Wegner, *The Farrell-Jones conjecture for virtually solvable groups*, J. Topol. **8** (2015), no. 4, 975–1016, DOI 10.1112/jtopol/jtv026. MR3431666

[Wi09] Daniel T. Wise, *Research announcement: the structure of groups with a quasiconvex hierarchy*, Electron. Res. Announc. Math. Sci. **16** (2009), 44–55, DOI 10.3934/era.2009.16.44. MR2558631

[Wi12a] D. Wise, *The structure of groups with a quasi-convex hierarchy*, 189 pages, preprint (2012), downloaded on October 29, 2012 from `http://www.math.mcgill.ca/wise/papers.html`

[Wi12b] Daniel T. Wise, *From riches to raags: 3-manifolds, right-angled Artin groups, and cubical geometry*, CBMS Regional Conference Series in Mathematics, vol. 117, Published for the Conference Board of the Mathematical Sciences, Washington, DC; by the American Mathematical Society, Providence, RI, 2012. MR2986461

[Wi14] Daniel T. Wise, *Cubular tubular groups*, Trans. Amer. Math. Soc. **366** (2014), no. 10, 5503–5521, DOI 10.1090/S0002-9947-2014-06065-0. MR3240932

Fakultät für Mathematik, Universität Regensburg, 93040 Regensburg, Germany
Email address: `sfriedl@gmail.com`

Mathematisches Institut der Universität Bonn, Endenicher Allee 60, 53115 Bonn, Germany
Email address: `wolfgang.lueck@him.uni-bonn.de`
URL: `http://www.him.uni-bonn.de/lueck`

School of Mathematics and Statistics, The University of Sydney, Sydney NSW 2006 Australia
Email address: `stephan.tillmann@sydney.edu.au`

Proceedings of Symposia in Pure Mathematics
Volume **102**, 2019
https://doi.org/10.1090/pspum/102/01817

Functions on surfaces and constructions of manifolds in dimensions three, four and five

David T. Gay

Abstract. We offer a new proof that two closed oriented 4-manifolds are cobordant if their signatures agree, in the spirit of Lickorish's proof [**6**] that all closed oriented 3-manifolds bound 4-manifolds. Where Lickorish uses Heegaard splittings we use trisections. In fact we begin with a subtle recasting of Lickorish's argument: Instead of factoring the gluing map for a Heegaard splitting as a product of Dehn twists, we encode each handlebody in a Heegaard splitting in terms of a Morse function on the surface and build the 4-manifold from a generic homotopy between the two functions. This extends up a dimension by encoding a trisection of a closed 4-manifold as a triangle (circle) of functions and constructing an associated 5-manifold from an extension to a 2-simplex (disk) of functions. This borrows ideas from Hatcher and Thurston's proof [**3**] that the mapping class group of a surface is finitely presented.

1. Introduction

The central theme of this paper is that manifolds of dimension $n+3$ can be described by generic n-parameter families of functions on 2-manifolds (at least for $n \leq 2$). After some preparatory discussion and definitions, this is formulated as our main theorem, Theorem 5. The following well known theorems are then proved as corollaries:

Theorem 1 (Rohlin [**7**], Thom [**11**]). *For every closed, oriented 3-manifold M there is a compact, oriented 4-manifold X with $\partial X = M$.*

Theorem 2 (Pontrjagin, Rohlin [**8**]). *For every closed, oriented 4-manifold X there is a compact, oriented 5-manifold Z with $\partial Z = X \amalg (\amalg^p \mathbb{C}P^2) \amalg (\amalg^q \overline{\mathbb{C}P^2})$, for some $p, q \geq 0$.*

Theorem 1 is compactly stated as $\Omega_3 = 0$, i.e. the oriented 3-dimensional cobordism group is trivial. Theorem 2 is usually stated by saying that $\Omega_4 = \mathbb{Z}$, i.e. that the oriented 4-dimensional cobordism group is $\mathbb{Z}$, with generator $\mathbb{C}P^2$; this follows from our statement and the fact that the signature is a 4-dimensional cobordism invariant [**11**].

Our proof of Theorem 1 is very similar in spirit to Lickorish's proof [**6**], but our proof of Theorem 2 is quite different from other proofs, and depends on the existence of trisections of 4-manifolds [**2**], in the same way that Lickorish's proof

2010 *Mathematics Subject Classification.* Primary 57R99; Secondary 55N22, 57R45.

depends on the existence of Heegaard splittings of 3-manifolds. In a nutshell, the idea is to use generic B^n- or S^n-parameterized families of smooth functions on closed, connected, oriented surfaces to construct $(n+3)$-manifolds. We work only in the cases where $n \leq 2$, but leave open the interesting question of to what extent these ideas extend to higher dimensions.

The author is aware of two other papers which relate the construction and existence of cobordisms with "higher Morse theory", but in subtly different ways: In [**1**], 4-manifolds bounded by given 3-manifolds are constructed using information coming from generic (stable) smooth maps from 3-manifolds to $\mathbb{R}^2$. In [**9**], it is shown how to reconstruct a 4-manifold up to 5-dimensional cobordism from data associated with stable smooth maps from 4-manifolds to 3-manifolds; from this one sees that $\mathbb{C}P^2$ generates Ω_4.

Throughout this paper all manifolds are smooth and oriented. Unless explicitly stated otherwise, all surfaces (2-manifolds) are closed, connected and oriented, but manifolds in other dimensions may have nonempty boundary. (In fact, our manifolds with boundary may have corners as well, and a "diffeomorphism" between two manifolds with boundary is allowed to bend at the corners. It should be clear from context when this is happening and that this does not raise any significant issues.)

2. Generic families of functions

We will work with smooth parameterized families of real-valued functions on a surface Σ; when the parameter space is A, we will denote such a family as a function $F : A \to C^\infty(\Sigma)$, always with the understanding that this really means that the associated function $A \times \Sigma \to \mathbb{R}$ is smooth. Our parameter spaces will always be spheres or balls. Given $F : A \to C^\infty(\Sigma)$ we will use the notation $\partial F : \partial A \to C^\infty(\Sigma)$ to refer to $F|_{\partial A}$.

Recall [**10**], when the dimension of A is 0, 1 or 2, that such a parameterized family of functions $F : A \to C^\infty(\Sigma)$ is *stable* if for each $p \in A$ and q in Σ we can find coordinates near p, q and $F(p,q) \in \mathbb{R}$ with respect to which $F : A \times \Sigma \to \mathbb{R}$ has one of the following models, listed according to the dimension of A (where (x,y) are the coordinates on Σ):

(0) When $\dim(A) = 0$ we have two options:
   Regular point: $(x,y) \mapsto x$
   Morse critical point: $(x,y) \mapsto \pm x^2 \pm y^2$

(1) When $\dim(A) = 1$ we have three options (here t is the coordinate on A):
   Path of regular points: $(t,x,y) \mapsto x$
   Path of Morse critical points: $(t,x,y) \mapsto \pm x^2 \pm y^2$
   Birth/death event: $(t,x,y) \mapsto x^3 + tx \pm y^2$, with $t = 0$ not on ∂A.

(2) When $\dim(A) = 2$ we have four options (here s and t are the coordinates on A):
   Disk of regular points: $(s,t,x,y) \mapsto x$
   Disk of Morse critical points: $(s,t,x,y) \mapsto \pm x^2 \pm y^2$
   Path of birth/death events: $(s,t,x,y) \mapsto x^3 + tx \pm y^2$, with $\{t = 0\}$ transverse to ∂A.
   Swallowtail: $(s,t,x,y) \mapsto x^4 + sx^2 + tx \pm y^2$, with $s = t = 0$ not on ∂A.

Furthermore, given such a family $F : A \to C^\infty(\Sigma)$ we can also think of F as a function $F : A \times \Sigma \to A \times \mathbb{R}$, being id_A on the first factor, and consider the locus of critical points $\mathrm{Crit}(F) \subset A \times \Sigma$; this is a smooth codimension 2 submanifold mapped by F to $A \times \mathbb{R}$. We then impose the following transversality conditions, again listed according to the dimension of A:

(0) If $\dim(A) = 0$ then, for each $p \in A$, we require that F is injective on $\mathrm{Crit}(F)$, i.e. all Morse critical points have distinct critical values

(1) If $\dim(A) = 1$, then $\mathrm{Crit}(F) = \mathrm{Crit}_0(F) \cup \mathrm{Crit}_1(F)$, where $\mathrm{Crit}_0(F)$ is the locus of critical points that arise in dimension 0, i.e. Morse critical points, and $\mathrm{Crit}_1(F)$ is the locus of birth/death events. Thus $\mathrm{Crit}_0(F)$ is 1-dimensional and $\mathrm{Crit}_1(F)$ is 0-dimensional. Here we require that:
   (a) F is injective on $\mathrm{Crit}_1(F)$,
   (b) the image of $\mathrm{Crit}_1(F)$ is disjoint from the image of $\mathrm{Crit}_0(F)$ and from $\partial A \times \mathbb{R}$, and
   (c) $\mathrm{Crit}_0(F)$ is immersed with at worst transverse double point self-intersections, which should occur away from $\partial A \times \mathbb{R}$. (We call this event, which occurs at isolated parameter values, a *critical value height switch*, or simply a *height switch*.)

(2) If $\dim(A) = 2$ then $\mathrm{Crit}(F) = \mathrm{Crit}_0(F) \cup \mathrm{Crit}_1(F) \cup \mathrm{Crit}_2(F)$, where $\mathrm{Crit}_0(F)$ is the locus of critical points that arise in dimension 0, $\mathrm{Crit}_1(F)$ is the locus of birth/death events, and $\mathrm{Crit}_2(F)$ is the locus of swallowtails. Now $\mathrm{Crit}_0(F)$ is 2-dimensional, $\mathrm{Crit}_1(F)$ is 1-dimensional and $\mathrm{Crit}_2(F)$ is 0-dimensional. We now require that:
   (a) F is injective on $\mathrm{Crit}_2(F)$,
   (b) the image of $\mathrm{Crit}_2(F)$ is disjoint from the images of $\mathrm{Crit}_1(F)$ and $\mathrm{Crit}_0(F)$ and from $\partial A \times \mathbb{R}$,
   (c) the image of $\mathrm{Crit}_1(F)$ is embedded and transverse to the image of $\mathrm{Crit}_0(F)$ and to $\partial A \times \mathbb{R}$, and
   (d) $\mathrm{Crit}_0(F)$ is immersed with at worst arcs of transverse double points (the arcs being transverse to $\partial A \times \mathbb{R}$) and isolated transverse triple points (away from $\partial A \times \mathbb{R}$). The arcs of transverse double points are thus 1-parameter families of critical value height switches and the isolated triple points correspond to three critical values switching heights in a 2-parameter fashion, to be discussed in more detail later.

We will abuse the term "generic" in reference to such families.

DEFINITION 3. If $A \in \{B^0, S^0, B^1, S^1, B^2\}$, an A-parameterized family of functions on a surface Σ, namely $F : A \to C^\infty(\Sigma)$, is *generic* if it is stable and satisfies the appropriate transversality condition on $\mathrm{Crit}(F)$ described above. For purely formal completeness, if $A = S^{-1}$ we declare there to be a unique S^{-1}-parametrized family, and declare it to be *generic*.

All we really need in this paper are the appropriate (and standard) relative existence statements: Every generic $G : \partial A \to C^\infty(\Sigma)$ extends to a generic $F : A \to C^\infty(\Sigma)$ with $\partial F = G$.

DEFINITION 4. A *manifold-surface pair* is a pair (Y, Σ), where Y is a manifold of some dimension ≥ 2 and $\Sigma \subset Y$ is a surface submanifold of Y. If $\partial Y \neq \emptyset$, we require that $\Sigma \subset \partial Y$. We define the *boundary of a manifold-surface pair* $\partial(Y, \Sigma)$ to be either $(\emptyset, \emptyset)$ if $\partial Y = \emptyset$ or to be $(\partial Y, \Sigma)$ if $\partial Y \neq \emptyset$). An embedding (resp.

diffeomorphism) $\phi : Y \to Y'$ is an *embedding (resp. diffeomorphism) of pairs* $(Y, \Sigma) \to (Y', \Sigma')$ if $\phi(\Sigma) = \Sigma'$, and is an embedding (resp. diffeomorphism) *rel* Σ if $\Sigma = \Sigma'$ and $\phi|_\Sigma$ is the identity map. The *dimension* of a pair (Y, Σ) is the dimension of Y.

3. The main theorem

We now present the main idea of this paper in the form of a single theorem, and then we will prove Theorems 1 and 2 as immediate corollaries.

THEOREM 5. *For each* $A \in \{S^{-1}, B^0, S^0, B^1, S^1, B^2\}$, *with* $\dim(A) = n \in \{-1, 0, 1, 2\}$, *there exists a function* $\mathcal{P}_A = \mathcal{P}$ *(we generally drop the subscript) from the set of generic A-parameterized functions on surfaces to the set of* $(n+3)$*-dimensional manifold-surface pairs, satisfying the following properties:*

(1) *If* $A = S^{-1} = \emptyset$ *and* $F : A \to C^\infty(\Sigma)$ *is the unique A-parameterized family of functions on a fixed surface* Σ, *then* $\mathcal{P}(F) = (\Sigma, \Sigma)$.
(2) *For* $A \in \{S^{-1}, B^0, S^0, B^1, S^1\}$, $\partial\mathcal{P}(F) = \mathcal{P}(\partial F)$. *(When* $\partial A = \emptyset$, $\mathcal{P}(\partial F) = \emptyset$.*)*
(3) *For* $A = B^2$, *for every* Σ *and every generic* $F : A \to C^\infty(\Sigma)$ *there are non-negative integers p and q such that* $\partial\mathcal{P}(F) = \mathcal{P}(\partial F) \amalg (\amalg^p(\mathbb{C}P^2, \emptyset)) \amalg (\amalg^q(\overline{\mathbb{C}P^2}, \emptyset))$.
(4) *For every closed connected* $(n+3)$*-manifold* M, *where* $n \in \{-1, 0, 1\}$, *there is a surface* $\Sigma \subset M$ *and a generic* $F : S^n \to C^\infty(\Sigma)$ *such that* (M, Σ) *is diffeomorphic rel.* Σ *to* $\mathcal{P}(F)$.

In fact $\mathcal{P}_A$ can be constructed for $A = S^2$ using our techniques but we have no use for this case in this paper, and cannot state an existence result like item 4 above, so we leave this for others to investigate more thoroughly, along with higher dimensional parameter spaces.

PROOF OF THEOREM 1. Given a closed connected 3-manifold M, choose a surface $\Sigma \subset M$ and a generic $F : S^0 \to C^\infty(\Sigma)$ such that (M, Σ) is diffeomorphic rel. Σ to $\mathcal{P}(F)$. Extend F to a generic $G : B^1 \to C^\infty(\Sigma)$ with $G|_{\partial B^1} = F$, and let $(X, \Sigma) = \mathcal{P}(G)$. Then X is a 4-manifold and $\partial X = M$. □

PROOF OF THEOREM 2. Given a closed connected 4-manifold X, choose a surface $\Sigma \subset X$ and a generic $F : S^1 \to C^\infty(\Sigma)$ such that (X, Σ) is diffeomorphic rel. Σ to $\mathcal{P}(F)$. Extend F to a generic $G : B^2 \to C^\infty(\Sigma)$ with $G|_{\partial B^2} = F$, and let $(Z, \Sigma) = \mathcal{P}(G)$. Then Z is a 5-manifold and $\partial Z = X \amalg (\amalg^p \mathbb{C}P^2) \amalg (\amalg^q \overline{\mathbb{C}P^2})$. □

PROOF OF THEOREM 5. This proof occupies the remainder of the paper but we give the outline here, to be filled in by means of explicit constructions and lemmas later. There is nothing to prove for item 1, as this constitutes the definition of the function $\mathcal{P}_{S^{-1}}$. Items 2 and 3 will all be proved as part of the construction of each $\mathcal{P}_A$ for $A \in \{B^0, S^0, B^1, S^1, B^2\}$. Item 4 is trivial in the case $n = -1$, is explained for $n = 0$ in the construction of $\mathcal{P}_{S^0}$ as a simple consequence of the existence of Heegaard splittings of 3-manifolds, and will be proved for $n = 1$ via a sequence of lemmas after the constructions of $\mathcal{P}_{B^1}$ and $\mathcal{P}_{S^1}$, ultimately as a consequence of the existence of trisections of 4-manifolds. □

4. Handlebodies from single functions and Heegaard splittings

DEFINITION 6. A *regular level component* of a smooth real-valued function f on a surface Σ is a connected component C of a level set $f^{-1}(z)$ of f such that C contains no critical points of f. Note that $z \in \mathbb{R}$ may be a critical value, but none of its critical preimages may be in the component C. A *critical level component* of f is a connected component of a level set containing at least one critical point.

Of course, if Σ is closed then all regular level components are smooth circles. Critical level components may be arbitrarily complicated depending on the nature of the singularities involved.

DEFINITION 7. Given a surface Σ, a *handlebody filling* of Σ is a pair (H, Σ), where H is a 3-dimensional handlebody with $\Sigma = \partial H$.

LEMMA 8. *Given a surface Σ and a Morse function $f : \Sigma \to \mathbb{R}$ which is injective on critical points, there is a handlebody filling (H, Σ) of Σ with the property that every regular level component of f bounds a disk in H. Any two such handlebody fillings of Σ are diffeomorphic rel. Σ, and any handlebody filling of Σ arises in this way from some such Morse function on Σ.*

PROOF. The only ambiguity in the definition of H comes from critical level components of f. Suppose $p \in \Sigma$ is a critical point of f, with $f(p) = a$, and let $\epsilon > 0$ be such that $[a-\epsilon, a+\epsilon]$ contains no critical values other than a. Let Γ be the component of $f^{-1}(a)$ containing p and let N be the component of $f^{-1}([a-\epsilon, a+\epsilon])$ containing p. All boundary components of N are regular level components of f and thus bound disks in H_f; if we can show that N is a subsurface of Σ of genus 0, then all simple closed curves on N will automatically bound in H_f and thus there will in fact be no ambiguity in the definition of H_f. Because f is injective on critical points, p is the only critical point in N, and Γ is the only critical level set in N. If p is a critical point of index 0 or 2, N is a disk and thus has genus 0. If p has index 1 then Γ is a figure eight graph. Because Σ is oriented and because Γ separates N into two halves $N \cap f^{-1}([0, a+\epsilon])$ and $N \cap f^{-1}([a-\epsilon, 0])$, we see that N is a pair of pants, with either two boundary components in $f^{-1}(a-\epsilon)$ and one in $f^{-1}(a+\epsilon)$ or vice versa. Thus N again has genus 0.

To see that every handlebody filling of Σ arises this way, note that for each g there exists a standard genus g handlebody H_g embedded in $\mathbb{R}^3$ such that the standard height function is a Morse function on ∂H_g with regular level components bounding disks in H_g. Given a particular handlebody filling (H, Σ), choose a diffeomorphism of H with H_g and pull back the height function on ∂H_g to Σ. □

Note that this lemma can fail if f is not injective on critical points, even if f is Morse, because components of critical level sets containing several critical points may have higher genus neighborhoods.

CONSTRUCTION OF $\mathcal{P}_{B^0}$: A generic B^0-parameterized family F is nothing more than a single Morse function $f = F(0)$ on a surface Σ which is injective on critical points. Define $\mathcal{P}(F) = \mathcal{P}_{B^0}(F)$ to be the handlebody filling (H, Σ) uniquely characterized by f as in the preceding lemma. □

CONSTRUCTION OF $\mathcal{P}_{S^0}$: Given a family of functions $F : S^0 = \{-1, 1\} \to C^\infty(\Sigma)$, let $\mathcal{P}_{S^0}(F)$ be the closed 3-manifold obtained by gluing the two handlebodies $\mathcal{P}(F_{-1})$ and $\mathcal{P}(F_1)$ together along their common boundary Σ. More precisely, with orientations in mind:

$$\mathcal{P}(F) = -\mathcal{P}(F|_{-1}) \cup_\Sigma \mathcal{P}(F|_1)$$

Here it is important that the boundaries of these handlebodies are both *equal* to Σ, not just diffeomorphic to Σ, so that we can really glue by the identity map. The fact (item 4 in Theorem 5) that every closed 3-manifold arises this way is simply the fact that every 3-manifold has a Heegaard splitting together with the part of Lemma 8 asserting that all handlebody fillings come from Morse functions. □

5. 1-parameter families and 4-manifolds

Next we consider generic S^1- and B^1-parametrized families. Note that, in a generic family $F : A \to C^\infty(\Sigma)$, with $A = S^1$ or $A = B^1$, A can be subdivided into subintervals $A = A_1 \cup \ldots \cup A_n$ such that each $F|_{A_i}$ has the following behavior: (1) $F|_{\partial A_i}$ is generic, i.e. a pair of Morse functions injective on critical points. (2) In the interior of each A_i there is at most one value of t at which $F(t)$ is not Morse and injective on critical points; if there is such a value, label it c_i. (3) For each c_i, as t varies from $c_i - \epsilon$ to $c_i + \epsilon$ for some small $\epsilon > 0$, either $F(t)$ experiences a birth or death of a single cancelling pair of critical points (so that $F(c_i)$ is not Morse) or exactly two critical points switch heights (so that $F(c_i)$ is Morse but not injective on critical points).

DEFINITION 9. Each such subinterval A_i in a subdivision of A as above will be called an *elementary subinterval* and we say that $F|_{A_i}$ is an *elementary* 1*-parameter family.*

In preparation for the upcoming discussion of $\mathcal{P}_{S^0}$ and $\mathcal{P}_{B^1}$ applied to elementary 1-parameter families, we name some specific manifold-surface pairs. Fix integers $g \geq k \geq 0$.

(1) The 3-dimensional pair $(\#^k S^1 \times S^2, \#^k S^1 \times S^1)$ refers to the standard genus k Heegaard splitting of $\#^k S^1 \times S^2$. This can be explicitly described by identifying $\#^k S^1 \times S^2$ with $\partial([-1,1] \times (\natural^k S^1 \times B^2))$, in which case $\#^k S^1 \times S^1$ is really $\{0\} \times (\#^k S^1 \times S^1)$.
(2) The corresponding 4-dimensional pair is $(\natural^k S^1 \times B^3, \#^k S^1 \times S^1)$, with boundary equal to $(\#^k S^1 \times S^2, \#^k S^1 \times S^1)$.
(3) The 3-dimensional pair $(\#^k S^1 \times S^2, \#^g S^1 \times S^1)$ refers to the result of stabilizing (in the sense of Heegaard splittings) $(\#^k S^1 \times S^2, \#^k S^1 \times S^1)$ exactly $g - k$ times.
(4) The corresponding 4-dimensional pair is $(\natural^k S^1 \times B^3, \#^g S^1 \times S^1)$, with boundary equal to $(\#^k S^1 \times S^2, \#^g S^1 \times S^1)$.

LEMMA 10. *Suppose the surface Σ is a surface of genus g, and that $F : A \to C^\infty(\Sigma)$ is an elementary* 1*-parameter family. Then $\mathcal{P}(\partial F)$ is diffeomorphic to either $(\#^g S^1 \times S^2, \#^g S^1 \times S^1)$ or $(\#^{g-1} S^1 \times S^2, \#^g S^1 \times S^1)$.*

PROOF. If we can show that the handlebodies $\mathcal{P}(F|_{\{-1\}})$ and $\mathcal{P}(F|_{\{1\}})$ are diffeomorphic rel. Σ, then we know that $\mathcal{P}(\partial F) \cong (\#^g S^1 \times S^2, \#^g S^1 \times S^1)$. We will show this in all but one case.

When $F(t)$ is Morse and injective on critical points for all t in A, the regular level components of $F(t)$ move by an isotopy of Σ, so that the handlebodies $\mathcal{P}(F|_{\{-1\}})$ and $\mathcal{P}(F|_{\{1\}})$ are diffeomorphic rel. Σ.

Otherwise, let $t_* \in A$ be the parameter value at which the birth/death or height switch occurs, and let $f_* = F(t_*) : \Sigma \to \mathbb{R}$.

In the case of a birth/death event, we claim again that $\mathcal{P}(F|_{\{-1\}})$ and $\mathcal{P}(F|_{\{1\}})$ are diffeomorphic rel. Σ. To see this, let Γ be the critical level component of f_* containing the birth/death point p, assume that $f_*(p) = 0$, and let N be the component of $f_*^{-1}[-\epsilon, \epsilon]$ containing p. (Choose ϵ small enough so that N contains no other critical points, of course.) From the local model of a birth/death, we see that Γ is a circle (smooth except at p) and N is a smooth cylinder with ∂N equal to two regular level components of f_*. We may assume that the regular level components of $F(t)$ do not change (except by isotopy) outside P as t ranges from $t < t_*$ to $t > t_*$ and thus, since the genus of N is 0, we do not need to know anything about regular level components in N to know the handlebody filling.

When the singular event is a critical value crossing, we have further cases to consider. Let the two critical points whose critical values cross be labelled p and q. If p and q occur in different critical level components of their common level set for f_*, then changing their relative heights can only change regular level components by isotopies and thus does not change the handlebodies, and once again $\mathcal{P}(F|_{\{-1\}})$ and $\mathcal{P}(F|_{\{1\}})$ are diffeomorphic rel. Σ. If either p or q has index 0 or 2, this certainly happens. If p and q both have index 1 and occur in the same component with, say, $f_*(p) = f_*(q) = 0$, let N be the component of $f_*^{-1}[-\epsilon, \epsilon]$ containing p and q, where $[-\epsilon, \epsilon]$ contains no other critical values, and let Γ be the component of $f_*^{-1}(0)$ containing p and q. Then N is a regular neighborhood of the graph Γ, which has two 4-valent vertices p and q. To construct all possible pairs (N, Γ) like this, we can start with one edge between p and q and construct the various possibilities from there. Figure 1 enumerates the possibilities, bearing in mind the constraints that N should be orientable and that Γ should separate N into a positive half $N \cap f_*^{-1}[0, \epsilon]$ and a negative half $N \cap f_*^{-1}[-\epsilon, 0]$. We end up with four options, three of which have genus 0 and one of which is genus 1. In the genus 0 case, we

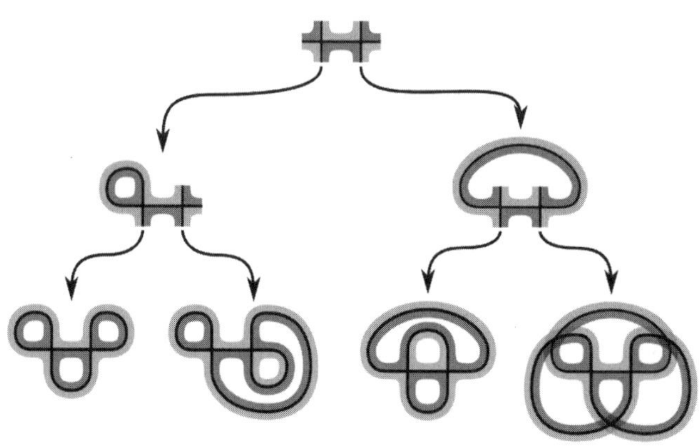

FIGURE 1. Enumerating possible regular neighborhoods of a critical level set containing two index 1 critical points.

again claim, for the same reason as in the birth/death case, that $\mathcal{P}(F|_{\{-1\}})$ and $\mathcal{P}(F|_{\{1\}})$ are diffeomorphic rel. Σ. This is because we can assume that $F(-1)$ and $F(1)$ agree outside N (up to isotopy), and thus the circles on $\Sigma \setminus N$ which bound disks in $\mathcal{P}(F|_{\{-1\}})$ also bound disks in $\mathcal{P}(F|_{\{1\}})$ and vice versa. Since N

has genus 0, specifying the curves that bound in $\Sigma \setminus N$ unambiguously determines a handlebody filling of Σ up to diffeomorphism rel. Σ.

In the center in Figure 2 we have redrawn the genus 1 case, with the two critical points labelled p and q. The genus 1 surface is obtained from the hexagon by identifying opposite edges. Note that N has two boundary components, $f_*^{-1}(\pm\epsilon)$.

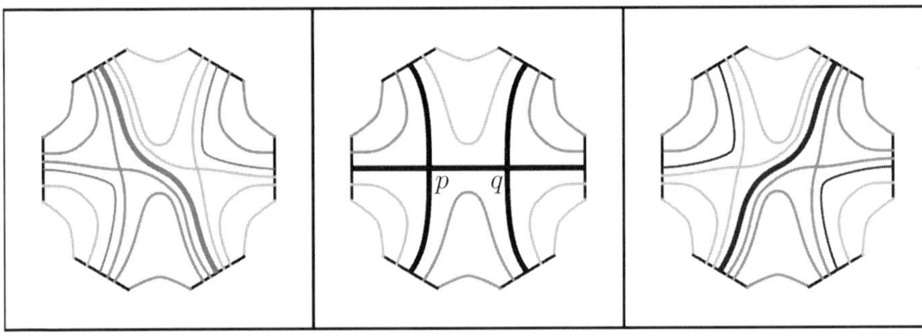

FIGURE 2. A movie of the genus 1 case for a critical value crossing.

To the left and right we have drawn various regular and critical level sets as we perturb f_* to a nearby function f so as to either make $f(p)$ positive and $f(q)$ negative or vice versa. Positive level sets are orange while negative level sets are pink; we highlight the 0 level set in red on the left, black in the middle and blue on the right. Thus these are three frames in a movie of $F(t)$, as t ranges from $t_* - \epsilon$ to $t_* + \epsilon$. Let $H_{\pm 1} = \mathcal{P}(F|_{\{\pm 1\}})$, the "before" and "after" handlebody fillings of Σ. Note that the thicker red and blue level components, if drawn on the same surface, intersect at a point; these are colored red and blue to suggest α and β curves in a Heegaard diagram. The upshot is that Σ contains an embedded genus 1 surface N with two boundary components such that:

(1) Every curve in $\Sigma \setminus N$ which bounds a disk in H_{-1} also bounds a disk in H_1, and vice versa.
(2) Both boundary components of N bound disks in both H_{-1} and H_1.
(3) There are essential simple closed curves C_{-1} and C_1 in N which intersect transversely at one point such that C_{-1} bounds a disk in H_{-1} and C_1 bounds a disk in H_1.

Note that since both boundary components of N bound disks in $H_{\pm 1}$, any band sum in N of these two boundary components bounds a disk in $H_{\pm 1}$, and thus we may replace N with a genus 1 surface with one boundary component, and still keep items (1) and (3) from the list above. From this we see that $\mathcal{P}(\partial F)$ comes with a Heegaard splitting $(-H_{-1}) \cup_\Sigma H_1$ which is the result of stabilizing a genus $g-1$ splitting in which every curve that bounds a disk in one half of the splitting bounds in the other half as well. This implies that $\mathcal{P}(\partial F) \cong (\#^{g-1} S^1 \times S^2, \#^g S^1 \times S^1)$. $\square$

DEFINITION 11. An elementary sub-interval A_i, or elementary 1-parameter family $F|_{A_i}$, is of *type 0* if, in the above theorem, we have $\mathcal{P}(F|_{\partial A_i}) \cong (\#^g S^1 \times S^2, \#^g S^1 \times S^1)$, and is of *type 1* if $\mathcal{P}(F|_{\partial A_i}) \cong (\#^{g-1} S^1 \times S^2, \#^g S^1 \times S^1)$.

CONSTRUCTION OF $\mathcal{P}_{B^1}$: Given a generic $F : B^1 \to C^\infty(\Sigma)$, subdivide $B^1 = [-1, 1]$ into elementary subintervals $A_i = [t_{i-1}, t_i]$, with $t_0 = -1 < t_1 < \ldots < t_n = 1$. Choose $\epsilon > 0$ such that, on each interval A_i, the parameter value c_i at which F

fails to be Morse and injective on critical points is in the interval $(t_{i-1}+\epsilon, t_i-\epsilon)$. We begin the construction of $\mathcal{P}(F)$ with $X_* = [-\epsilon, 0] \times [-1,1] \times \Sigma$. For each t_i, let $(H_i, \Sigma) = \mathcal{P}(F|_{\{t_i\}})$ be the handlebody filling of Σ determined by the Morse function $F(t_i) : \Sigma \to \mathbb{R}$. Let $X_0 = [t_0, t_0+\epsilon] \times H_0$, let $X_n = [t_n - \epsilon, t_n] \times H_n$, and for each $i = 1, \ldots, n-1$ let $X_i = [t_i - \epsilon, t_i + \epsilon] \times H_i$. Let X_{**} be the result of gluing each X_i to X_* by identifying $\{t\} \times \Sigma \subset H_i$ with $\{-\epsilon\} \times \{t\} \times \Sigma \subset X_*$.

Note that now ∂X_{**} has $n+1$ boundary components, each obtained by gluing two handlebody fillings of Σ together. One of them is $\mathcal{P}(F|_{\partial B^1})$, and the other n are $\mathcal{P}(F|_{\partial A_i})$ for $i = 1, \ldots, n$. Since each A_i is an elementary subinterval we can fill in each $\mathcal{P}(F|_{\partial A_i})$ with either $\natural^g S^1 \times B^3$ (in the type 0 case) or $\natural^{g-1} S^1 \times B^3$ (in the type 1 case); the result is our 4-manifold X, with $\mathcal{P}(F) = (X, \Sigma)$, where Σ is $\{(0,0)\} \times \Sigma \subset X_* \subset X$. Note that these final fillings are unique due to Laudenbach and Poenaru's result [5] that every self-diffeomorphism of $\#^n S^1 \times S^2$ extends to $\natural^n S^1 \times B^3$.

A finer subdivision of B^1 into elementary subintervals only results in more type 0 subintervals, and since $\natural^g S^1 \times B^3 \cong [0,1] \times (\natural^g S^1 \times B^2)$, it is easy to see that creating more type 0 subintervals does not change the resulting 4-manifold-surface pair in our construction. □

Construction of $\mathcal{P}_{S^1}$. Given a generic $F : S^1 \to C^\infty(\Sigma)$, simply split S^1 into two B^1's on which F remains generic (i.e. split at parameter values where F is Morse and injective on critical points), as $S^1 = B^1_- \cup B^1_+$, and glue $-\mathcal{P}(F|_{B^1_-})$ to $\mathcal{P}(F|_{B^1_+})$ along their common boundary $\mathcal{P}(F|_{S^0})$, where the S^0 is $\partial B^1_+ = -\partial B^1_-$. (Note that we are gluing manifold-surface pairs, so the gluings should match the Σ on one side with the Σ on the other side, via the identity map. Since Σ splits each 3-dimensional boundary into two handlebodies, and since a self-diffeomorphism of a handlebody is determined up to isotopy by its restriction to the boundary, the gluings here are well defined.) Item 2 of Theorem 5 is clear, but the important thing that remains in this dimension is to prove item 4, namely that all closed 4-manifolds can be produced this way. This happens in the following lemmas. □

Lemma 12. *Let $X = \natural^k S^1 \times B^3$ and let Σ be a genus $g \geq k$ Heegaard surface in $\#^k S^1 \times S^2 = \partial X$. Then there exists a generic B^1-parameterized family of functions F on Σ such that (X, Σ) is diffeomorphic rel. Σ to $\mathcal{P}(F)$.*

Proof. Consider the standard genus g Heegaard diagram (Σ, α, β) for $\#^k S^1 \times S^2 = \partial X$ shown in Figure 3, and let $f_0 : \Sigma \to \mathbb{R}$ be the (Morse) height function as indicated in the figure, so that all the α curves are regular level components of f_0. For $i = k+1, \ldots, g$, the dual pair (α_i, β_i) lies in a twice-punctured torus T_i with both components of ∂T_i being regular level components of f_0, and with no other α_j or β_j intersecting T_i. Let f_t, $t \in [0,1]$, be the generic family such that, for $i = k+1, \ldots, g$, for $t \in [(i-k-1)/(g-k), (i-k)/(g-k)]$, f_t does not vary with t outside T_i and, inside T_i, f_t undergoes a critical value crossing that changes from having α_i as a regular level component to having β_i as a regular level component. In other words, the associated family $F : [0,1] \to C^\infty(\Sigma)$ is elementary of type 1 on each subinterval $[(i-k-1)/(g-k), (i-k)/(g-k)]$, switching α_i to β_i as regular level components, exactly as in Figure 2.

We now claim that, for this F, we have that $\mathcal{P}(F)$ is diffeomorphic rel. Σ to (X, Σ). To see this, think of $\mathcal{P}(F)$ as a cobordism (with sides) from the 3-dimensional handlebody $\mathcal{P}(F|_0)$ to the handlebody $\mathcal{P}(F|_1)$ built by stacking the

FIGURE 3. Standard genus g Heegaard diagram for $\#^k S^1 \times S^2$, with the vertical height function so that the α curves (red) are all regular level components. (Here $g = 5$ and $k = 2$.)

elementary cobordisms corresponding to each elementary subinterval $[t_{i-1}, t_i]$, with $t_i = (i-k)/(g-k)$. Each of these elementary cobordisms is simply a 4-dimensional 2-handle attached to $[t_{i-1}, t_{i-1} + \epsilon] \times H$, where H is a genus g handlebody, and because the surgery corresponding to this 2-handle attachment results in switching from α_i bounding a disk to β_i bounding a disk, we see that this 2-handle simply cancels the 4-dimensional 1-handle in $[t_{i-1}, t_{i-1} + \epsilon] \times H$ dual to α_i. Thus, the full cobordism contains $(g-k)$ 2-handles each cancelling one of the 1-handles in $[0, \epsilon] \times (\natural^g S^1 \times B^2)$, and this is precisely X. $\square$

LEMMA 13. *Consider generic $G, G' : B^0 \to C^\infty(\Sigma)$ such that the handlebodies $\mathcal{P}(G)$ and $\mathcal{P}(G')$ are diffeomorphic rel. Σ. Then there exists a generic family $F : B^1 \to C^\infty(\Sigma)$, with $F|_{-1} = G$ and $F|_1 = G'$ such that any subdivision of B^1 into elementary subintervals for F involves only subintervals of type 0. In particular, $\mathcal{P}(F)$ is a product, i.e. if $\mathcal{P}(G) = (H, \Sigma)$ then $\mathcal{P}(F)$ is diffeomorphic rel. Σ to $([-1,1] \times H, 0 \times \Sigma)$.*

PROOF. Let $f = G(0)$ and $f' = G(0)$, generic Morse functions on Σ with the property that there is a single handlebody filling H of Σ such that all the regular level components of f and f' bound disks in H. We will construct a generic 1-parameter family f_t, $t \in [-1,1]$, with $f_{-1} = f$ and $f_1 = f'$, together with a subdivision of $[-1,1]$ into subintervals $[t_{i-1}, t_i]$ such that, on each interval $[t_{i-1}, t_i]$, f_t does not vary with t outside a genus 0 subsurface of Σ bounded by regular level components. This suffices to prove the lemma since f_t changing with t inside a genus 0 subsurface cannot change the associated handlebody filling.

Let C be a cut system for H made of level components for f, and let C' be a similar cut system for f'. Let $C_0 = C, C_1, C_2, \ldots, C_n = C'$ be a sequence of cut systems for H, each related to the next by a single handle slide [**4**]. Then there exist Morse functions $f_1, f_2, \ldots, f_n$ such that both C_0 and C_1 are collections of regular level components for f_1, both C_1 and C_2 are collections of levels for f_2, and so on, until C_{n-1} and C_n are levels for f_n. This reduces our problem to having to interpolate from $f = f_0$ to f_1, from f_1 to f_2, and so on, in each case interpolating between two functions which have a single common cut system made of level components for both functions.

So now suppose f and f' are given and there is a single cut system C such that each curve in C is a level component for both f and f'. After a generic homotopy supported in a neighborhood of each component of C (i.e. only modifying functions in genus 0 subsurfaces) we may assume f and f' agree on a neighborhood of C. Since C is a cut system, the complement of this neighborhood is also genus 0, so we may now homotope f to f' via a generic homotopy supported on this genus 0 subsurface, and we are done. □

PROOF OF ITEM 4 OF THEOREM 5, WHEN $n = 1$. We need to prove that, for every closed 4-manifold X, there is a surface $\Sigma \subset X$ and a generic S^1-parameterized family $F : S^1 \to C^\infty(\Sigma)$ such that (X, Σ) is diffeomorphic rel. Σ to $\mathcal{P}(F)$.

Let $X = X_1 \cup X_2 \cup X_3$ be a (g,k)-*trisection* of X; this is a decomposition such that each X_i is diffeomorphic to $\natural^k S^1 \times B^3$, each $X_i \cap X_j$ is diffeomorphic to the 3-dimensional handlebody $\natural^g S^1 \times B^2$, and $X_1 \cap X_2 \cap X_3$ is a genus g surface. The existence of trisections on all closed 4-manifolds is proved in [**2**]. Let $\Sigma = X_1 \cap X_2 \cap X_3$. Applying Lemma 12 to each X_i, we get three families $F_1, F_2, F_3 : B^1 \to C^\infty(\Sigma)$ such that (X_i, Σ) is diffeomorphic rel. Σ to $\mathcal{P}(F_i)$. We cannot glue these together to form a S^1-parameterized family because we do not know, for instance, that $F_1|_1 = F_2|_{-1}$ or that $F_2|_1 = F_3|_{-1}$. We do know, however, that $\mathcal{P}(F_i|_1)$ is diffeomorphic rel. Σ to $\mathcal{P}(F_{i+1}|_{-1})$, both being diffeomorphic rel. Σ to the handlebody $(X_i \cap X_{i+1}, \Sigma)$. Thus, by Lemma 13 we can interpolate between $F_i|_1$ and $F_{i+1}|_{-1}$ to create a S^1-parameterized family F which is obtained by gluing the X_i together with extra collars on the handlebodies between them, and the result is diffeomorphic rel. Σ to (X, Σ). □

6. 2-parameter families and 5-manifolds

Finally we consider 2-parameter families; in this paper we are only concerned with the case of B^2-parameterized families. Note that when F is a generic B^2-parameterized family of functions on Σ, there is a decomposition of B^2 into polygonal disks $B^2 = P_1 \cup \ldots \cup P_n$ which meet in pairs along intervals (edges) and in triples at points (vertices) such that:

(1) F is generic on each polygon, along each edge and at each vertex.
(2) For each edge e, $F|_e$ is an elementary 1-parameter family.
(3) For each polygon P_i, there is at most one point $p_i \in P_i$ at which one of the following occurs:
   (a) A swallowtail.
   (b) A transverse triple point of self-intersection of the image of $\mathrm{Crit}_0(F)$ in $B^2 \times \mathbb{R}$, i.e. three Morse critical points with the same critical value.
   (c) A transverse intersection of $\mathrm{Crit}_1(F)$ with $\mathrm{Crit}_0(F)$, i.e. a birth/death and a Morse critical point with the same critical value.

DEFINITION 14. Each such polygon P_i in a subdivision of A as above will be called an *elementary polygon* and we say that $F|_{A_i}$ is an *elementary 2-parameter family*.

We now prepare to understand the 4-manifolds obtained by applying $\mathcal{P}_{S^1}$ to the boundary of an elementary 2-parameter family. To do this, we first describe and name some model 4-manifold-surface pairs and associated 5-manifold-surface pairs.

(1) Given a genus g, let $\Sigma_g = \#^g S^1 \times S^1$, which is the boundary of the handlebody $H_g = \natural^g S^1 \times B^2$. Let $Z_g = D^2 \times H_g$, identify Σ_g with

$\{0\} \times \Sigma_g \subset Z_g$, and let $X_g = \partial Z_g$. This gives us pairs $(X_g, \Sigma_g) = \partial(Z_g, \Sigma_g)$, with $X_g \cong \#^g S^1 \times S^3$ and $Z_g \cong \natural^g S^1 \times B^4$.

(2) Recall that the 4-manifold-surface pair associated to a type 1 elementary 1-parameter family is $(\natural^{g-1} S^1 \times B^3, \#^g S^1 \times S^1)$, where the surface $\#^g S^1 \times S^1$ is the once-stabilized Heegaard surface in $\#^{g-1} S^1 \times S^2 = \partial \natural^{g-1} S^1 \times B^3$. From this we construct a 5-manifold-surface pair (Z_{g-1}, Σ_g) where $Z_{g-1} = [-1,1] \times \natural^{g-1} S^1 \times B^3$ and $\Sigma_g = \{0\} \times \#^g S^1 \times S^1$, and we let $(X_{g-1}, \Sigma_g) = \partial(Z_{g-1}, \Sigma_g)$. Note that (X_{g-1}, Σ_g) can also naturally be seen as the double of $(\natural^{g-1} S^1 \times B^3, \#^g S^1 \times S^1)$, and that $X_{g-1} \cong \#^{g-1} S^1 \times S^3$ and $Z_{g-1} \cong \natural^{g-1} S^1 \times B^4$.

(3) Let $T = \{[z_0 : z_1 : 1] \mid |z_0| = |z_1| = 1\}$ be the standard Lagrangian torus in $\mathbb{C}P^2$, giving us the 4-manifold-surface pair $(\mathbb{C}P^2, T)$ and consider the 5-manifold-surface pair $([0,1] \times \mathbb{C}P^2, T = \{0\} \times T)$. Note that $\partial([0,1] \times \mathbb{C}P^2, \{0\} \times T) = (\mathbb{C}P^2, T) \amalg (\mathbb{C}P^2, \emptyset)$.

(4) Finally, let $X_g^\pm = X_{g-1} \#(\pm\mathbb{C}P^2)$, let $\Sigma_g^\pm = \Sigma_{g-1} \# T$ and let $Z_g^\pm = Z_{g-1} \natural([0,1] \times (\pm\mathbb{C}P^2))$, where the boundary connected sum occurs along the boundary component $\{0\} \times (\pm\mathbb{C}P^2)$ in the $[0,1] \times \mathbb{C}P^2$ summand. This yields the pairs $(X_g^\pm, \Sigma_g^\pm)$ and $(Z_g^\pm, \Sigma_g^\pm)$, with $\partial(Z_g^\pm, \Sigma_g^\pm) = (X_g^\pm, \Sigma_g^\pm) \amalg (\pm\mathbb{C}P^2, \emptyset)$.

LEMMA 15. *If P_i is an elementary polygon for a generic $F : B^2 \to C^\infty(\Sigma)$, with Σ a surface of genus g, then $\mathcal{P}(F|_{\partial P_i})$ is diffeomorphic to either (X_g, Σ_g), (X_{g-1}, Σ_g) or $(X_g^\pm, \Sigma_g^\pm)$.*

PROOF. To repeat, there is at most one point $p_i \in P_i$ at which one of the following occurs:

(1) A swallowtail.
(2) A transverse triple point of self-intersection of the image of $\mathrm{Crit}_0(F)$ in $B^2 \times \mathbb{R}$, i.e. three Morse critical points with the same critical value. We will call this a *triple critical value switch.*
(3) A transverse intersection of $\mathrm{Crit}_1(F)$ with $\mathrm{Crit}_0(F)$, i.e. a birth/death and a Morse critical point with the same critical value. We will call this a *birth-Morse crossing*

First we consider the case where there is no such point p_i. Then either $F|_t$ is Morse and injective on critical points for all $t \in \partial P_i$ or exactly two critical value switches occur in $F|_{\partial P_i}$. In both cases, $\mathcal{P}(F|_{\partial P_i})$ is a double of $\mathcal{P}(F|_e)$ where e is one edge of P_i, and thus the result is either (X_g, Σ_g) or (X_{g-1}, Σ_g).

When there is such a point p_i, first consider a swallowtail. From the local model of a swallowtail, the critical level component containing the swallowtail is a figure eight graph just like the critical level component containing a Morse critical point in the B^0-parameterized case. (Note that the swallowtail is the only critical point in this level set.) In particular, the critical level component has a genus 0 neighborhood, so the handlebody filling is the same for any small perturbation of the swallowtail and thus $\mathcal{P}(F|_{\partial P_i})$ is (X_g, Σ_g).

In a birth-Morse crossing, we have a Morse critical value which a birth/death event crosses in a 2-parameter sense. At the parameter value when the Morse critical point and the birth/death point are at the same level, even if they are in the same component of the same level, that component is still a figure eight (with one non-smooth point, the birth/death), and it's neighborhood is thus genus 0.

Therefore the birth-Morse crossing can be taken to be a 2-parameter family of functions which does not change outside a genus 0 subsurface, so that $\mathcal{P}(F|_{\partial P_i})$ is (X_g, Σ_g).

Thus the only thing interesting that can happen happens with a triple critical value switch. To analyze this, we consider the critical level component of $f_* = F|_{\{p_i\}}$ containing three index 1 Morse critical points. (If we do not have all three critical points in the same component, we are again in one of the two easy cases (X_g, Σ_g) or (X_{g-1}, Σ_g). If any have index 0 or 2, they cannot all lie in one component.) This is a 4-valent graph Γ with three vertices p, q and r (the three critical points) and an oriented regular neighborhood N such that Γ separates N into two components $N_\pm$. Euler characteristic considerations show that P can either be genus 0 with 5 boundary components or genus 1 with 3 boundary components and, again, we are only interested in the genus 1 case. (Genus 0 means that the handlebody filling is the same for all perturbations of the function, so that we get (X_g, Σ_g).) The polygon P_i must be a hexagon, since there are six ways to order the three critical points p, q, r by height, and each edge of the hexagon represents a 1-parameter critical value switch. Figure 4 illustrates a neighborhood of the singular point p_i at the center of P_i, labelled by *ordered partitions* of the set $\{p, q, r\}$ of critical points. The orderings are given by the Morse function f_t for $t \in P_i$, and are abbreviated in the figure with the following notation: $pq|r$ means that $f_t(p) = f_t(q) < f_t(r)$, or pqr means $f_t(p) = f_t(q) = f_t(r)$, or $r|q|p$ means $f_t(r) < f_t(q) < f_t(p)$. Thus the Morse function at the center of the figure has all three critical points at the same level, and along the six black rays two critical points are at the same level while one is above or below that level, and between the rays all three critical points have distinct critical values. The hexagon P_i is dual to the decomposition shown; the edges of P_i are perpendicular to the six rays shown. (This hexagon is otherwise known as the *permutahedron* on three letters, a clue for how one might approach higher dimensional cases.)

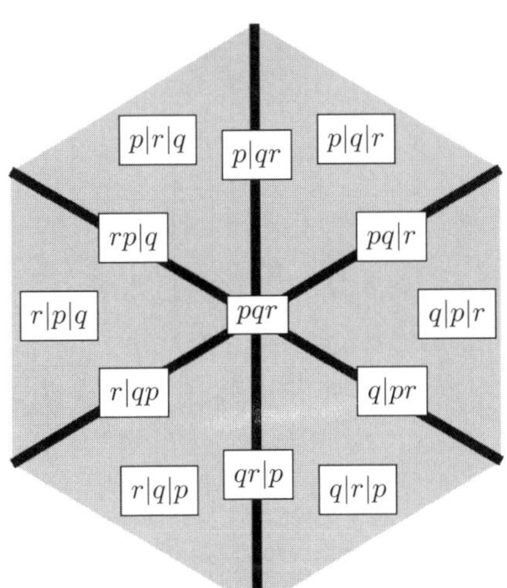

FIGURE 4. The permutahedron on 3 letters, representing a neighborhood in parameter space of a triple critical value switch.

Thus $F|_{\partial P_i}$ is divided into six elementary 1-parameter families; we claim now that at most three of these can be of type 1. If one believes this claim, then the handlebody filling associated to a point on ∂P_i, as one moves around ∂P_i, changes at most three times, and those changes (in the sense of which curves bound disks in the handlebody) all occur in a fixed genus 1 subsurface P of Σ bounded by curves that themselves bound disks in the handlebody. Up to diffeomorphism, then, we either have no changes, or two changes, switching from a meridian bounding a disk to a longitude bounding a disk and then back, or three changes, switching from a meridian to a longitude to a $(1, \pm 1)$-curve on the torus. Zero changes gives (X_g, Σ_g) and two changes gives (X_{g-1}, Σ_g), exactly as before. Three changes means that $\mathcal{P}(F|_{\partial P_i})$ is described by a trisection diagram $(\Sigma, \alpha, \beta, \gamma)$ in which α, β and γ only differ in one T^2 summand, which gives $(X_g^{\pm}, \Sigma_g^{\pm})$ (see the trisection diagrams for $S^1 \times S^3$ and $\pm \mathbb{C}P^2$ in [**2**]).

It remains to prove that we have at most three type 1 critical value switches around ∂P_i. In fact this claim follows from a simpler claim: Suppose that $f_t : P \to [-1, 1]$ is a generic 1-parameter family of functions on a surface P of genus 1 with 3 boundary components, with $f_t^{-1}(1)$ equal to two boundary components of P and $f_t^{-1}(-1)$ equal to one boundary component. Furthermore, suppose that f_t has exactly three index 1 critical points, and no other critical points, for all t, that there is exactly one critical value switch at $t = -1/2$ and one critical value switch at $t = 1/2$, and f_t is a generic Morse function for all other values of t, and finally that the two critical points that switch heights at $t = -1/2$ are not the same two critical points that switch at $t = 1/2$. Then f_t can not be of type 1 on both intervals $[-1, 0]$ and $[0, 1]$. This is because, up to diffeomorphism, f_0 can be taken to be the vertical height function on one of the three genus 1 surfaces embedded in $\mathbb{R}^3$ illustrated in Figure 5. If, for example, $f_0(p) = 1/2$, $f_0(q) = 0$, and p and q are to switch heights in a type 1 switch, the component of $f_0^{-1}([-\epsilon, 1/2 + \epsilon])$ containing p and q needs to have genus 1, but this can clearly not be true for both the pairs (p, q) and (q, r).

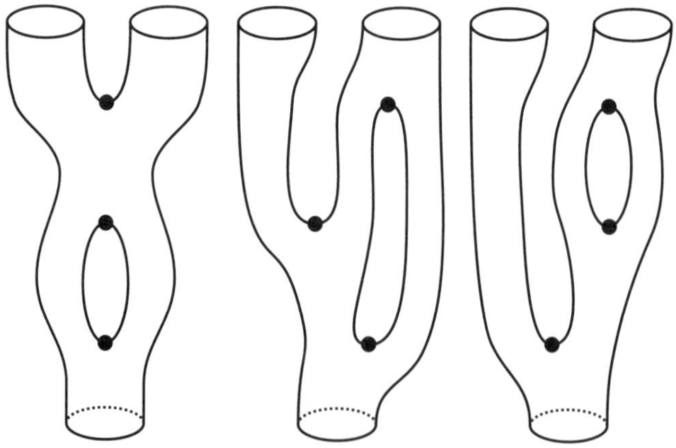

FIGURE 5. Three Morse functions on a genus 1 surface with 3 boundary components.

□

DEFINITION 16. An elementary polygon P_i, or an elementary 2-parameter family $F = F|_{P_i}$ is of *type 0* if, in the above theorem, we have $\mathcal{P}(F) \cong (X_g, \Sigma_g)$, is of *type 1* if $\mathcal{P}(F) \cong (X_{g-1}, \Sigma_g)$, and is of *type 2* if $\mathcal{P}(F) \cong (X_g^{\pm}, \Sigma_g^{\pm})$.

CONSTRUCTION OF $\mathcal{P}_{B^2}$: We parallel the construction of $\mathcal{P}_{B^1}$ as closely as possible. Given a generic $F : B^2 \to C^\infty(\Sigma)$, subdivide B^2 into elementary polygons $B^2 = P_1 \cup \ldots \cup P_n$. Let $v_1, \ldots, v_l$ be the vertices and let $e_1, \ldots, e_m$ be the edges of this subdivision. Thicken the 0- and 1-skeleton of this subdivision in the following sense: Let V_i be a small disk neighborhood of each vertex v_i (or a half-disk neighborhood when $v_i \in \partial B^2$). Shrink each e_j so as to intersect the V_i only at the endpoints of e_j and let E_j be a collar neighborhood of each e_j (i.e. E_j is identified with $I \times e_j$ where $I = [-\epsilon, \epsilon]$ or $[-\epsilon, 0] \times e_j$, according to whether e_j is in the interior or boundary of B^2) which intersects the V_i only at $I \times \partial e_j$. Finally shrink each P_k so that P_k intersects the union of the V_i and the E_j exactly along ∂P_k.

Let $Z_* = [-\epsilon, 0] \times B^2 \times \Sigma$. For each vertex v_i, let $(H_i, \Sigma) = \mathcal{P}(F|_{\{v_i\}})$, and let $Z_{v_i} = V_i \times H_i$. Let Z_{**} be the result of attaching each Z_{v_i} to Z_* by identifying $\{t\} \times \Sigma \subset Z_{v_i}$ with $\{-\epsilon\} \times \{t\} \times \Sigma \subset Z_*$.

Now, for each edge e_j, let $(X_j, \Sigma) = \mathcal{P}(F|_{e_j})$ and let $Z_{e_j} = I \times X_j$ (where $I = [-\epsilon, \epsilon]$ or $[-\epsilon, 0]$ as above, according to whether e_j is in the interior or boundary of B^2). Recall from the construction of $\mathcal{P}_{B^1}$ that X_j is built beginning with $[-\epsilon, 0] \times e_j \times \Sigma$, and thus for each $t \in e_j$ we have a particular copy of Σ in X_j, namely $\{0\} \times \{t\} \times \Sigma$. Let Z_{***} be the result of attaching each Z_{e_j} to Z_{**} as follows: For each $x \in I$ and each $t \in e_j$, first identify $\{x\} \times \{0\} \times \{t\} \times \Sigma$ in X_j with $\{-\epsilon\} \times \{(x,t)\} \times \Sigma$ in $Z_* \subset Z_{**}$. (The last (x,t) is a point in $I \times e_j$ which is identified with $E_j \subset B^2$.) Also, recall that each E_j intersects two vertex neighborhoods V_i and $V_{i'}$ in intervals which can be naturally identified with $I \times \{v_i\}$ and $I \times \{v_{i'}\}$. Since X_j contains at one end of its boundary a copy of the handlebody H_i and at the other end a copy of the handlebody $H_{i'}$, we can next identify $I \times H_i$ in Z_{e_j} with $I \times H_i$ in $Z_{v_i} \subset Z_{**}$ and also identify $I \times H_{i'}$ in Z_{e_j} with $I \times H_{i'}$ in $Z_{v_{i'}} \subset Z_{**}$. This completes the construction of Z_{***}.

Now note that Z_{***} has one "outer" boundary component, diffeomorphic to $\mathcal{P}(F|_{\partial B^2})$, and one "inner" boundary component for each elementary polygon P_k, diffeomorphic to $\mathcal{P}(F|_{\partial P_k})$. Since we have classified the manifolds that arise as $\mathcal{P}(F|_{\partial P_k})$, we know how to fill these extra boundary components in: In the type 0 case fill in with $\natural^g S^1 \times B^4$, in the type 1 case fill in with $\natural^{g-1} S^1 \times B^4$, and in the type 2 case fill in with $(\natural^{g-1} S^1 \times B^4) \natural ([0,1] \times (\pm \mathbb{C}P^2))$. In the type 2 case we are left with $\pm \mathbb{C}P^2$ boundary components, and in the other cases we completely cap off these boundaries. □

References

[1] Francesco Costantino and Dylan Thurston, *3-manifolds efficiently bound 4-manifolds*, J. Topol. **1** (2008), no. 3, 703–745, DOI 10.1112/jtopol/jtn017. MR2417451

[2] David Gay and Robion Kirby, *Trisecting 4-manifolds*, Geom. Topol. **20** (2016), no. 6, 3097–3132, DOI 10.2140/gt.2016.20.3097. MR3590351

[3] A. Hatcher and W. Thurston, *A presentation for the mapping class group of a closed orientable surface*, Topology **19** (1980), no. 3, 221–237, DOI 10.1016/0040-9383(80)90009-9. MR579573

[4] Klaus Johannson, *Topology and combinatorics of 3-manifolds*, Lecture Notes in Mathematics, vol. 1599, Springer-Verlag, Berlin, 1995. MR1439249

[5] François Laudenbach and Valentin Poénaru, *A note on 4-dimensional handlebodies*, Bull. Soc. Math. France **100** (1972), 337–344. MR0317343

[6] W. B. R. Lickorish, *A representation of orientable combinatorial* 3*-manifolds*, Ann. of Math. (2) **76** (1962), 531–540, DOI 10.2307/1970373. MR0151948

[7] V. A. Rohlin, *A three-dimensional manifold is the boundary of a four-dimensional one* (Russian), Doklady Akad. Nauk SSSR (N.S.) **81** (1951), 355–357. MR0048808

[8] V. A. Rohlin, *New results in the theory of four-dimensional manifolds* (Russian), Doklady Akad. Nauk SSSR (N.S.) **84** (1952), 221–224. MR0052101

[9] Osamu Saeki, *Singular fibers and 4-dimensional cobordism group*, Pacific J. Math. **248** (2010), no. 1, 233–256, DOI 10.2140/pjm.2010.248.233. MR2734174

[10] V. V. Sharko, *Functions on manifolds*, Translations of Mathematical Monographs, vol. 131, American Mathematical Society, Providence, RI, 1993. Algebraic and topological aspects; Translated from the Russian by V. V. Minachin [V. V. Minakhin]. MR1248167

[11] R. Thom, *Quelques propriétés des variétés-bords* (French), Colloque de Topologie de Strasbourg, 1951, no. V., La Bibliothèque Nationale et Universitaire de Strasbourg, 1952, pp. 10. MR0051511

Euclid Lab, 160 Milledge Terrace, Athens, Georgia 30606; and Department of Mathematics, University of Georgia, Athens, Georgia 30602

Email address: d.gay@euclidlab.org

Proceedings of Symposia in Pure Mathematics
Volume **102**, 2019
https://doi.org/10.1090/pspum/102/01816

On braided, banded surfaces and ribbon obstructions

J. Elisenda Grigsby

ABSTRACT. We discuss how to apply work of L. Rudolph to braid conjugacy class invariants to obtain potentially effective obstructions to a slice knot being ribbon. We then apply these ideas to a family of braid conjugacy class invariants [**6**] coming from Khovanov-Lee theory and explain why we do not obtain effective ribbon obstructions in this case.

1. Introduction

Recall that a knot $K \subseteq (S^3 = \partial B^4)$ is said to be *slice*[1] if K bounds a properly imbedded disk in B^4. K is said to be *ribbon* if K bounds an immersed disk in S^3 whose singularities are all of ribbon type, as in Figure 1. An immersed surface in S^3 whose singularities are all of ribbon type is said to be *ribbon-immersed.*

By pushing neighborhoods of the ribbon singularities into B^4, one sees that every ribbon knot is slice. The converse statement is a long-standing conjecture of Fox [**4**]:

CONJECTURE 1 (Slice-Ribbon conjecture). If K is slice, then K is ribbon.

There are a number of proposed counterexamples, cf. [**5**] and [**3**].

The current note has two modest goals:

(1) to advertise how work of L. Rudolph can be applied to braid conjugacy class invariants to in principle obstruct a slice knot from being ribbon,
(2) to explain why this will not work when the ideas in (1) are applied to the conjugacy class invariant defined and studied in [**6**],

We are disappointed about (2) but hope that (1) may still prove useful.

2. Obtaining ribbon obstructions from braid conjugacy class invariants

Let $\mathfrak{B}_n$ denote the n–strand braid group, with standard Artin generators $\sigma_1, \dots \sigma_{n-1}$. In this paper, braid words will be read from left to right, and braid diagrams will be read (and oriented) from top to bottom, with the first strand on the left. Let $i_{n,n+k} : \mathfrak{B}_n \to \mathfrak{B}_{n+k}$ denote the standard inclusion of the n–strand braid group into the $(n+k)$–strand braid group that adjoins k trivial strands on

2010 *Mathematics Subject Classification.* Primary 57M27; Secondary 81R50.
This work was partially supported by NSF CAREER award DMS-1151671 and a grant from the Simons Foundation (#396324 to J. Elisenda Grigsby).

[1]We work in the smooth category throughout.

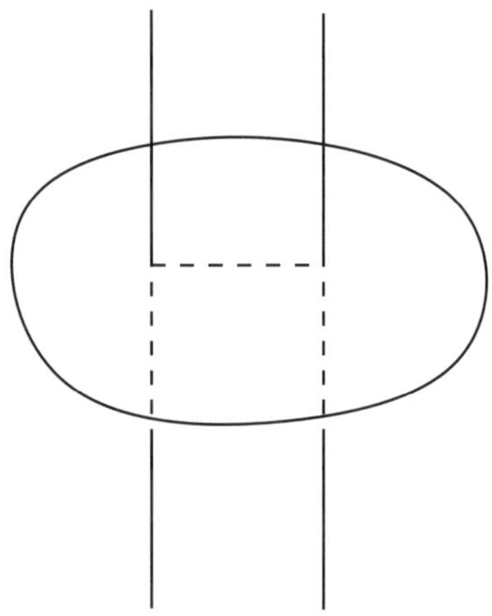

FIGURE 1. A *ribbon type singularity* of an immersed surface $f : S \looparrowright S^3$ with $\partial S \neq \emptyset$ is an imbedded arc of self-intersection, $a \subset S^3$, whose preimage, $f^{-1}(a) \subset S$, has two imbedded connected components, one with boundary in ∂S and one with boundary in $S \setminus \partial S$.

the right. We adopt Rudolph's conventions in [**10**] and denote $i_{n,n+k}(\beta) \in \mathfrak{B}_{n+k}$ by $\beta^{(k)}$. Let $\mathfrak{B} := \bigcup_n \mathfrak{B}_n$.

For $\beta \in \mathfrak{B}_n$, we use $[\beta]$ to denote its conjugacy class in $\mathfrak{B}_n$ and will denote its positive (resp., negative) Markov stabilization, $i_{n,n+1}(\beta)\sigma_n$ (resp., $i_{n,n+1}(\beta)\sigma_n^{-1}$), by β^+ (resp., by β^-).

For $\beta \in \mathfrak{B}_n$, we use $\widehat{\beta}_{S^3}$ to denote its *braid-oriented* closure as a link in S^3 and $\widehat{\beta}_{S^3 \setminus B}$ to denote its braid-oriented closure as a link in the complement of the braid axis, B. Of course, if $[\beta] = [\beta']$, then $\widehat{\beta}_{S^3} = \widehat{\beta}'_{S^3}$ as links in S^3, but the converse is not generally true. On the other hand, $[\beta] = [\beta']$ iff $\widehat{\beta}_{S^3 \setminus B} = \widehat{\beta}'_{S^3 \setminus B}$.

We begin with a few definitions (cf. [**10**]) in preparation for reminding the reader of two theorems of L. Rudolph.

DEFINITION 2. A positive (resp., negative) *band* in $\mathfrak{B}_n$ is any braid $\beta \in \mathfrak{B}_n$ for which $[\beta] = [\sigma_1]$ (resp., $[\beta] = [\sigma_1^{-1}]$).

REMARK 3. Any band can be written in the form $\omega\sigma_1\omega^{-1}$, where ω is a word in the Artin generators. Note that $\sigma_i^{\pm}$ is a band for each i, via the conjugating word

$$\omega = (\sigma_{i-1}\sigma_i)(\sigma_{i-2}\sigma_{i-1}) \cdots (\sigma_1\sigma_2).$$

DEFINITION 4. A *band presentation* of a braid β is an explicit decomposition of β as a product of bands:

$$\beta = \prod_{j=1}^{c} \omega_j \sigma_1^{\pm} \omega_j^{-1},$$

where, in addition, an explicit word in the Artin generators has been chosen for each of the conjugating braids, ω_j. The *band rank* of a braid β, denoted $\mathrm{rk}_n(\beta)$, is the minimal $c \in \mathbb{Z}^{\geq 0}$ such that β can be expressed as a product of c bands. Note that $\mathrm{rk}_n(\beta)$ is a braid conjugacy class invariant.

The nomenclature is justified by Rudolph's observation that if $\beta \in \mathfrak{B}_n$ can be written as a product of c bands, then the closure of β bounds an obvious ribbon-immersed surface in S^3 with a disk (0–handle) for each strand of the braid and a band (1–handle) for each term in the product. This surface is in general not imbedded, since each band may intersect the n disks, as well as itself. However, the resulting singularities will always be of ribbon type. By appropriately pushing neighborhoods of the ribbon singularities into B^4 one therefore obtains a properly-imbedded surface, Morse with respect to the radial function, with no maxima.

DEFINITION 5. A *braided, banded surface* is a ribbon-immersed oriented surface in S^3 bounded by the closure of a band-presented braid $\beta \in \mathfrak{B}_n$ as described above. Precisely, let $\beta = \prod_{j=1}^{c} \omega_j \sigma_1^{\pm} \omega_j^{-1}$ be a band presentation of β. Begin with the n braid-oriented disjoint disks $D_1, \ldots, D_n$ bounded by the closure of the n–strand trivial braid $\mathbb{1}_n$ and attach c half-twisted bands $b_1, \ldots, b_c$ to $D_1, \ldots, D_n$ as follows.[2]

(1) One of the two ends of b_j is attached to D_1. A neighborhood of this end looks like the standard oriented band bounded by $\sigma_1^{\pm}$ (with orientation induced by the orientation on $\sigma_1^{\pm}$).
(2) Construct the remainder of b_j by extending the standard band above, reading the word ω_j backwards one letter at a time and forming the associated nested subconjugates of $\omega_j \sigma_1^{\pm} \omega_j^{-1}$. Note that as the band is extended in this way, it may intersect itself and the interiors of the disks D_i, but all of these intersections will be of ribbon type.
(3) Let $\pi : \mathfrak{B}_n \to S_n$ be the standard projection of the braid group to the symmetric group on the same number of letters, and let π_j be the image of 2 under the permutation $\pi(\omega_j^{-1})$. Then the other end of b_j will be attached to D_{π_j}.

See Figure 2.

Theorems 6 and 8 and their corollaries are due to L. Rudolph.

THEOREM 6. [**9**, Sec. 3] *Let $\Sigma \subset S^3$ be a ribbon-immersed orientable surface. Then Σ is isotopic to a* braided, banded surface.

COROLLARY 7. *If K is ribbon, then for some n there exists a braid $\beta \in \mathfrak{B}_n$ with $\widehat{\beta}_{S^3} = K$ for which $rk_n(\beta) = n - 1$.*

PROOF. If K is ribbon, then it bounds a ribbon-immersed orientable disk in S^3. By Theorem 6, this ribbon-immersed disk is isotopic to a braided, banded surface S for some $\beta \in \mathfrak{B}_n$ with $\widehat{\beta}_{S^3} = K$. This implies that β has a band presentation with $n-1$ bands, so $\mathrm{rk}_n(\beta) \leq n - 1$. But β cannot be presented with fewer than $n-1$ bands, since this would force $\widehat{\beta}_{S^3}$ to be disconnected. □

THEOREM 8. [**10**, Prop. 6.11] *Let $\widehat{\beta}_{S^3}$ be a closed braid bounding a properly-imbedded oriented surface $S \subset B^4$. Then there exists $q \in \mathbb{Z}^{\geq 0}$ and some band presentation of $\beta^{(q)}$ for which the push-in to B^4 of the associated braided, banded surface is isotopic to S with q open disks removed.*

[2] The jth band, b_j, is a ribbon-immersed 2–dimensional 1–handle. We shall refer to the two attaching regions of each band as its *ends*. This should not be confused with the ends of the braid word $\omega_j \sigma_1^{\pm} \omega_j^{-1}$ defining the jth band. The geometric band is constructed by reading the braid word out from the center, rather than from left to right.

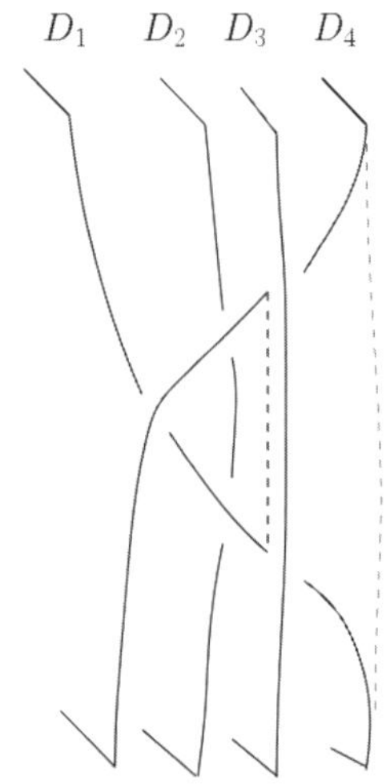

FIGURE 2. The positively half-twisted band associated to the conjugate $\omega\sigma_1\omega^{-1} \in \mathfrak{B}_4$, where $\omega = \sigma_3^{-1}\sigma_2$. One end of the band is attached to D_1, the other to D_4, and the band intersects D_3 in a ribbon singularity.

COROLLARY 9. *If K is slice, then for some n, q there exists a braid $\beta \in \mathfrak{B}_n$ with $\widehat{\beta}_{S^3} = K$ for which*

$$rk_{n+q}\left(\beta^{(q)}\right) \leq (n-1) + 2q.$$

PROOF. Let K be slice, and choose $\beta \in \mathfrak{B}_n$ with $\widehat{\beta}_{S^3} = K$. Then $\widehat{\beta}_{S^3}$ bounds a slice disk in B^4. Now Theorem 8 tells us there exists a q for which $\widehat{\beta}^{(q)}_{S^3}$ bounds a braided, banded surface of Euler characteristic $1-q$ (its push-in is the slice disk with q open disks removed). The associated band presentation of $\widehat{\beta}^{(q)}_{S^3}$ therefore has b bands, where

$$(n+q) - b = 1 - q.$$

The result follows. □

PROPOSITION 10. *Let $\phi : \mathfrak{B} \to \mathbb{R}$ be a real-valued braid invariant with the following properties.*

(1) $\phi(\beta) = \phi(\beta')$ *if* $[\beta] = [\beta']$,
(2) $\phi(\beta^{\pm}) = \phi(\beta) + 1$,
(3) $\phi(\beta) \leq rk_n([\beta])$,
(4) *For some n, k, $\exists\, \beta \in \mathfrak{B}_n$ satisfying $\phi\left(\beta^{(k)}\right) \neq \phi(\beta) + 2k$,*
(5) *There exists a knot K with a braid representative $\beta \in \mathfrak{B}_n$ satisfying $\phi(\beta) > n-1$.*

Then $\phi(\beta) - n$ is an oriented link invariant that has the potential to obstruct a slice knot from being ribbon.

PROOF. The Alexander and Markov theorems (cf. [**2**]) tell us that any (oriented) link is isotopic to a (braid-oriented) closed braid and that any two closed braid representatives of the same link are related by a finite sequence of closed braid isotopies (conjugations) and Markov de/stabilizations. Properties (1) and (2) then imply that $\phi(\beta) - n$ is an oriented link invariant.

Property (5) tells us that there exists a knot K with a braid representative $\beta \in \mathfrak{B}_{n_0}$ (for some n_0) satisfying $n_0 - 1 < \phi(\beta)$. Properties (1) and (2) and the Markov theorem then tell us that for *every* braid representative $\beta \in \mathfrak{B}_n$ of K (for *every* n), $\phi(\beta) > n - 1$. Property (3) and Corollary 7 then tells us that $K = \widehat{\beta}_{S^3}$ cannot be ribbon.

Finally, we must show that it's possible for ϕ to obstruct a *slice* knot from being ribbon. Let K be a slice knot and $\beta \in \mathfrak{B}_n$ a braid for which $\widehat{\beta}_{S^3} = K$. Then Corollary 9 tells us that for some q we have $\mathrm{rk}_{n+q}\left(\widehat{\beta}^{(q)}_{S^3}\right) \leq (n-1) + 2q$. Property (3) then tells us that $\phi(\beta^{(q)}) \leq (n-1) + 2q$, and Property (4) tells us that it's possible for $\phi(\beta) > n - 1$. □

3. The annular Rasmussen invariant does not give new ribbon obstructions

In [**6**] we define, for each braid $\beta \in \mathfrak{B}$ a family, $d_t(\widehat{\beta}_{S^3 \setminus B})$ for $t \in [0,1]$, of real-valued braid conjugacy class invariants.[3] These braid conjugacy class invariants are defined using the structure of the annular Khovanov-Lee complex of $\widehat{\beta}_{S^3}$ as a $(\mathbb{Z} \oplus \mathbb{Z})$–filtered chain complex.[4] We recall here some of its relevant properties.

THEOREM 11. [**6**, Thm. 1, Cor. 3, Thm. 6] *Let $\beta \in \mathfrak{B}_n$ be an n-strand braid equipped with the braid-like orientation. Viewed as a function from $[0,1]$ to $\mathbb{R}$, $d_t(\widehat{\beta}_{S^3 \setminus B})$ satisfies the following properties:*

(1) *$d_t(\widehat{\beta}_{S^3 \setminus B})$ is a conjugacy class invariant,*
(2) *$d_t(\widehat{\beta}_{S^3 \setminus B})$ is continuous and piecewise linear, with slopes*
$$m_t(\widehat{\beta}_{S^3 \setminus B}) \in \{-n, -n+2, \ldots, n-2, n\},$$
(3) *$d_0(\widehat{\beta}_{S^3 \setminus B}) = s(\widehat{\beta}_{S^3}) - 1$,*
(4) *$d_1(\widehat{\beta}_{S^3 \setminus B}) = w(\widehat{\beta}_{S^3 \setminus B})$*
(5) *$|d_t(\widehat{\beta}_{S^3 \setminus B}) - d_t(\widehat{\mathbb{1}}_n)| \leq rk_n(\widehat{\beta}_{S^3 \setminus B})$*
(6) *$d_t(\widehat{\mathbb{1}}_n) = n(t-1)$*

In the above, $s(\widehat{\beta}_{S^3})$ is Beliakova-Wehrli's generalization [**1**] of Rasmussen's invariant of knots [**8**] to an invariant of oriented links in S^3, and $w(\widehat{\beta}_{S^3 \setminus B})$ is the writhe of $\widehat{\beta}_{S^3 \setminus B}$ equipped with the braid-orientation. In the literature, $w(\widehat{\beta}_{S^3 \setminus B})$ is sometimes called the *exponent sum*.

Recall that we're trying to apply the lower bound on band rank coming from the invariant $d_t(\widehat{\beta}_{S^3 \setminus B})$ to find an obstruction to a slice knot K being ribbon. Rudolph's Corollary 7 tells us that if we can find a lower bound on the band rank that is well-behaved enough under Markov de/stabilization, it might yield an effective ribbon obstruction.

[3] In [**6**], $d_t(\widehat{\beta}_{S^3 \setminus B})$ is defined for $t \in [0,2]$. However, the relevant interval is $t \in [0,1]$ because the invariant is symmetric about $t = 1$ (cf. [**6**, Thm. 1]).

[4] In [**6**], d_t is defined as an invariant of oriented annular links, but it is noted there that the closure of a braid comes equipped with a standard orientation that we call the braid-like orientation, see [**6**, Sec. 3.2], and two braid-oriented braid closures are isotopic as annular links iff they are conjugate in $\mathfrak{B}$.

Unfortunately, Theorem 12 below tells us that the best lower bound on band rank coming from the d_t invariant occurs at either $t = 0$ or $t = 1$. Moreover, Lemmas 15 and 16 tell us that neither of these invariants is strong enough to obstruct a slice knot from being ribbon.

Indeed, none of these invariants is strong enough to obstruct a knot of *finite concordance order* from being ribbon. Recall that the concordance order of an oriented knot K is the minimal $n \in \mathbb{Z}^{\geq 0}$ for which the connected sum of n copies of K with itself is slice. Theorem 12 combined with Lemmas 15 and 16 tell us that if $d_t(\widehat{\beta}_{S^3\setminus B})$ obstructs $\widehat{\beta}_{S^3} = K$ from being ribbon, then K has infinite order in the smooth knot concordance group.

THEOREM 12. *Let $\beta \in \mathfrak{B}_n$ be an n-strand braid whose closure, $K = \widehat{\beta}_{S^3}$, has finite concordance order. Then*

$$\max_{t\in[0,1]} |d_t(\widehat{\beta}_{S^3\setminus B}) - d_t(\widehat{\mathbb{1}}_n)|$$

occurs at either $t = 0$ or $t = 1$.

PROOF. Recall that Theorem 11, Property (6), tells us that $d_t(\widehat{\mathbb{1}}_n) = nt - n$ on $[0,1]$. So knowing that $d_t(\widehat{\beta}_{S^3\setminus B})$ is continuous and piecewise linear on $[0,1]$ with slopes bounded between $-n$ and n tells us that

$$f(t) := d_t(\widehat{\beta}_{S^3\setminus B}) - d_t(\widehat{\mathbb{1}}_n)$$

is continuous and piecewise linear on $[0,1]$ with slopes bounded between $-2n$ and 0.

Remarks 13 and 14 below hold for any continuous and piecewise-linear $f : [0,1] \to \mathbb{R}$ with slopes bounded between $-2n$ and 0. They are both immediate from the fact that such a function is non-increasing on the interval $[0,1]$.

REMARK 13. Let $[t_0, t_1] \subseteq [0,1]$ be an interval on which $f(t) \geq 0$ (resp., on which $f(t) \leq 0$). Then $|f(t)|$ is non-increasing on $[t_0, t_1]$ (resp., $|f(t)|$ is non-decreasing on $[t_0, t_1]$). Hence, $\max_{t\in[t_0,t_1]}|f(t)|$ occurs at t_0 (resp., $\max_{t\in[t_0,t_1]}|f(t)|$ occurs at t_1).

REMARK 14. $f(t)$ changes sign at most once on the interval $[0,1]$. In particular, if $f(t_0) \leq 0$ for some $t_0 \in [0,1)$, then $f(t) \leq 0$ for all $t \in [t_0, 1]$.

Now consider the function $f(t) := d_t(\widehat{\beta}_{S^3\setminus B}) - d_t(\widehat{\mathbb{1}}_n)$ for $\widehat{\beta}_{S^3}$ with finite concordance order. Since $\widehat{\beta}_{S^3}$ has finite concordance order, $s(\widehat{\beta}_{S^3}) = 0$. Hence Theorem 11, Property (3) tells us that $d_0(\widehat{\beta}_{S^3\setminus B}) = s(\widehat{\beta}_{S^3}) - 1 = -1$. Moreover, $d_t(\widehat{\mathbb{1}}_n) = nt - n$ on $[0,1]$, so $f(0) = n - 1$.

Now assume that $\max_{t\in[0,1]}|f(t)|$ does not occur at $t = 0$.

Let $t' > 0$ be the first time the maximum value occurs. That is, $|f(t')| = \max_{t\in[0,1]}|f(t)|$ and $|f(t)| < |f(t')|$ for all $t \in [0, t')$.

We know that $f(t') < 0$, since Remark 14 and the fact that $f(0) > 0$ implies that if $f(t') > 0$ then $f(t) > 0$ for all $t \in [0, t']$. But if this were the case then Remark 13 would tell us $\max_{t\in[0,t']}|f(t)|$ occurs at $t = 0$, a contradiction.

But since $f(t') < 0$, Remark 14 tells us that $f(t) < 0$ for all $t \in [t', 1]$. Hence, by Remark 13, $\max_{t\in[t',1]}|f(t)|$ occurs at $t = 1$. Since $|f(t)| < |f(t')|$ for all $t \in [0, t']$, $\max_{t\in[0,1]} |f(t)|$ occurs at $t = 1$, as desired. □

LEMMA 15. *Let $\beta \in \mathfrak{B}_n$ be an n-strand braid whose closure, $\widehat{\beta}_{S^3}$, is a knot K. If K has finite concordance order, then $|d_0(\widehat{\beta}_{S^3\setminus B}) - d_0(\widehat{\mathbb{1}}_n)| = n - 1$.*

PROOF. If K has finite concordance order, then $s(K) = d_0(\widehat{\beta}_{S^3\setminus B}) + 1 = 0$, by [**8**]. The result then follows from Theorem 11, Property (6). $\square$

LEMMA 16. *Let $\beta \in \mathfrak{B}_n$ be an n-strand braid whose closure, $\widehat{\beta}_{S^3}$, is a knot K. If K has finite concordance order, then $|d_1(\widehat{\beta}_{S^3\setminus B}) - d_1(\widehat{\mathbb{1}}_n)| \leq n-1$.*

PROOF. This follows immediately from Theorem 11, Properties (4) and (6), and [**7**, Lem. 2.1], which says that the absolute value of the *writhe* of an n–braid whose closure has finite concordance order cannot exceed $n-1$. Lisca's lemma in turn relies on the slice-Bennequin inequality [**11**]. $\square$

Acknowledgments

This note is based on a talk the author delivered at the Georgia Topology Conference, 2017. She thanks Mark Hughes, whose questions following the talk prompted a deeper investigation, Kyle Hayden, with whom she studied the constructions of Rudolph described here, and Joan Birman, Tony Licata, and Stephan Wehrli for interesting conversations. The author is also grateful to Adam Saltz for carefully reading a previous draft and pointing out a gap (now filled) in the proof of Theorem 12.

References

[1] Anna Beliakova and Stephan Wehrli, *Categorification of the colored Jones polynomial and Rasmussen invariant of links*, Canad. J. Math. **60** (2008), no. 6, 1240–1266, DOI 10.4153/CJM-2008-053-1. MR2462446

[2] Joan S. Birman, *Braids, links, and mapping class groups*, Princeton University Press, Princeton, N.J.; University of Tokyo Press, Tokyo, 1974. Annals of Mathematics Studies, No. 82. MR0375281

[3] P. Cahn and A. Kjuchukova, *Singular branched covers of four-manifolds*, 2017. math.GT/1710.11562.

[4] R. H. Fox, *Some problems in knot theory*, Topology of 3-manifolds and related topics (Proc. The Univ. of Georgia Institute, 1961), Prentice-Hall, Englewood Cliffs, N.J., 1962, pp. 168–176. MR0140100

[5] Robert E. Gompf, Martin Scharlemann, and Abigail Thompson, *Fibered knots and potential counterexamples to the property 2R and slice-ribbon conjectures*, Geom. Topol. **14** (2010), no. 4, 2305–2347, DOI 10.2140/gt.2010.14.2305. MR2740649

[6] J. E. Grigsby, A.M. Licata, and S. M. Wehrli, *Annular Khovanov-Lee homology, braids, and cobordisms*, 2016. math.GT/1612.05953.

[7] Paolo Lisca, *On 3-braid knots of finite concordance order*, Trans. Amer. Math. Soc. **369** (2017), no. 7, 5087–5112, DOI 10.1090/tran/6888. MR3632561

[8] Jacob Rasmussen, *Khovanov homology and the slice genus*, Invent. Math. **182** (2010), no. 2, 419–447, DOI 10.1007/s00222-010-0275-6. MR2729272

[9] Lee Rudolph, *Braided surfaces and Seifert ribbons for closed braids*, Comment. Math. Helv. **58** (1983), no. 1, 1–37, DOI 10.1007/BF02564622. MR699004

[10] Lee Rudolph, *Special positions for surfaces bounded by closed braids*, Rev. Mat. Iberoamericana **1** (1985), no. 3, 93–133, DOI 10.4171/RMI/16. MR836285

[11] Lee Rudolph, *Quasipositivity as an obstruction to sliceness*, Bull. Amer. Math. Soc. (N.S.) **29** (1993), no. 1, 51–59, DOI 10.1090/S0273-0979-1993-00397-5. MR1193540

DEPARTMENT OF MATHEMATICS, BOSTON COLLEGE, 522 MALONEY HALL, CHESTNUT HILL, MASSACHUSETTS 02467

Email address: grigsbyj@bc.edu

Proceedings of Symposia in Pure Mathematics
Volume **102**, 2019
https://doi.org/10.1090/pspum/102/01810

A remark on the geography problem in Heegaard Floer homology

Jonathan Hanselman, Çağatay Kutluhan, and Tye Lidman

Abstract. We give new obstructions to the module structures arising in Heegaard Floer homology. As a corollary, we characterize the possible modules arising as the Heegaard Floer homology of an integer homology sphere with one-dimensional reduced Floer homology. Up to absolute grading shifts, there are only two. We use this corollary to show that the chain complex depicted by Ozsváth, Stipsicz, and Szabó to argue that there is no algebraic obstruction to the existence of knots with trivial ϵ invariant and non-trivial Υ invariant cannot be realized as the knot Floer complex of a knot.

1. Introduction

Heegaard Floer homology is a collection of three-manifold invariants defined by Ozsváth and Szabó which were inspired by the Seiberg–Witten equations in gauge theory [**24**]. The most refined of these invariants is HF^+, which is a graded module over $\mathbb{F}[U]$, where U is an endomorphism of degree -2, and $\mathbb{F}$ denotes the field $\mathbb{Z}/2\mathbb{Z}$. The simplest example is the Heegaard Floer homology of the 3-sphere for which $HF^+(S^3) = \mathcal{T}^+_{(0)}$, where $\mathcal{T}^+_{(d)}$ denotes $\mathbb{F}[U, U^{-1}]/U \cdot \mathbb{F}[U]$ with $gr(1) = d$. More interesting examples include $HF^+(\pm\Sigma(2,3,5)) = \mathcal{T}^+_{(\mp 2)}$ and $HF^+(\Sigma(2,3,7)) = \mathcal{T}^+_{(0)} \oplus \mathbb{F}_{(0)}$, while $HF^+(-\Sigma(2,3,7)) = \mathcal{T}^+_{(0)} \oplus \mathbb{F}_{(-1)}$ where a positive orientation refers to the orientation induced on the boundary of a positive-definite plumbing. In fact, for every Spin^c rational homology sphere $(Y, \mathfrak{s})$, we have a (non-canonical) splitting $HF^+(Y,\mathfrak{s}) = \mathcal{T}^+_{(d)} \oplus HF_{red}(Y,\mathfrak{s})$, where $HF_{red}(Y,\mathfrak{s})$ is a finitely generated torsion module and $d \in \mathbb{Q}$. If Y is an integer homology sphere, then there is a unique Spin^c structure.

The d-invariant is an invariant of Spin^c rational homology cobordism, and has become pervasive in applications to singularity theory, knot concordance, and unknotting numbers of knots (see for instance [**1, 20, 25**]). On the other hand, if $HF_{red}(Y) = \bigoplus_{\mathfrak{s}} HF_{red}(Y,\mathfrak{s}) = 0$, then Y cannot admit a co-orientable taut foliation [**22**]. The interplay between the d-invariants and the reduced Floer homology

2010 *Mathematics Subject Classification.* Primary 57M27, 57R58.
The first author was supported in part by NSF grants DMS-1148490 and DMS-1711926.
The second author was supported in part by NSF grant DMS-1360293.
The third author was supported in part by NSF grants DMS-1148490 and DMS-1709702.

is also quite powerful; this was used to prove the Dehn surgery characterization of the unknot in S^3 [**11**] (see also [**6, 22**]).

In this note, we give new restrictions on the module structure of the Heegaard Floer homology of rational homology spheres.

THEOREM 1. *Let Y be a rational homology sphere and $\mathfrak{s}$ a self-conjugate Spin^c structure. If $HF_{red}(Y, \mathfrak{s})$ is supported only in degrees strictly greater than $d(Y, \mathfrak{s})$, then $\dim_{\mathbb{F}} HF_{red}(Y, \mathfrak{s})$ is even. The same statement holds if $HF_{red}(Y, \mathfrak{s})$ is supported in degrees strictly less than $d(Y, \mathfrak{s}) - 1$.*

Note that every three-manifold admits at least one self-conjugate Spin^c structure and the unique Spin^c structure on an integer homology sphere is tautologically self-conjugate. Theorem 1 immediately yields a characterization of the modules with one-dimensional reduced Floer homology.

COROLLARY 2. *Let Y be a rational homology sphere equipped with a self-conjugate Spin^c structure $\mathfrak{s}$. If $\dim_{\mathbb{F}} HF_{red}(Y, \mathfrak{s}) = 1$ then $HF^+(Y, \mathfrak{s}) = \mathcal{T}^+_{(d)} \oplus \mathbb{F}_{(d)}$ or $HF^+(Y, \mathfrak{s}) = \mathcal{T}^+_{(d)} \oplus \mathbb{F}_{(d-1)}$.*

By the computations of $HF^+(\pm\Sigma(2,3,7))$ stated above, we see that the two possible relatively graded modules with $\dim_{\mathbb{F}} HF_{red} = 1$ are realized.

One may also use the above restrictions to obstruct a chain complex from being realized as the knot Floer complex of a knot inside an integer homology spheres. The following corollary demonstrates this utility of Theorem 1 and Corollary 2.

COROLLARY 3. *The ϵ-trivial complex C with non-trivial Υ_C depicted in* [**21**] *cannot be realized as the knot Floer complex of a knot inside an integer homology sphere.*

The argument for the proof of Theorem 1 will be a result of the isomorphisms with monopole Floer homology (see Theorem 4) and its relationship, via the Gysin sequence of Lin [**18**], with the Pin(2)-monopole Floer homology. For the reader with a distaste for gauge theory, we point out that the arguments only use the formal properties of these theories. We briefly review these properties in Section 2, and provide a proof of Theorem 1 in Section 3. We prove Corollary 3 in Section 4.

We hope that this note encourages further work utilizing the strengths of both Heegaard Floer homology and Seiberg-Witten theory in conjunction.

2. Background

In order to prove Theorem 1, we use the Pin(2)-symmetry of solutions of the Seiberg-Witten equations to rule out certain graded module structures in Heegaard Floer homology. To be more precise, we use the Pin(2)-monopole Floer homology as defined by Lin in [**18**]. The latter is a Morse–Bott version of Kronheimer and Mrowka's monopole Floer homology (see [**12**]). In this article, we will not need the definitions of either the monopole Floer homology or the Pin(2)-monopole Floer homology. It suffices to work with their formal properties, which we review next.

First, to make connection with the Seiberg-Witten equations, we appeal to the isomorphism between Heegaard Floer homology and monopole Floer homology as is proved in [**13–17**].

THEOREM 4 (Main Theorem in [**13**]). *Let Y be a closed oriented three-manifold and $\mathfrak{s}$ be a Spin^c structure on Y. Then, $HF^+(Y, \mathfrak{s})$ (respectively, $HF^-(Y, \mathfrak{s})$ and*

$HF^\infty(Y,\mathfrak{s})$) and $\widecheck{HM}_*(Y,\mathfrak{s},c_b)$ (respectively, $\widehat{HM}_*(Y,\mathfrak{s},c_b)$ and $\overline{HM}_*(Y,\mathfrak{s},c_b)$) are isomorphic as relatively graded $\mathbb{F}[U]$-modules.

Here, $\mathring{HM}(Y,\mathfrak{s},c_b)$ denotes the monopole Floer homology of $(Y,\mathfrak{s})$ with a balanced perturbation. In the case of a torsion Spin^c structure, this is the same as the standard monopole Floer homology (see [**12**, §30]). Otherwise, one would work with completions of these modules, denoted $\mathring{HM}_\bullet(Y,\mathfrak{s},c_b)$, with respect to the variable U, making them modules over $\mathbb{F}[[U]]$. According to [**12**, Theorem 31.1.1], we have the following isomorphisms:

$$\widecheck{HM}_\bullet(Y,\mathfrak{s},c_b)\cong\widecheck{HM}_\bullet(Y,\mathfrak{s}),\ \widehat{HM}_\bullet(Y,\mathfrak{s},c_b)\cong\widehat{HM}_\bullet(Y,\mathfrak{s}),\ \overline{HM}_\bullet(Y,\mathfrak{s},c_b)\cong\overline{HM}_\bullet(Y,\mathfrak{s}).$$

In any case, completion does not affect the chain complex that defines $\widecheck{HM}$ since every element of $\widecheck{HM}$ is annihilated by some finite power of U.

The part of the isomorphism between Heegaard Floer homology and monopole Floer homology relevant to this article also follows from combined works of Taubes [**28**–**32**] and Colin, Ghiggini, and Honda [**2**–**4**]. In fact, together with works of Cristofaro-Gardiner [**5**], Huang and Ramos [**10**], and Ramos [**27**], it follows that for a rational homology sphere Y, $HF^+(Y,\mathfrak{s})$ and $\widecheck{HM}_*(Y,\mathfrak{s})$ are isomorphic as absolutely $\mathbb{Q}$-graded $\mathbb{F}[U]$-modules.

Like in Heegaard Floer and monopole Floer homologies, the Pin(2)-monopole Floer homology comes in a variety of flavors: $\widecheck{HS}_\bullet$, $\widehat{HS}_\bullet$, and $\overline{HS}_\bullet$. However, the Pin(2)-monopole Floer homology comes equipped with a more interesting module structure; this is the same structure which enables the more refined invariants leading to Manolescu's disproof of the Triangulation Conjecture in dimensions $\geqslant 5$ [**19**] (see also [**18**]). For a closed oriented three-manifold Y equipped with a self-conjugate Spin^c structure $\mathfrak{s}$, the invariants $\widecheck{HS}_\bullet(Y,\mathfrak{s}), \widehat{HS}_\bullet(Y,\mathfrak{s}), \overline{HS}_\bullet(Y,\mathfrak{s})$ take the form of $\mathbb{Q}$-graded modules over $\mathcal{R}=\mathbb{F}[[V]][Q]/Q^3$, where V and Q are endomorphisms of degrees -4 and -1, respectively. Note that we can also naturally equip any $\mathbb{F}[[U]]$-module with an $\mathcal{R}$-module structure, by having Q act by 0 and V by U^2.

The following proposition displays the clear analogy between the flavors of the monopole Floer and Pin(2)-monopole Floer homologies.

PROPOSITION 5 (Proposition 4.6 in Chapter 4 of [**18**]). *Let $\mathfrak{s}$ be a self-conjugate Spin^c structure on a rational homology sphere Y. Then, up to an absolute grading shift, $\overline{HS}_\bullet(Y,\mathfrak{s})\cong\mathbb{F}[[V,V^{-1}]][Q]/Q^3$.*

This is analogous to the fact that $\overline{HM}_\bullet(Y,\mathfrak{s})\cong\mathbb{F}[[U,U^{-1}]$ for any Spin^c rational homology sphere. Recall that this implies, using the long exact sequence relating $\widecheck{HM}_\bullet$, $\widehat{HM}_\bullet$, and $\overline{HM}_\bullet$, that in sufficiently large gradings, the dimension of $\widecheck{HM}_\bullet(Y,\mathfrak{s})$ alternates between one and zero. Likewise, Proposition 5 has the following consequence with regard to the rank of $\widecheck{HS}$ of rational homology spheres in sufficiently large gradings.

LEMMA 6. *Let $\mathfrak{s}$ be a self-conjugate Spin^c structure on a rational homology sphere Y. Then $\dim_{\mathbb{F}}\widecheck{HS}_k(Y,\mathfrak{s})\leqslant 1$ for $k\gg 0$.*

PROOF. This follows readily from the definition of the groups $\widecheck{HS}_\bullet$, $\widehat{HS}_\bullet$, $\overline{HS}_\bullet$, the long exact sequence relating them, and Proposition 5. To be more explicit, by definition $\widehat{HS}_k(Y,\mathfrak{s})$ is zero for all sufficiently large $k \gg 0$. Then the long exact sequence,

$$\cdots \xrightarrow{j*} \widehat{HS}_{k+1}(Y,\mathfrak{s}) \xrightarrow{p*} \overline{HS}_k(Y,\mathfrak{s}) \xrightarrow{i*} \widecheck{HS}_k(Y,\mathfrak{s}) \xrightarrow{j*} \widehat{HS}_k(Y,\mathfrak{s}) \xrightarrow{p*} \cdots,$$

implies that $\overline{HS}_k(Y,\mathfrak{s}) \cong \widecheck{HS}_k(Y,\mathfrak{s})$ as vector spaces over $\mathbb{F}$ for all sufficiently large $k \gg 0$. On the other hand, Proposition 5 implies that $\overline{HS}_k(Y,\mathfrak{s})$ has rank at most 1 for any $k \in \mathbb{Z}$. This gives the desired result. □

The key fact which allows us to transport information from $\widecheck{HS}_\bullet$ to $\widecheck{HM}_\bullet$ is the following Gysin sequence.

PROPOSITION 7 (Proposition 3.10 in Chapter 4 of [**18**]). *Let Y be a closed oriented three-manifold equipped with a self-conjugate Spin^c structure $\mathfrak{s}$. Then there exists a long exact sequence:*

$$\ldots \to \widecheck{HS}_{k+1}(Y,\mathfrak{s}) \overset{e_{k+1}}{\to} \widecheck{HS}_k(Y,\mathfrak{s}) \overset{\iota_k}{\to} \widecheck{HM}_k(Y,\mathfrak{s}) \overset{\pi_k}{\to} \widecheck{HS}_k(Y,\mathfrak{s}) \overset{e_k}{\to} \widecheck{HS}_{k-1}(Y,\mathfrak{s}) \to \ldots$$

Further, the maps in this long exact sequence respect the $\mathcal{R}$-module structures.

With the preceding understood, we are ready to prove Theorem 1.

3. Proof of Theorem 1

In order to prove Theorem 1, we will simply show that an $\mathbb{F}[[U]]$-module of the form $\mathcal{T}^+_{(d)} \oplus N$ where N is an r-dimensional torsion module supported in degrees greater than d with r an odd integer cannot fit into the Gysin sequence with an $\mathcal{R}$-module satisfying Lemma 6. As explained momentarily, a duality argument rules out the case where N is supported in degrees less than $d-1$. This will imply that such an $\mathbb{F}[[U]]$-module cannot occur as the monopole Floer homology of a rational homology sphere with a self-conjugate Spin^c structure. The $\mathcal{R}$-module structure will be key.

Note that the isomorphisms of Theorem 4 are only relatively graded. However, from the proof, it will be clear that the absolute grading does not play a role. We therefore assume for notational simplicity that $d = 0$ throughout.

Meanwhile, by [**12**, Proposition 28.3.4], $\widecheck{HM}_\bullet(Y,\mathfrak{s}) \cong \widehat{HM}^\bullet(-Y,\mathfrak{s})$ via an isomorphism sending elements in grading k to elements in grading $-(k+1)$. Working with coefficients in the field $\mathbb{F}$, we also have $\widehat{HM}^\bullet(-Y,\mathfrak{s}) \cong \widehat{HM}_\bullet(-Y,\mathfrak{s})$. Hence, if $\widecheck{HM}_\bullet(Y,\mathfrak{s}) \cong \mathcal{T}^+_{(0)} \oplus HM_{red}(Y,\mathfrak{s})$ where $HM_{red}(Y,\mathfrak{s})$ is r-dimensional and is supported in degrees less than -1 with r an odd integer, then $\widecheck{HM}_\bullet(-Y,\mathfrak{s}) \cong \mathcal{T}^+_{(0)} \oplus HM_{red}(-Y,\mathfrak{s})$ where $HM_{red}(-Y,\mathfrak{s})$ is r-dimensional and is supported in degrees greater than 0. Therefore, it suffices to prove the non-realizability of $\mathcal{T}^+_{(0)} \oplus HM_{red}$ where HM_{red} is r-dimensional and is supported in degrees greater than 0 with r an odd integer.

With the preceding understood, suppose that $\widecheck{HM}_\bullet(Y,\mathfrak{s}) \cong \mathcal{T}^+_{(0)} \oplus HM_{red}(Y,\mathfrak{s})$ with $HM_{red}(Y,\mathfrak{s})$ supported only in positive degree.

LEMMA 8. $\widecheck{HS}_k(Y,\mathfrak{s}) = 0$ *for* $k < 0$ *and* $\widecheck{HS}_0(Y,\mathfrak{s}) = \mathbb{F}$.

PROOF. Since $\widecheck{HM}_k(Y,\mathfrak{s}) = 0$ for $k < 0$, the Gysin sequence in Proposition 7 gives isomorphisms $\widecheck{HS}_k(Y,\mathfrak{s}) \cong \widecheck{HS}_{k-1}(Y,\mathfrak{s})$ for all $k < 0$. Thus $\widecheck{HS}_k(Y,\mathfrak{s})$ is isomorphic for all $k < 0$. But $\widecheck{HS}_k(Y,\mathfrak{s}) = 0$ for sufficiently negative k by definition, so we must have that $\widecheck{HS}_k(Y,\mathfrak{s}) = 0$ for all $k < 0$. Finally, $\widecheck{HM}_0(Y,\mathfrak{s}) = \mathbb{F}$ (since HM_{red} is 0 in degree 0), and the exactness of

$$(1) \qquad \cdots \to \widecheck{HS}_0(Y,\mathfrak{s}) \xrightarrow{i_0} \widecheck{HM}_0(Y,\mathfrak{s}) \xrightarrow{\pi_0} \widecheck{HS}_0(Y,\mathfrak{s}) \to 0$$

implies that π_0 is an isomorphism and $\widecheck{HS}_0(Y,\mathfrak{s}) = \mathbb{F}$. □

Given $k \geqslant 0$ even, let $\bar{\pi}_k$ denote the restriction of the map $\pi_k : \widecheck{HM}_k(Y,\mathfrak{s}) \to \widecheck{HS}_k(Y,\mathfrak{s})$ in the Gysin sequence in Proposition 7 to the part of the tower $\mathcal{T}^+_{(0)}$ with grading k. Although the splitting of $\widecheck{HM}(Y,\mathfrak{s}) \cong \mathcal{T}^+_{(0)} \oplus HM_{red}(Y,\mathfrak{s})$ is non-canonical, we can identify $\mathcal{T}^+_{(0)}$ canonically as a submodule of $\widecheck{HM}(Y,\mathfrak{s})$ by considering the image of U^ℓ for $\ell \gg 0$. Thus, the restriction of π_k to $\mathcal{T}^+_{(0)}$ is well-defined.

LEMMA 9. *For each $i \geqslant 0$, $\bar{\pi}_{4i}$ is nontrivial and $\bar{\pi}_{4i+2}$ is trivial.*

PROOF. Suppose that $\bar{\pi}_i$ is nontrivial for some even i. We deduce that $\bar{\pi}_{i+4}$ is also nontrivial from the fact that the Gysin sequence respects the module structures on $\widecheck{HM}_\bullet$ and $\widecheck{HS}_\bullet$. In particular $\bar{\pi}_i \circ U^2 = V \circ \bar{\pi}_{i+4}$. Since U^2 gives a nontrivial map between (the restrictions to $\mathcal{T}^+_{(0)}$ of) $\widecheck{HM}_{i+4}(Y,\mathfrak{s})$ and $\widecheck{HM}_i(Y,\mathfrak{s})$ and $\bar{\pi}_i$ is nontrivial, we must have that $\bar{\pi}_{i+4}$ is nontrivial. That $\bar{\pi}_0$ is nontrivial follows from Equation (1) (note that since HM_{red} is trivial in degree 0, we have $\bar{\pi}_0 = \pi_0$). By induction, it follows that $\bar{\pi}_{4i}$ is nontrivial for all $i \geqslant 0$.

Now suppose that $\bar{\pi}_{4i+2}$ is also nontrivial for some $i \geqslant 0$. Then by the argument above $\bar{\pi}_{4j+2}$ is nontrivial for all $j \geqslant i$, and so $\bar{\pi}_{2k}$ is nontrivial for all $k \geqslant 2i$. For sufficiently high degrees (in particular, higher than the support of HM_{red}), the Gysin sequence breaks into pieces of the form

$$0 \to \widecheck{HS}_{2k+1}(Y,\mathfrak{s}) \to \widecheck{HS}_{2k}(Y,\mathfrak{s}) \xrightarrow{0} \mathbb{F} \overset{\bar{\pi}_{2k}}{\hookrightarrow} \widecheck{HS}_{2k}(Y,\mathfrak{s}) \to \widecheck{HS}_{2k-1}(Y,\mathfrak{s}) \to 0.$$

Thus we have isomorphisms $\widecheck{HS}_{2k+1}(Y,\mathfrak{s}) \cong \widecheck{HS}_{2k}(Y,\mathfrak{s})$ and $\dim_{\mathbb{F}} \widecheck{HS}_{2k}(Y,\mathfrak{s})$ is strictly larger than $\dim_{\mathbb{F}} \widecheck{HS}_{2k-1}(Y,\mathfrak{s})$. It follows that $\dim_{\mathbb{F}} \widecheck{HS}_k(Y,\mathfrak{s})$ grows without bound for sufficiently large k, violating Lemma 6. Thus $\bar{\pi}_{4i+2}$ must be trivial for all $i \geqslant 0$. □

With these lemmas, we are ready to complete the proof of Theorem 1. For sufficiently large k, $\widecheck{HM}_{4k+1}(Y,\mathfrak{s})$ and $\widecheck{HM}_{4k+3}(Y,\mathfrak{s})$ are zero. Consider the portions of the Gysin sequence centered around $\widecheck{HM}_{4k+2}(Y,\mathfrak{s})$ and $\widecheck{HM}_{4k}(Y,\mathfrak{s})$:

$$0 \to \widecheck{HS}_{4k+3}(Y,\mathfrak{s}) \to \widecheck{HS}_{4k+2}(Y,\mathfrak{s}) \twoheadrightarrow \mathbb{F} \xrightarrow{0} \widecheck{HS}_{4k+2}(Y,\mathfrak{s}) \to \widecheck{HS}_{4k+1}(Y,\mathfrak{s}) \to 0,$$

$$0 \to \widecheck{HS}_{4k+1}(Y,\mathfrak{s}) \to \widecheck{HS}_{4k}(Y,\mathfrak{s}) \xrightarrow{0} \mathbb{F} \hookrightarrow \widecheck{HS}_{4k}(Y,\mathfrak{s}) \to \widecheck{HS}_{4k-1}(Y,\mathfrak{s}) \to 0.$$

Note that the non-triviality (respectively triviality) of the map $\bar{\pi}_{4k} : \mathbb{F} \to \widecheck{HS}_{4k}(Y,\mathfrak{s})$ (respectively $\bar{\pi}_{4k+2} : \mathbb{F} \to \widecheck{HS}_{4k+2}(Y,\mathfrak{s})$) is determined by Lemma 9. For $k \gg 0$, this gives the following isomorphisms:

$$\widecheck{HS}_{4k+1}(Y,\mathfrak{s}) \cong \widecheck{HS}_{4k}(Y,\mathfrak{s}),$$
$$\widecheck{HS}_{4k+2}(Y,\mathfrak{s}) \cong \widecheck{HS}_{4k+1}(Y,\mathfrak{s}),$$
$$\widecheck{HS}_{4k+3}(Y,\mathfrak{s}) \oplus \mathbb{F} \cong \widecheck{HS}_{4k+2}(Y,\mathfrak{s}),$$
$$\widecheck{HS}_{4k+4}(Y,\mathfrak{s}) \cong \widecheck{HS}_{4k+3}(Y,\mathfrak{s}) \oplus \mathbb{F}.$$

In particular, since $\widecheck{HS}_{4k+4}(Y,\mathfrak{s}) \cong \widecheck{HS}_{4k+3}(Y,\mathfrak{s}) \oplus \mathbb{F}$ and by Lemma 6 we have that $\dim_{\mathbb{F}} \widecheck{HS}_{4k+4}(Y,\mathfrak{s}) \leqslant 1$, it follows that $\widecheck{HS}_{4k+3}(Y,\mathfrak{s}) = 0$ for all sufficiently large k.

Fix some large k such that $4k+3$ is larger than the maximum degree in the support of $HM_{red}(Y,\mathfrak{s})$ and $\widecheck{HS}_{4k+3}(Y,\mathfrak{s}) = 0$. Consider the Gysin sequence between $\widecheck{HM}_{4k+3}(Y,\mathfrak{s}) = 0$ and $\widecheck{HM}_{-1}(Y,\mathfrak{s}) = 0$. By exactness, the sum of dimensions of each group in this sequence must be even. The groups that appear in this sequence are $\widecheck{HM}_i(Y,\mathfrak{s})$ for each $0 \leqslant i \leqslant 4k+2$, $\widecheck{HS}_{4k+3}(Y,\mathfrak{s}) = 0$, $\widecheck{HS}_{-1}(Y,\mathfrak{s}) = 0$, and two copies of $\widecheck{HS}_i(Y,\mathfrak{s})$ for each $0 \leqslant i \leqslant 4k+2$. It follows that

$$\sum_{i=0}^{4k+2} \dim_{\mathbb{F}} \widecheck{HM}_i(Y,\mathfrak{s}) = 2(k+1) + \sum_{i=0}^{4k+2} \dim_{\mathbb{F}} HM_{red,i}(Y,\mathfrak{s})$$

is even, where the first term on the right is the contribution to the tower from degrees zero to $4k+2$, and the second term is simply $\dim_{\mathbb{F}} HM_{red}(Y,\mathfrak{s})$ since by assumption $HM_{red}(Y,\mathfrak{s})$ is supported in positive degrees at most $4k+2$. Therefore, $\dim_{\mathbb{F}} HM_{red}(Y,\mathfrak{s})$ must be even. This completes the proof of Theorem 1.

4. Proof of Corollary 3

By way of background, Hom introduced a concordance invariant, denoted ϵ, that takes values in the set $\{-1,0,1\}$ [**7**]. This invariant is zero for smoothly slice knots, and it was used by Hom to prove the existence of an infinite rank summand in the subgroup of the smooth concordance group generated by topologically slice knots [**8**]. Later on, Ozsváth, Stipsicz, and Szabó introduced another concordance invariant, denoted Υ, which is a piecewise linear real-valued function on $[0,2]$ [**21**]. Hom showed in [**9**] that there exist knots K with $\Upsilon_K \equiv 0$ but $\epsilon(K) \neq 0$. Conversely, Ozsváth, Stipsicz, and Szabó presented a chain complex C that is a free $\mathbb{F}[U,U^{-1}]$-module with five generators and differential described by the following diagram. They show in [**21**, Proposition 9.4] that the chain complex C depicted in Figure 1

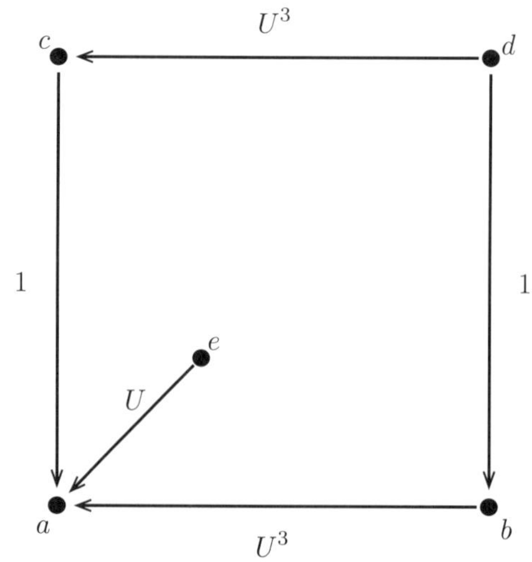

FIGURE 1. The Alexander gradings of the five generators are as follows: $\mathcal{A}(a) = \mathcal{A}(d) = \mathcal{A}(e) = 0$, $\mathcal{A}(b) = -3$, and $\mathcal{A}(c) = 3$.

is ϵ-trivial, equivalently $\epsilon(C) = 0$, but it has non-trivial Υ_C. They use this to argue that there is no algebraic obstruction to existence of knots K with $\epsilon(K) = 0$ but $\Upsilon_K \not\equiv 0$. We show that the chain complex C cannot be realized as the knot Floer complex of a knot.

PROOF. To start the proof, suppose to the contrary that such a knot $K \subset Y$ exists. Then $CFK^\infty(Y, K)$ is as in Figure 2a. By [**23**, Theorem 4.4] (also see [**26**, Theorem 2.3]), $HF^+(Y_p(K), s) \cong H_*(A_0^+)$ for large odd values of $p > 0$ where s is the unique self-conjugate Spin^c-structure on $Y_p(K)$ and A_0^+ is the chain complex consisting of elements in $CFK^\infty(Y, K)$ with $max(i, j) \geqslant 0$ (see Figure 2b.) It is easy to check that $H_*(A_0^+) = (\mathbb{F}[U, U^{-1}]/U \cdot \mathbb{F}[U]) \cdot [U^3 b] \oplus \mathbb{F} \cdot [Ue]$. Note also that the relative grading between Ue and $U^3 b$ is more than 1; in fact, $gr(Ue, U^3 b) = 4$. Therefore, $H_*(A_0^+) \cong \mathcal{T}^+_{(-4)} \oplus \mathbb{F}_{(0)}$, as relatively-graded groups, which contradicts Corollary 2. As a result, there does not exist any $K \subset Y$ with $CFK^\infty(Y, K)$ as described by the diagram in Figure 1. □

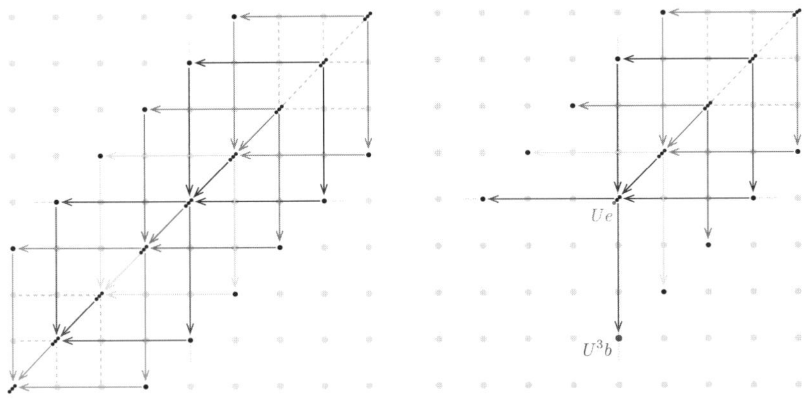

FIGURE 2. (A) The complex $CFK^\infty(Y, K)$ in the (i, j)-plane. Different U-translates of the diagram in Figure 1 are depicted in different colors. (B) The complex A_0^+.

REMARK 1. We would like to note here that if the chain complex C were modified so as to make the diagonal arrow have twice the length then the resulting chain complex would still be ϵ-trivial with non-trivial Υ invariant, but the $\mathbb{F}[U]$-torsion in the homology of the resulting A_0^+ complex would have even rank over $\mathbb{F}$. Therefore, we are not able to use the above argument to obstruct that complex from being the knot Floer complex of a knot in an integer homology sphere.

References

[1] Maciej Borodzik and Charles Livingston, *Heegaard Floer homology and rational cuspidal curves*, Forum Math. Sigma **2** (2014), e28, 23, DOI 10.1017/fms.2014.28. MR3347955

[2] Vincent Colin, Paolo Ghiggini, and Ko Honda, *Equivalence of Heegaard Floer homology and embedded contact homology via open book decompositions*, Proc. Natl. Acad. Sci. USA **108** (2011), no. 20, 8100–8105, DOI 10.1073/pnas.1018734108. MR2806645

[3] Vincent Colin, Paolo Ghiggini, and Ko Honda, *The equivalence of Heegaard Floer homology and embedded contact homology via open book decompositions II*, preprint (2012), available at arXiv:1208.1077.

[4] Vincent Colin, Paolo Ghiggini, and Ko Honda, *The equivalence of Heegaard Floer homology and embedded contact homology III: from hat to plus*, preprint (2012), available at arXiv:1208.1526.

[5] Daniel Cristofaro-Gardiner, *The absolute gradings on embedded contact homology and Seiberg-Witten Floer cohomology*, Algebr. Geom. Topol. **13** (2013), no. 4, 2239–2260, DOI 10.2140/agt.2013.13.2239. MR3073915

[6] Fyodor Gainullin, *Heegaard Floer homology and knots determined by their complements*, Algebr. Geom. Topol. **18** (2018), no. 1, 69–109, DOI 10.2140/agt.2018.18.69. MR3748239

[7] Jennifer Hom, *The knot Floer complex and the smooth concordance group*, Comment. Math. Helv. **89** (2014), no. 3, 537–570, DOI 10.4171/CMH/326. MR3260841

[8] Jennifer Hom, *An infinite-rank summand of topologically slice knots*, Geom. Topol. **19** (2015), no. 2, 1063–1110, DOI 10.2140/gt.2015.19.1063. MR3336278

[9] Jennifer Hom, *A note on the concordance invariants epsilon and upsilon*, Proc. Amer. Math. Soc. **144** (2016), no. 2, 897–902, DOI 10.1090/proc/12706. MR3430863

[10] Yang Huang and Vinicius G. B. Ramos, *An absolute grading on Heegaard Floer homology by homotopy classes of oriented 2-plane fields*, J. Symplectic Geom. **15** (2017), no. 1, 51–90, DOI 10.4310/JSG.2017.v15.n1.a2. MR3652073

[11] P. Kronheimer, T. Mrowka, P. Ozsváth, and Z. Szabó, *Monopoles and lens space surgeries*, Ann. of Math. (2) **165** (2007), no. 2, 457–546, DOI 10.4007/annals.2007.165.457. MR2299739

[12] Peter Kronheimer and Tomasz Mrowka, *Monopoles and three-manifolds*, New Mathematical Monographs, vol. 10, Cambridge University Press, Cambridge, 2007. MR2388043

[13] Çağatay Kutluhan, Yi-Jen Lee, and Clifford H. Taubes, *$HF = HM$ I : Heegaard Floer homology and Seiberg–Witten Floer homology*, preprint (2010), available at arXiv:1007.1979.

[14] Çağatay Kutluhan, Yi-Jen Lee, and Clifford H. Taubes, *$HF = HM$ II : Reeb orbits and holomorphic curves for the ech/Heegaard Floer homology correspondence*, preprint (2010), available at arXiv:1008.1595.

[15] Çağatay Kutluhan, Yi-Jen Lee, and Clifford H. Taubes, *$HF = HM$ III : Holomorphic curves and the differential for the ech/Heegaard Floer homology correspondence*, preprint (2010), available at arXiv:1010.3456.

[16] Çağatay Kutluhan, Yi-Jen Lee, and Clifford H. Taubes, *$HF = HM$ IV : The Seiberg–Witten Floer homology/ech correspondence*, preprint (2011), available at arXiv:1107.2297.

[17] Çağatay Kutluhan, Yi-Jen Lee, and Clifford H. Taubes, *$HF = HM$ V : Seiberg–Witten Floer homology and handle addition*, preprint (2012), available at arXiv:1204.0115.

[18] Francesco Lin, *A Morse-Bott approach to monopole Floer homology and the triangulation conjecture*, Mem. Amer. Math. Soc. **255** (2018), no. 1221, v+162, DOI 10.1090/memo/1221. MR3827053

[19] Ciprian Manolescu, *Pin(2)-equivariant Seiberg-Witten Floer homology and the triangulation conjecture*, J. Amer. Math. Soc. **29** (2016), no. 1, 147–176, DOI 10.1090/jams829. MR3402697

[20] Ciprian Manolescu and Brendan Owens, *A concordance invariant from the Floer homology of double branched covers*, Int. Math. Res. Not. IMRN **20** (2007), Art. ID rnm077, 21, DOI 10.1093/imrn/rnm077. MR2363303

[21] Peter S. Ozsváth, András I. Stipsicz, and Zoltán Szabó, *Concordance homomorphisms from knot Floer homology*, Adv. Math. **315** (2017), 366–426, DOI 10.1016/j.aim.2017.05.017. MR3667589

[22] Peter Ozsváth and Zoltán Szabó, *Holomorphic disks and genus bounds*, Geom. Topol. **8** (2004), 311–334, DOI 10.2140/gt.2004.8.311. MR2023281

[23] Peter Ozsváth and Zoltán Szabó, *Holomorphic disks and knot invariants*, Adv. Math. **186** (2004), no. 1, 58–116, DOI 10.1016/j.aim.2003.05.001. MR2065507

[24] Peter Ozsváth and Zoltán Szabó, *Holomorphic disks and topological invariants for closed three-manifolds*, Ann. of Math. (2) **159** (2004), no. 3, 1027–1158, DOI 10.4007/annals.2004.159.1027. MR2113019

[25] Peter Ozsváth and Zoltán Szabó, *Knots with unknotting number one and Heegaard Floer homology*, Topology **44** (2005), no. 4, 705–745, DOI 10.1016/j.top.2005.01.002. MR2136532

[26] Peter S. Ozsváth and Zoltán Szabó, *Knot Floer homology and integer surgeries*, Algebr. Geom. Topol. **8** (2008), no. 1, 101–153, DOI 10.2140/agt.2008.8.101. MR2377279

[27] Vinicius Gripp Barros Ramos, *Absolute gradings on ECH and Heegaard Floer homology*, Quantum Topol. **9** (2018), no. 2, 207–228, DOI 10.4171/QT/107. MR3812797
[28] Clifford Henry Taubes, *Embedded contact homology and Seiberg-Witten Floer cohomology I*, Geom. Topol. **14** (2010), no. 5, 2497–2581, DOI 10.2140/gt.2010.14.2497. MR2746723
[29] Clifford Henry Taubes, *Embedded contact homology and Seiberg-Witten Floer cohomology II*, Geom. Topol. **14** (2010), no. 5, 2583–2720, DOI 10.2140/gt.2010.14.2583. MR2746724
[30] Clifford Henry Taubes, *Embedded contact homology and Seiberg-Witten Floer cohomology III*, Geom. Topol. **14** (2010), no. 5, 2721–2817, DOI 10.2140/gt.2010.14.2721. MR2746725
[31] Clifford Henry Taubes, *Embedded contact homology and Seiberg-Witten Floer cohomology IV*, Geom. Topol. **14** (2010), no. 5, 2819–2960, DOI 10.2140/gt.2010.14.2819. MR2746726
[32] Clifford Henry Taubes, *Embedded contact homology and Seiberg-Witten Floer cohomology V*, Geom. Topol. **14** (2010), no. 5, 2961–3000, DOI 10.2140/gt.2010.14.2961. MR2746727

Department of Mathematics, Princeton University, Princeton, New Jersey 08544
Email address: jh66@princeton.edu

Department of Mathematics, University at Buffalo, Buffalo, New York 14260
Email address: kutluhan@buffalo.edu

Department of Mathematics, North Carolina State University, Raleigh, North Carolina 27695
Email address: tlid@math.ncsu.edu

Proceedings of Symposia in Pure Mathematics
Volume **102**, 2019
https://doi.org/10.1090/pspum/102/01804

A note on knot concordance and involutive knot Floer homology

Kristen Hendricks and Jennifer Hom

ABSTRACT. We prove that if two knots are concordant, then their involutive knot Floer complexes satisfy a certain type of stable equivalence.

1. Introduction

The knot Floer homology package of Ozsváth-Szabó [**OS04**] and Rasmussen [**Ras03**] has many applications to concordance. For example, many different smooth concordance invariants can be extracted from the filtered chain homotopy type of the knot Floer complex, such as τ [**OS03**], $\Upsilon(t)$ [**OSS17**], and ν^+ [**HW16**]. Furthermore, the second author [**Hom14**] showed that, modulo an appropriate equivalence relation, the set of knot Floer complexes forms a group, and that there is a homomorphism from the knot concordance group to this group. In [**Hom17**, Theorem 1], she showed that if two knots are concordant, then their knot Floer complexes satisfy a certain type of stable equivalence.

Recently, Manolescu and the first author [**HM17a**] used the conjugation symmetry on Heegaard Floer complexes to define involutive Heegaard Floer homology. They similarly considered the conjugation action on the knot Floer complex. Zemke [**Zem17**] showed that, under an appropriate equivalence relation, the set of knot Floer complexes together with the extra structure given by the conjugation action form a group, and that there is a homomorphism from the knot concordance group to this group. The aim of this note is to prove an involutive analog of [**Hom17**, Theorem 1]. Throughout, $\mathbb{F} = \mathbb{Z}/2\mathbb{Z}$.

THEOREM 1. *If K is slice, then $(CFK^\infty(K), \iota_K)$ is filtered chain homotopic to $(\mathbb{F}[U, U^{-1}], \mathrm{id}) \oplus (A, \iota_A)$, where A is acyclic, i.e., $H_*(A) = 0$.*

COROLLARY 2. *If K_1 and K_2 are concordant, then we have the following filtered chain homotopy equivalence*

$$(CFK^\infty(K_1), \iota_{K_1}) \oplus (A_1, \iota_{A_1}) \simeq (CFK^\infty(K_2), \iota_{K_2}) \oplus (A_2, \iota_{A_2}),$$

where A_1, A_2 are acyclic, i.e., $H_*(A_1) = H_*(A_2) = 0$.

2010 *Mathematics Subject Classification.* Primary 57M27, 57M25.

The first author was partially supported by NSF grant DMS-1663778.

The second author was partially supported by NSF grant DMS-1552285 and a Sloan Research Fellowship.

2. Background

In 2013, Manolescu introduced a $Pin(2)$-equivariant version of Seiberg-Witten Floer homology and used it to resolve the Triangulation Conjecture [**Man16**]. Since then, several authors have given applications of this invariant, especially to the homology cobordism group [**Man14, Lin15b, Sto15b, Sto15a, Sto16**]. F. Lin also gave a reformulation to monopole Floer homology, and deduced various applications [**Lin14, Lin15a, Lin16c, Lin16b, Lin16a**].

Two years later, Manolescu and the first author introduced a shadow of $Pin(2)$-equivariant Seiberg-Witten Floer homology, called *involutive Heegaard Floer homology* [**HM17b**], in Ozsváth-Szabó's Heegaard Floer homology [**OSz04**]. Involutive Heegaard Floer homology has had a number of applications, again mainly to the homology cobordism group [**HMZ17, BH16, DM17, Zem17, HL17**].

Like ordinary Heegaard Floer homology, involutive Heegaard Floer homology has a version for knots: Manolescu and the first author associate to a knot K an order-four symmetry ι_K on the knot Floer complex $CFK^\infty(K)$, and extract various concordance invariants from this data [**HM17b**]. In [**Zem17**], Zemke studies the behavior of these complexes and the associated involutions under connected sum. In this section, we recap some of his definitions and results, in preparation for proving Theorem 1 in Section 3. We begin with the following definition, which is a specialization of [**Zem17**, Definition 2.2].

DEFINITION 2.1. We say that $(C, \partial, B, \iota_C)$ is an *ι_K-complex* if

- (C, ∂) is a finitely-generated, free, $\mathbb{Z}$-graded, $(\mathbb{Z} \oplus \mathbb{Z})$-filtered, $\mathbb{F}[U, U^{-1}]$-complex with a filtered basis B;
- Given an element $x \in B$, $\partial x = \sum_{y \in B} U^{n_y} y$ for some set of integers $n_y \geq 0$;
- The action of U lowers homological grading by 2 and each filtration level by 1;
- There is an isomorphism $H_*(C, \partial) \cong \mathbb{F}[U, U^{-1}]$;
- ι_C is a skew-filtered U-equivariant endomorphism of C;
- $\iota_C^2 \simeq \mathrm{id} + \Phi_B \circ \Psi_B$, where $\Phi_B \colon C \to C$ and $\Psi_B \colon C \to C$ are formal derivatives of ∂.

(For more on the definition of the maps Φ_B and Ψ_B, see [**Zem17**, p. 7].)

Typically we omit the differential and basis from the notation. This definition is not quite Zemke's; one can think of our ι_K-complexes as the part of his ι_K-complexes concentrated in Alexander grading zero [**Zem17**, Remark 2.3]. If K is a knot, then $(CFK^\infty(K), \iota_K)$ can be made into an ι_K-complex by picking a basis for $CFK^\infty(K)$. The following notion of equivalence between two ι_K-complexes is particularly useful for studying concordance.

DEFINITION 2.2. [**Zem17**, Definition 2.4] Two ι_K-complexes (C_1, ι_{C_1}) and (C_2, ι_{C_2}) are said to be *locally equivalent* if there are filtered, grading-preserving $\mathbb{F}[U, U^{-1}]$-equivariant chain maps

$$F \colon C_1 \to C_2 \qquad G \colon C_2 \to C_1$$

such that F and G induce isomorphisms on homology and furthermore

$$F \circ \iota_{C_1} \simeq \iota_{C_2} \circ F \qquad G \circ \iota_{C_2} \simeq \iota_{C_1} \circ G$$

via skew-filtered U-equivariant chain homotopy equivalences. (If in addition $F \circ G \simeq \mathrm{id}$ and $G \circ F \simeq \mathrm{id}$ via filtered U-equivariant chain homotopy equivalences, the ι_K-complexes are said to be *homotopy equivalent.*)

One can define two possible products on the set of ι_K-complexes, denoted $\times_1$ and $\times_2$, and given by

$$(C_1, \iota_{C_1}) \times_1 (C_2, \iota_{C_2}) = (C_1 \otimes C_2, \iota_{C_1} \otimes \iota_{C_2} + (\Phi_{B_1} \otimes \Psi_{B_2}) \circ (\iota_{C_1} \otimes \iota_{C_2}))$$
$$(C_1, \iota_{C_1}) \times_2 (C_2, \iota_{C_2}) = (C_1 \otimes C_2, \iota_{C_1} \otimes \iota_{C_2} + (\Psi_{B_1} \otimes \Phi_{B_2}) \circ (\iota_{C_1} \otimes \iota_{C_2}))$$

Zemke shows that $(C_1, \iota_{C_1}) \times_1 (C_2, \iota_{C_2})$ is filtered chain-homotopy equivalent to $(C_1, \iota_{C_1}) \times_2 (C_2, \iota_{C_2})$. Following similar work in [**Sto15b**] and [**HMZ17**], Zemke further shows that either of these products makes the set of ι_K-complexes up to the relationship of local equivalence into an abelian group $\mathfrak{I}_K$ [**Zem17**, Proposition 2.6]. One then obtains a homomorphism from $\mathcal{C}$ the smooth knot concordance group to $\mathfrak{I}_K$ as follows.

PROPOSITION 2.3. [**Zem17**, Theorem 1.5] *Let $\mathcal{C}$ be the smooth knot concordance group. The map*

$$\begin{aligned} \mathcal{C} &\to \mathfrak{I}_K \\ K &\mapsto [CFK^\infty(K), \iota_K] \end{aligned}$$

is a well-defined group homomorphism.

Zemke [**Zem17**, Definition 2.5 and Proposition 2.6] shows that the inverse of $[C, \iota_C]$ is $[C^*, \iota_C^*]$, where $C^* = \mathrm{Hom}_{\mathbb{F}[U,U^{-1}]}(C, \mathbb{F}[U, U^{-1}])$, the map ι^* is the dual of ι, and B^* is a dual basis to B. The identity element of $\mathfrak{I}_K$ is $[\mathbb{F}[U, U^{-1}], \mathrm{id}]$.

3. Proof of Theorem

Since Zemke [**Zem17**, Theorem 1.5] showed that concordant knots have locally equivalent ι_K-complexes, Theorem 1 and Corollary 2 follow immediately from the following proposition and corollary.

PROPOSITION 3.1. *If (C, ι_C) is locally equivalent $(\mathbb{F}[U, U^{-1}], \mathrm{id})$, then (C, ι_C) is filtered chain homotopy equivalent to*

$$(\mathbb{F}[U, U^{-1}], \mathrm{id}) \oplus (A, \iota_A),$$

where A is some acyclic complex, i.e., $H_*(A) = 0$.

Here (A, ι_A) is not technically an ι_K complex, because $H_*(A)$ is trivial rather than $\mathbb{F}[U, U^{-1}]$. The notation should be taken to indicate that (A, ι_A) satisfies all other structural requirements in the definition of an ι_K-complex.

COROLLARY 3.2. *Two ι_K-complexes (C_1, ι_{C_1}) and (C_2, ι_{C_2}) are locally equivalent if and only if there is a filtered chain homotopy equivalence*

$$(C_1, \iota_{C_1}) \oplus (A_1, \iota_{A_1}) \simeq (C_2, \iota_{C_2}) \oplus (A_2, \iota_{A_2}),$$

for some acyclic complexes A_1 and A_2.

PROOF OF COROLLARY 3.2. The reverse direction is straightfoward: $(C_1, \iota_{C_1}) \oplus (A_1, \iota_{A_1})$ is locally equivalent to (C_1, ι_{C_1}), so if $(C_1, \iota_{C_1}) \oplus (A_1, \iota_{A_1})$ is filtered homotopy equivalent to $(C_2, \iota_{C_2}) \oplus (A_2, \iota_{A_2})$, or even merely locally equivalent, then (C_1, ι_{C_1}) is locally equivalent to (C_2, ι_{C_2}).

We now turn to the forward direction. If (C_1, ι_{C_1}) and (C_2, ι_{C_2}) are locally equivalent, then by [**Zem17**, Proposition 2.6] $(C_1, \iota_{C_1}) \times (C_2^*, \iota_{C_2}^*)$ is locally equivalent to $(\mathbb{F}[U, U^{-1}], \mathrm{id})$, where $\times$ denotes either $\times_1$ or $\times_2$. Then by Proposition 3.1, $(C_1, \iota_{C_1}) \times (C_2^*, \iota_{C_2}^*)$ is filtered chain homotopy equivalent to $(\mathbb{F}[U, U^{-1}], \mathrm{id}) \oplus (A, \iota_A)$.

Consider $(C_1, \iota_{C_1}) \times (C_2^*, \iota_{C_2}^*) \times (C_2, \iota_{C_2})$. By [**Zem17**, Theorem 1.1], the product $\times$ respects splittings and $(\mathbb{F}[U, U^{-1}], \mathrm{id})$ is the identity element with respect to $\times$. Then

$$\begin{aligned}((C_1, \iota_{C_1}) \times (C_2^*, \iota_{C_2}^*)) \times (C_2, \iota_{C_2}) &\simeq \big((\mathbb{F}[U, U^{-1}], \mathrm{id}) \oplus (A, \iota_A)\big) \times (C_2, \iota_{C_2}) \\ &\simeq (C_2, \iota_{C_2}) \oplus (A', \iota'_A),\end{aligned}$$

where $(A', \iota'_A) = (A, \iota_A) \times (C_2, \iota_{C_2})$. Similarly, for some acyclic complex D, we have

$$\begin{aligned}(C_1, \iota_{C_1}) \times ((C_2^*, \iota_{C_2}^*) \times (C_2, \iota_{C_2})) &\simeq (C_1, \iota_{C_1}) \times \big((\mathbb{F}[U, U^{-1}], \mathrm{id}) \oplus (D, \iota_D)\big) \\ &\simeq (C_1, \iota_{C_1}) \oplus (D', \iota'_D),\end{aligned}$$

where $(D', \iota'_D) = (D, \iota_D) \times (C_2, \iota_{C_2})$. This concludes the proof of the corollary. □

Using the language of local equivalence, we reprove [**Hom17**, Theorem 1].

LEMMA 3.3. *If (C, ι_C) is locally equivalent to $(\mathbb{F}[U, U^{-1}], \mathrm{id})$, then C is filtered chain homotopic to $\mathbb{F}[U, U^{-1}] \oplus A$ for some acyclic complex A.*

PROOF OF LEMMA 3.3. Since (C, ι_C) and $(\mathbb{F}[U, U^{-1}], \mathrm{id})$ are locally equivalent, there exist grading-preserving, filtered chain maps

$$F \colon \mathbb{F}[U, U^{-1}] \to C$$
$$G \colon C \to \mathbb{F}[U, U^{-1}]$$

that induce isomorphisms on homology. Since $\mathbb{F}[U, U^{-1}]$ is isomorphic to its homology, G is surjective and $G \circ F = \mathrm{id}$. Then a standard algebra argument shows that C is filtered isomorphic to $\mathbb{F}[U, U^{-1}] \oplus \ker G$. Namely, $\Phi \colon \mathbb{F}[U, U^{-1}] \oplus \ker G \to C$ given by $(x, y) \mapsto x + y$ and $\Psi \colon C \to \mathbb{F}[U, U^{-1}] \oplus \ker G$ given by $z \mapsto (F \circ G(z), z + F \circ G(z))$ provide the necessary isomorphisms, where we identify $\mathbb{F}[U, U^{-1}]$ with $\operatorname{im} F$. □

Notice that in general ι_C does not respect the splitting in the above lemma. However, we will show that ι_C is homotopic to a map that does split.

PROOF OF PROPOSITION 3.1. By Lemma 3.3, we may assume C is of the form $\mathbb{F}[U, U^{-1}] \oplus A$. Since (C, ι_C) and $(\mathbb{F}[U, U^{-1}], \mathrm{id})$ are locally equivalent, there exist grading-preserving, filtered chain maps

$$F \colon (\mathbb{F}[U, U^{-1}], \mathrm{id}) \to (C, \iota_C)$$
$$G \colon (C, \iota_C) \to (\mathbb{F}[U, U^{-1}], \mathrm{id})$$

such that $F \circ \mathrm{id} \simeq \iota_C \circ F$ via a skew-filtered chain homotopy H_F and $G \circ \iota_C \simeq \mathrm{id} \circ G$ via a skew-filtered chain homotopy H_G.

We consider the splitting given in Lemma 3.3. Let $p_i \colon \mathbb{F}[U, U^{-1}] \oplus A \to \mathbb{F}[U, U^{-1}] \oplus A$ denote projection onto the ith factor. We have that $p_1 = F \circ G$ and $p_2 = \mathrm{id} + F \circ G$.

Define

$$\iota'_C(x,y) = (x,0) + p_2 \circ \iota_C(0,y).$$

We claim that $\iota_C \simeq \iota'_C$ via the homotopy $J = H_F \circ G + F \circ H_G \circ p_2$. Indeed,

$$\begin{aligned}
\iota_C(x,y) &+ \iota'_C(x,y) \\
&= \iota_C(x,0) + \iota_C(0,y) + (x,0) + p_2 \circ \iota_C(0,y) \\
&= \iota_C(x,0) + \iota_C(0,y) + (x,0) + \iota_C(0,y) + F \circ G \circ \iota_C(0,y) \\
&= \iota_C \circ F \circ G(x,y) + F \circ \mathrm{id} \circ G(x,y) + F \circ G \circ \iota_C(0,y) + F \circ \mathrm{id} \circ G(0,y) \\
&= \partial \circ H_F \circ G(x,y) + H_F \circ \partial \circ G(x,y) + F \circ \partial \circ H_G(0,y) + F \circ H_G \circ \partial(0,y) \\
&= \partial \circ J(x,y) + J \circ \partial(x,y).
\end{aligned}$$

It is straightforward to check that ι'_C respects the splitting $\mathbb{F}[U, U^{-1}] \oplus A$ and that it is the identity on the first factor, as desired. □

Acknowledgments

We thank Tye Lidman, Charles Livingston, Ciprian Manolescu, Matt Stoffregen, and Ian Zemke for helpful conversations. We also thank the anonymous referee for helpful comments.

References

[BH16] Maciej Borodzik and Jennifer Hom, *Involutive Heegaard Floer homology and rational cuspidal curves*, Proc. Lond. Math. Soc (Published online July 26, 2018), DOI 10.1112/plms.12179.

[DM17] Irving Dai and Ciprian Manolescu, *Involutive Heegaard Floer homology and plumbed three-manifolds*, J. Inst. Math. Jussieu (Published online September 4, 2017), DOI 10.1017/S1474748017000329.

[HL17] Kristen Hendricks and Robert Lipshitz, *Involutive bordered Floer homology*, 2017, preprint, arXiv:1706.06557. To appear in Trans. Amer. Math. Soc.

[HM17a] Kristen Hendricks and Ciprian Manolescu, *Involutive Heegaard Floer homology*, Duke Math. J. **166** (2017), no. 7, 1211–1299, DOI 10.1215/00127094-3793141. MR3649355

[HM17b] Kristen Hendricks and Ciprian Manolescu, *Involutive Heegaard Floer homology*, Duke Math. J. **166** (2017), no. 7, 1211–1299, DOI 10.1215/00127094-3793141. MR3649355

[HMZ17] Kristen Hendricks, Ciprian Manolescu, and Ian Zemke, *A connected sum formula for involutive Heegaard Floer homology*, Selecta Math. (2017), 1–63.

[Hom14] Jennifer Hom, *The knot Floer complex and the smooth concordance group*, Comment. Math. Helv. **89** (2014), no. 3, 537–570, DOI 10.4171/CMH/326. MR3260841

[Hom17] Jennifer Hom, *A survey on Heegaard Floer homology and concordance*, J. Knot Theory Ramifications **26** (2017), no. 2, 1740015, 24, DOI 10.1142/S0218216517400156. MR3604497

[HW16] Jennifer Hom and Zhongtao Wu, *Four-ball genus bounds and a refinement of the Ozváth-Szabó tau invariant*, J. Symplectic Geom. **14** (2016), no. 1, 305–323, DOI 10.4310/JSG.2016.v14.n1.a12. MR3523259

[Lin14] Francesco Lin, *A Morse-Bott approach to monopole Floer homology and the triangulation conjecture*, Mem. Amer. Math. Soc. **255** (2018), no. 1221, v+162, DOI 10.1090/memo/1221. MR3827053

[Lin15a] Francesco Lin, *The surgery exact triangle in* Pin(2)*-monopole Floer homology*, Algebr. Geom. Topol. **17** (2017), no. 5, 2915–2960, DOI 10.2140/agt.2017.17.2915. MR3704248

[Lin15b] Jianfeng Lin, *Pin(2)-equivariant KO-theory and intersection forms of spin 4-manifolds*, Algebr. Geom. Topol. **15** (2015), no. 2, 863–902, DOI 10.2140/agt.2015.15.863. MR3342679

[Lin16a] Francesco Lin, *Khovanov homology in characteristic two and involutive monopole Floer homology*, 2016, preprint, arXiv:1610.08866. To appear in Math. Ann.

[Lin16b] ———, *Manolescu correction terms and knots in the three-sphere*, 2016, preprint, arXiv:1607.05220.

[Lin16c] Francesco Lin, Pin(2)-*monopole Floer homology, higher compositions and connected sums*, J. Topol. **10** (2017), no. 4, 921–969, DOI 10.1112/topo.12027. MR3705144

[Man14] Ciprian Manolescu, *On the intersection forms of spin four-manifolds with boundary*, Math. Ann. **359** (2014), no. 3-4, 695–728, DOI 10.1007/s00208-014-1010-1. MR3231012

[Man16] Ciprian Manolescu, *Pin(2)-equivariant Seiberg-Witten Floer homology and the triangulation conjecture*, J. Amer. Math. Soc. **29** (2016), no. 1, 147–176, DOI 10.1090/jams829. MR3402697

[OS03] Peter Ozsváth and Zoltán Szabó, *Knot Floer homology and the four-ball genus*, Geom. Topol. **7** (2003), 615–639, DOI 10.2140/gt.2003.7.615. MR2026543

[OS04] Peter Ozsváth and Zoltán Szabó, *Holomorphic disks and knot invariants*, Adv. Math. **186** (2004), no. 1, 58–116, DOI 10.1016/j.aim.2003.05.001. MR2065507

[OSS17] Peter S. Ozsváth, András I. Stipsicz, and Zoltán Szabó, *Concordance homomorphisms from knot Floer homology*, Adv. Math. **315** (2017), 366–426, DOI 10.1016/j.aim.2017.05.017. MR3667589

[OSz04] Peter Ozsváth and Zoltán Szabó, *Holomorphic disks and topological invariants for closed three-manifolds*, Ann. of Math. (2) **159** (2004), no. 3, 1027–1158, DOI 10.4007/annals.2004.159.1027. MR2113019

[Ras03] Jacob Andrew Rasmussen, *Floer homology and knot complements*, ProQuest LLC, Ann Arbor, MI, 2003. Thesis (Ph.D.)–Harvard University. MR2704683

[Sto15a] Matthew Stoffregen, *Manolescu invariants of connected sums*, Proc. Lond. Math. Soc. (3) **115** (2017), no. 5, 1072–1117, DOI 10.1112/plms.12060. MR3733559

[Sto15b] Matthew Henry Stoffregen, *Applications of Pin(2)-equivariant Seiberg-Witten Floer Homology*, ProQuest LLC, Ann Arbor, MI, 2017. Thesis (Ph.D.)–University of California, Los Angeles. MR3697767

[Sto16] Matthew Stoffregen, *A remark on* Pin(2)-*equivariant Floer homology*, Michigan Math. J. **66** (2017), no. 4, 867–884, DOI 10.1307/mmj/1508810818. MR3720328

[Zem17] Ian Zemke, *Connected sums and involutive link Floer homology*, 2017, preprint, arXiv:1705.01117. To appear in Proc. Lond. Math. Soc.

Department of Mathematics, Michigan State University, East Lansing, Michigan 48824

Email address: `hendricks@math.msu.edu`

School of Mathematics, Georgia Institute of Technology, Atlanta, Georgia 30332

Email address: `hom@math.gatech.edu`

Proceedings of Symposia in Pure Mathematics
Volume **102**, 2019
https://doi.org/10.1090/pspum/102/01811

A Heegaard Floer analog of algebraic torsion

Çağatay Kutluhan, Gordana Matić, Jeremy Van Horn-Morris, and Andy Wand

ABSTRACT. This article provides the origin story for *spectral order*, an invariant of contact structures on closed 3-manifolds defined by the authors. This invariant can be thought of as an assignment of a 'degree of tightness' to a contact structure. The principal motivation for the development of this invariant was work of Hutchings establishing a filtration on the differential in embedded contact homology. We explain how this filtration may be translated into the language of Heegaard Floer homology.

1. Introduction

In the '80s, Bennequin [**Ben83**] initiated and Eliashberg [**Eli87, Eli88**] modernized the study of contact structures on 3-manifolds, demonstrating first that they fall into two categories: *overtwisted* and *tight*. Overtwisted contact structures on closed 3-manifolds contain a specific embedded submanifold, an *overtwisted disk*, which enables complete 'flexiblity.' On a closed 3-manifold, each overtwisted contact structure is uniquely determined up to contact deformation by its structure as an oriented 2-plane field up to homotopy [**Eli89**], and every oriented 2-plane field is homotopic to an overtwisted contact structure [**Mar71**]. Tight contact structures, on the other hand, exhibit 'rigidity': they do not exist in every homotopy class of oriented 2-plane fields, and when they do, they are not necessarily unique [**Eli96**] [**LM96**].

A fundamental problem in 3-dimensional contact topology is the classification of tight contact structures: *which 3-manifolds carry tight contact structures, and for those that do, how do we distinguish distinct tight contact structures?* Eliashberg showed that the contact 3-manifold that naturally arises on the boundary of a compact symplectic 4-manifold is tight [**Eli90**] (see also [**Gro85**]), which yielded a wealth of constructions. However, not every tight contact structure arises this

2010 *Mathematics Subject Classification.* Primary 57M17, 57R58.

The first author was supported in part by NSF grants DMS-1128155 and DMS-1360293, and a Simons Foundation grant No. 519352.

The second author was supported in part by Simons Foundation grant No. 246461.

The third author was supported by Simons Foundation grant No. 279342 and NSF grant DMS-1612412.

The fourth author was supported in part by ERC grant GEODYCON and EPSRC EP/P004598/1.

way, [**EH02**, **LS03**].[1] Such phenomena sparked a search for invariants of contact structures that could detect tightness and fillability properties.

The foundational invariant of contact manifolds is *symplectic field theory* (SFT) of Eliashberg, Givental, and Hofer [**EGH00**]. It conjecturally detects tight contact structures on closed 3-manifolds. In [**LW11**], Latschev and Wendl investigated a property of the SFT algebra, its *algebraic torsion*, which measures the torsion of a certain element in the ground ring. They found out that if SFT algebraic torsion is finite, it obstructs the existence of symplectic fillings. Furthermore, it is non-decreasing under Legendrian surgery, making it a numerical measurement that further partitions the class of non-fillable contact structures.

Similar in construction to symplectic field theory is Hutchings's *embedded contact homology* (ECH) [**Hut10**] of closed contact 3-manifolds. In [**LW11**, Appendix], Hutchings described a method to define an analog of the SFT algebraic torsion in the framework of embedded contact homology. In the ECH analog of algebraic torsion, Hutchings uses a normalized Euler characteristic of the pseudoholomorphic curves counted in defining the ECH differential to introduce a filtration on it. He then uses this filtration to measure the 'degree of vanishing' of a canonical element in ECH.

In [**KMVWc**], the authors define the *spectral order* of a closed contact 3-manifold. It arises as an analog of the SFT algebraic torsion in the context of *Heegaard Floer homology*, a topological invariant of closed oriented 3-manifolds introduced by Ozsváth and Szabó [**OS04**], which is shown to be isomorphic to ECH [**CGH12b**, **CGH12c**, **CGH12a**]. Given a contact structure ξ on a closed 3-manifold Y, Ozsváth and Szabó specify a distinguished homology class $\hat{c}(\xi)$ (well defined up to ± 1) in the Heegaard Floer homology group $\widehat{HF}(Y)$. This class, known as the *Ozsváth–Szabó contact class*, yields an invariant of contact structures on closed 3-manifolds. The vanishing or non-vanishing of this class has been a powerful instrument to detect or obstruct tightness and fillability properties of contact structures on closed 3-manifolds. Ozsváth and Szabó proved that $\hat{c}(\xi)$ vanishes for overtwisted contact structures and is non-vanishing for symplectically fillable ones. However, it does not always detect tightness even in its strongest form (see, for example, [**Ghi06**]). Spectral order attempts to refine the information that one could extract from the Ozsváth–Szabó contact class.

The definition of spectral order is algebraic in nature. It uses a version of the usual Heegaard Floer chain complex associated to an open book decomposition with twisted coefficients resulting from the normalized Euler characteristic used by Hutchings. It takes values in the set $\mathbb{Z}_{\geq 0} \cup \{\infty\}$ and it has many desirable properties, perhaps foremost that it vanishes for overtwisted contact structures and is non-decreasing under Legendrian surgery and contact connected sum. These latter two properties are, for example, sufficient for it to give a very general obstruction to existence of Stein cobordisms between two closed contact 3-manifolds, answering [**LW11**, Question 1]. Furthermore, in [**KMVWc**], the authors give examples of families of contact 3-manifolds for which $\hat{c}(\xi) = 0$ but with non-zero spectral order and realizing infinitely many distinct positive integer values. These examples prove

[1] Indeed, there are many different flavors of 'fillings' of contact 3-manifolds by compact symplectic 4-manifolds. For ease of reading, in most cases throughout the article 'symplectically fillable' will mean 'strongly symplectically fillable.'

that spectral order detects non-trivial information about contact structures that the Ozsváth–Szabó contact class could not detect.

The goal of this article is to provide the origin story for spectral order. In this article, we describe in detail how a filtration on the ECH differential as defined by Hutchings can be ported to the setting of Heegaard Floer homology, details missing from the paper [**KMVWc**]. The main tool is the isomorphism between the latter and a variant of embedded contact homology for certain stable Hamiltonian structures, denoted *ech*, by Lee, Taubes, and the first named author (see [**KLT10a**, **KLT10b**, **KLT10c**], or [**Kut13**], for an outline).

Organization. In Section 2, we outline how an open book decomposition supporting a contact structure is used to construct an auxiliary manifold with a stable Hamiltonian structure on it, and how this is used to define *ech*. In Section 3, we describe Hutchings's filtration adapted to *ech* and define an analog of Latschev and Wendl's algebraic torsion in the context of *ech*. In Section 4, we translate the latter to Heegaard Floer homology.

2. The *ech* chain complex

In order to define a Heegaard Floer analog of Latschev and Wendl's algebraic torsion, we first define a filtered chain complex using the *ech* chain complex starting from an open book decomposition. In this regard, let M be a closed, connected, and oriented 3-manifold with a co-oriented contact structure ξ, and (S, ϕ) an abstract open book decomposition of M supporting ξ. Here, S (the *page*) is a compact oriented surface of genus g with B boundary components, and ϕ (the *monodromy*) is an orientation preserving diffeomorphism of S which restricts to identity in a neighborhood of the boundary. The manifold M is then diffeomorphic to $S \times [0,1]/\sim$ where $(p,1) \sim (\phi(p),0)$ for any $p \in S$ and $(p,t) \sim (p,t')$ for any $p \in \partial S$ and $t, t' \in [0,1]$. Let $\mathrm{G} = 2g+\mathrm{B}-1$, and fix a self-indexing Morse function on S that has a single index-2 critical point, no index-0 critical points, and attains its minimum along ∂S. Then a suitably chosen pseudo-gradient vector field for the latter defines a basis of arcs $\mathbf{a} = \{a_1, \dots, a_{\mathrm{G}}\}$ on S; i.e. a disjoint collection of properly embedded arcs which cut S into a polygon. Each arc belonging to $\mathbf{a}$ is the descending manifold of an index-1 critical point of the Morse function for the chosen pseudo-gradient vector field. As in [**HKM09b**, §3.1], this basis together with the monodromy ϕ determines a Heegaard diagram $(\Sigma, \{\beta_1, \dots, \beta_{\mathrm{G}}\}, \{\alpha_1, \dots, \alpha_{\mathrm{G}}\})$ for $-M$. To be more explicit, let $b = \{b_1, \dots, b_{\mathrm{G}}\}$ be a collection of arcs on S such that each b_i is obtained from a_i by an isotopy pushing each endpoint slightly along ∂S in the direction of the standard (counter-clockwise) orientation (see Figure 1). Then we have $\Sigma = S \times \{\frac{1}{2}\} \cup_{\partial S} -S \times \{0\}$, $\alpha_i = a_i \times \{\frac{1}{2}\} \cup a_i \times \{0\}$, and $\beta_i = b_i \times \{\frac{1}{2}\} \cup \phi(b_i) \times \{0\}$.

The pseudo-gradient vector field on S and the monodromy ϕ can be used to define a self-indexing Morse function f and a pseudo-gradient vector field v on $-M$ which yield the Heegaard diagram $(\Sigma, \{\beta_1, \dots, \beta_{\mathrm{G}}\}, \{\alpha_1, \dots, \alpha_{\mathrm{G}}\})$. Each α-curve appears as the intersection of Σ with the descending manifold of an index-1 critical point of f for v, while each β-curve appears as the intersection of Σ with the ascending manifold of an index-2 critical point of f for v. Therefore, the canonical pairing of the α- and β-curves indicated by their indexing induces a canonical pairing of the index-1 and index-2 critical points of f. With this understood, we construct a new manifold Y out of M by attaching *handles* following [**KLT10b**, §1a]. Each handle is a smooth manifold diffeomorphic to $[-1,1] \times S^2$ and it is

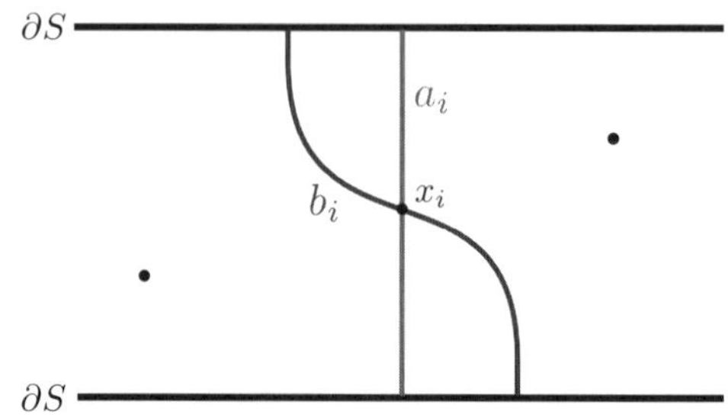

FIGURE 1. The arcs a_i and b_i on the surface S.

attached along its boundary to the complement of some regular coordinate balls centered at a distinguished pair of critical points of f. The resulting manifold Y is diffeomorphic to $M\#_{G+1}S^1 \times S^2$ via an orientation preserving diffeomorphism[2]. To be more explicit, one handle, denoted $\mathcal{H}_0$, is attached along the index-0 and index-3 critical points of f, and G additional handles are attached along pairs of index-1 and index-2 critical points of f. We denote the latter kind of handles by $\mathcal{H}_i$ for $i \in \{1, \dots, G\}$.

Next fix a basepoint $z \in \Sigma \setminus \bigcup_{i\in\{1,\dots,G\}}(\alpha_i \cup \beta_i)$ according to the convention in [**HKM09b**, §3.1], namely, fix a point in $S \times \{\frac{1}{2}\}$ outside the thin strips between a_i and b_i in Figure 1. By changing the monodromy in its isotopy class, one can make sure that the pointed Heegaard diagram $(\Sigma, \{\beta_1, \dots, \beta_G\}, \{\alpha_1, \dots, \alpha_G\}, z)$ is *strongly admissible* for any given *Spin*^{c} structure. As stated in [**KLT10b**, Lemma 1.1], this means that there exists an area form w_Σ on Σ with total area equal to 2, and the signed area of each periodic domain $\mathcal{P}$ on this Heegaard diagram is equal to the pairing $\langle c_1(\mathfrak{s}_\xi), \mathcal{H}(\mathcal{P})\rangle$, where $\mathcal{H}(\mathcal{P}) \in H_2(M;\mathbb{Z})$ is the homology class corresponding to $\mathcal{P}$. Use v and w_Σ, as in [**KLT10b**, §1a-1d], to construct a stable Hamiltonian structure on Y. The latter is a pair (a, w) where a is a smooth 1-form and w is a smooth closed 2-form such that $da = hw$ for some smooth function $h : Y \to \mathbb{R}$, and $a \wedge w$ is nowhere zero. Associated to this stable Hamiltonian structure is a vector field R satisfying $w(R, \cdot) = 0$ and $a(R) = 1$, which agrees with the pseudo-gradient vector field v on $Y \setminus \bigcup_{i\in\{0,1,\dots,G\}} \mathcal{H}_i$. Of interest are periodic orbits of the vector field R. A complete description of periodic orbits of R is provided in [**KLT10b**, §2]. In particular, there is a unique embedded periodic orbit passing through the basepoint z, denoted by γ_z. This orbit intersects every cross-sectional sphere of $\mathcal{H}_0$ exactly once. Meanwhile, for each $i \in \{1, \dots, G\}$, there are exactly two distinct embedded periodic orbits inside $\mathcal{H}_i$, denoted by γ_i^+ and γ_i^-, which are hyperbolic and homologically trivial.

Periodic orbits of the Reeb vector field and pseudo-holomorphic curves in $\mathbb{R} \times Y$ for suitably generic almost complex structure are used in [**KLT10b**, Appendix A] to define a variant of Hutchings's embedded contact homology for certain homology classes $\Gamma \in H_1(Y;\mathbb{Z})$. First, we recall some basic definitions. An *orbit set* is a finite collection of the form $\{(\gamma, m)\}$ where γ are distinct embedded periodic orbits of the Reeb vector field and m is a positive integer. An orbit set $\{(\gamma, m)\}$ with the

[2]Note that this construction is equivalent to attaching 4-dimensional 1-handles on the outgoing boundary of the product cobordism $I \times M$, and Y is the outgoing boundary of the resulting cobordism.

extra requirement that $m = 1$ when γ is hyperbolic is called an *admissible orbit set*. The homology class of an orbit set $\Theta = \{(\gamma, m)\}$ is defined by $\sum m[\gamma]$. Of interest here are admissible orbit sets with homology class $\Gamma \in H_1(Y;\mathbb{Z})$ such that $\langle PD(\Gamma), [S^2]\rangle = 0$ if $[S^2]$ is a generator of $H_2(\mathcal{H}_0;\mathbb{Z})$, $\langle PD(\Gamma), [S^2]\rangle = 1$ if $[S^2]$ is the positive generator of $H_2(\mathcal{H}_i;\mathbb{Z})$ for $i \in \{1, \dots, \text{G}\}$, and $\mathfrak{s} - \mathfrak{s}_\xi = PD(\Gamma|_M)$ where $\mathfrak{s}_\xi$ is the canonical Spin^c structure on M defined by the contact structure ξ, and $\mathfrak{s}$ is a Spin^c structure on M. Note that the homology class Γ is uniquely determined by these properties once a Spin^c structure $\mathfrak{s}$ on M is fixed. For such $\Gamma \in H_1(Y;\mathbb{Z})$, it follows from [**KLT10b**, §2] that any γ that belongs to an admissible orbit set with homology class Γ is hyperbolic. Hence, we may drop the integer m from the notation. The free $\mathbb{Z}$-module generated by admissible orbit sets with homology Γ will be denoted by $\widehat{ecc}(Y, \Gamma)$.

Having fixed an almost complex structure J on $\mathbb{R}\times Y$ satisfying the conditions in [**KLT10b**, §3a], the *ECH index* of a J-holomorphic curve in $\mathbb{R}\times Y$ asymptotic to orbit sets representing the same homology class at $\pm\infty$ is an additive quantity that depends only on the relative homology class of the curve, and it is equal to the Fredholm index for embedded curves (see [**Hut02**]). The differential $\widehat{\partial}_{ech}$ on $\widehat{ecc}(Y, \Gamma)$ is defined to be the endomorphism of $\widehat{ecc}(Y, \Gamma)$ sending a generator Θ_+ to

$$\sum \sigma(\Theta_+, \Theta_-)\Theta_-,$$

where $\sigma(\Theta_+, \Theta_-)$ is a signed count, modulo $\mathbb{R}$-translation, of ECH index-1 J-holomorphic curves asymptotic to Θ_+ at $+\infty$ and to Θ_- at $-\infty$, and disjoint from $\mathbb{R}\times\gamma_z$. Note that $\mathbb{R}\times\gamma_z$ is a J-holomorphic cylinder. Hence, by additivity and positivity of intersections of J-holomorphic curves, it follows from [**KLT10c**] (cf. [**HT07, HT09**]) that $\widehat{\partial}_{ech}\circ\widehat{\partial}_{ech} = 0$, and

$$\widehat{ech}(Y, \Gamma) := H_*(\widehat{ecc}(Y, \Gamma), \widehat{\partial}_{ech}).$$

A detailed description of J-holomorphic curves in $\mathbb{R}\times Y$ is given in [**KLT10b**, §3 and §4]. In what follows we specialize to a subcomplex $\widehat{ecc}_\circ(Y, \Gamma)$ generated by a smaller set of admissible orbit sets. To describe the latter, note that by [**KLT10b**, Proposition 2.8], there exists a 1–1 correspondence between the set of admissible orbit sets and the set $\mathcal{Z}_{\text{HF}} \times \prod_{i\in\{1,\dots,\text{G}\}}(\mathbb{Z}\times \text{O})$ where $\mathcal{Z}_{\text{HF}}$ denotes G-tuples of integral curves of the pseudo-gradient vector field v connecting the index-1 critical points of f to its index-2 critical points in such a way that each critical point appears as the limit of a unique integral curve, $\text{O} = \{0, +1, -1, \{+1, -1\}\}$, and $i \in \{1, \dots, \text{G}\}$ corresponds to the handle $\mathcal{H}_i$. Then $\widehat{ecc}_\circ(Y, \Gamma)$ is the subcomplex generated by admissible orbit sets whose restriction to each $\mathcal{H}_i$ correspond to the pair $(0, 0) \in \mathbb{Z}\times\text{O}$. The fact that $\widehat{ecc}_\circ(Y, \Gamma)$ is a subcomplex of $\widehat{ecc}(Y, \Gamma)$ follows from [**KLT10c**].

3. Hutchings's filtration in *ech*

Given orbit sets Θ_+ and Θ_- with the same homology class, Hutchings defines variants of his ECH index for relative homology classes in $H_2(Y, \Theta_+, \Theta_-)$, denoted by J_0, J_+, and J_- (see [**Hut**, §6]). Among other things, these are additive in the sense that for $Z_1 \in H_2(Y, \Theta_+, \Theta)$ and $Z_2 \in H_2(Y, \Theta, \Theta_-)$, we have

$$J_\circ(Z_1 + Z_2) = J_\circ(Z_1) + J_\circ(Z_2),$$

where $\circ \in \{0, +, -\}$ (see [**Hut**, Proposition 6.5]). If C is an embedded J-holomorphic curve in $\mathbb{R} \times Y$ with positive ends at an admissible orbit set Θ_+ and negative ends at an admissible orbit set Θ_-, then positive, respectively negative, ends of C are at distinct embedded periodic orbits of R, since an admissible orbit set consists only of hyperbolic orbits. Hence,

$$J_+(C) = -\chi(C) + |\Theta_+| - |\Theta_-|,$$

where $|\cdot|$ denotes the cardinality of an admissible orbit set. Decomposing into connected components, we can rewrite the above formula as

$$J_+(C) = \sum_{C_j \subset C} (2g_j - 2 + 2|\Theta_{j+}|), \tag{1}$$

where each C_j denotes a connected component of C, g_j denotes the genus of C_j, and each $\Theta_{j+} \subset \Theta_+$ is the collection of periodic orbits of the Reeb vector field that comprise the positive ends of C_i. Clearly, $2 \mid J_+(C)$.

LEMMA 1. *Let Θ_+ and Θ_- be generators of $\widehat{ecc}_\circ(Y, \Gamma)$ and C be an embedded J-holomorphic curve C with positive ends at Θ_+ and negative ends at Θ_-, and with no closed components. Then $J_+(C) \geq 0$.*

PROOF. Note that no subcollection of periodic orbits from an admissible orbit set Θ can be homologically trivial unless Θ contains one or more of the homologically trivial γ_i^+, $\gamma_i^- \subset \mathcal{H}_i$. Therefore, if Θ_+ and Θ_- are generators of $\widehat{ecc}_\circ(Y, \Gamma)$, then no non-empty subcollection of periodic orbits from Θ_+ or Θ_- can be homologically trivial, and every connected component of an embedded J-holomorphic curve C with positive ends at Θ_+ and negative ends at Θ_- has non-empty positive and negative ends by what is said in [**KLT10b**, §3 and §4]. It then follows from (1) that $J_+(C) \geq 0$. □

Following [**LW11**, Appendix], we decompose the differential $\widehat{\partial}_{ech}$ as

$$\widehat{\partial}_{ech} = \partial_0 + \partial_1 + \cdots + \partial_k + \cdots,$$

where ∂_k counts admissible ECH index-1 J-holomorphic curves C with $J_+(C) = 2k$ and empty intersection with $\mathbb{R} \times \gamma_z$. Since J_+ is additive under gluing of J-holomorphic curves, we can define a filtered chain complex $\widehat{ecc}_\circ(Y, \Gamma) \otimes_{\mathbb{Z}} \mathbb{Z}[t, t^{-1}]$ with differential $\sum_{k=0}^{\infty} \partial_k \otimes t^{-k}$. The spectral sequence associated to this filtered chain complex has pages $E^k(S, \phi, \mathbf{a}; J)$.

Note that there is a distinguished point x_i in $\alpha_i \cap \beta_i$ that lies in $S \times \{\frac{1}{2}\}$ for each $i = 1, \ldots, \text{G}$ as depicted in Figure 1. The set of distinguished points $\mathbf{x}_\xi = \{x_1, \ldots, x_{\text{G}}\}$ specifies a unique element of $\mathcal{Z}_{\text{HF}}$. This in turn specifies a distinguished generator Θ_ξ of $\widehat{ecc}_\circ(Y, \Gamma_\xi)$ where Γ_ξ is the unique homology class satisfying $\langle PD(\Gamma), [S^2] \rangle = 0$ if $[S^2]$ is a generator of $H_2(\mathcal{H}_0; \mathbb{Z})$, $\langle PD(\Gamma), [S^2] \rangle = 1$ if $[S^2]$ is the positive generator of $H_2(\mathcal{H}_i; \mathbb{Z})$ for $i \in \{1, \ldots, \text{G}\}$, and $\Gamma|_M = 0$. The generator Θ_ξ is a disjoint union of G embedded periodic orbits of the Reeb vector field. In fact, each of these orbits represents a positive generator of $H_1(S^1 \times S^2; \mathbb{Z})$ for a distinguished copy of $S^1 \times S^2$ in a connected sum decomposition of $Y \simeq M \#_{\text{G}+1} S^1 \times S^2$ determined by the unique collection of integral curves of the pseudo-gradient vector field v passing through the points $\{z, x_1, \ldots, x_{\text{G}}\}$. To define the *ech* analog of algebraic torsion, we keep track of the generator Θ_ξ in the spectral sequence associated to this filtered chain complex $\widehat{ecc}_\circ(Y, \Gamma_\xi) \otimes_{\mathbb{Z}} \mathbb{Z}[t, t^{-1}]$.

DEFINITION 2. Let $o(S, \phi, \mathbf{a}; J)$ be the smallest non-negative integer k such that the orbit set Θ_ξ represents the trivial class in $E^{k+1}(S, \phi, \mathbf{a}; J)$.

4. Porting Hutchings's filtration to Heegaard Floer homology

Next we translate the J_+ filtration in *ech* to Heegaard Floer homology. In order to do so, we make use of the construction of the isomorphism between *ech* and Heegaard Floer homology [**KLT10b**, **KLT10c**]:

$$\widehat{ech}(Y, \Gamma) \cong \widehat{HF}(-M, \mathfrak{s}_\xi + PD(\Gamma|_M)) \otimes \mathbb{Z}^{2^{\mathrm{G}}}.$$

The construction of this isomorphism exploits the cylindrical reformulation of Heegaard Floer homology due to Lipshitz [**Lip06**]. In Lipshitz's reformulation, the Heegaard Floer chain complex $\widehat{CF}(-M, \mathfrak{s})$ is generated by G-tuples of Reeb chords of the form $\{x_1, \ldots, x_{\mathrm{G}}\} \times [0,1] \subset \Sigma \times [0,1]$ where each x_i is a point in $\alpha_i \cap \beta_{\sigma(i)}$ for some $\sigma \in S_{\mathrm{G}}$. In comparison, $\{x_1, \ldots, x_{\mathrm{G}}\} \in Sym^{\mathrm{G}}(\Sigma)$ is a generator of $\widehat{CF}(-M, \mathfrak{s})$ in the original formulation of Heegaard Floer homology by Ozsváth and Szabó [**OS04**]. Each such G-tuple of Reeb chords corresponds to a unique element of $\mathcal{Z}_{\mathrm{HF}}$. As such, the set of such G-tuples of Reeb chords is in 1–1 correspondence with the set $\mathcal{Z}_{\mathrm{HF}}$. Meanwhile, given an almost complex structure J_{HF} on $\Sigma \times [0,1] \times \mathbb{R}$ satisfying conditions (**J1**)–(**J5**) in [**Lip06**, §1], there exists an almost complex structure J on $\mathbb{R} \times Y$ satisfying the conditions in [**KLT10b**, §3a]. In the present context, where we restrict ourselves to the subcomplex $\widehat{ecc}_\circ(Y, \Gamma)$, there is a 1–1 correspondence between ECH index-1 J-holomorphic curves in $\mathbb{R} \times Y$ with ends at generators of this subcomplex and Fredholm index-1 J_{HF}-holomorphic curves in $\Sigma \times [0,1] \times \mathbb{R}$ that satisfy conditions (**M0**)–(**M6**) in [**Lip06**, §1]. Therefore, $(\widehat{ecc}_\circ(Y, \Gamma), \widehat{\partial}_{ech})$ and $(\widehat{CF}(-M, \mathfrak{s}), \widehat{\partial}_{\mathrm{HF}})$ are canonically isomorphic as chain complexes via the aforementioned 1–1 correspondence between generators (see [**KLT10c**] for details).

Now let C_L be a Fredholm index-1 J_{HF}-holomorphic curve in $\Sigma \times [0,1] \times \mathbb{R}$ satisfying conditions (**M0**)–(**M6**) in [**Lip06**, §1], and $C \subset \mathbb{R} \times Y$ be the corresponding ECH index-1 J-holomorphic curve with positive ends at an admissible orbit set Θ_+ and negative ends at an admissible orbit set Θ_-, both of which are generators of $\widehat{ecc}_\circ(Y, \Gamma)$. Due to the aforementioned 1–1 correspondence, we can write $J_+(C_L)$ for $J_+(C)$. Topologically, C is obtained from C_L by adding 2-dimensional 1-handles in such a way that for each $i = 1, \ldots, \mathrm{G}$ a 1-handle is attached along a point in α_i and a point in β_i (See Figure 2). Therefore, $-\chi(C) = -\chi(C_L) + \mathrm{G}$, and

$$J_+(C_L) = -\chi(C_L) + \mathrm{G} + |\Theta_+| - |\Theta_-|. \tag{2}$$

Lipshitz proves that $\chi(C_L)$ is determined by the relative homology class $[C_L] \in \widehat{\pi}_2(\mathbf{x}_+, \mathbf{x}_-)$ (see [**Lip06**, Proposition 4.2], cf. [**Lip14**, Proposition 4.2′]); more specifically,

$$\chi(C_L) = \mathrm{G} - n_{\mathbf{x}_+}(\mathcal{D}([C_L])) - n_{\mathbf{x}_-}(\mathcal{D}([C_L])) + e(\mathcal{D}([C_L])). \tag{3}$$

Here, $\mathcal{D}([C_L])$ denotes the domain in the Heegaard diagram associated to C_L, $\mathbf{x}_+$ denotes the Heegaard Floer generator that corresponds to Θ_+, $\mathbf{x}_-$ denotes the Heegaard Floer generator that corresponds to Θ_-, $n_p(\mathcal{D}([C_L]))$ denotes the *point measure*, namely, the average of the coefficients of $\mathcal{D}([C_L])$ for the four regions with corners at $p \in \alpha_i \cap \beta_j$, and $e(\mathcal{D}([C_L]))$ is the *Euler measure* of $\mathcal{D}([C_L])$. Meanwhile, keeping the labeling of the α and β curves in mind, each Heegaard Floer generator

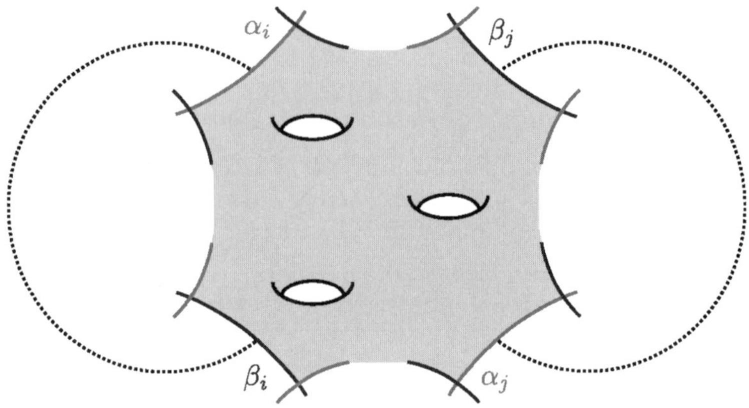

FIGURE 2. An embedded J_{HF}-holomorphic curve C_L in $\Sigma \times [0,1] \times \mathbb{R}$ satisfying conditions (**M0**)–(**M6**) in [**Lip06**, §1] shaded, and cores of some 2-dimensional 1-handles attached to C_L are indicated with dotted lines.

yields an element of the symmetric group, and $|\Theta_\pm|$ is equal to the number of cycles in the element of that symmetric group associated to $\mathbf{x}_\pm$, respectively. We will denote the latter quantity also by $|\cdot|$. With the preceding understood, combine (2) and (3) to obtain the following formula:

$$J_+([C_L]) = n_{\mathbf{x}_+}(\mathcal{D}([C_L])) + n_{\mathbf{x}_-}(\mathcal{D}([C_L])) - e(\mathcal{D}([C_L])) + |\mathbf{x}_+| - |\mathbf{x}_-|. \tag{4}$$

Note that the right hand side of (4) depends only on $\mathcal{D}([C_L])$, and in particular, since a class in $\widehat{\pi}_2(\mathbf{x}_+, \mathbf{x}_-)$ is specified by its domain, $J_+([C_L])$ depends only on the relative homology class $[C_L]$. In fact, we may define J_+ for any domain $\mathcal{D}$ in the Heegaard diagram by the right hand side of (4), which by [**Lip06**, Corollary 4.10] could also be written as

$$J_+(\mathcal{D}) := \mu(\mathcal{D}) - 2e(\mathcal{D}) + |\mathbf{x}_+| - |\mathbf{x}_-|, \tag{5}$$

where $\mu(\mathcal{D})$ is the *Maslov index* of $\mathcal{D}$. If there exists an embedded J_{HF}-holomorphic curve C_L representing $\mathcal{D}$, then the Maslov index of $\mathcal{D}$ agrees with the Fredholm index of C_L ([**Lip14**, Proposition 4.8′]) .

PROPOSITION 3. *The assignment $J_+(\mathcal{D})$ to domains in the Heegaard diagram is additive under concatenation of domains, and it is non-negative on Maslov index-1 domains represented by embedded pseudo-holomorphic curves.*

PROOF. The additivity claim follows from additivity of the Euler measure and the Maslov index under concatenation of domains. As for the non-negativity claim, this follows from Lemma 1 and by the correspondence between moduli spaces of pseudo-holomorphic curves to define the differentials for Heegaard Floer and *ech* chain complexes in [**KLT10c**]. □

For Maslov index-1 domains, it is easier to use the formula

$$J_+(\mathcal{D}) = 2[n_{\mathbf{x}_+}(\mathcal{D}) + n_{\mathbf{x}_-}(\mathcal{D})] - 1 + |\mathbf{x}_+| - |\mathbf{x}_-|. \tag{6}$$

With the preceding understood, we give an obstruction to overtwistedness:

THEOREM 4. *Let* ξ_{OT} *be an overtwisted contact structure on a closed* 3*-manifold* M. *Then there exist an open book decomposition* (S, ϕ) *supporting* ξ, *a basis of arcs* $\mathbf{a}$ *on* S, *and a generic split complex structure* J_{HF} *in the sense of* [**Lip06**] *such that* $o(S, \phi, \mathbf{a}; J_{HF}) = 0$.

PROOF. Note that an overtwisted contact structure is supported by an open book decomposition (S, ϕ) where the monodromy ϕ is not right-veering [**HKM07**, Theorem 1.1]. One can find a basis of arcs $\mathbf{a}$ on S so that in the corresponding Heegaard diagram $\widehat{\partial}_{\mathrm{HF}}\mathbf{y} = \mathbf{x}_{\xi_{\mathrm{OT}}}$ where $\mathbf{y} = \{y_1, x_2, \ldots, x_{\mathrm{G}}\}$ and there is exactly one Maslov index-1 holomorphic domain $\mathcal{D}$, a bigon, that contributes to the differential [**HKM09b**, Lemma 3.2] as defined by a split complex structure J_{HF} on $\Sigma \times [0,1] \times \mathbb{R}$. Therefore, $n_{\mathbf{y}}(\mathcal{D}) = \frac{1}{4}$, $n_{\mathbf{x}_{\xi_{\mathrm{OT}}}}(\mathcal{D}) = \frac{1}{4}$, $|\mathbf{y}| = \mathrm{G}$, and $|\mathbf{x}_{\xi_{\mathrm{OT}}}| = \mathrm{G}$. Applying (6), we find $J_+(\mathcal{D}) = 0$. As a result, $o(S, \phi, \mathbf{a}; J_{HF}) = 0$. □

We see this as the beginning of investigations of a new family of computable invariants of contact structures and related objects via Heegaard Floer homology. In the sequel [**KMVWc**], we adopt Lipshitz's cylindrical reformulation of Heegaard Floer homology, and use the above definition and Proposition 3 along with equation (6) to define the spectral order of a closed contact 3-manifold, then investigate its many properties. Our main result there can be stated as follows:

THEOREM 5. *There exists an invariant of contact structures on closed oriented* 3*-manifolds, denoted* $\mathbf{o}$, *taking values in* $\mathbb{Z}_{\geq 0} \cup \{\infty\}$ *with the following properties:*

- $\mathbf{o}(M, \xi) = 0$ *if* (M, ξ) *is overtwisted.*
- $\mathbf{o}(M, \xi)$ *is non-decreasing under Legendrian surgery.*
- $\mathbf{o}(M, \xi) = \infty$ *if* (M, ξ) *is Stein fillable.*

In later sequels, we will investigate spectral order's ability to detect tightness [**KMVWa**], and introduce a version of spectral order for compact contact manifolds with convex boundary [**KMVWb**] applying Definition 2 to partial open book decompositions with protected boundary. There will be a number of interesting applications of this version of spectral order. For example, in [**JK18**], Juhász and Kang independently observe that the definition of spectral order can be generalized to partial open book decompositions and state that, just like the maps on sutured Floer homology defined in [**HKM09a**], the *sutured* version of $\mathbf{o}$ behaves well under inclusion of contact manifolds—$\mathbf{o}$ of a contact submanifold is an upper bound for $\mathbf{o}$ of the ambient contact manifold. Applying this to the case of a Giroux torsion domain, they give an upper bound of 2 for the spectral order of a contact 3-manifold with positive Giroux torsion.

REMARK. During the course of this project we learned that John Baldwin and David Shea Vela-Vick have independently been working on an invariant of contact 3-manifolds in the framework of Heegaard Floer homology similar in spirit to what is outlined in this article. The definition of their invariant can be found in [**BSVV18**, §1.3].

Acknowledgments

This project was initiated at the "*Interactions between contact symplectic topology and gauge theory in dimensions 3 and 4*" workshop at Banff International Research Station (BIRS) in 2011. We would like to thank BIRS for their hospitality and for creating a wonderful environment for collaboration. We also thank Michael Hutchings for helpful correspondence.

References

[Ben83] Daniel Bennequin, *Entrelacements et équations de Pfaff* (French), Third Schnepfenried geometry conference, Vol. 1 (Schnepfenried, 1982), Astérisque, vol. 107, Soc. Math. France, Paris, 1983, pp. 87–161. MR753131

[BSVV18] John Baldwin and David Vela-Vick, *A note on the knot Floer homology of fibered knots*, Algebr. Geom. Topol. **18** (2018), no. 6, 3669–3690, DOI 10.2140/agt.2018.18.3669. MR3868231

[CGH12a] Vincent Colin, Paolo Ghiggini, and Ko Honda, *The equivalence of Heegaard Floer homology and embedded contact homology III: from hat to plus*, preprint (2012), available at arXiv:1208.1526.

[CGH12b] ———, *The equivalence of Heegaard Floer homology and embedded contact homology via open book decompositions I*, preprint (2012), available at arXiv:1208.1074.

[CGH12c] ———, *The equivalence of Heegaard Floer homology and embedded contact homology via open book decompositions II*, preprint (2012), available at arXiv:1208.1077.

[EGH00] Y. Eliashberg, A. Givental, and H. Hofer, *Introduction to symplectic field theory*, Geom. Funct. Anal. **Special Volume** (2000), 560–673, DOI 10.1007/978-3-0346-0425-3_4. GAFA 2000 (Tel Aviv, 1999). MR1826267

[EH02] John B. Etnyre and Ko Honda, *Tight contact structures with no symplectic fillings*, Invent. Math. **148** (2002), no. 3, 609–626, DOI 10.1007/s002220100204. MR1908061

[Eli87] Ya. M. Eliashberg, *A theorem on the structure of wave fronts and its application in symplectic topology* (Russian), Funktsional. Anal. i Prilozhen. **21** (1987), no. 3, 65–72, 96. MR911776

[Eli88] Ya. M. Eliashberg, *The structure of 1-dimensional wave fronts, nonstandard Legendrian loops and Bennequin's theorem*, Topology and geometry—Rohlin Seminar, Lecture Notes in Math., vol. 1346, Springer, Berlin, 1988, pp. 7–12, DOI 10.1007/BFb0082768. MR970069

[Eli89] Y. Eliashberg, *Classification of overtwisted contact structures on 3-manifolds*, Invent. Math. **98** (1989), no. 3, 623–637, DOI 10.1007/BF01393840. MR1022310

[Eli90] Yakov Eliashberg, *Filling by holomorphic discs and its applications*, Geometry of low-dimensional manifolds, 2 (Durham, 1989), London Math. Soc. Lecture Note Ser., vol. 151, Cambridge Univ. Press, Cambridge, 1990, pp. 45–67. MR1171908

[Eli96] Yasha Eliashberg, *Unique holomorphically fillable contact structure on the 3-torus*, Internat. Math. Res. Notices **2** (1996), 77–82, DOI 10.1155/S1073792896000074. MR1383953

[Ghi06] Paolo Ghiggini, *Infinitely many universally tight contact manifolds with trivial Ozsváth-Szabó contact invariants*, Geom. Topol. **10** (2006), 335–357, DOI 10.2140/gt.2006.10.335. MR2224460

[Gro85] M. Gromov, *Pseudo holomorphic curves in symplectic manifolds*, Invent. Math. **82** (1985), no. 2, 307–347, DOI 10.1007/BF01388806. MR809718

[HKM07] Ko Honda, William H. Kazez, and Gordana Matić, *Right-veering diffeomorphisms of compact surfaces with boundary*, Invent. Math. **169** (2007), no. 2, 427–449, DOI 10.1007/s00222-007-0051-4. MR2318562

[HKM09a] Ko Honda, William H. Kazez, and Gordana Matić, *The contact invariant in sutured Floer homology*, Invent. Math. **176** (2009), no. 3, 637–676, DOI 10.1007/s00222-008-0173-3. MR2501299

[HKM09b] Ko Honda, William H. Kazez, and Gordana Matić, *On the contact class in Heegaard Floer homology*, J. Differential Geom. **83** (2009), no. 2, 289–311. MR2577470

[HT07] Michael Hutchings and Clifford Henry Taubes, *Gluing pseudoholomorphic curves along branched covered cylinders. I*, J. Symplectic Geom. **5** (2007), no. 1, 43–137. MR2371184

[HT09] Michael Hutchings and Clifford Henry Taubes, *Gluing pseudoholomorphic curves along branched covered cylinders. II*, J. Symplectic Geom. **7** (2009), no. 1, 29–133. MR2491716

[Hut] Michael Hutchings, *The embedded contact homology index revisited*, New perspectives and challenges in symplectic field theory, CRM Proc. Lecture Notes, vol. 49, Amer. Math. Soc., Providence, RI, 2009, pp. 263–297. MR2555941

[Hut02] Michael Hutchings, *An index inequality for embedded pseudoholomorphic curves in symplectizations*, J. Eur. Math. Soc. (JEMS) **4** (2002), no. 4, 313–361, DOI 10.1007/s100970100041. MR1941088

[Hut10] Michael Hutchings, *Embedded contact homology and its applications*, Proceedings of the International Congress of Mathematicians. Volume II, Hindustan Book Agency, New Delhi, 2010, pp. 1022–1041. MR2827830

[JK18] András Juhász and Sungkyung Kang, *Spectral order for contact manifolds with convex boundary*, Algebr. Geom. Topol. **18** (2018), no. 6, 3315–3338, DOI 10.2140/agt.2018.18.3315. MR3868222

[KLT10a] Cagatay Kutluhan, *Floer homology and symplectic forms on S1 x M3*, ProQuest LLC, Ann Arbor, MI, 2009. Thesis (Ph.D.)–University of Michigan. MR2713879

[KLT10b] ______, *$HF = HM$ II : Reeb orbits and holomorphic curves for the ech/Heegaard Floer homology correspondence*, preprint (2010), available at arXiv:1008.1595.

[KLT10c] ______, *$HF = HM$ III : Holomorphic curves and the differential for the ech/Heegaard Floer homology correspondence*, preprint (2010), available at arXiv:1010.3456.

[KMVWa] Çağatay Kutluhan, Gordana Matić, Jeremy Van Horn-Morris, and Andy Wand, *Detecting tightness via spectral order*, in preparation.

[KMVWb] ______, *Spectral order for sutured contact manifolds*, in preparation.

[KMVWc] ______, *Filtering the Heegaard Floer contact invariant*, preprint (2016), available at arXiv:1603.02673.

[Kut13] Çağatay Kutluhan, *Lectures on the equivalence of Heegaard Floer and Seiberg-Witten Floer homologies*, Proceedings of the Gökova Geometry-Topology Conference 2012, Int. Press, Somerville, MA, 2013, pp. 1–42. MR3203356

[Lip06] Robert Lipshitz, *A cylindrical reformulation of Heegaard Floer homology*, Geom. Topol. **10** (2006), 955–1096, DOI 10.2140/gt.2006.10.955. Paging previously given as 955–1097. MR2240908

[Lip14] Robert Lipshitz, *Correction to the article: A cylindrical reformulation of Heegaard Floer homology [MR2240908]*, Geom. Topol. **18** (2014), no. 1, 17–30, DOI 10.2140/gt.2014.18.17. MR3158771

[LM96] Paolo Lisca and Gordana Matić, *On homotopic, non-isomorphic tight contact structures on* 3*-manifolds*, Turkish J. Math. **20** (1996), no. 1, 37–45. MR1392661

[LS03] Paolo Lisca and András I. Stipsicz, *An infinite family of tight, not semi-fillable contact three-manifolds*, Geom. Topol. **7** (2003), 1055–1073, DOI 10.2140/gt.2003.7.1055. MR2026538

[LW11] Janko Latschev and Chris Wendl, *Algebraic torsion in contact manifolds*, Geom. Funct. Anal. **21** (2011), no. 5, 1144–1195, DOI 10.1007/s00039-011-0138-3. With an appendix by Michael Hutchings. MR2846386

[Mar71] J. Martinet, *Formes de contact sur les variétés de dimension* 3 (French), Proceedings of Liverpool Singularities Symposium, II (1969/1970), Springer, Berlin, 1971, pp. 142–163. Lecture Notes in Math., Vol. 209. MR0350771

[OS04] Peter Ozsváth and Zoltán Szabó, *Holomorphic disks and topological invariants for closed three-manifolds*, Ann. of Math. (2) **159** (2004), no. 3, 1027–1158, DOI 10.4007/annals.2004.159.1027. MR2113019

Department of Mathematics, University at Buffalo, Buffalo, New York 14260
Email address: kutluhan@buffalo.edu

Department of Mathematics, University of Georgia, Athens, Georgia, 30602
Email address: gordana@math.uga.edu

Department of Mathematical Sciences, University of Arkansas, Fayetteville, Arkansas 72701
Email address: jvhm@uark.edu

School of Mathematics and Statistics, University of Glasgow, Glasgow G12 8QW, United Kingdom
Email address: andy.wand@glasgow.ac.uk

Proceedings of Symposia in Pure Mathematics
Volume **102**, 2019
https://doi.org/10.1090/pspum/102/01815

Realization problems for diffeomorphism groups

Kathryn Mann and Bena Tshishiku

Abstract. We discuss recent results and open questions on the broad theme of (Nielsen) realization problems. Beyond realizing subgroups of mapping class groups, there are many other natural instances where one can ask if a surjection from a group of diffeomorphisms of a manifold to another group admits a section over particular subgroups. This survey includes many open problems, and some short proofs of new results that are illustrative of key techniques; drawing attention to parallels between problems arising in different areas.

1. Introduction

One way to understand the algebraic structure of a large or complicated group is to surject it to a simpler group, then measure what is lost. In other words, given G, we take an exact sequence $1 \to K \to G \to H \to 1$ and study how much H differs from G. This difference can be measured through the obstructions to a (group-theoretic) *section* $\phi : H \to G$, as well as the obstructions to sections over subgroups of H.

In the case where G is the group of homeomorphisms or diffeomorphisms of a manifold M, this question has a long history and important interpretations, both from the topological point of view of flat bundles, and the dynamical point of view through group actions of M. This article is intended as an invitation to section problems for diffeomorphism groups; illustrating applicable techniques and listing open problems. To introduce this circle of ideas, we begin with the classical case of surface bundles and diffeomorphism groups of surfaces.

The flatness problem for surface bundles. Let Σ be a smooth, oriented closed surface and let $E \to B$ be a fiber bundle with fiber Σ, base a manifold B, and structure group the group $\mathrm{Diff}^r(\Sigma)$ of orientation-preserving C^r diffeomorphisms of Σ, for some fixed $r \geq 0$. Such a bundle is said to be *flat* or *foliated* of class C^r if it admits a C^r foliation transverse to the fibers whose leaves project to the base as covering spaces. Equivalently, $E \to B$ is flat if its structure group can be taken to be $\mathrm{Diff}^r(\Sigma)$ with the discrete topology. In a flat bundle, parallel transport along the leaves of the foliation defines a holonomy representation $\phi : \pi_1(B) \to \mathrm{Diff}^r(\Sigma)$.

It is a basic problem to determine when a surface bundle is flat; and even in the case where the base is a surface, it is a open question whether every surface bundle

2010 *Mathematics Subject Classification.* Primary 57S25; Secondary 55R40, 57M60.
The first author was partially supported by NSF grant DMS 1606254.
The second author was partially supported by NSF grant DMS 1502794.

admits a flat structure. A foundational result of Earle–Eells [**EE67**] states that, when $\chi(\Sigma) < 0$, a Σ-bundle $E \to B$ is determined up to bundle isomorphism by its *monodromy representation* $\rho : \pi_1(B) \to \mathrm{Mod}(\Sigma)$. Here $\mathrm{Mod}(\Sigma) := \pi_0(\mathrm{Diff}^r(\Sigma))$ denotes the *mapping class group* of Σ (which is independent of r). Thus, $E \to B$ admits a C^r flat structure if and only if its monodromy lifts to $\mathrm{Diff}^r(\Sigma)$, as in the diagram below. In this case, the lift agrees (up to conjugacy) with the holonomy ϕ mentioned above.

$$\begin{array}{ccc} & \tilde{\rho} \nearrow & \mathrm{Diff}^r(\Sigma) \\ & & \downarrow \pi \\ \pi_1(B) & \xrightarrow[\rho]{} & \mathrm{Mod}(\Sigma) \end{array}$$

The flat bundle question motivates the following problem. Let $\Sigma_{g,m}^b$ denote a compact surface of genus g with b boundary components and m marked points. Define $\mathrm{Diff}^r(\Sigma_{g,m}^b)$ as the group of orientation-preserving diffeomorphisms of Σ that preserve the marked points as a set, and define $\mathrm{Diff}^r_\partial(\Sigma_{g,m}^b) < \mathrm{Diff}^r(\Sigma_{g,m}^b)$ as the subgroup of diffeomorphisms that restrict to the identity on the boundary $\partial\Sigma$. The mapping class group $\mathrm{Mod}_{g,m}^b := \pi_0 \mathrm{Diff}^r_\partial(\Sigma_{g,m}^b)$ can be viewed as the group of diffeomorphism of $\Sigma_{g,m}^b$ modulo isotopies that fix the boundary and the punctures[1]. It is a classical fact, stemming from work of Munkres, Smale and Whitehead in the 50s, that $\pi_0(\mathrm{Diff}^r(\Sigma_{g,m}^b))$ is independent of the regularity r. See discussion in [**FM12**, §1.3.2].

PROBLEM 1.1 (Generalized Nielsen realization). Let Γ be a finitely-generated group, $0 \leq r \leq \infty$, and let $\rho : \Gamma \to \mathrm{Mod}_{g,m}^b$ be a homomorphism. Does there exist a homomorphism $\tilde{\rho} : \Gamma \to \mathrm{Diff}^r_\partial(\Sigma_{g,m}^b)$ so that $\pi \circ \tilde{\rho} = \rho$?

If the answer is yes, we say ρ is *realized by C^r diffeomorphisms*, or by *homeomorphisms* in the case $r = 0$. (We will use both $\mathrm{Diff}^0(M)$ and $\mathrm{Homeo}(M)$ as notation for the homeomorphism group of a manifold M.) The map $\tilde{\rho}$ is called a *section* of $\mathrm{Diff}^r(\Sigma_{g,m}^b) \to \mathrm{Mod}(\Sigma_{g,m}^b)$ over ρ, so we will also often refer to this and related questions as *section problems*. Earle–Eells' result generalizes to surfaces with boundary and marked points [**ES70**, **Hat11**], and from the bundle perspective, marked points correspond to preferred sections.

As is well known, Problem 1.1 was answered positively for finite, cyclic groups by Nielsen [**Nie43**]. Nielsen's original motivation was different from ours – the question he answered (positively) was: *if f is a diffeomorphism of Σ such that f^n is isotopic to the identity, is f isotopic to some $g \in \mathrm{Diff}^\infty(\Sigma)$ with $g^n = \mathrm{id}$?* Problem 1.1 for finite groups $\Gamma \subset \mathrm{Mod}_g$ eventually became known as Nielsen's problem. Fenchel [**Fen48**] gave a positive answer for finite *solvable* groups, and the general result for finite groups is due to Kerckhoff [**Ker83**], with alternate proofs by Tromba, Gabai and Wolpert [**Tro96**, **Gab91**, **Wol87**] appearing later. Going forward in this paper we will focus on realization problems for infinite groups.

The case $\Gamma = \mathrm{Mod}_g$ for $\tilde{\rho}$ with image in $\mathrm{Homeo}(\Sigma_g)$ was posed by Thurston in Kirby's problem list [**Kir97**, Prob. 2.6]. This follows an earlier theorem of Morita [**Mor87**], who showed that (for $g \geq 18$), there is no realization of Mod_g (or any finite index subgroup of Mod_g) in $\mathrm{Diff}^2(\Sigma_g)$. Later, Franks–Handel [**FH09**] showed that there is no realization of Mod_g in $\mathrm{Diff}^1(\Sigma_g)$; this does not generalize to finite index

[1] following standard convention, we denote $\mathrm{Mod}_{g,0}^0$ by Mod_g

subgroups, but covers several other cases of large infinite subgroups. We sketch both Morita's and the Franks–Handel proof later on. In [**Mar07**, **MS08**] Markovic and Markovic–Saric answered Thurston's question, showing that there is no lift of Mod_g to $\text{Homeo}(\Sigma_g)$ provided $g \geq 2$. However, their proof uses torsion elements in an essential way, so does not generalize to finite index subgroups or to many of the cases covered by Franks–Handel. Indeed, for "most" infinite subgroups $\Gamma < \text{Mod}_{g,m}^b$ Problem 1.1 remains completely open (see discussion in [**Far06a**, §6.3]).

One source of difficulty in Problem 1.1 is our poor understanding of finitely-generated subgroups of diffeomorphism groups. For instance, the following problem is open.

PROBLEM 1.2. Give an example of a finitely-generated, torsion free group Γ, and a surface S, such that Γ is not isomorphic to a subgroup of $\text{Homeo}(S)$.

Of course, it suffices to give an example for $\text{Homeo}_\partial(\mathbb{D}^2)$ as this embeds in $\text{Homeo}(S)$ – and in fact in the identity component $\text{Homeo}_0(S)$ – for any other surface. Indeed, the difficulty in the problem is understanding the algebraic structure of $\text{Homeo}_0(S)$; replacing $\text{Homeo}(S)$ by $\text{Mod}(S)$ renders it a reasonable exercise. Problem 1.2 is also completely open for homeomorphism groups of manifolds of higher dimension. In general the regularity of diffeomorphisms can make a great difference, and Problem 1.2 has been recently solved for C^2 diffeomorphisms with the solution to the Zimmer conjecture [**BFH16**], with examples given by higher rank lattices. However, we are far from a complete understanding of the algebraic structure of groups that can and cannot act.

Nielsen realization problems in higher dimensions. Problem 1.1 can be posed for manifolds M with $\dim M \geq 3$. Does $\text{Diff}^r(M) \to \pi_0\big(\text{Diff}^r(M)\big)$ admit a section? Which finitely-generated subgroups can be realized? Unlike in the surface case, the group $\pi_0(\text{Diff}^r(M))$ is not always finitely generated and its algebraic structure may depend on r. Spheres are a particularly interesting class of examples:

QUESTION 1.3 (A. Kupers). For $n \geq 5$, the group $\pi_0(\text{Diff}^\infty(S^n))$ is the group of homotopy $(n+1)$-spheres, a finite abelian group. Can this group be realized by diffeomorphisms of S^n?

Some cases are covered by Schultz [**Sch79**]. Other instances of Nielsen realization problems for finite subgroups of mapping class groups are discussed by Block–Weinberger in [**BW08**], who prove several non-realizability results and give a good survey of the state of the art. See also [**Zim85**]. Examples of non-realizability results for *infinite*, finitely-generated subgroups for higher-dimensional manifolds are given in [**Tsh15**]; see Theorem 2.10 below.

Beyond mapping class groups. When M is a closed manifold, the path-component of the identity $\text{Diff}_0^r(M) \subset \text{Diff}^r(M)$ is simple, provided that $r \neq \dim(M)+1$. For $r > 0$, this is a deep result of Mather [**Mat74**, **Mat75**], and for $r = 0$ it follows from [**And58**] and [**EK71**] (it is open for $r = \dim(M)+1$). Thus, realization problems for subgroups of mapping class groups are essentially the only section problems for these groups. However, when M has boundary, there is an obvious map $\text{Diff}^r(M) \xrightarrow{R} \text{Diff}^r(\partial M)$. This may not be surjective, but is when restricted to the identity components $\text{Diff}_0^r(M) \to \text{Diff}_0^r(\partial M)$. In [**Ghy91**], Ghys asked in which cases this map admits a section. In parallel with Problem 1.1, we ask:

PROBLEM 1.4. Let Γ be a finitely-generated group, $0 \leq r \leq \infty$, and let $\rho : \Gamma \to \mathrm{Diff}_0^r(\partial M)$ be a homomorphism. In which cases does there exist a homomorphism $\tilde{\rho} : \Gamma \to \mathrm{Diff}_0^r(M)$ so that $R \circ \tilde{\rho} = \rho$?

Rephrased in terms of flat bundles, Problem 1.4 asks whether a flat (∂M)-bundle E is the boundary of a flat M-bundle.

As a further variant on Problems 1.1 and 1.4, one can also restrict to *subgroups* of $\mathrm{Diff}^r(M)$ and pose the same family of questions. For real-analytic diffeomorphisms, Problem 1.1 for $\Gamma = \mathrm{Mod}_g$ was answered negatively in [**CC08**]. Bowden [**Bow11**] discusses Problem 1.4 for the boundary map on the group of surface symplectomorphisms $\mathrm{Symp}(\Sigma_g^1) \to \mathrm{Diff}_0(S^1)$, and Epstein–Markovic discuss the boundary map from quasi-conformal homeomorphisms of the disk to quasi-symmetric homeomorphisms of the circle. Interestingly, this latter problem is also related to Nielsen realization for the mapping class group of a surface; see Section 3.4 below.

Germs and holonomy of foliations. Considering foliations rather than foliated bundles leads to a class of related problems. Let $\mathrm{Diff}^r(\mathbb{R}^k, 0)$ denote the group of diffeomorphisms of $\mathbb{R}^k$ fixing the origin, and $\mathcal{G}^r(k)$ the group of germs at the origin of elements of $\mathrm{Diff}^r(\mathbb{R}^k, 0)$. Analogous to the holonomy of a flat bundle, when a manifold M admits a transversely-C^r codimension-k foliation $\mathcal{F}$, each leaf L has a holonomy representation $\pi_1(L) \to \mathcal{G}^r(k)$. Thus, understanding the structure of $\mathcal{G}^r(k)$ and its finitely-generated subgroups plays an important role in the study of the structure of foliations. However, the algebraic structure $\mathcal{G}^r(k)$ is not well understood. For instance, the following problem is open.

QUESTION 1.5. Suppose $\phi : \mathcal{G}^r(n) \to \mathcal{G}^s(m)$ is an isomorphism. Is it necessarily true that $m = n$, $r = s$, and ϕ is an inner automorphism?

The corresponding (positive) result for isomorphisms $\mathrm{Diff}^r(M) \to \mathrm{Diff}^s(M)$, where M is a compact manifold, is a theorem of Filipkiewicz [**Fil82**].

Finitely-generated subgroups of $\mathcal{G}^r(n)$ are also poorly understood. Navas [**DNR14**] asked whether every finitely-generated subgroup of $\mathcal{G}^0(2)$ is isomorphic to a subgroup of $\mathrm{Diff}^0(\mathbb{R}^2)$. While it is expected that the answer is negative, the question was motivated by a positive result in the case of $\mathbb{R}^1$; see [**DNR14**, §1.1.3]. Generalizing Navas' question above, we ask the following related (broad) question and realization problem:

PROBLEM 1.6. Which finitely-generated groups are subgroups of $\mathcal{G}^r(n)$? What are obstructions to lifting a representation $\rho : \Gamma \to \mathcal{G}^r(n)$ to $\mathrm{Diff}^r(\mathbb{R}^n, 0)$?

Further motivation for this problem comes from Thurston's *stability theorem* [**Thu74**] and the study of foliations. Thurston's theorem is one answer to the first part of the question above for $\mathcal{G}^1(n)$; it states that finitely-generated subgroups of $\mathcal{G}^1(n)$ with trivial derivative at 0 necessarily have the property that they admit surjective homomorphisms to $\mathbb{Z}$. (In other words, $\mathcal{G}^1(n)$ is a *locally indicable* group.) Thurston's motivation for this result was a generalization of the Reeb stability theorem for foliations; and indeed foliation theory and the structure of groups of germs are intimately linked, as representations into $\mathcal{G}^r(n)$ arise naturally from the holonomy of leaves of codimension n, transversely C^r foliations. For a further introduction to the relationship between groups of diffeomorphisms, foliations, germs, and their classifying spaces see Tsuboi's survey paper [**Tsu84**].

While foliation theory is not the focus of this paper, we will see numerous applications of Thurston stability in other contexts – including approaches to Problem 1.1 – below in §3. One is optimistic that any further answer to the first part of Problem 1.6 would have a similarly wide range of applications.

Organization of paper. In order to draw parallels between the various problems described above, we have organized this paper by *technique* (broadly categorized as dynamical or cohomological) rather than by results. In §2 we discuss examples where cohomology of groups and classifying spaces has been successfully used to obstruct lifting problems. §3 contains a discussion of dynamical techniques used to provide other obstructions. In §4 we give some positive results for Problem 1.1, i.e. interesting groups that *can* be realized by diffeomorphisms. Questions and problems have been attributed to their authors when known, although we did not make an exhaustive attempt to trace them to their original sources.

2. Cohomological techniques

In this section we take an algebro-topological approach to lifting problems. In its most basic form, the general approach is as follows: Given a surjection $\pi : G \to H$ and a homomorphism $\rho : \Gamma \to H$, a lift $\tilde{\rho}$ as in the figure on the left below induces maps on group cohomology as indicated on the right.

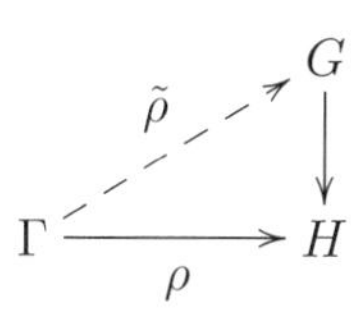

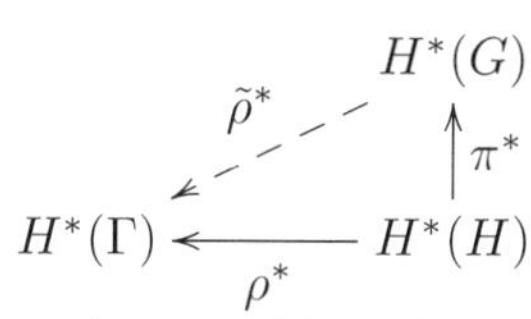

Classes in the kernel of π^* serve as obstructions to lifts. Specifically, if $c \in \ker \pi^*$, and ρ is a representation such that $\rho^*(c) \neq 0 \in H^*(\Gamma)$, then no map $\tilde{\rho}$ can make the diagram commute.

While this approach (or variations of it) can be successful, in general it can be difficult to find a class c that works, and the absence of cohomological obstructions is not sufficient to guarantee the existence of a lift. For an example, compare Theorems 2.16 and 3.9 below. However, one benefit of the cohomological approach (when it works) is that often one can conclude that no realization exists even after passing to a finite-index subgroup of Γ.

2.1. The Euler class as an obstruction. We begin with a simple example of the strategy introduced above. Fix $h \geq 2$. Then $\pi_1(\Sigma_h)$ is a discrete, subgroup of $\mathrm{PSL}(2, \mathbb{R})$, which acts on S^1 by Möbius transformations. Let $\rho : \pi_1(\Sigma_h) \to \mathrm{PSL}(2, \mathbb{R}) \to \mathrm{Homeo}^+(S^1)$ denote this homomorphism.

PROPOSITION 2.1. *Fix $g \geq 1$, and let Σ_g^1 be a genus g surface with 1 boundary component. Then the action ρ of $\pi_1(\Sigma_h)$ on $\partial \Sigma_g^1 \simeq S^1$ does not extend to a representation into* $\mathrm{Homeo}_0(\Sigma_g^1)$.

Here, as always, Homeo_0 denotes the identity component of the group. As we remark after the proof, the situation is much more subtle (and the proposition false) if Homeo_0 is replaced by Homeo in the statement above.

The proof of Proposition 2.1 is easy when framed in the language of characteristic classes and classifying spaces. For the reader unacquainted with this theory, we summarize the main properties we need before giving the proof. See [**Mor01**, Ch. 4] for more details.

Classifying space basics. Every topological group G has a classifying space BG that classifies principal G-bundles. If $G < \mathrm{Homeo}(X)$, then BG also classifies X-bundles with structure group G. If G has the discrete topology (indicated by G^δ), then $BG^\delta \sim K(G,1)$ is an *Eilenberg–Maclane space* for G (a connected CW-complex with fundamental group G and contractible universal cover). In particular $B\,\mathrm{Diff}(\Sigma_g)$ classifies Σ_g-bundles and $B\,\mathrm{Diff}(\Sigma_g)^\delta$ classifies flat Σ_g-bundles.

The classifying space construction $G \mapsto BG$ is functorial: a continuous homomorphism $G \to H$ induces a continuous map $BG \to BH$. Furthermore, if $G \to H$ is a homotopy equivalence, then $BG \to BH$ is also a homotopy equivalence, and a short exact sequence $1 \to K \to G \to H \to 1$ induces a fibration $BK \to BG \to BH$.

In the special case where $G = \mathrm{Homeo}(S^1)$, the inclusion $\mathrm{SO}(2) \hookrightarrow \mathrm{Homeo}_+(S^1)$ induces a homotopy equivalence, so the classifying space $B\,\mathrm{Homeo}_+(S^1)$ is homotopy equivalent to $B\,\mathrm{SO}(2) \sim \mathbb{C}P^\infty$. As such $H^*(B\,\mathrm{Homeo}(S^1);\mathbb{Z})$ is a polynomial ring $\mathbb{Z}[e]$ generated by a class $e \in H^2(B\,\mathrm{Homeo}(S^1))$ which has the following property: a circle bundle $E \to B$ classified by a map $f_E : B \to B\,\mathrm{Homeo}(S^1)$ admits a continuous section if and only if $f_E^*(e) = 0$.

Proof of Proposition 2.1. If ρ extends to $\tilde{\rho} : \pi_1(\Sigma_h) \to \mathrm{Homeo}_0(\Sigma_g^1)$, then there is a diagram that commutes up to homotopy:

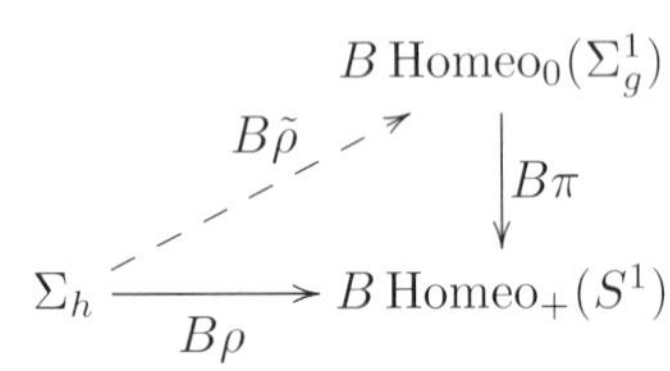

Let $e(\rho) := B\rho^*(e) \in H^2(\Sigma_h)$ be the Euler class of ρ. On the one hand, $e(\rho)$ is the Euler class of the unit tangent bundle $T^1\Sigma_h \to \Sigma_h$, which is nonzero because $\chi(\Sigma_h) \neq 0$. On the other hand, $\mathrm{Homeo}_0(\Sigma_g^1)$ is contractible, so $B\,\mathrm{Homeo}_0(\Sigma_g^1)$ is also contractible, and this implies that $\pi^*(e) = 0$. Thus $\tilde{\rho}$ does not exist. $\square$

The above argument shows that no action $\rho : \pi_1(\Sigma_h) \to \mathrm{Homeo}_+(S^1)$ with $e(\rho) \neq 0$ extends to $\mathrm{Homeo}_0(\Sigma_g^1)$ for $g \geq 1$. There are many such actions, even among representations into $\mathrm{PSL}(2,\mathbb{R})$. However, one can produce actions $\rho : \pi_1(\Sigma_h) \to \mathrm{Homeo}(S^1)$ with $e(\rho) \neq 0$ that *do* extend to $\mathrm{Homeo}(\Sigma_g^1)$ for any $g \geq 2$. This uses a "stabilization" argument [**Bow11**]: start with any $\rho' : \pi_1(\Sigma_{h'}) \to \mathrm{Homeo}(S^1)$, present $\pi_1(\Sigma_{h'}) = \langle a_1, \ldots, b_{h'} \mid \prod_{i=1}^{h'}[a_i, b_i] = 1\rangle$, and choose $\tilde{\rho}'(a_i)$, $\tilde{\rho}'(b_i) \in \mathrm{Homeo}(\Sigma_g^1)$ extending $\rho'(a_i)$ and $\rho'(b_i)$ on $\partial\Sigma_g^1$ (technically we want extensions that are a trivial product on a collar neighborhood of $\partial\Sigma_g^1$). Then $f := \prod_{i=1}^{h}[\tilde{\rho}'(a_i), \tilde{\rho}'(b_i)]$ belongs to the group $\mathrm{Homeo}_{N(\partial)}(\Sigma_g^1)$ of homeomoprhisms that are the identity on a neighborhood of $\partial\Sigma_g^1$. This group is perfect [**Fis60**] [**FM12**, §5.1], so we can write $f^{-1} = \prod_{j=1}^{n}[x_j, y_j]$. Let $h = h' + n$, and extend ρ'

to

$$\rho : \pi_1(\Sigma_h) = \langle a_i, b_i, x_j, y_j \mid \prod_{i=1}^{h'}[a_i, b_i]\prod_{j=1}^{n}[x_j, y_j] = 1\rangle \to \mathrm{Homeo}_+(S^1)$$

by mapping x_j and y_j to the identity. By design, $e(\rho) = e(\rho')$, and ρ extends to $\mathrm{Homeo}(\Sigma_g^1)$. Note that $e(\rho) \in \mathbb{Z}$ determines the class $[f] \in \pi_0\big[\mathrm{Homeo}_{N(\partial)}(\Sigma_g^1) \cap \mathrm{Homeo}_0(\Sigma_g^1)\big] \simeq \mathbb{Z}$. In particular $x_i, y_i \in \mathrm{Homeo}_0(\Sigma_g^1)$ is possible only if $e(\rho) = 0$. This is consistent with Proposition 2.1.

QUESTION 2.2. Given a representation $\rho : \pi_1(\Sigma_{h'}) \to \mathrm{Homeo}_+(S^1)$ as above, what is the minimal stabilization genus h required to extend the action to $\mathrm{Homeo}_+(\Sigma_g^1)$? Are there examples where this answer is different in the C^0 and C^∞ cases?

The Milnor-Wood inequality (Theorem 2.6 below) gives bounds on the value of the Euler class $\rho^*(e) \in H^2(\Sigma_h; \mathbb{Z})$ for a flat bundle. Since none of the examples obtained by stabilization are maximal, we ask:

QUESTION 2.3. Does there exist a representation $\rho : \pi_1(\Sigma_h) \to \mathrm{Homeo}(S^1)$ with maximal Euler class that lifts to $\mathrm{Homeo}(\Sigma_g^1)$?

Since $\mathrm{Diff}_0^r(\mathbb{D}^2) \to \mathrm{Diff}_0^r(S^1)$ is a homotopy equivalence for all r, the argument in the proof of Proposition 2.1 gives no obstruction to extending any action on S^1 to an action on the disk. (Indeed for $r = 0$, any group action can be extended by "coning off" the circle to the disk. However, we will see examples of group actions by C^1 diffeomorphisms that cannot be lifted to $\mathrm{Diff}^1(\mathbb{D}^2)$ in Section 3.) Geometric actions of surface groups provide interesting examples to study:

PROBLEM 2.4. The action of $\mathrm{PSL}(2, \mathbb{R})$ by Möbius transformations on S^1 extends to a smooth action on $\mathbb{D}^2$, so the representations $\pi_1(\Sigma_h) \to \mathrm{PSL}(2, \mathbb{R}) \to \mathrm{Diff}^\infty(S^1)$ discussed above extend to $\mathrm{Diff}^\infty(\mathbb{D}^2)$.

(a) Classify all extensions of these actions. (Are there any exotic ones?)
(b) If $k > 1$ divides $2h - 2$, then $\pi_1(\Sigma_h)$ embeds as a lattice in the central extension of $\mathrm{PSL}(2, \mathbb{R})$ by a cyclic group of order k, which also acts naturally on the circle. (This action is simply the lift of the natural Möbius action of $\mathrm{PSL}(2, \mathbb{R})$ to the k-fold cover of S^1, which is also a circle.) Does such an action of $\pi_1(\Sigma_h)$ on S^1 extend to $\mathrm{Diff}^\infty(\mathbb{D}^2)$?

The actions described in (b) are called *geometric* and share many properties with Fuchsian actions, including strong topological rigidity [**Man15b, MW17**]. It would be interesting to know if they have obstructions to extensions over the disc.

Surface braid groups and point-pushing. There is a surjection $\mathrm{Mod}_{g,m}^b \to \mathrm{Mod}_g^b$ that forgets the marked points. The kernel is described by an exact sequence

$$\pi_1(\mathrm{Diff}(\Sigma)) \to \mathrm{Br}_m(\Sigma) \xrightarrow{P} \mathrm{Mod}_{g,m}^b \to \mathrm{Mod}_g^b \to 1, \tag{2.1}$$

where $\mathrm{Br}_m(\Sigma)$ is the *surface braid group*, defined as the fundamental group $\pi_1(\mathrm{Conf}_m(\Sigma))$ of the space of unordered configurations of m points on Σ. The homomorphism P is called the *point-pushing homomorphism*; see [**FM12**] for further details.

THEOREM 2.5 (Bestvina–Church–Souto [**BCS13**]). *The homomorphism P in (2.1) is not realized by C^1 diffeomorphisms when Σ is closed, $g \geq 2$ and $m \geq 1$.*

We sketch Bestvina–Church–Souto's argument, which relies on the fact that the Euler class introduced above is a *bounded* cohomology class [**Mil58, Woo71**].

THEOREM 2.6 (Milnor-Wood inequality). *Let Σ_g be a closed oriented surface of genus $g \geq 1$, and let $[\Sigma_g] \in H_2(\Sigma_g;\mathbb{Z})$ be the fundamental class. For any $\rho : \pi_1(\Sigma_g) \to \mathrm{Homeo}(S^1)$, the Euler class $e(\rho) \in H^2(\Sigma_g;\mathbb{Z}) \simeq \mathbb{Z}$ satisfies*

$$2 - 2g \leq e(\rho) \leq 2g - 2.$$

Furthermore, if ρ factors through $r : \mathrm{GL}_2^+(\mathbb{R}) \to \mathrm{Homeo}(S^1)$ acting on rays through the origin in $\mathbb{R}^2$, then $|e(\rho)| \leq g - 1$.

PROOF OF THEOREM 2.5. We sketch a proof in the case $m = 1$. In this case $\mathrm{Br}_1(\Sigma_g) \simeq \pi_1(\Sigma_g)$. If P is realized by diffeomorphisms we have the following diagram.

$$\begin{array}{ccccc} & & \mathrm{Diff}^1(\Sigma_{g,1}) & \xrightarrow{D} & \mathrm{GL}_2^+(\mathbb{R}) \\ & \tilde{P} \nearrow & \downarrow & & \downarrow r \\ \pi_1(\Sigma_g) & \xrightarrow[P]{} & \mathrm{Mod}_{g,1} & \xrightarrow[\alpha]{} & \mathrm{Homeo}(S^1) \end{array} \tag{2.2}$$

Here D is induced by the action on the tangent space at the marked point and r is the homomorphism from Theorem 2.6. The homomorphism α was originally defined by Nielsen; it comes from the isomorphism $\mathrm{Mod}_{g,1} \simeq \mathrm{Aut}(\pi_1\Sigma_g)$ together with the action on the Gromov boundary $\partial\pi_1(\Sigma_g) \simeq S^1$ (or equivalently the boundary of the compactification of the universal cover $\tilde{\Sigma} \simeq \mathbb{H}^2$). See §3 for a description. The left-hand triangle commutes, but the right-hand square does not. However, it does commute up to homotopy on the level of classifying spaces [**BCS13**, §3]. The proof concludes by comparing the image of the Euler class $e \in H^2(B\,\mathrm{Homeo}(S^1))$ when pulled back to $H^2(\Sigma_g) \simeq \mathbb{Z}$ along the outer paths in the diagram.

On the one hand, the composition $\alpha \circ P$ factors as $\pi_1(\Sigma_g) \to \mathrm{PSL}(2,\mathbb{R}) \to \mathrm{Homeo}(S^1)$, and it follows that $e(\alpha \circ P) \in H^2(\Sigma_g)$ is the Euler number of the unit tangent bundle $T^1\Sigma_g \to \Sigma_g$ (as in the proof of Proposition 2.1). Thus $e(\alpha \circ P) = 2 - 2g$. On the other hand, by the Milnor-Wood inequality (see Theorem 2.6), $|e(r \circ D \circ \tilde{P})| \leq g - 1$. This contradiction implies that $\tilde{P}$ does not exist. □

Topologically, $P : \pi_1(\Sigma_g) \to \mathrm{Mod}_{g,1}$ is the monodromy of the trivial bundle with respect to the diagonal section $\Delta \subset \Sigma \times \Sigma$. Theorem 2.5 implies that this bundle has no flat connection such that the diagonal is *parallel*, i.e. a leaf of the foliation. See also [**MV08**].

QUESTION 2.7 (Realizing point-pushing by homeomorphisms). Can $P : \pi_1(\Sigma_g) \to \mathrm{Mod}_{g,1}$ be realized by homeomorphisms? What about by homeomorphisms of the blow up of $\Sigma_{g,1}$ at the marked point (i.e. by homeomorphisms of Σ_g^1 that may be nontrivial on the boundary). If one tries to run the above proof of Theorem 2.5, one must replace $\mathrm{GL}_2^+(\mathbb{R})$ with $\mathrm{Homeo}(S^1)$ in (2.2), and then the Milnor–Wood inequality does not provide any contradiction.

QUESTION 2.8 (Disc pushing). If we replace the marked point by a boundary component, there is a "disk-pushing" subgroup $D : \pi_1(T^1\Sigma_g) \hookrightarrow \mathrm{Mod}_g^1$ analogous to the point-pushing subgroup $P : \pi_1(\Sigma_g) \hookrightarrow \mathrm{Mod}_{g,1}$. Is D realized by diffeomorphisms? The proof of Theorem 2.5 does not adapt to this case because the Euler class on Mod_g^1 is trivial.

QUESTION 2.9 (Handle pushing). Consider an inclusion $\Sigma_g^1 \hookrightarrow \Sigma_{g+1}$ obtained by gluing on a genus-1 handle. This induces a homomorphism $\mathrm{Mod}_g^1 \to \mathrm{Mod}_{g+1}$. Is the composition $H : \pi_1(T^1\Sigma_g) \xrightarrow{D} \mathrm{Mod}_g^1 \to \mathrm{Mod}_{g+1}$ realized by diffeomorphisms? This homomorphism can be interpreted as "handle-pushing". This problem is harder than disk-pushing since a realization does not have an obvious invariant submanifold (e.g. a point or a disk as before).

The second author proved the following generalization of Theorem 2.5 to many other locally symmetric manifolds [**Tsh15**].

THEOREM 2.10 (Tshishiku). *Let G be a real simple noncompact Lie group. Assume G is not isogenous to* $\mathrm{SO}(n,1)$ *for any $n \geq 3$. Then there exists a lattice $\Gamma < G$ so that for the associated locally symmetric space $M = \Gamma\backslash G/K$, the point-pushing homomorphism $P : \pi_1(M) \to \mathrm{Mod}(M,*)$ is not realized by diffeomorphisms.*

The outline is the same as the proof of Theorem 2.5, but it uses different obstructions, including Pontryagin classes, Chern–Weil theory, and Margulis superrigidity.

QUESTION 2.11. Let M be a finite volume hyperbolic 3-manifold. Is the point-pushing homomorphism $P : \pi_1(M) \to \mathrm{Mod}(M,*)$ realized by diffeomorphisms?

2.2. The MMM classes and Morita's nonlifting theorem. In this section, we discuss the question of whether or not the surjection $\pi : \mathrm{Diff}_\partial(\Sigma_{g,m}^b) \to \mathrm{Mod}_{g,m}^b$ is split, i.e. whether or not the entire mapping class group is realized by diffeomorphisms (this is a special case of Problem 1.1). There are a few low-complexity examples when π does split; for example when $\Sigma = \Sigma_1$ is a closed torus or when $\Sigma = \Sigma_{0,3}^1$ is a disk with 3 marked points; these examples will be discussed more in §4. The first negative result was proved by Morita [**Mor87**], originally with the assumption $g \geq 18$.

THEOREM 2.12 (Morita nonlifting). *Let Σ_g be a closed surface of genus at least 2. Then the surjection $\pi : \mathrm{Diff}^r(\Sigma_g) \to \mathrm{Mod}_g$ does not split for any $r \geq 0$. Furthermore, if $g \geq 10$ then π has no C^2 splitting over any finite-index subgroup of* Mod_g.

We sketch Morita's proof using the following theorem on the cohomology of Mod_g. See [**Wah13**] for an exposition of the proof.

THEOREM 2.13 (Morita, Harer stability, Madsen–Weiss). *In degrees $* \leq 2(g-1)/3$, the cohomology $H^*(\mathrm{Mod}_g;\mathbb{Q})$ is a polynomial algebra generated by certain* tautological classes $e_i \in H^{2i}(\mathrm{Mod}_g;\mathbb{Q})$. *As a characteristic class of a Σ_g-bundle $E \to M$ over a manifold M, the class e_i is defined as*

$$e_i(E) = \int_\Sigma e(T^vE)^{i+1} \in H^{2i}(M),$$

where $\mathbb{R}^2 \to T^vE \to E$ is the vertical tangent bundle, *$e(T^vE)$ is the* Euler class, *and $\int_\Sigma : H^*(E) \to H^{*-2}(M)$ is the* fiber-integration *(or Gysin) map.*

Morita [**Mor87**] defined the characteristic classes $e_i \in H^*(\mathrm{Mod}_g)$, and showed they are often nonzero (by constructing interesting Σ_g-bundles). Harer [**Har85**] proved that they are *stably* nonzero, and Madsen–Weiss [**MW07**] proved that these classes account for all of the cohomology in low degree (their techniques also give an alternate proof that $e_i \neq 0$ for g large).

PROOF OF THEOREM 2.12. Let $\mathrm{Diff}^2(\Sigma_g)^\delta$ denote $\mathrm{Diff}^2(\Sigma_g)$ with the discrete topology. We may view π as a composition of continuous maps $\mathrm{Diff}^2(\Sigma_g)^\delta \xrightarrow{\alpha} \mathrm{Diff}^2(\Sigma_g) \xrightarrow{\beta} \mathrm{Mod}_g$. To prove the theorem, it suffices to show that

$$\alpha^* \circ \beta^* : H^*(B\,\mathrm{Mod}_g;\mathbb{Q}) \to H^*(B\,\mathrm{Diff}^2(\Sigma_g);\mathbb{Q}) \to H^*(B\,\mathrm{Diff}^2(\Sigma_g)^\delta;\mathbb{Q})$$

has nontrivial kernel. The result of Earle–Eells [**EE67**] cited in the previous section is that $\mathrm{Diff}^2(\Sigma_g) \to \mathrm{Mod}_g$ is a homotopy equivalence, so by classifying space theory the induced map on classifying spaces is as well, hence β^* is an isomorphism. Thus, we need to show that α^* is not injective. Since $B\,\mathrm{Diff}^2(\Sigma)^\delta$ classifies flat surface bundles, $\ker\alpha^*$ is the ideal of characteristic classes of Σ_g-bundles that vanish for flat bundles. Using Bott's vanishing theorem (a topological obstruction to existence of certain foliations), which applies because of the C^2 hypothesis, one checks that if $E \to M$ is flat, then $e_i(E) = 0$ for $i \geq 3$. By the Theorem 2.13, $e_3 \in H^*(\mathrm{Mod}_g;\mathbb{Q})$ is nonzero if $g \geq 10$. Thus e_3 is a *nonzero* element of $\ker\alpha^*$.

If $j : \Gamma \hookrightarrow \mathrm{Mod}_g$ is finite index, then the induced map $j^* : \mathbb{H}^*(\mathrm{Mod}_g;\mathbb{Q}) \to H^*(\Gamma;\mathbb{Q})$ is injective by transfer, so $j^*(e_3) \neq 0$ and the same argument shows Γ is not realized by diffeomorphisms. □

The proof above shows Mod_g is not realized by C^2 diffeomorphisms when $g \geq 10$. As mentioned in the introduction, there are now many different proofs of Theorem 2.12 [**BCS13, FH09, ST16, MS08**] that show Mod_g is not realized by C^1 or C^0 diffeomorphisms (with varying improvements on the genus bound); in particular, both Theorem 2.5 and Theorem 3.9 have Theorem 2.12 as a consequence. However, none of these alternate proofs give information about finite-index subgroups of Mod_g. In particular, Markovic–Saric's proof of non-realization by homeomorphisms uses torsion elements in a crucial way, so does not answer the following.

PROBLEM 2.14 (Robust non-lifting). Show that no finite-index subgroup of Mod_g is realized by homeomorphisms for $g \geq 2$.

A seemingly more difficult problem is to show non-lifting for the *Torelli group.*

PROBLEM 2.15. Show that $\mathcal{I}_g$, the kernel of $\mathrm{Mod}_g \to \mathrm{Sp}_{2g}(\mathbb{Z})$, is not realized by diffeomorphisms.

The question of realizing mapping class groups $\mathrm{Mod}_{g,m}^b$ when $m, b \geq 1$ is also interesting. For example, the mapping class group $\mathrm{Mod}_{0,m}^1$ of a disk $\mathbb{D}^2$ with m marked points $\mathbf{z}_m \subset \mathbb{D}^2$ is the classical braid group B_m. The following theorem of Nariman [**Nar18**] states that there is no cohomological obstruction to splitting $\mathrm{Diff}_\partial(\mathbb{D}^2 - \mathbf{z}_m) \to B_m$. Nevertheless, in contrast, the mapping class group of a disk is not realized by C^1 diffeomorphisms – see Theorem 3.9 below.

THEOREM 2.16 (Nariman). *Consider the surjection* $\mathrm{Diff}^\infty(\mathbb{D}^2 - \mathbf{z}_m) \to B_m$. *The induced map on group cohomology* $H^*(B_m; A) \to H^*(\mathrm{Diff}^\infty(\mathbb{D}^2 - \mathbf{z}_m); A)$ *is split injective in all cohomological degrees and for all abelian groups* A.

Nariman's work also considers another question related to subgroups of mapping class groups. There is a natural "geometric" map $\psi : \mathrm{Br}_{2g+2} \hookrightarrow \mathrm{Mod}_g^2$ induced by lifting mapping classes to a double cover Σ_g^2 of the disk $\mathbb{D}^2$ ramified over the points of $\mathbf{z}_{2g+2}$. One description of this map is as follows: Each $f \in \mathrm{Diff}(\mathbb{D}^2, \mathbf{z}_{2g+2})$ has a canonical lift to a homeomorphism of the cover Σ_g^2; this is the lift that fixes both boundary components pointwise. This gives an injective map $\Psi : \mathrm{Diff}(\mathbb{D}^2, \mathbf{z}_{2g+2}) \to \mathrm{Homeo}(\Sigma_g^2, \partial\Sigma_g^2)$, and the induced map on isotopy classes is ψ. Nariman asked if Ψ can be "smoothed": is there a homomorphism $\mathrm{Diff}(\mathbb{D}^2, \mathbf{z}_{2g+2}) \to \mathrm{Diff}(\Sigma_g^2, \partial\Sigma_g^2)$ that induces ψ on mapping class groups? Building on the techniques used in his proof of Theorem 2.16, Nariman shows that there is no cohomological obstruction.

THEOREM 2.17 (Nariman [**Nar18**]). *Let $g \geq 1$. There is a space Y, homologically equivalent to the classifying space $B\,\mathrm{Diff}^\delta(\Sigma_g^2, \partial\Sigma_g^2)$ and a map Φ such that the following diagram is homotopy commutative*

$$\begin{array}{ccc} B\,\mathrm{Diff}^\delta(\mathbb{D}^2, \mathbf{z}_{2g+2}) & \xrightarrow{\Phi} & Y \\ \downarrow & & \downarrow \\ B\,\mathrm{Mod}(\mathbb{D}^2, \mathbf{z}_{2g+2}) & \xrightarrow[B\psi]{} & B\,\mathrm{Mod}(\Sigma_g^2) \end{array}$$

We give an alternative proof in §4 by constructing an explicit lift. However, the following question remains open.

QUESTION 2.18. Can the "geometric" braid group $\psi : \mathrm{Br}_{2g+2} \hookrightarrow \mathrm{Mod}_g^2$ described above be realized in $\mathrm{Diff}(\Sigma_g^2)$?

3. Dynamical obstructions to realizations

This section showcases the use of dynamical tools in realization problems. In many contexts, one can use the algebraic structure of a group Γ to force some dynamical behavior – such as the existence of fixed points or invariant sets – when Γ acts by diffeomorphisms on a manifold M. If the action comes from a section of $\mathrm{Diff}_0^r(M) \to H$ for some group H, properties of H often furnish additional dynamical information.

The proofs of non-realizability of Mod_g by homeomorphisms due to Markovic and Markovic–Saric [**Mar07, MS08**] are the most sophisticated instance of this phenomenon. The heart of Markovic's idea is to show that the realization of Dehn twists about simple closed curves have support on sets that, roughly, look like the annuli on which the Dehn twists are defined. Once this is appropriately formalized, one uses relations in Mod_g to derive a contradiction. As the argument is quite involved, we will not be able to do it justice here and instead refer the reader to Le Cavez's work [**Cal12**], which contains a nice exposition as well as some alternative techniques towards the proof.

Here we focus on simpler examples that illustrate a tension between the algebraic structure of a group Γ and the dynamics of a realization. As a first basic illustration, we discuss non-realizability for groups of germs.

3.1. Non-realizability of groups of germs. Recall that $\mathcal{G}^r(n)$ denotes the group of germs at 0 of orientation-preserving C^r diffeomorphisms of $\mathbb{R}^n$ fixing 0 (i.e. the quotient of $\mathrm{Diff}^r(\mathbb{R}^n, 0)$ by the subgroup of diffeomorphisms that restrict to the identity in a neighborhood of 0), and $\Gamma \subset \mathcal{G}^r(n)$ is *realizable by diffeomorphisms* if there is a section of $\mathrm{Diff}^r(\mathbb{R}^n, 0) \to \mathcal{G}^r(n)$ over Γ. Here, we prove the following non-realizability result, as a first step towards Problem 1.6.

THEOREM 3.1. *There exists an (explicit) finitely-generated, torsion-free subgroup* Γ *of* $\mathcal{G}^0(n)$ *that cannot be realized in* $\mathrm{Diff}^0(\mathbb{R}^n, 0)$.

As we remark in the proof, the subgroup Γ can in fact be taken to lie in $\mathcal{G}^\infty(n)$.

Theorem 3.1 was proved in [**Man15a**] for $n = 1$. Here, we give a simplified argument that proves the general case. The argument uses essentially only one dynamical tool (which will make a later appearance in Section 3.3). This is the *basin of attraction.*

DEFINITION 3.2. Let f be a homeomorphism of a manifold M. A point $p \in M$ is *attracting* if there exists a neighborhood W of p such that $\bigcap_{n>0} f^n(\overline{W}) = p$. In this case the *basin of attraction* of f at p is the set $\{x \in M \mid f^n(x) \to p\}$.

It follows from the definition that the basin of attraction is f-invariant, and equal to $\bigcup_{n>0} f^n(W)$ for any sufficiently small neighborhood W of the attracting fixed point – so in particular is open. Furthermore, if f and g commute, then g permutes the attracting fixed points of f and their basins of attraction.

PROOF OF THEOREM 3.1. We first define Γ by specifying homeomorphisms of which we take the germs, then prove that there is no realization. The key algebraic properties of Γ are that it contains a Baumslag-Solitar subgroup $BS(1,2) = \langle a, b \mid aba^{-1} = b^2\rangle$, as well as several commuting elements.

Parametrize $\mathbb{R}^n \setminus \{0\} = S^{n-1} \times \mathbb{R}$, with the sets $S^{n-1} \times \{t\}$ approaching 0 as $t \to -\infty$. All of the homeomorphisms in Γ will be defined radially, so we start by working on $\mathbb{R}$. A $BS(1,2)$ subgroup of $\mathrm{Diff}^\infty(\mathbb{R})$ is given by $a : x \mapsto 2x$ and $b : x \mapsto x - 1$. To define the other elements, let $Y \subset X$ be two a-invariant disjoint unions of intervals in $(-\infty, 0)$ and let f_1 be a diffeomorphism supported on X that agrees with b on Y in some neighborhood of $-\infty$. Let $f_2 = f_1^{-1} \circ b$, which, in a neighborhood of $-\infty$, is supported on the a-invariant set $\mathbb{R} \setminus Y$. Let g_i be arbitrary diffeomorphisms of $\mathbb{R}$ that commute with a in a neighborhood of $-\infty$ and have support equal to $\mathrm{fix}(f_i)$.

For each $h \in \{a, b, f_1, f_2, g_1, g_2\}$ the map $(s,t) \mapsto (s, h(t))$ of $S^{n-1} \times \mathbb{R}$ extends to a homeomorphism of $\mathbb{R}^2$ fixing 0. (By first conjugating a and b by a strong contraction at $-\infty$ as in Construction 4.3, one may in fact take these to be germs of smooth diffeomorphisms, infinitely tangent to the identity at 0; in which case f_i and g_i may be chosen so that these define diffeomorphisms as well.) Abusing notation slightly, we identify h with the germ of the homeomorphism so defined. These germs satisfy the relations $aba^{-1} = b^2$, $f_1 f_2 = b$, and $[f_i, a] = [f_i, g_i] = 1$.

Let $\Gamma < \mathcal{G}^0(n)$ be the subgroup generated by a, b, f_1, f_2, g_1 and g_2, and suppose that $\phi : \Gamma \to \mathrm{Homeo}(\mathbb{R}^n, 0)$ is a realization. Let B denote the attracting basin for b. The relation $aba^{-1} = b^2$ implies that B is a-invariant. Note that, if U is a sufficiently small ball about the origin, then B is the increasing union of the sets $\bigcup_{n>0} a^{-1}(U)$; each a homeomorphic image of a ball. Using this (and the annulus theorem), we can parametrize $B \setminus 0$ by $S^{n-1} \times \mathbb{R}$, on which $\phi(b)$ acts by $(s,t) \mapsto (s, t-1)$. We now claim that the attracting basin for $\phi(a)$ has closure contained in B, and will show this by finding an a-invariant set properly contained in B. Take a fundamental domain $S^{n-1} \times [0,1]$ for $\phi(b)$; the relation $aba^{-1} = b^2$ implies that the image of any fundamental domain for b under a is a fundamental domain for b^2. Now $\phi(a)(S^{n-1} \times [0,1])$ is compact so contained in $S^{n-1} \times [-M, M]$ for some M, hence $\phi(a)^{-1}(S^{n-1} \times [-2M, 2M])$ is a fundamental domain for b^{2M}

containing $[0,1]$, thus contained in $S^{n-1} \times [-2M, 2M]$. This gives an a-invariant compact set with closure contained in B, as desired.

Finally, since ϕ is a realization, we known the germ of $\phi(g_i)$ at 0: in a neighborhood of 0 the homeomorphism $\phi(g_i)$ is supported on the disjoint union of an annulus $A \cong S^{n-1} \times I$ with its images under $\phi(a)^n$ for $n \in \mathbb{N}$. Since g_i and a commute, the boundary of the basin of attraction of a is equal to the set of accumulation points of the iterates $a^{-n}(A)$, for $n > 0$. Since g_i and f_i commute, this set of accumulation points is f_i-invariant. But $f_1 f_2 = b$, contradicting the fact that there is no b-invariant set contained in $B \setminus \{0\}$. □

Despite the elementary nature of this example, the algebraic structure of groups of germs is far from understood. As well as the problems raised in the introduction, we ask (expecting the answer to be positive):

QUESTION 3.3. Can $\mathcal{G}^r(n)$ and $\mathcal{G}^r(m)$ be distinguished by their finite subgroups? What about finitely-generated, torsion-free subgroups? *Realizable* subgroups?

QUESTION 3.4. Fix $n > 1$. Can $\mathcal{G}^r(n)$ and $\mathrm{Diff}^r(\mathbb{R}^n)$ be distinguished by their finitely-generated, torsion-free subgroups?

3.2. C^1 techniques: Linear representations and Thurston stability. In [**Ghy91**], Ghys proved the following non-realization result for boundary maps.

THEOREM 3.5 (Ghys). *There is no section of the natural restriction map* $\mathrm{Diff}^1_0(\mathbb{D}^n) \to \mathrm{Diff}^1_0(S^{n-1})$.

In the case when n is even, the proof actually gives a finitely-generated subgroup $\Gamma \subset \mathrm{Diff}^\infty_0(S^{n-1})$ which has no realization in $\mathrm{Diff}^1_0(\mathbb{D}^n)$, we give a sketch below.

PROOF SKETCH – EVEN DIMENSIONAL CASE. Supposing $n = 2k$, identify S^{n-1} with the unit sphere $\{(z_1, ... z_k) \in \mathbb{C}^k \mid \sum |z_i|^2 = 1\}$. Let $\mu_1, \dots, \mu_k$ be distinct p^{th} roots of unity for some prime $p > n$, and define an order p diffeomorphism $f : (z_1, \dots, z_k) \mapsto (w_1 z_1, \dots, w_k z_k)$. One can explicitly write f as a product of commutators $f = [a_1, b_1] \dots [a_j, b_j]$ such that a_i and b_i all commute with f. Let $\Gamma = \langle f \rangle \times \langle a_1, \dots, b_j \rangle$. Supposing that ϕ is an extension, $\phi(f)$ is a finite order element of $\mathrm{Diff}^1_0(\mathbb{D}^n)$ with no fixed points on S^{n-1} hence has an interior fixed point (by Brouwer). Smith theory says that the action of $\phi(f)$ on the doubled disc (i.e. S^n) has fixed set equal to a mod-p homology sphere; since this fixed set is disconnected, it must be a 0-sphere. One concludes that $\phi(f)$ has a single fixed point, say $x \in \mathbb{D}^n$, preserved by $\phi(a_i)$ and $\phi(b_i)$ since these commute with $\phi(f)$. Since x is an isolated fixed point, the derivative $D\phi(f)_x$ is a linear map of order p and full rank. However, the centralizer of such a map in $\mathrm{GL}(n, \mathbb{R})$ is abelian, so $D\phi(f)_x = \prod [D\phi(a_i)_x, D\phi(b_i)_x] = 1$, contradiction. □

The tension here comes from forcing a global fixed point (through Smith theory and the algebraic structure of Γ), then using the constraints from the algebraic structure of Γ on the linear representation given by taking derivatives at that point. It would be interesting to see if this line of argument can be improved to even dimensional spheres – because there is no fixed-point free finite order map of S^{2k}, Ghys' argument in this case is more involved and uses simplicity of the full diffeomorphism group. Torsion elements also play an essential role in the argument, so the following question remains open in all cases except $n = 2$.

QUESTION 3.6. Does there exist a torsion-free, finitely-generated $G \subset \mathrm{Diff}_0(S^{n-1})$ with no section to $\mathrm{Diff}_0(\mathbb{D}^n)$?

The $n = 2$ is covered in [**Man13**], and parallels the strategy in Ghys' proof above. *Distorted elements* in G are used as a substitute for torsion elements, and work of Franks and Handel on distorted surface diffeomorphisms is used as a substitute for Smith theory to force an isolated fixed point. To complete the second half of the argument, one needs an element with nontrivial, finite-order, derivative at the fixed point; this is accomplished using rotation numbers.

In the absence of a nontrivial linear representation, *Thurston's stability theorem* gives an alternative – and more powerful – tool whenever one has a C^1 action with global fixed point.

THEOREM 3.7 (Thurston stability [**Thu74**]). *Let M be a manifold, $x \in M$, and let G be finitely-generated group of germs of C^1 diffeomorphisms fixing x and with trivial derivative at x. Then there is a surjective homomorphism $G \to \mathbb{Z}$.*

The proof uses a clever renormalization procedure to compare growth of higher order terms. A nice exposition can be found in [**CC00**, §6.2].

APPLICATION 3.8. A typical application of Theorem 3.7 is as follows: suppose Σ is a surface and x an accumulation point of fixed points for $G \subset \mathrm{Diff}^1(\Sigma)$. Then the linear representation of G by taking derivatives at x either gives a homomorphism to $\mathbb{Z}$ via determinant, or via conjugation to the unipotent subgroup $\left(\begin{smallmatrix} 1 & * \\ 0 & 1 \end{smallmatrix}\right)$, or is trivial. In the last case, Thurston stability produces a homomorphism to $\mathbb{Z}$. Thus, to apply this reasoning to a non-realization problem in $\mathrm{Diff}^1(\Sigma)$, one aims to force the existence of accumulation points of fixed points for a finitely-generated subgroup G with $H^1(G; \mathbb{Q}) = 0$.

As an illustrative example, here is a special case of a result of Salter-Tshishiku.

THEOREM 3.9 (Salter-Tshishiku [**ST16**]). *Let $n \geq 5$. There is no realization of $\mathrm{Diff}^1(\mathbb{D}^2, \mathbf{z}_n) \to B_n$.*

PROOF. Suppose $\phi : B_n \to \mathrm{Diff}^1(\mathbb{D}^2, \mathbf{z}_n)$ is a realization. The commutator subgroup of B_n is finitely generated, and by a theorem of Gorin–Lin [**GL69**], is perfect provided that $n \geq 5$. Let x be a point in the frontier of $\mathrm{fix}(\phi([B_n, B_n]))$, which is nonempty since $\phi(B_n)$ pointwise fixes $\partial\mathbb{D}^2$. Now use the argument from Application 3.8 to derive a contradiction. □

Salter–Tshishiku actually prove the stronger result that, for any surface $\Sigma_{g,n}^b$ (possibly with $b = 0$) the surface braid subgroup of $\mathrm{Mod}(\Sigma_{g,n}^b)$ is not realized by C^1 diffeomorphisms whenever $n \geq 6$, then use the same technique to re-prove Morita's non-lifting Theorem 2.12 in class C^1 for surfaces of genus $g \geq 2$. The idea is as follows: Supposing that a realization $\mathrm{Mod}_g \to \mathrm{Diff}^1(\Sigma)$ exists, let $f \in \mathrm{Mod}_g$ be the hypereliptic involution. The Lefschetz fixed point theorem implies that $\phi(f)$ has $2g + 2$ fixed points; these are permuted by the centralizer of f. Using Dehn twists about explicit curves, one produces a subgroup of the centralizer isomorphic to a quotient of B_{2g+2}, acting with a global fixed point, and applies a Thurston stability argument to derive a contradiction.

We will show in §4 that B_3 can be realized by (smooth) diffeomorphisms of a disc with three marked points; however, the following are open.

QUESTION 3.10. Can B_4 be realized by diffeomorphisms?

QUESTION 3.11. For $n > 3$, can B_n be realized by homeomorphisms of the marked disc?

QUESTION 3.12. The argument of [**ST16**] does not generalize to finite index subgroups. For $n \geq 4$, is there a finite index subgroup of B_n that *can* be realized by diffeomorphisms? What about by homeomorphisms? As a particular case, what about the *pure braid group* $P_n := \ker[B_n \to S_n]$?

3.3. Franks–Handel nonlifting. Franks and Handel have a different proof of a general version of Morita's nonlifting Theorem 2.12 for genus ≥ 3.

THEOREM 3.13 (Franks–Handel [**FH09**]). *Let $f \in \mathrm{Mod}_g$ preserve a genus $h \geq 1$ subsurface S, and be pseudo-Anosov on S and isotopic to the identity on $\Sigma_g \setminus S$. Let $H \cong \mathrm{Mod}(\Sigma_g \setminus S) \subset \mathrm{Mod}_g$ be the subgroup of mapping classes of diffeomorphisms pointwise fixing S; this centralizes f. Provided that $\Sigma_g \setminus S$ has genus at least 2, $\langle f, H\rangle$ cannot be realized in* $\mathrm{Diff}^1(\Sigma_g)$.

The proof uses two of the dynamical tools we have discussed: Thurston stability and basins of attraction. The additional ingredients are a) the theory of *prime ends* – a compactification method compatible with group actions on open, connected, simply connected domains in the plane – and b) the Thurston–Nielsen classification of mapping classes and the dynamics of their boundary maps, which we describe next. Notably, the proof does not use torsion in $\mathrm{Mod}(\Sigma)$.

The Thurston–Nielsen classification. Building on work of Nielsen [**Nie44**], Thurston [**Thu88**] gave a classification of elements of the mapping class group of a surface. Each $\phi \in \mathrm{Mod}_g$ has a representative homeomorphism which is either finite order, pseudo-Anosov, or reducible. A homeomorphism f being *Pseudo-Anosov* means that the surface admits two f-invariant transverse measured foliations, whose transverse measures are expanded by λ and $1/\lambda$ (respectively), for some $\lambda > 0$. The foliations may have finitely many singularities, each a p-pronged saddle, for some p. *Reducible* means that there is a finite collection $\mathcal{A}$ of disjoint simple closed curves invariant under f such that, after passing to a power that fixes each curve, f preserves a small tubular neighborhood $N(\mathcal{A})$ of $\mathcal{A}$, and acts on each connected component of $\Sigma \setminus N(\mathcal{A})$ separately either by the identity, a pseudo-Anosov homeomorphism, or possibly a Dehn twist if the component is an annulus. The structure of reducible elements is discussed more in §4.2.

Any homeomorphism f of Σ can be lifted to a homeomorphism of the universal cover $\tilde{\Sigma} \cong \mathbb{H}^2$. The lift is a *quasi-isometry* of $\mathbb{H}^2$ so extends to a homeomorphism of the boundary S^1 of the Poincaré disc. Thinking of the mapping class group as the group of outer automorphisms of $\pi_1(\Sigma)$; a choice of lift of f fixes a choice of representative *automorphism* corresponding to the mapping class of f, and the action of the lift on S^1 agrees with the action of the automorphism on the Gromov boundary of $\pi_1(\Sigma)$. Thus, if f and g are isotopic, two isotopic lifts determine the same boundary action; and the Thurston–Nielsen classification above gives a helpful way to understand the boundary action of mapping classes, as will be used in the proof below. Thurston's survey [**Thu88**] is a good introduction to this theory; more details can be found in [**FM12**, §13] and [**FLP79**].

Using this machinery, we now sketch the proof of Franks and Handel's theorem. Their exposition is focused on the dynamics of homeomorphisms of the disc with fixed points; we focus on the dynamics of pseudo-Anosov mapping classes applicable to this situation.

PROOF SKETCH OF THEOREM 3.13. Suppose for contradiction that $\phi : G = \langle f, H \rangle \to \mathrm{Diff}^1(\Sigma_g)$ is a realization. One may check directly from a presentation of H that $H^1(H;\mathbb{Z}) = 0$, so the goal is to find an accumulation point of global fixed points for $\phi(H)$ and apply Thurston stability.

We first lift ϕ to an action on the universal cover, and study its action by homeomorphisms on the compactification of $\widetilde{\Sigma_g} \cong \mathbb{H}^2$ as $\mathbb{D}^2$. Let a be a pseudo-Anosov representative of f on S. After passing to a power of a (and f) if needed, we may assume a fixes a point $x \in S$, and (after a further power) also fixes all leaves through x. Choose a connected component $\widetilde{S}$ of the preimage of S in $\mathbb{H}^2$ and take a lift $\tilde{a}$ fixing $\tilde{x} \in \widetilde{S}$. Then the boundary action of $\tilde{a}$ has a finite, even number of fixed points, alternately attracting and repelling, corresponding to the endpoints of the leaves of the (lifted) foliations that pass through $\tilde{x}$. If x is a non-singular point, there will be two attractors and two repellers, otherwise, for a p-pronged singularity, there are p of each. Let $\tilde{f}$ be the lift of $\phi(f)$ with boundary action agreeing with $\tilde{a}$; these lifts have the property that $\tilde{f}$ and $\tilde{a}$ are isotopic. There is also a canonical choice of lift $\tilde{H}$ of $\phi(H)$ pointwise fixing $\partial\widetilde{S} \subset \partial\mathbb{H}^2$. This will commute with $\tilde{f}$, giving a well-defined lift of ϕ.

The next step is to understand the dynamics of the action of $\tilde{f}$ on the interior of the disc. Although f is only isotopic to a, so *a priori* it can have very different dynamics, we claim that the attracting points of $\tilde{f}$ on the boundary really are attractors, in the sense that there exists an open neighborhood $W \subset \mathbb{D}^2$ of each such point p with the property that $\tilde{f}^n(W)$ converges to p. To see this, we use the fact that f is a quasi-isometry of $\mathbb{D}^2$, so there exists K such that the image of each geodesic under f is a K-quasi-geodesic. Also, since $\tilde{f}$ and $\tilde{a}$ are isotopic, there exists K' such that $d(\tilde{f}(x), \tilde{a}(x)) < K'$ (in the hyperbolic distance) for all $x \in \mathbb{H}^2$. Recall also that a expands length along the singular leaf L ending at p by some $\lambda > 1$. Let γ be a bi-infinite geodesic perpendicular to a passing through L at distance t from $\tilde{x}$. Then $\tilde{f}(\gamma)$ is a K quasi-geodesic passing through a point at distance K' from a point distance λt from $\tilde{x}$ along L, and it is easy to verify that whenever t is sufficiently large, $\tilde{f}(\gamma)$ is contained inside the hemisphere cut out by γ. This suffices to show that p is an attracting point.

Let U denote the basin of attraction of $\tilde{f}$ at p (as defined in Section 3.1). Since $\tilde{f}$ and $\tilde{H}$ commute, and H fixes p, the basin U is $\tilde{H}$-invariant. An easy improvement of this argument shows that, in fact, for every $\tilde{h} \in \tilde{H}$, there exists $n \in N$ such that for each attractor or repeller z of $\tilde{f}$ in $\partial\mathbb{H}^2$, the map $\tilde{f}^n\tilde{h}$ has z as an attractor (or repeller) with the same basin at z as f.

Now we find a point $y \in \mathbb{H}^2$ fixed by any such homeomorphism $\tilde{f}^n \circ \tilde{h}$. As an easy case, suppose U were an open ball with boundary topologically an embedded circle C. Let y be the closest point of $\mathrm{fix}(\tilde{f}^n \circ \tilde{h}) \cap C$ to the right of the repelling point q adjacent to p, as indicated in the figure. Using the fact that f and H commute, one shows this "first point" for $\tilde{f}^n \circ \tilde{h}$ is actually independent of h, and so fixed by each such $\tilde{f}^n\tilde{h}$. When $h = \mathrm{id}$, we may take $n = 0$, so this shows it is also a fixed point of $\tilde{f}$, and hence of $\tilde{H}$ as well. In general, U is not necessarily a homeomorphic disc with circle boundary, but one can show that U is open, connected, and simply connected,

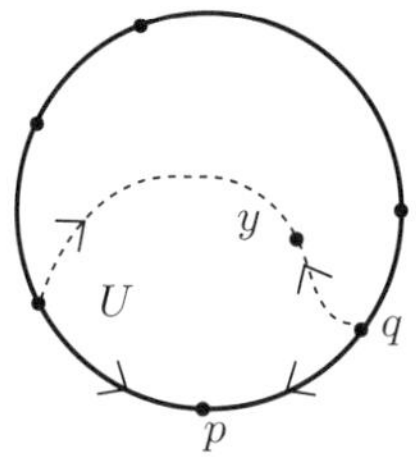

FIGURE 1. Dynamics of $\tilde{f}$ on $\overline{\mathbb{H}^2}$

and thus replace ∂U in the argument above with the *prime end compactification* of U to finish the argument in an analogous manner.

This gives us a single fixed point for $\phi(H)$ in Σ_g. To produce infinitely many more, one repeats the argument above using other powers of a. It is a standard fact (a consequence of Poincaré recurrence) that pseudo-Anosov diffeomorphisms have infinitely many periodic points – in fact, a dense set in the surface. See [**FM12**, §14]. These are partitioned into (infinitely many distinct) *Nielsen classes*, where x and y are said to be equivalent if there exists a lift of a^n to $\mathbb{H}^2$ that fixes both x and y. The argument we gave above produced a global fixed point y for $\phi(H)$ in the same Nielsen class as the original periodic point x of a. Thus, starting with points from different Nielsen classes for a will produce infinitely many distinct fixed points for $\phi(H)$, which must accumulate somewhere on Σ. This is what we needed to show. □

3.4. Obstructions to realizing non-discrete groups: Forcing continuity. We return to discussing realization problems for large (non finitely-generated) subgroups under the "restrict-to-boundary" map. Groups of *quasiconformal* homeomorphisms are an interesting case. Quasiconformal mappings are those with "bounded deviation" from conformal maps: as a special case, if f is a differentiable map defined on an open domain $U \subset \mathbb{C}$, f is quasiconformal if $\sup_{p \in U} \frac{|f_z(p)|+|f_{\bar{z}}(p)|}{|f_z(p)|-|f_{\bar{z}}(p)|}$ is finite. There are other formulations of the definition which apply to homeomorphisms of arbitrary metric spaces. The pseudo-Anosov maps from the Thurston–Nielsen classification above give examples of quasiconformal homeomorphisms between Riemann surfaces; indeed one approach to proving the classification (due to Bers [**Ber78**]) is through the theory of quasiconformal maps.

A quasiconformal homeomorphism of a surface with boundary restricts to what is called a *quasi-symmetric* homeomorphism of S^1. Thus, there is a natural surjection $\mathrm{QC}(\mathbb{D}^2) \to \mathrm{QS}(S^1)$. The section problem for this map was answered negatively by Epstein and Markovic.

THEOREM 3.14 (Epstein–Markovic [**EM07**]). *There is no section of the restriction map* $\mathrm{QC}(\mathbb{D}^2) \to \mathrm{QS}(S^1)$.

As explained in [**EM07**] (and perhaps part of their original motivation) there is a nice connection between this problem and the Nielsen realization problem for Mod_g. If $\mathrm{QC}(\mathbb{D}^2) \to \mathrm{QS}(S^1)$ had a section *with the additional property* that the extension over $\mathrm{PSL}(2,\mathbb{R})$ were the identity, then the mapping class group $\mathrm{Mod}(\Sigma_g)$

could be realized by homeomorphisms, as follows: Let G be the group of all boundary mappings of lifts of homeomorphisms of Σ_g to $\mathbb{D}^2 = \overline{\mathbb{H}^2}$. Since the representative diffeomorphisms from the Thurston–Nielsen classification of mapping-classes are quasiconformal, these boundary maps are all quasi-symmetric. There is an exact sequence $\pi_1(\Sigma_g) \to G \to \mathrm{Mod}(\Sigma_g)$. Assuming a section as above, let ϕ be a lift of G to $\mathrm{QC}(\mathbb{D}^2)$; then $\pi_1(\Sigma_g)$ acts by Möbius transformations. The quotient of $\phi(G)$ by $\phi(\pi_1(\Sigma_g))$ then gives a realization of $\mathrm{Mod}(\Sigma_g)$ in the group of quasiconformal homeomorphisms of Σ_g.

A major step in Epstein and Markovic's proof of Theorem 3.14 is to show that any such realization would be forced to be *continuous*, and so they ask whether every section of the boundary map $\mathrm{Homeo}(\mathbb{D}^2) \to \mathrm{Homeo}(S^1)$ is necessarily continuous. Note that at least one section exists, namely from "coning off" the circle to the disc. More generally, one can ask the following.

QUESTION 3.15. Let M be a manifold with boundary, fix $r > 0$, and suppose that $\mathrm{Diff}_0^r(M) \to \mathrm{Diff}_0^r(\partial M)$ admits a section. Is that section necessarily continuous?

Only two results are known in this direction, one in C^∞ and one in C^0 regularity. Both prove something much stronger than the answer to the question above, but both use their respective assumptions on regularity in an essential way.

THEOREM 3.16 (Hurtado [**Hur15**]). *Let M and N be closed manifolds, and $\phi : \mathrm{Diff}_0^\infty(M) \to \mathrm{Diff}_0^\infty(N)$ any homomorphism. Then ϕ is continuous.*

THEOREM 3.17 ([**Man16**]). *Let M be a closed manifold and G a separable topological group. Any homomorphism $\phi : \mathrm{Homeo}_0(M) \to G$ is continuous.*

We conclude this section with two additional open questions.

QUESTION 3.18 (Epstein–Markovic). Fix $n > 2$. Does the map $\mathrm{QC}(\mathbb{D}^n) \to \mathrm{Homeo}(S^{n-1})$ have a section over its image?

QUESTION 3.19. Let M be a manifold with boundary. If $\mathrm{Diff}_0^r(M) \to \mathrm{Diff}_0^r(\partial M)$ admits a section, what can one say about the topology of M? Are there examples where the action of $\mathrm{Diff}_0^r(\partial M)$ does not preserve a foliation?

For Diff_0^∞ this question was asked by Ghys [**Ghy91**], who also asks specifically whether the map admits a section when M is a handlebody.

4. Positive results

Up to now we have only discussed *negative* results and obstructions to realization. In this section we give some examples of large groups of mapping classes that *are* realized by diffeomorphisms and techniques to realize them.

4.1. Automorphisms, blow-ups and smoothings. The mapping class group of the torus $\mathrm{Mod}_{1,1} \cong \mathrm{Mod}_1 \cong \mathrm{SL}(2,\mathbb{Z})$ is realized by linear diffeomorphisms of $\mathbb{T}^2 := \mathbb{R}^2/\mathbb{Z}^2$ (with 0 as a marked point). Automorphism groups of translation surfaces are a natural generalization of this example.

A *translation surface* is a surface with the additional structure of a holomorphic 1-form; equivalently, it is a surface obtained by taking a finite collection of polygons in $\mathbb{R}^2$ and identifying parallel sides in pairs by translations. Such a surface inherits a singular Euclidean structure with finitely many cone singularities, each with angle a positive multiple of 2π. An (orientation preserving) *affine automorphism* of a translation surface X is a homeomorphism of X that permutes its singularities and is locally in $\mathrm{Aff}^+(\mathbb{R}^2)$ at all other points. These form a group, denoted $\mathrm{Aut}^+(X)$.

While the automorphism group of the square torus includes both $\mathrm{SL}(2, \mathbb{Z})$ and all translations in $\mathbb{R}^2/\mathbb{Z}^2$, when a translation surface X has at least one singularity, the map $\mathrm{Aut}^+(X) \to \mathrm{Mod}(X)$ is always injective. By definition $\mathrm{Aut}^+(X) < \mathrm{Mod}(X)$ is realized by homeomorphisms, but not obviously by *diffeomorphisms*. In this section, we discuss techniques that allow one to promote an action by homeomorphisms to a smooth action, and use this to realize affine automorphisms and other groups.

Realizing $\mathrm{Aut}^+(X)$ **by diffeomorphisms.** Let X be a translation surface with holomorphic 1-form ω. Assume X is not a flat torus, so has at least one singularity. Taking the linear parts of the defining affine maps gives a homomorphism $\mathrm{Aut}^+(X) \to \mathrm{SL}(2, \mathbb{R})$ whose image, denoted $\mathrm{SL}(X, \omega)$, is a discrete subgroup. This subgroup may be finite covolume or not, but is never cocompact. (For a gentle introduction, see [**Sch11**].) We show how to realize $\mathrm{Aut}^+(X)$ by diffeomorphisms, under the simplifying assumption that $\mathrm{SL}(X, \omega)$ is finitely generated.

In this case, $\mathrm{SL}(X, \omega)$ has a presentation of the form

$$\langle a_1, b_1, \ldots, a_g, b_g, c_1, \ldots, c_m, d_1, \ldots, d_n \mid [a_1, b_1] \cdots [a_g, b_g] c_1 \cdots c_m d_1 \cdots d_n, (c_i)^{q_i} \rangle$$

where $q_i \in \mathbb{N}$, $n \geq 1$ and $g, m \geq 0$. Let $p_1, \ldots, p_k$ denote the singularities of X. The stabilizer of p_i is a finite index subgroup of $\mathrm{SL}(X, \omega)$, so finitely generated and of the form above. The open surface $X \setminus \{p_1, \ldots, p_k\}$ can be compactified by adding a circle of "straight line directions" at each end, giving a surface $\overline{X}$ with k boundary components, to which the action of $\mathrm{Aut}^+(X)$ extends.

If p_1 has cone angle $2\pi M$, its stabilizer in $\mathrm{SL}(X, \omega)$ acts on the circle of directions by lifts of the linear action of $\mathrm{SL}(2, \mathbb{R})$ on the circle of rays from the origin in $\mathbb{R}$ (topologically a circle) to the M-fold cover of this circle. Other elements of the stabilizer in $\mathrm{Aut}^+(X)$, if any, act by rotations of order k/M, commuting with the lifted linear action.

This action can be smoothly isotoped to one where, using the presentation above, all the a_i, b_i, and $d_1, \ldots, d_{n-1}$ act trivially, and the c_i and d_n lie in $\mathrm{SO}(2)$ (we do not change the action of elements that map to the identity in $\mathrm{SL}(X, \omega)$). To do this, work within the cyclic extension of $\mathrm{SL}(2, \mathbb{R})$ by $\mathbb{Z}/M\mathbb{Z}$ in which these elements lie, and isotope a_i, b_i, and d_i, smoothly towards the identity, while isotoping each c_i (through the space of order q_i elements) to rotations. The product $[a_1, b_1] \ldots [a_g, b_g] c_1 \ldots c_m d_1 \ldots d_{n-1}$ varies smoothly through the isotopy, so can be taken to define an isotopy of d_n^{-1}. This isotopy of the action allows one to "cone off" the circle to a disc, extending the action of the stabilizer smoothly over the disc. A variation of this argument is given in detail in the next example.

If $\mathrm{Aut}^+(X)$ acts transitively on the singularities, this is enough to determine a smooth action of $\mathrm{Aut}^+(X)$ on a closed surface obtained by gluing discs to each boundary component of $\overline{X}$ (which is topologically the surface X). If not, repeat the procedure for each orbit of the action. □

A realization of B_3 **by diffeomorphisms.** Thurston observed that the braid group B_3 could be realized by homeomorphisms. Building on our work above, we describe his construction and prove the following.

THEOREM 4.1. *There is a section of* $\mathrm{Diff}^\infty_\partial(\mathbb{D}^2, \mathbf{z}_3) \to \mathrm{Mod}(\Sigma^1_{0,3}) = B_3$.

PROOF OF THEOREM 4.1. Consider the linear action of $\mathrm{SL}(2,\mathbb{Z})$ on $\mathbb{T}^2 = \mathbb{R}^2/\mathbb{Z}^2$. The center $\{\pm 1\}$ of $\mathrm{SL}(2,\mathbb{Z})$ acts on $\mathbb{T}^2$ as the hyperelliptic involution; it has four fixed points $\{0\} \cup \{p_1, p_2, p_3\}$, which $\mathrm{SL}(2,\mathbb{Z})$ permutes (later we will use that these points are $p_1 = (0, \frac{1}{2})$, $p_2 = (\frac{1}{2}, \frac{1}{2})$, $p_3 = (\frac{1}{2}, 0)$, when viewed in $\mathbb{R}^2$). The point 0 is a global fixed point and $\mathrm{SL}(2,\mathbb{Z})$ acts on $\mathbf{p} = \{p_1, p_2, p_3\}$ transitively. The quotient $\mathrm{PSL}(2,\mathbb{Z}) = \mathrm{SL}(2,\mathbb{Z})/\{\pm 1\}$ acts on the quotient $\mathbb{T}^2/\{\pm 1\}$, which is a sphere with four cone points, each of angle π. (In the language of the previous section, this sphere is a *half translation surface*, or Riemann surface equipped with a *quadratic differential.*)

Since 0 is a global fixed point, we can compactify $\mathbb{T}^2/\{\pm 1\} \setminus \{0\}$ by adding a circle of directions at 0, obtaining an action of $\mathrm{PSL}(2,\mathbb{Z})$ on a disk with 3 marked points $\Sigma^1_{0,3} = (\mathbb{D}^2, \mathbf{p})$.

To relate B_3 to $\mathrm{PSL}(2,\mathbb{Z})$, note that there is an exact sequence $1 \to \mathbb{Z} \to B_3 \to \mathrm{PSL}_2(\mathbb{Z}) \to 1$. On the level of mapping class groups, the projection $B_3 \to \mathrm{PSL}(2,\mathbb{Z})$ is equal to the map

$$B_3 \simeq \mathrm{Mod}^1_{0,3} \to \mathrm{Mod}^0_{0,4} \simeq (\mathbb{Z}/2\mathbb{Z} \times \mathbb{Z}/2\mathbb{Z}) \rtimes \mathrm{PSL}(2,\mathbb{Z}) \to \mathrm{PSL}(2,\mathbb{Z})$$

induced by the map $\Sigma^1_{0,3} \to \Sigma^0_{0,4}$ that collapses the boundary component of to a point (see [**FM12**, Prop. 2.7]). On the group level $\mathrm{PSL}(2,\mathbb{Z}) \simeq \mathbb{Z}/2\mathbb{Z} * \mathbb{Z}/3\mathbb{Z} = \langle x, y : x^2 = 1 = y^3 \rangle$, and B_3 has a presentation $\langle a, b : a^2 = b^3 \rangle$, with $B_3 \to \mathrm{PSL}(2,\mathbb{Z})$ the obvious homomorphism. Thus, we can view the $\mathrm{PSL}(2,\mathbb{Z})$ action on $(\mathbb{D}^2, \mathbf{p})$ as a non-faithful action of B_3. To obtain a smooth realization of B_3, we will (1) isotope the action of $x, y \in \mathrm{PSL}_2(\mathbb{Z})$ on the $\partial\mathbb{D}^2$ to the identity, preserving the relation $x^2 = y^3$ to get an action by homeomorphisms, and (2) glue discs in at singularities, as we did for translation surfaces above, to smooth the action.

For (1), attach an annulus $A \simeq S^1 \times [0,1]$ to $\partial\mathbb{D}^2$ along $S^1 \times \{0\}$, enlarging the disc. Under suitable parameterization, $a \in B_3$ acts on $\partial\mathbb{D}^2$ by a standard order two rotation, and b by an order 3 projectively linear map. Let b_t, $0 \le t \le 1/2$ be a smooth path of conjugates of b through $\mathrm{PSL}(2,\mathbb{R})$ such that $b_0 = b$, and $b_{1/2} \in \mathrm{SO}(2)$. Now extend this to a smooth path through $\mathrm{SO}(2)$ for $1/2 \le t \le 1$, with $b_{1-\epsilon} = \mathrm{id}$ for all small ϵ. Let a_t be a smooth path in $\mathrm{SO}(2)$ from $a_0 = a$ to $a_1 = \mathrm{id}$ such that $a_t^2 = b_t^3$ for all t. Defining a, b on $S^1 \times \{t\} \subset A$ to agree with a_t, b_t gives an extension of the action to a smooth action on the annulus that is identity in a neighborhood of $S^1 \times \{1\}$.

For (2), as in the argument for translation surfaces, compactify $\mathbb{D}^2 \setminus \mathbf{p}$ by a circle C_i of straight line directions at each p_i. Observe that the action of $\mathrm{PSL}(2,\mathbb{Z})$ on $C_1 \cup C_2 \cup C_3$ is the action of $\mathrm{PSL}(2,\mathbb{Z})$ on the projectivized unit tangent circles at p_1, p_2, p_3 (which we may identify using the natural trivialization of $T\mathbb{D}^2$). Under our identification we have

$$\begin{aligned} &a : C_2 \to C_2 \text{ is the identity, } a : C_1 \to C_3 \text{ agrees with } \left(\begin{smallmatrix} 0 & -1 \\ 1 & 0 \end{smallmatrix}\right) \in \mathrm{PSL}(2,\mathbb{R}), \\ &b : C_i \to C_{i+1} \text{ agrees with } \left(\begin{smallmatrix} 0 & -1 \\ 1 & 1 \end{smallmatrix}\right) \in \mathrm{PSL}(2,\mathbb{R}). \end{aligned} \tag{4.1}$$

Glue a disc D_i to each C_i and fix smooth collar neighborhoods of $C_i \subset D_i$ parametrized by $C_i \times [0,1]$. To extend to this collar, we can choose smooth isotopies of a and b from the maps in (4.1) to the identity, and extend the action to $\cup_i C_i \times \{t\}$ by a_t, b_t, so that a_0, b_0 agree with rigid translations on $\cup_i C_i \times \{0\}$. Then we can extend the action to the rest of D_i by rigid translations. The result is an action by diffeomorphisms of a and b on a disc, fixing the boundary pointwise, and permuting the centers of D_1, D_2 and D_3, which we take as three marked points. The reader may easily check that the mapping classes of a and b agree with the standard generators of B_3, as required. □

Building on this line of argument, we also give a positive answer to Nariman's question mentioned at the end of Section 2.

THEOREM 4.2. *Let* $\mathbf{z} = \mathbf{z}_{2g+2}$ *for some* $g \geq 1$. *There is a continuous homomorphism* $\psi : \mathrm{Diff}(\mathbb{D}^2, \mathbf{z}) \to \mathrm{Diff}(\Sigma_g^2, \partial\Sigma_g^2)$ *that induces the geometric homomorphism* $B_{2g+2} \to \mathrm{Mod}_g^2$ *on mapping class groups.*

While Nariman showed there is no cohmological obstruction to this lift, the existence of the realization is somewhat surprising, given a few of the results we have already mentioned. Specifically, Theorem 3.9 implies that ψ cannot be obtained by a map that factors through B_{2g+2}, and Theorem 3.16 (which has a variation for manifolds that are not closed) also implies that such a map ψ should essentially be continuous, with further work of Hurtado implying that its restriction to the subgroup $\mathrm{Diff}_c(\mathbb{D}^2, \mathbf{z})$ of diffeomorphisms fixing a neighborhood of $\mathbf{z}$ (which we know to be nontrivial by [**ST16**]) must be obtained by embedding copies of covers of the open, punctured disc into Σ_g^2. This suggests, at least vaguely, that ψ would have to be obtained by branching the punctured disc over $\mathbf{z}$, an inherently non-smooth construction.

The proof uses two constructions, one a trick for smooth gluing, and the other a classical "blow up" procedure similar to the technique above to deal with singularities of translation surfaces.

CONSTRUCTION 4.3 (Smoothing actions glued on a codimension 1 submanifold). Let G be a group acting by smooth diffeomorphisms on manifolds S_1 and S_2. Let X_1 and X_2 be diffeomorphic connected components of ∂S_1 and ∂S_2 respectively, and let S be the manifold obtained by gluing S_1 and S_2 by a diffeomorphism $X_1 \to X_2$. If, for each $g \in G$, the action of g on X_1 agrees with that on X_2 under the identification used in the gluing, then there is an obvious induced action of G on S by homeomorphisms. However, this is *conjugate* to an action by *smooth diffeomorphisms* on S. The conjugacy can be obtained by a map $f : S \to S$ which is the identity outside a tubular neighborhood of the glued boundary components, and in the tubular neighborhood (identified with $X \times [-1, 1]$, with the glued boundary components at $X \times \{0\}$) is locally a very strong contraction at 0; taken strong enough so that fgf^{-1} becomes infinitely tangent to the identity in the direction transverse to the boundary at $X \times \{0\}$. Details are worked out in [**Par15**] using the local contraction $(x, y) \mapsto (x, e^{\frac{-1}{e^{-1/|y|}}})$.

CONSTRUCTION 4.4 (Blow-up). Let $\mathbf{p} = \{p_1, p_2, ..., p_k\}$ be a finite set of points in a manifold M^n. The (oriented) *blowup of* M *at* $\mathbf{p}$ is a smooth manifold $\hat{M}$ obtained from M by replacing each $x \in \mathbf{p}$ with the space of directions $T_x^1 M \simeq S^{n-1}$ in its tangent space. There is a natural projection $\Phi : \hat{M} \to M$ that is a diffeomorphism away from $\mathbf{p}$, and a natural injection $\Phi^* : \mathrm{Diff}^r(M, \mathbf{p}) \to \mathrm{Diff}^{r-1}(\hat{M})$ for $r \geq 1$. In particular, any C^1 action on $(M, \mathbf{p})$ induces an action on $\hat{M}$ by homeomorphisms.

PROOF OF THEOREM 4.2. Our strategy is to first build a map $\phi : \mathrm{Diff}(\mathbb{D}^2, \mathbf{z}) \to \mathrm{Diff}(\mathbb{D}^2, \mathbf{z})$, which has image in a subgroup that acts on a given neighborhood of $\mathbf{z}$

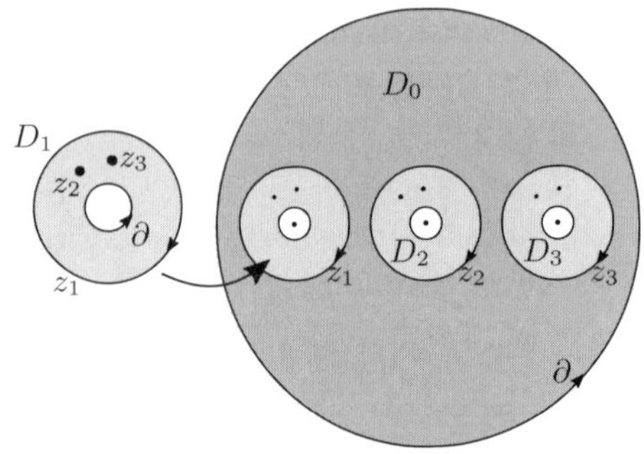

FIGURE 2. Gluing copies of $\mathbb{D}^2$ blown up at one point of $\mathbf{z}$ into a copy of $\mathbb{D}^2$ blown up at $\mathbf{z}$

by translations, and also induces the identity map $B_n \to B_n$. Building this map is the bulk of the construction. Given such a map ϕ, the diffeomorphisms in its image can then be lifted to *diffeomorphisms* of a cover branched over $\mathbf{z}$ as described above.

To start, apply Construction 4.4 to blow up $\mathbb{D}^2$ at the set $\mathbf{z}$. The new surface obtained (call it D_0) has $n+1$ boundary components, one corresponding to the original boundary ∂ of the disc, and the others corresponding to the blown up points.

Enumerate $\mathbf{z} = \{z_1, z_2, ..., z_n\}$ and for $i = 1, 2, ..., n$, let D_i be a blow-up of $\mathbb{D}^2$ at $\{z_i\}$. Glue each D_i to D_0 along the blow-up of z_i, using the identity map on the space of tangent directions at z_i. The result is an $(n+1)$-holed sphere. Now embed this $(n+1)$-holed sphere into $\mathbb{D}^2$ with the boundary component ∂ mapping to $\partial\mathbb{D}^2$. The result is pictured in Figure 2; boundary components of the D_i are labeled by their images under the map from the blow-up construction.

Let $C_1, C_2, ...C_n$ be the connected components of the complement of the image of the embedding (shown in white on the figure). We may arrange the embedding so that each C_i is a round disc of radius ϵ, centered around the marked point z_i on the original disc $\mathbb{D}^2$ (shown as midpoints of the white regions of the figure). Construction 4.4 gives a natural homomorphism from $\text{Diff}(\mathbb{D}^2, \mathbf{z})$ to $\text{Diff}(D_i)$. These actions of $\text{Diff}(\mathbb{D}^2, \mathbf{z})$ on the various D_i for $i = 0, 1, ..., n$ agree on their glued boundary components, so Construction 4.3 produces a homomorphism from $\text{Diff}(\mathbb{D}^2, \mathbf{z})$ to the diffeomorphisms of the $n+1$ holed sphere that was obtained by gluing the D_i together. We identify this surface with the image of its embedding in $\mathbb{D}^2$. Since elements of $\text{Diff}(\mathbb{D}^2, \mathbf{z})$ fix a neighborhood of $\partial\mathbb{D}^2$ pointwise, we may also arrange the embedding so that this action permutes the boundaries of the complementary discs C_i by *rigid translations.* Thus, the action naturally extends to an action on $\mathbb{D}^2$ by diffeomorphisms, permuting the discs C_i by translations. In particular, the set of midpoints of the C_i is preserved, so this action is by elements of $\text{Diff}(\mathbb{D}^2, \mathbf{z})$. Let $\phi : \text{Diff}(\mathbb{D}^2, \mathbf{z}) \to \text{Diff}(\mathbb{D}^2, \mathbf{z})$ denote this action.

Finally we check that ϕ induces the identity map on the quotient $B_n = \text{Diff}(\mathbb{D}^2, \mathbf{z}) / \text{Diff}_0(\mathbb{D}^2, \mathbf{z})$. By design of our blow-up, gluing, and embedding, if $f(z_i) = z_j$, then $\phi(f)$ maps D_i to D_j, hence maps C_i to C_j, and its center point z_i to z_j.

We may now quickly finish the proof. As explained in §2, there is an injective map $\Psi : \mathrm{Diff}(\mathbb{D}^2, \mathbf{z}) \to \mathrm{Homeo}_\partial(\Sigma_g^2)$. Consider the map $\Psi \circ \phi : \mathrm{Diff}(\mathbb{D}^2, \mathbf{z}) \to \mathrm{Homeo}_\partial(\Sigma_g^2)$, which agrees with Ψ on mapping class groups. Each diffeomorphism in the image of ϕ has trivial (i.e. constant $\equiv id$) derivative in a neighborhood of each $z \in \mathbf{z}$, so its image under Ψ is smooth everywhere. Thus, $\Psi \circ \phi$ gives the desired map $\mathrm{Diff}(\mathbb{D}^2, \mathbf{z}) \to \mathrm{Diff}_\partial(\Sigma_g^2)$. □

4.2. Realizing free abelian and right-angled Artin groups. Recall from §3.3 the Nielsen–Thurston classification of elements of $\mathrm{Mod}(\Sigma)$ (finite order, reducible, pseudo-Anosov). We now discuss reducible elements in more detail. By definition, a reducible $\phi \in \mathrm{Mod}(\Sigma)$ admits a *reduction system*, a finite collection $\mathcal{A}$ of disjoint isotopy classes of simple closed curves that is invariant under ϕ. The intersection of all *maximal* reduction systems is called the *canonical reduction system* $\mathcal{A}_\phi$.

Birman–Lubotzky–McCarthy [**BLM83**] use reduction systems to show that all solvable subgroups of $\mathrm{Mod}(\Sigma_{g,n}^b)$ are virtually abelian, and give a (sharp) bound on the torsion-free rank of abelian subgroups. Farb [**Far06b**, Ch. 2, §6.3] remarks (without proof) that their perspective can be used to show that all free abelian subgroups $\mathbb{Z}^n < \mathrm{Mod}_{g,m}^b$ can be realized by homeomorphisms. In essence, since the image of a reduction system $\mathcal{A}$ for ϕ under some element ψ is a reduction system for $\psi\phi\psi^{-1}$, whenever $\Gamma < \mathrm{Mod}(\Sigma_{g,n}^b)$ is abelian, then the union $\bigcup_{\phi \in \Gamma} \mathcal{A}_\phi$ is a reduction system for each element of Γ, and restrictions of commuting elements to subsurfaces are necessarily either powers of the same pseudo-Anosov, or of the same Dehn twist, or possibly the identity. From there (proceeding with some care on annular neighborhoods of the reducing system) one can choose commuting representative homeomorphisms for elements of Γ.

A natural next step is to generalize this strategy to *right-angled Artin subgroups* of $\mathrm{Mod}(\Sigma)$. A right-angled Artin group $A(G)$ is defined by a finite undirected graph $G = (V, E)$ via

$$A(G) = \langle x \in V \mid [x, y] = 1 \text{ if } \{x, y\} \in E \rangle.$$

Hence, free groups (which are always realizable) correspond to totally disconnected graphs, and free abelian groups to complete graphs. There are many other interesting RAAG subgroups of mapping class groups, see [**Kob17**, §8] and references therein.

QUESTION 4.5. Is every right-angled Artin subgroup $A(G) < \mathrm{Mod}(\Sigma)$ realized by homeomorphisms?

We remark that the reduction-system strategy above seems to go through in some special cases; for instance, if $A(G) < \mathrm{Mod}(\Sigma)$ is such that each $x \in V$ *fixes* each curve in $\mathcal{A}_x$. However, there are many RAAG subgroups that do not have this property.

Acknowledgments

The authors would like to thank B. Farb, N. Salter and A. Kupers for feedback on this work. Theorem 3.1 came up in discussion with F. Le Roux. Thanks also to the referee for carefully reading the paper and offering many helpful suggestions.

References

[And58] R. D. Anderson, *The algebraic simplicity of certain groups of homeomorphisms*, Amer. J. Math. **80** (1958), 955–963, DOI 10.2307/2372842. MR0098145

[BCS13] Mladen Bestvina, Thomas Church, and Juan Souto, *Some groups of mapping classes not realized by diffeomorphisms*, Comment. Math. Helv. **88** (2013), no. 1, 205–220, DOI 10.4171/CMH/283. MR3008918

[Ber78] Lipman Bers, *An extremal problem for quasiconformal mappings and a theorem by Thurston*, Acta Math. **141** (1978), no. 1-2, 73–98, DOI 10.1007/BF02545743. MR0477161

[BFH16] A. Brown, D. Fisher, and S. Hurtado. Zimmer's conjecture: Subexponential growth, measure rigidity, and strong property (T). preprint. arxiv:1608.04995, December 2016.

[BLM83] Joan S. Birman, Alex Lubotzky, and John McCarthy, *Abelian and solvable subgroups of the mapping class groups*, Duke Math. J. **50** (1983), no. 4, 1107–1120, DOI 10.1215/S0012-7094-83-05046-9. MR726319

[Bow11] Jonathan Bowden, *Flat structures on surface bundles*, Algebr. Geom. Topol. **11** (2011), no. 4, 2207–2235, DOI 10.2140/agt.2011.11.2207. MR2826937

[BW08] Jonathan Block and Shmuel Weinberger, *On the generalized Nielsen realization problem*, Comment. Math. Helv. **83** (2008), no. 1, 21–33, DOI 10.4171/CMH/116. MR2365406

[Cal12] P. Le Calvez. A periodicity criterion and the section problem on the mapping class group, 2012.

[CC00] Alberto Candel and Lawrence Conlon, *Foliations. I*, Graduate Studies in Mathematics, vol. 23, American Mathematical Society, Providence, RI, 2000. MR1732868

[CC08] Serge Cantat and Dominique Cerveau, *Analytic actions of mapping class groups on surfaces*, J. Topol. **1** (2008), no. 4, 910–922, DOI 10.1112/jtopol/jtn028. MR2461860

[DNR14] B. Deroin, A. Navas, and C. Rivas. Groups, orders, and dynamics. preprint. arXiv:1408.5805, October 2014.

[EE67] C. J. Earle and J. Eells, *The diffeomorphism group of a compact Riemann surface*, Bull. Amer. Math. Soc. **73** (1967), 557–559, DOI 10.1090/S0002-9904-1967-11746-4. MR0212840

[EK71] Robert D. Edwards and Robion C. Kirby, *Deformations of spaces of imbeddings*, Ann. Math. (2) **93** (1971), 63–88. MR0283802

[EM07] David Epstein and Vladimir Markovic, *Extending homeomorphisms of the circle to quasiconformal homeomorphisms of the disk*, Geom. Topol. **11** (2007), 517–595, DOI 10.2140/gt.2007.11.517. MR2302497

[ES70] C. J. Earle and A. Schatz, *Teichmüller theory for surfaces with boundary*, J. Differential Geometry **4** (1970), 169–185. MR0277000

[Far06a] Benson Farb, *Some problems on mapping class groups and moduli space*, Problems on mapping class groups and related topics, Proc. Sympos. Pure Math., vol. 74, Amer. Math. Soc., Providence, RI, 2006, pp. 11–55, DOI 10.1090/pspum/074/2264130. MR2264130

[Far06b] Benson Farb, *Some problems on mapping class groups and moduli space*, Problems on mapping class groups and related topics, Proc. Sympos. Pure Math., vol. 74, Amer. Math. Soc., Providence, RI, 2006, pp. 11–55, DOI 10.1090/pspum/074/2264130. MR2264130

[Fen48] W. Fenchel, *Estensioni di gruppi discontinui e trasformazioni periodiche delle superficie* (Italian), Atti Accad. Naz Lincei. Rend. Cl. Sci. Fis. Mat. Nat.(8) **5** (1948), 326–329. MR0029161

[FH09] John Franks and Michael Handel, *Global fixed points for centralizers and Morita's theorem*, Geom. Topol. **13** (2009), no. 1, 87–98, DOI 10.2140/gt.2009.13.87. MR2469514

[Fil82] R. P. Filipkiewicz, *Isomorphisms between diffeomorphism groups*, Ergodic Theory Dynamical Systems **2** (1982), no. 2, 159–171 (1983). MR693972

[Fis60] Gordon M. Fisher, *On the group of all homeomorphisms of a manifold*, Trans. Amer. Math. Soc. **97** (1960), 193–212, DOI 10.2307/1993298. MR0117712

[FLP79] *Travaux de Thurston sur les surfaces* (French), Astérisque, vol. 66, Société Mathématique de France, Paris, 1979. Séminaire Orsay; With an English summary. MR568308

[FM12] Benson Farb and Dan Margalit, *A primer on mapping class groups*, Princeton Mathematical Series, vol. 49, Princeton University Press, Princeton, NJ, 2012. MR2850125

[Gab91] David Gabai, *Convergence groups are Fuchsian groups*, Bull. Amer. Math. Soc. (N.S.) **25** (1991), no. 2, 395-402, DOI 10.1090/S0273-0979-1991-16082-9. MR1102752

[Ghy91] Étienne Ghys, *Prolongements des difféomorphismes de la sphère* (French), Enseign. Math. (2) **37** (1991), no. 1-2, 45–59. MR1115743

[GL69] E. A. Gorin and V. Ja. Lin, *Algebraic equations with continuous coefficients, and certain questions of the algebraic theory of braids* (Russian), Mat. Sb. (N.S.) **78 (120)** (1969), 579–610. MR0251712

[Har85] John L. Harer, *Stability of the homology of the mapping class groups of orientable surfaces*, Ann. of Math. (2) **121** (1985), no. 2, 215–249, DOI 10.2307/1971172. MR786348

[Hat11] A. Hatcher. A short exposition of the Madsen-Weiss theorem. preprint. arXiv:1103.5223, January 2011.

[Hur15] Sebastian Hurtado, *Continuity of discrete homomorphisms of diffeomorphism groups*, Geom. Topol. **19** (2015), no. 4, 2117–2154, DOI 10.2140/gt.2015.19.2117. MR3375524

[Ker83] Steven P. Kerckhoff, *The Nielsen realization problem*, Ann. of Math. (2) **117** (1983), no. 2, 235–265, DOI 10.2307/2007076. MR690845

[Kir97] Rob Kirby (ed.), *Problems in low-dimensional topology*, Geometric topology (Athens, GA, 1993), AMS/IP Stud. Adv. Math., vol. 2, Amer. Math. Soc., Providence, RI, 1997, pp. 35–473. MR1470751

[Kob17] T. Koberda. Actions of right-angled Artin groups in low dimensions, 2017. arXiv:1707.05958.

[Man13] Kathryn Mann, *Diffeomorphism groups of balls and spheres*, New York J. Math. **19** (2013), 583–596. MR3119099

[Man15a] Kathryn Mann, *Left-orderable groups that don't act on the line*, Math. Z. **280** (2015), no. 3-4, 905–918, DOI 10.1007/s00209-015-1455-2. MR3369358

[Man15b] Kathryn Mann, *Spaces of surface group representations*, Invent. Math. **201** (2015), no. 2, 669–710, DOI 10.1007/s00222-014-0558-4. MR3370623

[Man16] Kathryn Mann, *Automatic continuity for homeomorphism groups and applications*, Geom. Topol. **20** (2016), no. 5, 3033–3056, DOI 10.2140/gt.2016.20.3033. With an appendix by Frédéric Le Roux and Mann. MR3556355

[Mar07] Vladimir Markovic, *Realization of the mapping class group by homeomorphisms*, Invent. Math. **168** (2007), no. 3, 523–566, DOI 10.1007/s00222-007-0039-0. MR2299561

[Mat74] John N. Mather, *Commutators of diffeomorphisms*, Comment. Math. Helv. **49** (1974), 512–528, DOI 10.1007/BF02566746. MR0356129

[Mat75] John N. Mather, *Commutators of diffeomorphisms. II*, Comment. Math. Helv. **50** (1975), 33–40, DOI 10.1007/BF02565731. MR0375382

[Mil58] John Milnor, *On the existence of a connection with curvature zero*, Comment. Math. Helv. **32** (1958), 215–223, DOI 10.1007/BF02564579. MR0095518

[Mor87] Shigeyuki Morita, *Characteristic classes of surface bundles*, Invent. Math. **90** (1987), no. 3, 551–577, DOI 10.1007/BF01389178. MR914849

[Mor01] Shigeyuki Morita, *Geometry of characteristic classes*, Translations of Mathematical Monographs, vol. 199, American Mathematical Society, Providence, RI, 2001. Translated from the 1999 Japanese original; Iwanami Series in Modern Mathematics. MR1826571

[MS08] Vladimir Markovic, *Realization of the mapping class group by homeomorphisms*, Invent. Math. **168** (2007), no. 3, 523–566, DOI 10.1007/s00222-007-0039-0. MR2299561

[MV08] Yoshihiko Mitsumatsu and Elmar Vogt, *Foliations and compact leaves on 4-manifolds. I. Realization and self-intersection of compact leaves*, Groups of diffeomorphisms, Adv. Stud. Pure Math., vol. 52, Math. Soc. Japan, Tokyo, 2008, pp. 415–442, DOI 10.1142/e015. MR2509719

[MW07] Ib Madsen and Michael Weiss, *The stable moduli space of Riemann surfaces: Mumford's conjecture*, Ann. of Math. (2) **165** (2007), no. 3, 843–941, DOI 10.4007/annals.2007.165.843. MR2335797

[MW17] K. Mann and M. Wolff. Rigidity and geometricity for surface group actions on the circle. preprint arXiv:1710.04902 [math.GT], 2017.

[Nar18] Sam Nariman, *Braid groups and discrete diffeomorphisms of the punctured disk*, Math. Z. **288** (2018), no. 3-4, 1255–1271, DOI 10.1007/s00209-017-1933-9. MR3778996

[Nie43] Jakob Nielsen, *Abbildungsklassen endlicher Ordnung* (German), Acta Math. **75** (1943), 23–115, DOI 10.1007/BF02404101. MR0013306

[Nie44] Jakob Nielsen, *Surface transformation classes of algebraically finite type*, Danske Vid. Selsk. Math.-Phys. Medd. **21** (1944), no. 2, 89. MR0015791

[Par15] Kiran Parkhe, *Smooth gluing of group actions and applications*, Proc. Amer. Math. Soc. **143** (2015), no. 1, 203–212, DOI 10.1090/S0002-9939-2014-12231-3. MR3272745

[Sch79] Reinhard Schultz, *Isotopy classes of periodic diffeomorphisms on spheres*, Algebraic topology, Waterloo, 1978 (Proc. Conf., Univ. Waterloo, Waterloo, Ont., 1978), Lecture Notes in Math., vol. 741, Springer, Berlin, 1979, pp. 334–354. MR557176

[Sch11] Richard Evan Schwartz, *Mostly surfaces*, Student Mathematical Library, vol. 60, American Mathematical Society, Providence, RI, 2011. MR2809109

[ST16] Nick Salter and Bena Tshishiku, *On the non-realizability of braid groups by diffeomorphisms*, Bull. Lond. Math. Soc. **48** (2016), no. 3, 457–471, DOI 10.1112/blms/bdw016. MR3509906

[Thu74] William P. Thurston, *A generalization of the Reeb stability theorem*, Topology **13** (1974), 347–352, DOI 10.1016/0040-9383(74)90025-1. MR0356087

[Thu88] William P. Thurston, *On the geometry and dynamics of diffeomorphisms of surfaces*, Bull. Amer. Math. Soc. (N.S.) **19** (1988), no. 2, 417–431, DOI 10.1090/S0273-0979-1988-15685-6. MR956596

[Tro96] A. J. Tromba, *Dirichlet's energy on Teichmüller's moduli space and the Nielsen realization problem*, Math. Z. **222** (1996), no. 3, 451–464, DOI 10.1007/PL00004542. MR1400202

[Tsh15] Bena Tshishiku, *Cohomological obstructions to Nielsen realization*, J. Topol. **8** (2015), no. 2, 352–376, DOI 10.1112/jtopol/jtu028. MR3356765

[Tsu84] Takashi Tsuboi, *Homology of diffeomorphism groups, and foliated structures* (Japanese), Sūgaku **36** (1984), no. 4, 320–343. Translated in Sugaku Expositions **3** (1990), no. 2, 145–181. MR780156

[Wah13] Nathalie Wahl, *The Mumford conjecture, Madsen-Weiss and homological stability for mapping class groups of surfaces*, Moduli spaces of Riemann surfaces, IAS/Park City Math. Ser., vol. 20, Amer. Math. Soc., Providence, RI, 2013, pp. 109–138. MR3114683

[Wol87] Scott A. Wolpert, *Geodesic length functions and the Nielsen problem*, J. Differential Geom. **25** (1987), no. 2, 275–296. MR880186

[Woo71] John W. Wood, *Bundles with totally disconnected structure group*, Comment. Math. Helv. **46** (1971), 257–273, DOI 10.1007/BF02566843. MR0293655

[Zim85] Bruno Zimmermann, *Nielsensche Realisierungssätze für Seifertfaserungen* (German, with English summary), Manuscripta Math. **51** (1985), no. 1-3, 225–242, DOI 10.1007/BF01168354. MR788680

Department of Mathematics, Brown University, 151 Thayer Street, Providence, Rhode Island 02912

Email address: mann@math.brown.edu

Department of Mathematics, Harvard Univeristy, One Oxford Street, Cambridge, Massachusetts 02138

Email address: bena@math.harvard.edu

Proceedings of Symposia in Pure Mathematics
Volume **102**, 2019
https://doi.org/10.1090/pspum/102/01808

Problems, questions, and conjectures about mapping class groups

Dan Margalit

Abstract. We discuss a number of open problems about mapping class groups of surfaces. In particular, we discuss problems related to linearity, congruence subgroups, cohomology, pseudo-Anosov stretch factors, Torelli groups, and normal subgroups.

Beginning with the work of Max Dehn a century ago, the subject of mapping class groups has become a central topic in mathematics. It enjoys deep and varied connections to many other subjects, such as low-dimensional topology, geometric group theory, dynamics, Teichmüller theory, algebraic geometry, and number theory. The number of papers on mapping class groups recorded on MathSciNet in the last six decades has grown from 205 to 386 to 525 to 791 to 1,121 to 1,390. The subject seems to enjoy an endless supply of beautiful ideas, pictures, and theorems.

At the 2017 Georgia International Topology Conference, the author gave a lecture called "Problems and progress on mapping class groups." This paper is a summary of parts of that lecture. What follows is not a comprehensive list in any way, even among the topics it attempts to address. Rather it gives a mix of problems—from the famous and notoriously difficult to the eminently doable. There is little attempt to give background; the reader may find that in the book by Farb and the author [**59**] and in the other references therein.

There are other problem lists on mapping class groups. In fact Farb has edited an entire book of problem lists on mapping class groups [**53**]. Ivanov's problem list [**81**], which also appears in Farb's book, has been particularly influential on the author of this article. That problem list is an updated version of a problem list written in conjunction with the 1993 Georgia International Topology Conference, twenty-four years earlier. Some of Ivanov's problems also appeared in Kirby's problem list, compiled around the same time [**89**]. Joan Birman's classic book contains several problems on mapping class groups, some of which are still open [**15**].

For a surface S, its mapping class group $\mathrm{Mod}(S)$ is the group of homotopy classes of orientation-preserving homeomorphisms of S. If S has boundary and/or marked points, then the homeomorphisms (and homotopies) are required to fix the

---

2010 *Mathematics Subject Classification.* Primary 20F38.

This material is based upon work supported by the National Science Foundation under Grant No. DMS - 1057874.

marked points as a set and are required to fix the boundary of S pointwise. We denote the closed orientable surface of genus g by S_g.

1. Linearity

We begin with one of the most basic and famous questions about mapping class groups, which appears, for example, in Birman's book [**15**, Appendix, Problem 30]. In the statement, we say that a group is linear if it admits a faithful linear representation into some $\mathrm{GL}_n(k)$ where k is a field.

QUESTION 1.1. Is $\mathrm{Mod}(S_g)$ linear?

Dehn proved that $\mathrm{Mod}(T^2)$, the mapping class group of the torus, is isomorphic to $\mathrm{SL}_2(\mathbb{Z}) \subseteq \mathrm{GL}_2(\mathbb{C})$. In the case $g = 2$ the Birman–Hilden theory [**110**] gives a short exact sequence:

$$1 \to \langle \iota \rangle \to \mathrm{Mod}(S_2) \to \mathrm{Mod}(S_{0,6}) \to 1$$

where ι is the hyperelliptic involution of S_2. The group $\mathrm{Mod}(S_{0,6})$ is closely related to the braid group on 5 strands. As such, Bigelow–Budney and Korkmaz were able to prove that $\mathrm{Mod}(S_2)$ is linear, using the theorem of Krammer and Bigelow that braid groups are linear [**14, 91**]. For $g \geq 3$, Question 1.1 is wide open.

Linearity of the braid group. One might be tempted to think that $\mathrm{Mod}(S_g)$ is not linear, because if it were then we would already have stumbled across the representation. On the other hand, we should draw inspiration from the case of the braid group. The Burau representation of the braid group was introduced in 1935 [**33**]. It was shown to be unfaithful for $n \geq 6$ by Long–Paton and Moody [**101, 122**] in 1993 and later it was shown to be unfaithful for $n \geq 5$ by Bigelow [**12**]. (The case $n = 4$ is open; see Question 3.1 below.) The Lawrence–Krammer representation was introduced in 1990 by Lawrence [**95**] and was proved to be faithful by Bigelow and Krammer around the turn of the century. See the surveys by Birman and Brendle [**16**] and by Turaev [**153**].

Whereas the Burau representation considers the action of the braid group on the fundamental group of the punctured disk, the Lawrence–Krammer representation considers the action on the fundamental group of the space of configurations of two points in the punctured disk. The fundamental group of a space is the fundamental group of the space of configurations of a single point, and so in that sense the Lawrence–Krammer representation is a mild generalization. This gives hope that some mild generalization of a known representation of $\mathrm{Mod}(S_g)$ could also be faithful.

Lawrence–Krammer representations for mapping class groups. Let us define the kth Lawrence representation of $\mathrm{Mod}(S_g)$ to be the action of $\mathrm{Mod}(S_g)$ on the homology of the universal abelian cover of the configuration space of k distinct, ordered points in S_g, considered as a module over $H_1(S_g;\mathbb{Z})$. (This is not a perfect analog of the Lawrence–Krammer representation because there is no $\mathrm{Mod}(S_g)$-equivariant surjective map $H_1(S_g;\mathbb{Z}) \to \mathbb{Z}$.)

QUESTION 1.2. Is the kth Lawrence representation of $\mathrm{Mod}(S_g)$ faithful for any $k \geq 1$?

Some positive evidence[1] in the direction of Question 1.2 is given by the work of Moriyama [**130**], who showed that the kth term of the Johnson filtration for the once-punctured surface $S_{g,1}$ is given by the kernel of the action of $\mathrm{Mod}(S_{g,1})$ on the compactly supported cohomology of the configuration space of k distinct points in $S_{g,1}$.

Kontsevich's approach. Kontsevich has proposed a family of linear representations of the mapping class group, described in the online problem list for the Center for the Topology and Quantization of Moduli Spaces [**100**] (there is also a discussion on MathOverflow [**135**]). Kontsevich has conjectured that this approach gives a faithful linear representation of the mapping class group. Here is his description of the main idea, taken from the CTQM web site [**100**] and mildly edited for clarity.

According to Thurston, a generic element f in $\mathrm{Mod}(S_g)$ *is pseudo-Anosov and has a canonical representative that preserves two transverse laminations and multiplies the transverse measures by λ and $1/\lambda$, where $\lambda > 1$ is an algebraic integer. The number λ appears as the largest eigenvalue of the action of f on the first cohomology of the ramified double covering of S_g on which the quadratic differential associated to the laminations for f becomes an abelian differential. Moreover, one may assume that the laminations have $4g-4$ triple singular points. I would like to construct a representation in which f acts non-trivially, because there will be some eigenvalue of the corresponding matrix equal to a positive power of λ.*

A representation of $\mathrm{Mod}(S_g)$ *is the same as a local system on moduli space $\mathcal{M}_g$. Here is the description of it. The fiber of it at a complex curve C is a kind of middle (or total?) cohomology of the configuration space* $\mathrm{Conf}_{4g-4}(C) = C^{4g-4} \setminus$ *big diagonal (the configuration space of $4g-4$ distinct points $(x_1, \ldots, x_{4g-4})$ on C), with coefficients in the flat bundle, the first cohomology group of the double cover of C, doubly ramified at $(x_1, \ldots, x_{4g-4})$.*

Why I believe that this should work: Thurston's representative gives a map from $\mathrm{Conf}_{4g-4}(C)$ *(together with the local system) to itself, with a preferred marked point such that the action on the fiber has the largest eigenvalue equal to λ. Can one use Lefschetz fixed point formula, or maybe some dynamical reasoning, to see that this eigenvalue appears in the total cohomology of* $\mathrm{Conf}_{4g-4}(C)$ *with local coefficients as well?*

There are various challenges to making Kontsevich's idea work. For instance, it is not clear why one can assume that the laminations only have singular points of degree 3; it is true that any foliation is Whitehead equivalent to such a foliation, but then there is no pseduo-Anosov homeomorphism preserving the modified foliation. Even assuming that all pseudo-Anosov foliations have only singular points of degree 3, it is still not clear how to construct the local system; it seems that this would require a deformation retraction of $\mathrm{Conf}_{4g-4}(C)$ to some base point.

Question 1.3. Can Kontsevich's idea be fleshed out to give a faithful linear representation of the mapping class group, or not?

In unpublished work, Dunfield has performed some computer calculations suggesting that the Kontsevich approach will not result in a faithful representation, at least for the braid groups [**48**].

[1] According to the anonymous referee, Moriyama's result should be considered as negative evidence!

No poison. We end this section by mentioning one other result in the direction of Question 1.1. Let F_n denote the free group of rank n. Formanek–Procesi [**61**] showed that the automorphism group $\mathrm{Aut}(F_n)$ and the outer automorphism group $\mathrm{Out}(F_n)$ both fail to be linear for $n \geq 4$. They accomplish this by producing certain "poison subgroups" that are patently non-linear (the $n = 3$ cases are open). Brendle and Hamidi-Tehrani showed that there are no analogous poison subgroups in the mapping class group [**26**].

2. The congruence subgroup problem

For a given group, the congruence subgroup problem asks if every finite-index subgroup of a given group contains a congruence subgroup. The precise meaning of this problem depends on an appropriate choice of definition for a congruence subgroup.

The *principal congruence subgroup of level* m in the special linear group $\mathrm{SL}_n(\mathbb{Z})$ is the kernel of the reduction homomorphism $\mathrm{SL}_n(\mathbb{Z}) \to \mathrm{SL}_n(\mathbb{Z}/m)$. It is a theorem of Bass–Milnor–Serre [**8**] that $\mathrm{SL}_n(\mathbb{Z})$ has the congruence subgroup property. As above, this means that every subgroup of finite index in $\mathrm{SL}_n(\mathbb{Z})$ contains a principal congruence subgroup.

The most naïve analogue of the principal congruence subgroups for $\mathrm{Mod}(S_g)$ are the level m congruence subgroups $\mathrm{Mod}(S_g)[m]$, defined to be the kernels of the compositions

$$\mathrm{Mod}(S_g) \to \mathrm{Sp}_{2g}(\mathbb{Z}) \to \mathrm{Sp}_{2g}(\mathbb{Z}/m),$$

where $\mathrm{Sp}_{2g}(\mathbb{Z})$ and $\mathrm{Sp}_{2g}(\mathbb{Z}/m)$ are the symplectic groups over $\mathbb{Z}$ and $\mathbb{Z}/m$. This composition gives exactly the action of $\mathrm{Mod}(S_g)$ on the homology group $H_1(S_g;\mathbb{Z}/m)$. The Torelli group $\mathcal{I}(S_g)$ is the kernel of the map $\mathrm{Mod}(S_g) \to \mathrm{Sp}_{2g}(\mathbb{Z})$, and as such it is the intersection of all of the level m congruence subgroups for $m \geq 2$.

There are finite-index subgroups of $\mathrm{Mod}(S_g)$ not containing any congruence subgroup $\mathrm{Mod}(S_g)[m]$. A fortiori there are subgroups of finite index that do not contain the Torelli group. Here is how we can construct such a subgroup. Let H denote $H_1(S_g;\mathbb{Z})$. Morita [**124**, Theorem 6.2] defined a homomorphism

$$\tilde{\tau} : \mathrm{Mod}(S_g) \to \left(\wedge^3 H/H\right) \rtimes \mathrm{Sp}_{2g}(\mathbb{Z})$$

that extends the Johnson homomorphism $\tau : \mathcal{I}(S_g) \to \wedge^3 H/H$, in the sense that the restriction of $\tilde{\tau}$ to $\mathcal{I}(S_g)$ is equal to the post-composition of τ with the inclusion $\left(\wedge^3 H/H\right) \to \left(\wedge^3 H/H\right) \rtimes \mathrm{Sp}_{2g}(\mathbb{Z})$. Consider the subgroup $2 \wedge^3 H/2H$ where the inclusion $2H \to 2 \wedge^3 H$ is the restriction of the usual inclusion $H \to \wedge^3 H$. The action of $\mathrm{Sp}_{2g}(\mathbb{Z})$ on $\wedge^3 H$ preserves both $2 \wedge^3 H$ and $2H$, and so we obtain a subgroup $\left(2 \wedge^3 H/2H\right) \rtimes \mathrm{Sp}_{2g}(\mathbb{Z})$ inside $\left(\wedge^3 H/H\right) \rtimes \mathrm{Sp}_{2g}(\mathbb{Z})$. The preimage of this subgroup under $\tilde{\tau}$ is the desired subgroup.

We may vary this construction in the following way: if Γ is a subgroup of finite index in $\mathrm{Sp}_{2g}(\mathbb{Z})$ and Δ is a subgroup of finite index in $\wedge^3 H/H$ invariant under Γ, then $\Delta \rtimes \Gamma$ is a finite-index subgroup of $\mathrm{Mod}(S_g)$. In what follows we will refer to these as Morita subgroups.

Other examples of finite-index subgroups of $\mathrm{Mod}(S_g)$ that do not contain the Torelli group $\mathcal{I}(S_g)$ are given by Cooper [**42**], who constructs two different characteristic p analogues of the Johnson filtration.

The upshot is that if we want the mapping class group to have the congruence subgroup property then we shall require a richer class of subgroups of finite index.

To this end, consider a characteristic subgroup Γ of finite index in $\pi_1(S_g)$. There is a homomorphism $\text{Mod}(S_g) \to \text{Out}(\pi_1(S_g)/\Gamma)$. Since Γ has finite index the group $\pi_1(S_g)/\Gamma$, hence $\text{Out}(\pi_1(S_g)/\Gamma)$, is finite. The kernel of the homomorphism is thus a subgroup of finite index.

For instance if Γ is the kernel of the composition $\pi_1(S_g) \to H_1(S_g;\mathbb{Z}) \to H_1(S_g;\mathbb{Z}/m)$ then we recover the congruence subgroup $\text{Mod}(S_g)[m]$. Following Ivanov, we refer to any subgroup of $\text{Mod}(S_g)$ constructed in this manner as a congruence subgroup.

QUESTION 2.1. Is it true that every finite-index subgroup in $\text{Mod}(S_g)$ contains a congruence subgroup?

This is the first problem on Ivanov's list [**81**], and in fact he states it as a conjecture. More modestly, we may ask the following.

QUESTION 2.2. Do the Morita subgroups of $\text{Mod}(S_g)$ constructed above contain congruence subgroups?

It would even be helpful to clarify the relationship between the Morita subgroups and the Cooper subgroups.

We can reinterpret the construction of congruence subgroups in terms of covering spaces. Fix a characteristic subgroup Γ of finite index in $\pi_1(S_g)$ and consider the associated characteristic finite-sheeted cover of S_g. The entire mapping class group $\text{Mod}(S_g)$ is liftable in the sense of Birman–Hilden [**17, 110**], that is, each element lifts to a mapping class of the cover. As such, $\text{Mod}(S_g)$ acts by outer automorphisms on the deck group of the cover, and the kernel of this action is precisely the congruence subgroup associated to Γ.

This idea gives another source for subgroups of finite index. For a subgroup Γ of finite index in $\pi_1(S_g)$, we may consider the associated finite-sheeted cover of S_g. The resulting liftable mapping class group is again a subgroup of $\text{Mod}(S_g)$ of finite index. It is a straightforward exercise to show that these liftable mapping class groups all contain congruence subgroups as defined by Ivanov.

There are some surfaces for which Question 2.1 has been resolved. Diaz–Donagi–Harbater [**45**] answered it in the affirmative for a sphere with four punctures (although their proof applies to spheres with any number of punctures). Asada [**5**] answered the question in the affirmative for the sphere with any number of punctures and for the torus with any positive number of punctures (see also [**34, 50**]). Thurston gave a more elementary proof for punctured spheres, based on Asada's work; see the paper by McReynolds [**117**] for a presentation of Thurston's proof as well as some historical discussion. More recently, a new proof in the case of the punctured torus was given by Kent [**88**]. Question 2.1 was also resolved for surfaces of genus two by Boggi [**18, 20**]. (Boggi [**19**] had claimed a general solution to Question 2.1 but a gap was found by Abramovich, Kent, and Wieland in his Theorem 5.4.)

As mentioned by Ivanov [**81**], Voevodsky observed that a positive answer to Question 2.1 would give a proof of a conjecture of Grothendieck that a smooth algebraic curve defined over $\mathbb{Q}$ is determined by its algebraic fundamental group—which is isomorphic to the profinite completion of $\pi_1(S_g)$—together with the natural action of the absolute Galois group $\text{Gal}(\bar{\mathbb{Q}}/\mathbb{Q})$ on the latter (the conjecture was proved by Mochizuki [**119, 120**]).

3. The integral Burau representation

Denote by B_n the braid group on n strands. The Burau representation

$$\mathrm{B}_n \to \mathrm{GL}_n(\mathbb{Z}[t,t^{-1}])$$

is the most well-studied linear representation of B_n. It is easily seen to be faithful for $n = 3$. As discussed above, it is now known to be unfaithful for $n \geq 5$. This begs the following question, discussed, for example, in Birman's book [**15**].

QUESTION 3.1. Is the Burau representation faithful for $n = 4$?

Bigelow suggests that an element of the kernel of the Burau representation for $n = 4$ would give rise to a candidate knot for showing that the Jones polynomial does not detect the unknot [**13**]. Some experimental results on Question 3.1 were recently found by Fullarton and Shadrach [**64**]. Birman showed that Question 3.1 is equivalent to the question of whether the images of $\sigma_1\sigma_3^{-1}$ and $\sigma_2\sigma_3\sigma_1^{-1}\sigma_2^{-1}$ generate a free group of rank 2 [**15**, Theorem 3.19].

While the Burau representation is not faithful in general, we can still use it to effectively probe the braid group. For instance the kernel of the Burau representation gives an interesting normal subgroup of the braid group. Church and Farb showed that this group is not finitely generated [**38**]. Still, we have the following problem, implicit in Morita's survey [**125**, Problem 6.24].

PROBLEM 3.2. Describe the kernel of the Burau representation. Find a natural generating set.

In Morita's survey, he mentions that it is still open whether the Gassner representation of the pure braid group is faithful or not; see Gassner's paper [**68**] for the definition. Bachmuth [**6**] had claimed that it was faithful, but his argument was refuted by Abramenko–Müller [**134**]. We refer the reader to Birman's extensive review of Bachmuth's paper on MathSciNet.

Braid Torelli groups. We may also consider specializations of the Burau representation obtained by setting t to be some fixed complex number. There are two values of t that give specializations to $\mathrm{GL}_n(\mathbb{Z})$, namely $t = 1$ and $t = -1$. The specialization at $t = 1$ is nothing other than the standard representation to the symmetric group. The specialization at $t = -1$, called the integral Burau representation, is more interesting. (Specializations at other roots of unity have also been studied; see, e.g. [**7, 65, 67, 154**].)

For $n = 2g+1$ there is a map $\mathrm{B}_{2g+1} \to \mathrm{Mod}(S_g^1)$ obtained by lifting homeomorphisms through the branched double cover $S_g^1 \to D_{2g+1}$ (here S_g^1 is the compact surface of genus g with one boundary component and D_{2g+1} is the disk with $2g+1$ marked points). The integral Burau representation is the representation obtained by post-composing this map with the standard symplectic representation $\mathrm{Mod}(S_g^1) \to \mathrm{Sp}_{2g}(\mathbb{Z})$. For n even the integral Burau representation has a similar description, with S_g^1 replaced by S_g^2; see [**25**].

The kernel of the integral Burau representation is called the braid Torelli group $\mathcal{BI}_n$. Hain has identified this group with the fundamental group of the branch locus for the period mapping on Torelli space [**72**], and so this group is of interest in algebraic geometry. With Brendle and Putman, the author of this article showed that $\mathcal{BI}_n$ is generated by squares of Dehn twists about curves in D_{2g+1} that surround an odd number of marked points [**25**].

Smythe [**145**] showed that $\mathcal{BI}_4$ is isomorphic to F_∞, the free group on countably many generators. It follows from a theorem of Mess [**118**] that $\mathcal{BI}_5 \cong F_\infty \times \mathbb{Z}$ and then from a theorem of Brendle and the author [**29**] that $\mathcal{BI}_6 \cong F_\infty \ltimes F_\infty$. The following problem is suggested by Hain [**72**, Problem 7], although it was clearly of interest much earlier.

QUESTION 3.3. Is $\mathcal{BI}_n$ finitely generated for any $n \geq 7$? Is the abelianization finitely generated?

It is further of interest to find (possibly infinite) presentations of the groups $\mathcal{BI}_n$.

Level m congruence subgroups. We obtain finite-index analogs of $\mathcal{BI}_n$ by considering the mod m reductions of the integral Burau representation. The kernels of these representations are called the level m congruence subgroups of the braid group, and are denoted $\mathrm{B}_n[m]$. Arnol'd [**3**] proved that $\mathrm{B}_n[2]$ is nothing other than the pure braid group PB_n. Brendle and the author of this article [**31**] proved that $\mathrm{B}_n[4]$ is equal to the group generated by squares of Dehn twists in D_n and that both of these are equal to the group PB_n^2.

A'Campo proved that for $n = 2g+1$ the image of PB_n in $\mathrm{Sp}_{2g}(\mathbb{Z})$ is the level two congruence subgroup $\mathrm{Sp}_{2g}(\mathbb{Z})[2]$. It follows that for m even the image of $\mathrm{B}_n[m]$ in $\mathrm{Sp}_{2g}(\mathbb{Z})$ is $\mathrm{Sp}_{2g}(\mathbb{Z})[m]$.

PROBLEM 3.4. Describe the image of $\mathrm{B}_n[m]$ in $\mathrm{Sp}_{2g}(\mathbb{Z})$ for m odd.

The group $\mathrm{PB}_n = \mathrm{B}_n[2]$ is generated by Dehn twists and (as above) the group $\mathrm{B}_n[4]$ is generated by squares of Dehn twists [**31**].

PROBLEM 3.5. Find natural generating sets for the groups $\mathrm{B}_n[m]$.

One generating set for the case of m an odd prime was given by Styliankis [**147**]. The answer to Problem 3.5 will be more complicated in general than for the cases of $\mathrm{B}_n[2]$ and $\mathrm{B}_n[4]$. Those are the only two that contain all squares of Dehn twists. Since $\mathcal{BI}_n$ is generated by squares of Dehn twists about odd curves (as above), the group $\mathcal{BI}_n$ is automatically contained in their generating sets.

PROBLEM 3.6. Compute the abelianizations (and other homology groups) of the groups $\mathrm{B}_n[m]$.

Kordek points out the following application: if the ranks of the abelianizations of $\mathrm{B}_n[2m]$ are unbounded for fixed n and varying m, then $\mathcal{BI}_n$ is not finitely generated (in fact its abelianization has infinite rank). The proof of this uses the fact, due to Každan [**87**], that the congruence subgroups $\mathrm{Sp}_{2g}(\mathbb{Z})[2m]$ have trivial first integral cohomology.

Styliankis [**147**] has shown that for p an odd prime the quotient group $\mathrm{B}_n[p]/\mathrm{B}_n[2p]$ is isomorphic to the symmetric group Σ_n. This in particular implies that the cohomology of $\mathrm{B}_n[2p]$ is a representation of the symmetric group.

In forthcoming work, the first rational homology of $\mathrm{B}_n[4]$ is computed by Kordek and the author.

4. Generating with torsion

Almost 50 years ago, MacLachlan proved that $\mathrm{Mod}(S_g)$ is generated by torsion [**105**]. In the time since, there have been many (successful) attempts to sharpen this result: Which orders are possible? What is the smallest number of generators needed?

Most of the work in this direction has to do with involutions, elements of order two. McCarthy and Papadopoulos proved that $\mathrm{Mod}(S_g)$ is generated by involutions, and in fact it is normally generated by a single involution [**112**]. Luo showed that $\mathrm{Mod}(S_g)$ is generated by $12g+2$ involutions [**104**]. Brendle and Farb sharpened this further: for $g \geq 3$ only 6 involutions are needed, independent of g. Kassabov [**86**] goes even further: only four involutions are needed if $g \geq 7$.

We know that $\mathrm{Mod}(S_g)$ cannot be generated by two involutions, because any such group is virtually cyclic. So we are left with a tantalizing problem, suggested by Brendle–Farb and Kassabov.

Question 4.1. Is $\mathrm{Mod}(S_g)$ generated by three involutions or not?

Of course there are other orders besides two. Monden [**121**] showed for $g \geq 3$ that $\mathrm{Mod}(S_g)$ is generated by three elements of order 3 and also by four elements of order 4. Yoshihara [**157**] has recently shown that $\mathrm{Mod}(S_g)$ is generated by four elements of order 6 when $g \geq 5$.

What about order 5? One obstacle here is that $\mathrm{Mod}(S_3)$ does not even have an element of order 5. In fact there are subtle number theoretic conditions that determine whether $\mathrm{Mod}(S_g)$ contains an element of order k. So it seems at first that there is no hope of extending Monden's results to periodic elements of higher periods.

However, it is a theorem of Tucker [**152**] that for any $k \geq 2$, the group $\mathrm{Mod}(S_g)$ has elements of order k as long as $g \geq (k-1)(k-3)$ (in fact his result is stronger). Building on this, Lanier [**92**] proved that for any $k \geq 6$ the group $\mathrm{Mod}(S_g)$ is generated by three elements of order k when $g \geq (k-1)^2+1$; the new phenomenon here is that the number of generators is independent of both g as well as k. Moreover, Lanier's elements are all conjugate to each other.

Given Tucker's result, we can ask the analogue of Question 4.1.

Question 4.2. For which k can $\mathrm{Mod}(S_g)$ be generated by two elements of order k?

The maximum order of a periodic element of $\mathrm{Mod}(S_g)$ is $4g+2$, and Korkmaz showed that $\mathrm{Mod}(S_g)$ is generated by only two of these [**90**]. Here the order $4g+2$ varies with g, so this result is in a slightly different direction than Question 4.2

Lanier chooses his periodic elements very carefully: they are all realized by rotations of S_g in $\mathbb{R}^3$. What about other periodic elements? With Lanier, the author proved that every nontrivial periodic element of $\mathrm{Mod}(S_g)$—except for the hyperelliptic involution—is a normal generator for $\mathrm{Mod}(S_g)$. The following question is asked in our paper [**93**].

Problem 4.3. Given a periodic element of $\mathrm{Mod}(S_g)$, how many of its conjugates are needed to generate $\mathrm{Mod}(S_g)$? Can this number be taken to be independent of the element chosen?

5. Generators and relations for Torelli groups

As mentioned, the Torelli group $\mathcal{I}(S_g)$ is the kernel of the representation $\mathrm{Mod}(S_g) \to \mathrm{Sp}_{2g}(\mathbb{Z})$ given by the action of $\mathrm{Mod}(S_g)$ on $H_1(S_g;\mathbb{Z})$. We may think of the Torelli group as being the non-linear, more mysterious part of the mapping class group.

The Torelli group arises in algebraic geometry because it is the fundamental group of Torelli space, the space of Riemann surfaces with homology framings. This space is an infinite regular cover of moduli space and it is the natural domain of the period mapping, which maps a Riemann surface with homology framing to Siegel space; see, e.g. Mess' paper [**118**].

Topologists became interested in the Torelli group because of the following fact: every homology 3-sphere can be obtained from S^3 by cutting along a surface and regluing by the product of the original gluing map and an element of the Torelli group. In his survey paper, Johnson credits the resurgence of interest in the Torelli group to Birman [**83**].

In 1983 Johnson [**82**] proved that $\mathcal{I}(S_g)$ is finitely generated by bounding pair maps for $g \geq 3$. Mess proved that $\mathcal{I}(S_2)$ is isomorphic to a free group of infinite rank [**118**]. The following is the most fundamental open question about the Torelli group. It is asked by Birman in her book [**15**, Appendix, Problem 29].

QUESTION 5.1. Let $g \geq 3$. Is the Torelli group $\mathcal{I}(S_g)$ finitely presented?

The Torelli group has been studied for a century, going back to the work of Dehn, Nielsen, and Magnus. Still this question is wide open. Putman has found an infinite presentation for $\mathcal{I}(S_g)$ where all the relations come from geometric considerations [**136**], but there is little hope of turning this into a finite presentation. Morita and Penner [**123**] give a sort of presentation for $\mathcal{I}(S_g)$. Their presentation has a geometric origin, but the generating set consists of infinitely many copies of each element of the group and the list of relations contains all relations between all elements (and indeed infinitely many copies of each relation).

Generating sets. There is still more to be said about the finite generation of $\mathcal{I}(S_g)$. The number of elements in Johnson's original finite generating set for $\mathcal{I}(S_g)$ is exponential in g. Johnson also proved [**85**] that the abelianization of $\mathcal{I}(S_g)$ has rank $\binom{2g+1}{3}$, and he conjectured [**82, 83**] that there should be a generating set for $\mathcal{I}(S_g)$ whose size is cubic in g. This conjecture was proved by Putman, who found a generating set for $\mathcal{I}(S_g)$ with $57\binom{g}{3} + 2g + 1$ elements [**138**]. While this is cubic in g, it leaves open the following question, asked explicitly by Johnson [**82**].

QUESTION 5.2. Is there a generating set for $\mathcal{I}(S_g)$ with exactly $\binom{2g+1}{3}$ elements?

Johnson whittled down his original generating set for $\mathcal{I}(S_3)$ to one with $\binom{7}{3} = 35$ elements, thus answering his own question affirmatively in this case.

Johnson's proof that $\mathcal{I}(S_g)$ is finitely generated requires some difficult computations. His arguments essentially reduce the problem to the case $g = 3$, but all of the difficult computations are still required even for this case. Putman's paper [**138**] also gives a conceptual reduction to the case $g = 3$. Still we have the following problem.

PROBLEM 5.3. Find a (more) conceptual proof that $\mathcal{I}(S_3)$ is finitely generated.

One approach to the previous problem would be to simply elucidate Johnson's proof, for instance by making his computations more transparent (say, realizing them as relations among push maps). A totally different approach would be to use the fact that the normal closure of any hyperelliptic involution contains $\mathcal{I}(S_g)$ as a subgroup of index 2; see, e.g., the author's paper with Lanier [**93**].

Even though $\mathcal{I}(S_2)$ is infinitely generated, there is still a basic open question about its generation. Mess shows that $\mathcal{I}(S_2)$ is generated by Dehn twists, one for each symplectic splitting of $H_1(S_2;\mathbb{Z})$. He posed the following [**118**].

PROBLEM 5.4. Give an explicit infinite generating set for $\mathcal{I}(S_2)$.

Mess conjectured in his paper that one can construct a generating set by choosing a hyperbolic metric on S_2 and choosing for each symplectic splitting the Dehn twist about the shortest curve inducing that splitting.

We expect that one can solve Problem 5.4 using the action of $\mathcal{I}(S_2)$ on the complex of cycles $\mathcal{B}(S_2)$ of Bestvina, Bux, and the author [**10**].

Two-generator subgroups. Related to presentations of the Torelli group we also have the following question. Our inability to answer it underscores our lack of knowledge about the basic algebraic structure of the Torelli group.

QUESTION 5.5. Let $g \geq 3$. Suppose that f and h are elements of $\mathcal{I}(S_g)$. Is it true that f and h either commute or generate a free group?

With Leininger, the author answered the analogous question for pure braid groups in the affirmative [**97**]. The pure braid group is the subgroup of the braid group acting trivially on the homology of the punctured disk, and so we may think of the pure braid group as one analogue of the Torelli group (probably a better analogue is the braid Torelli group; see [**25**]). It is also interesting to consider the analog of Question 5.5 for pure surface braid groups or for the Johnson kernel.

Question 5.5 is true when f and h are Dehn twists (cf. [**59**, Theorem 3.14]). However it is open when f and h are both bounding pair maps, or when one is a bounding pair map and one is a Dehn twist.

Leininger has pointed out that when f and h are bounding pair maps with the property that each of the four curves intersects each of the two curves of the other bounding pair in a single point, the group generated by f and h is free. This is because the two bounding pairs lie inside a torus with two boundary components, and the corresponding bounding pair maps become trivial after one of the boundary components is filled in. In other words, the two bounding pair maps lie in a subgroup isomorphic to $\pi_1(S_{1,1}) \cong F_2$.

Homology of Torelli groups. Beyond Question 5.1 we also have the following basic question, discussed, for example, by Farb [**54**, Problem 5.11].

QUESTION 5.6. For which natural numbers g and k is $H_k(\mathcal{I}(S_g);\mathbb{Z})$ finitely generated?

By Mess' theorem, we have a complete answer for $g \leq 2$, and by Johnson's theorem we have a complete answer for $k = 1$. With Bestvina and Bux, the author showed [**10**] that the cohomological dimension of $\mathcal{I}(S_g)$ is $3g-5$ (the lower bound was previously given by Mess [**69**]). We also showed that $H_{3g-5}(\mathcal{I}(S_g);\mathbb{Z})$ is not finitely generated, sharpening an earlier result of Akita [**1**]. It is a theorem of Johnson–Millson–Mess that $H_3(\mathcal{I}(S_3);\mathbb{Z})$ is not finitely generated [**118**].

Very recently, Gaifullin [**66**] has proved that the groups $H_k(\mathcal{I}(S_g);\mathbb{Z})$ are infinitely generated for $2g-3 \leq k \leq 3g-6$. It was explicitly asked in our paper with Bestvina and Bux if this was true [**10**, Question 8.11].

If a group is finitely presented then its second integral homology group is finitely generated, and so the case of $k = 2$ in Question 5.6 is of particular interest.

There is a folk conjecture that for fixed k the group $H_k(\mathcal{I}(S_g); \mathbb{Z})$ is finitely generated once g is large enough; see, e.g. [**54**, Conjecture 5.12].

The Johnson kernel. There are many variants of the Torelli group, all interesting for different reasons. The most well-studied subgroup of the Torelli group is the Johnson kernel $\mathcal{K}(S_g)$, so called because it is the kernel of the Johnson homomorphism $\tau : \mathcal{I}(S_g) \to \wedge^3 H_1(S_g; \mathbb{Z})$. Johnson proved that this kernel is equal to the subgroup of $\mathcal{I}(S_g)$ generated by Dehn twists about separating curves [**84**].

Most of our questions about $\mathcal{I}(S_g)$ are open for $\mathcal{K}(S_g)$ as well. It was only very recently shown that $\mathcal{K}(S_g)$ is finitely generated. This was originally shown by Ershov and He [**51**] for $g \geq 12$ and later extended to $g \geq 4$ by Church, Ershov, and Putman [**37**]. Dimça and Papadima had shown a few years earlier that the torsion-free part of the abelianization was finitely generated [**47**]. Recently Morita, Sakasai, and Suzuki [**129**] used the work of Dimça, Hain, and Papadima [**46**] to give an explicit description of $H_1(\mathcal{K}(S_g); \mathbb{Q})$ (from their description it is possible to also calculate the rank of $H_1(\mathcal{K}(S_g); \mathbb{Z})$).

PROBLEM 5.7. Find an explicit finite generating set for $\mathcal{K}(S_g)$.

We can also ask about the size of a minimal generating set for $\mathcal{K}(S_g)$. We also have the following question.

QUESTION 5.8. Is $\mathcal{K}(S_3)$ finitely generated or not?

The following natural question is made more intriguing by the Church–Ershov–Putman result.

QUESTION 5.9. Is $\mathcal{K}(S_g)$ finitely presented or not?

Also, it is known [**10**] that the cohomological dimension of $\mathcal{K}(S_g)$ is $2g - 3$ but the following is still open for $g \geq 3$.

QUESTION 5.10. Is $H_{2g-3}(\mathcal{K}(S_g); \mathbb{Z})$ infinitely generated?

We expect that one can use the techniques used for the Torelli group to answer this question; see [**10**].

Other terms of the Johnson filtration. The Johnson filtration of $\mathcal{I}(S_g)$ is the sequence of groups $\mathcal{N}_k(S_g)$ defined by

$$\mathcal{N}_k(S_g) = \ker\left(\mathrm{Mod}(S_g) \to \mathrm{Out}(\pi/\pi^k)\right)$$

where $\pi = \pi_1(S_g)$ and π^k is the kth term of its lower central series. The groups $\mathcal{N}_2(S_g)$ and $\mathcal{N}_3(S_g)$ are equal to $\mathcal{I}(S_g)$ and $\mathcal{K}(S_g)$, respectively. Magnus proved that the intersection of the $\mathcal{N}_k(S_g)$ (as k varies) is trivial [**107**].

Church, Ershov, and Putman proved that each term of the Johnson filtration is finitely generated when $k \geq 3$ and $g \geq 2k - 1$.

PROBLEM 5.11. Find explicit finite generating sets for the terms of the Johnson filtration.

It is even an open problem to find simple, explicit, infinite generating sets. Here is one proposal. For any $f \in \mathcal{N}_k(S_g)$ and any curve c in S_g the twist difference $[T_c, f] = T_c T_{f(c)}^{-1}$ lies in $\mathcal{N}_k(S_g)$.

QUESTION 5.12. Are the terms of the Johnson filtration generated by twist differences?

This question is already interesting for the Johnson kernel, where it is not known. We can refine the question by asking if the terms of the Johnson filtration are generated by twist differences on subsurfaces of uniformly small Euler characteristic (that is, the Euler characteristic only depends on k). Church and Putman showed that there does exist some (infinite) generating set with this property [**40**].

It is known that the cohomological dimension of $\mathcal{N}_k(S_g)$ lies between $g-1$ and $2g-3$ for $k \geq 4$; see [**10**]. The lower bound comes from an explicit construction of a free abelian subgroup of rank $g-1$. The upper bound comes from the fact that the cohomological dimension of $\mathcal{K}(S_g)$ is $2g-3$. Since cohomological dimension is non-increasing under passage to subgroups, it follows that as we increase k (and fix g) the cohomological dimension must eventually stabilize.

PROBLEM 5.13. Compute the cohomological dimension of $\mathcal{N}_k(S_g)$ for $k \geq 4$ and $g \geq 3$.

As a first start we have the following problem.

PROBLEM 5.14. Compute the maximal rank of an abelian subgroup of $\mathcal{N}_k(S_g)$ for $k \geq 4$.

We conjecture that the maximal rank in the above problem is $g-1$, the lower bound given in Farb's problem list [**54**, Theorem 5.10]. One feature to note is that for $k \geq 4$ the groups $\mathcal{N}_k(S_g)$ contain no multitwists [**10**, Theorem A.1].

The Magnus filtration. There is a natural variant of the Johnson filtration, called the Magnus filtration. These groups are obtained by replacing π^k in the definition of the Johnson filtration with the kth term of the derived series of $\pi_1(S_g)$. This filtration was defined by McNeill [**116**] who showed, among other things, that the quotient of each term by the next is infinitely generated. Beyond her paper, these groups have been unexplored.

PROBLEM 5.15. Investigate the basic properties of the terms of the Magnus filtration. For instance what is a simple, explicit generating set? What are the abelianizations?

6. Virtual surjection onto the integers

The mapping class group $\text{Mod}(S_g)$ is perfect for $g \geq 3$, meaning that it is equal to its own commutator subgroup. For $g \leq 2$ the commutator subgroup of $\text{Mod}(S_g)$ has finite index. So for every g all homomorphisms $\text{Mod}(S_g) \to \mathbb{Z}$ are trivial. The following is one of the most fundamental open questions about the mapping class group.

QUESTION 6.1. Is there a finite-index subgroup of $\text{Mod}(S_g)$ that surjects onto $\mathbb{Z}$?

Since $\text{Mod}(S_1)$ is virtually free the answer is 'yes' for $g = 1$. Taherkhani [**148**] and McCarthy [**113**] proved that the answer is also 'yes' for $g = 2$. Thus, we may restrict attention to $g \geq 3$.

McCarthy [**113**] and Hain [**73**] answered Question 6.1 in the negative for all subgroups of finite index containing the Torelli group, and Putman [**137**] extended this result to all subgroups of finite index containing the Johnson kernel. Finally, Ershov and He extended these results to all terms of the Johnson filtration [**51**].

PROBLEM 6.2. Compute the abelianizations of the various Morita subgroups, congruence subgroups, and liftable subgroups of $\text{Mod}(S_g)$ defined in Section 2.

Morita's subgroups contain the Johnson kernel, and so by Putman's result these abelianizations are torsion.

Putman and Wieland [**139**] conjecture that the answer to Question 6.1 is 'no.' They further prove that if their conjecture holds for $g = 3$ then it holds for all $g \geq 3$.

Putman and Wieland also relate Question 6.1 to a question about the so-called Prym representations of $\text{Mod}(S_g)$. In short, a Prym representation is the linear representation given by the action of $\text{Mod}(S_g)$ on the first homology of some finite, characteristic cover (in order for this representation to be well-defined, we must fix a basepoint in S_g; see their paper for details). Prym representations have been studied by Looijenga [**103**], Grunewald–Lubotzky [**71**], and Grunewald–Larsen–Lubotzky–Malestein [**70**]. Putman and Wieland conjecture the following.

CONJECTURE 6.3. *For $g \geq 2$ all nonzero orbits of all Prym representations of $\text{Mod}(S_g)$ are infinite.*

Putman and Wieland proved that this conjecture is essentially equivalent to their conjecture that the answer to Question 6.1 is 'no.' In particular, to answer Question 6.1 in the negative, it is enough to prove Conjecture 6.3 for the case of a surface of genus two with a single boundary component.

Farb and Hensel proved that the analog of Conjecture 6.3 for graphs is true [**55**].

7. Ivanov's metaconjecture

The problems in the next two sections are motivated by the following general question:

What can we say about the typical (or general) normal subgroup of the mapping class group?

In this section we discuss the automorphism groups of normal subgroups. The story begins with a theorem of Ivanov [**80**], which states that

$$\text{Aut}\,\text{Mod}(S_g) \cong \text{Mod}^{\pm}(S_g)$$

for $g \geq 3$. Here $\text{Mod}^{\pm}(S_g)$ is the extended mapping class group, the group of homotopy classes of all homeomorphisms of S_g (including the orientation-reversing ones).

Ivanov's theorem gives the isomorphism explicitly: the natural map $\text{Mod}^{\pm}(S_g) \to \text{Aut}\,\text{Mod}(S_g)$ given by conjugation is an isomorphism. A key ingredient is the accompanying theorem that the automorphism group of the complex of curves $\mathcal{C}(S_g)$ is also isomorphic to $\text{Mod}^{\pm}(S_g)$. To compute $\text{Aut}\,\text{Mod}(S_g)$, Ivanov first shows that an automorphism of $\text{Mod}(S_g)$ must preserve powers of Dehn twists, and then uses this to reduce to his theorem about $\mathcal{C}(S_g)$.

After Ivanov's work, many other similar theorems were proved, each saying that the automorphism group of some normal subgroup of $\text{Mod}(S_g)$ or some simplicial complex associated to S_g is isomorphic to $\text{Mod}^{\pm}(S_g)$. In response to many of these results, Ivanov posed the following [**81**].

METACONJECTURE 7.1. *Every object naturally associated to a surface S and having a sufficiently rich structure has $\mathrm{Mod}^{\pm}(S_g)$ as its group of automorphisms. Moreover, this can be proved by a reduction to the theorem about the automorphisms of $\mathcal{C}(S)$.*

One type of "objects" to consider are the normal subgroups of $\mathrm{Mod}(S_g)$. Farb and Ivanov [**56**] proved that the Torelli group $\mathcal{I}(S_g)$ satisfies the metaconjecture. Brendle and the author did the same for the Johnson kernel [**27**, **28**, **30**]. Bridson, Pettet, and Souto [**32**] then announced a generalization: all terms of the Johnson filtration have automorphism group isomorphic to $\mathrm{Mod}^{\pm}(S_g)$ for $g \geq 4$. It is rather startling that these arbitrarily small subgroups of the mapping class group somehow remember the structure of the original group. (In fact much more is true: in each case it was further shown that the abstract commensurator group of the given subgroup is $\mathrm{Mod}^{\pm}(S_g)$.)

In the spirit of Ivanov, each of the above results is proved by showing that some appropriate combinatorial complex has automorphism group isomorphic to $\mathrm{Mod}^{\pm}(S_g)$: the complex of curves, the complex of separating curves and bounding pairs, the complex of separating curves, and the complex of shirts and straitjackets. In fact, there were many other papers proving that the automorphism group of some combinatorial complex associated to S_g is isomorphic to $\mathrm{Mod}^{\pm}(S_g)$. See the author's survey with Brendle [**23**] or the paper by McCarthy–Papadopoulos [**114**] for comprehensive surveys.

The result of Bridson–Pettet–Souto gives many examples of normal subgroups satisfying Ivanov's metaconjecture. One might hope that all nontrivial normal subgroups of $\mathrm{Mod}(S_g)$ satisfy the metaconjecture. However, this is certainly not the case. Dahmani–Guirardel–Osin [**44**] constructed examples of infinitely generated, free, normal subgroups of $\mathrm{Mod}(S_g)$; the automorphism groups of these groups are uncountable, and hence definitely not isomorphic to $\mathrm{Mod}^{\pm}(S_g)$.

The metaconjecture for normal subgroups. A recent theorem of the author with Brendle gives a large class of normal subgroups of $\mathrm{Mod}(S_g)$ that satisfy Ivanov's metaconjecture [**24**]. We will state a simplified version here.

Say that a subsurface of S_g is *small* if it is contained as a non-peripheral subsurface of a subsurface of S_g of genus k with connected boundary where $k = \lfloor g/3 \rfloor$. An element of $\mathrm{Mod}(S_g)$ has *small support* if it has a representative whose support is small.

The theorem we proved is that if N is a normal subgroup of $\mathrm{Mod}(S_g)$ that contains a nontrivial element of small support, then

$$\mathrm{Aut}\, N \cong \mathrm{Mod}^{\pm}(S_g).$$

Additionally, we prove that the abstract commensurator group of N is $\mathrm{Mod}^{\pm}(S_g)$. The examples of Dahmani–Guirardel–Osin are all pseudo-Anosov, which means that the support of every nontrivial element is the whole surface S_g. Our hypothesis exactly rules out these types of examples.

Our theorem implies the result announced by Bridson–Pettet–Souto about the terms of the Johnson filtration (only for $g \geq 7$). It also applies to many other normal subgroups, including the terms of the Magnus filtration, discussed above. Using our theorem one can define many other normal subgroups satisfying Ivanov's metaconjecture, for instance, the group generated by 100th powers of Dehn twists about separating curves of genus 7 (our result applies for $g \geq 25$).

In the statement of the theorem there is a $g/3$ that should morally be $g/2$; that is, half the surface instead of one-third. Say that a nonseparating subsurface of S_g is *not large* if it is homeomorphic to a proper subsurface of its complement. And say that an arbitrary subsurface is not large if it is contained in a nonseparating subsurface that is not large. With Brendle, we make the following conjecture [**24**].

CONJECTURE 7.2. *Suppose that N is a normal subgroup of* $\text{Mod}(S_g)$ *that contains an element whose support is not large. Then the automorphism group (and also the abstract commensurator group) of N is isomorphic to* $\text{Mod}^{\pm}(S_g)$.

Chen has pointed out that whenever we have a normal subgroup N of $\text{Mod}(S_g)$ with $\text{Aut}\, N \cong \text{Mod}^{\pm}(S_g)$ then every normal subgroup of $\text{Mod}(S_g)$ isomorphic to N is equal to N; see [**24**, Corollary 1.2]. We can ask if the same holds when the mystery subgroup is not assumed to be normal.

QUESTION 7.3. Is it true that if N is a normal subgroup of $\text{Mod}(S_g)$ with $\text{Aut}\, N \cong \text{Mod}^{\pm}(S_g)$ then every subgroup of $\text{Mod}(S_g)$ isomorphic to N is equal to N? Is this true for the Torelli group $\mathcal{I}(S_g)$?

The metaconjecture for curve complexes. Our theorem with Brendle is proved by showing that there is a large class of simplicial complexes associated to S_g whose members all have automorphism group isomorphic to $\text{Mod}^{\pm}(S_g)$. Our complexes have vertices corresponding to proper, connected subsurfaces of S_g and edges for disjointness. The hypotheses of our theorem require that the simplicial complexes are connected, that they have a small element, and that they admit no exchange automorphisms, which are automorphisms that interchange two vertices and fix all other vertices. There are a number of ways in which one might try to improve on our result.

PROBLEM 7.4. Improve on our theorem with Brendle about automorphisms of simplicial complexes associated to S_g in the following ways:

(1) replace "small" with "not large,"
(2) allow vertices corresponding to disconnected subsurfaces, and
(3) allow edges for configurations other than disjointness.

The last two items would allow for complexes such as the pants complex, whose vertices correspond to disconnected subsurfaces and whose edges do not correspond to disjointness.

In our paper [**24**] we conjecture that there is a stronger statement than the one suggested by item (1) in Problem 7.4. Specifically, we conjecture that the small assumption can be removed altogether, if we keep the assumption of connectivity.

CONJECTURE 7.5. *Let $\mathcal{C}_A(S_g)$ be a complex of regions that is connected and admits no exchange automorphisms. Then the natural map*

$$\text{Mod}^{\pm}(S_g) \to \text{Aut}\, \mathcal{C}_A(S_g)$$

is an isomorphism.

Beyond the metaconjecture. With Brendle, we also give the following generalization of Ivanov's metaconjecture [**23**].

GENERALIZED METACONJECTURE 7.6. *Suppose that Y is a nice space. Every object naturally associated to Y and having a sufficiently rich structure has* $\text{Out}\, \pi_1(Y)$ *as its group of automorphisms.*

This is indeed a generalization of Ivanov's metaconjecture because of the Dehn–Nielsen–Baer theorem [**59**, Theorem 8.1] which states that $\operatorname{Out}\pi_1(S_g) \cong \operatorname{Mod}^{\pm}(S_g)$. One example of a nice space Y is the connected sum of n copies of $S^2 \times S^1$. The fundamental group of this manifold is the free group F_n. Scott [**143**] has shown that many complexes associated to this manifold have automorphism group $\operatorname{Out}(F_n)$. Other specific complexes associated to this manifold have been studied by Aramayona–Souto [**2**], Bestvina–Bridson [**9**], and Pandit [**131, 132**].

8. Normal right-angled Artin subgroups

As mentioned, Dahmani–Guirardel–Osin constructed free normal subgroups in $\operatorname{Mod}(S_g)$. In their subgroups, all of the nontrivial elements are pseudo-Anosov [**44**]. Their subgroups are constructed by taking the normal closure of some finite collection of pseudo-Anosov mapping classes.

In joint work with Clay and Mangahas [**41**], we construct similar free groups with partial pseudo-Anosov elements instead of pseudo-Anosov elements. By our theorem with Brendle about automorphisms of normal subgroups above, the supports of these partial pseudo-Anosov elements must be large in the sense of the previous section. In other words the supports take up more than half the surface.

By varying our construction we also produce normal subgroups of $\operatorname{Mod}(S_g)$ of the following form:

$$\mathop{\ast}_{\infty}(F_\infty \times F_\infty) \quad \text{and} \quad \mathop{\ast}_{\infty}(F_\infty \times F_\infty \times \mathbb{Z}).$$

For surfaces with punctures, we also construct normal subgroups isomorphic to

$$\mathop{\ast}_{\infty}(F_\infty \times \mathbb{Z}).$$

To create a normal subgroup of the first type in $\operatorname{Mod}(S_g)$, we take g to be even and take the normal closure of a (high power of a) partial pseudo-Anosov element f of $\operatorname{Mod}(S_g)$ whose support is homeomorphic to $S^1_{g/2}$. The element f has a conjugate in $\operatorname{Mod}(S_g)$ supported in the complement; this is how the commuting elements arise. To construct an example of the second type of group, we take the normal closure of the same f and a power of T_c, where c is a curve in S_g of genus $g/2$ (that is, c cuts S_g in half).

Conjecture 8.1. *Any normal subgroup of* $\operatorname{Mod}(S_g)$ *isomorphic to a right-angled Artin group is isomorphic to a free product of groups from the following list:*

$$F_\infty, \quad \mathop{\ast}_{\infty}(F_\infty \times F_\infty), \quad \text{and} \quad \mathop{\ast}_{\infty}(F_\infty \times F_\infty \times \mathbb{Z}).$$

We also conjecture that these are the only examples of normal subgroups that do not have automorphism group the extended mapping class group.

Conjecture 8.2. *A normal subgroup of* $\operatorname{Mod}(S_g)$ *either has automorphism group* $\operatorname{Mod}^{\pm}(S_g)$ *or it is isomorphic to a right-angled Artin group.*

9. Cohomology of the mapping class group

In this section we discuss the following central open problem. Throughout we take our coefficients for homology to be $\mathbb{Q}$.

Problem 9.1. Compute $H^k(\operatorname{Mod}(S_g))$ for each g and k.

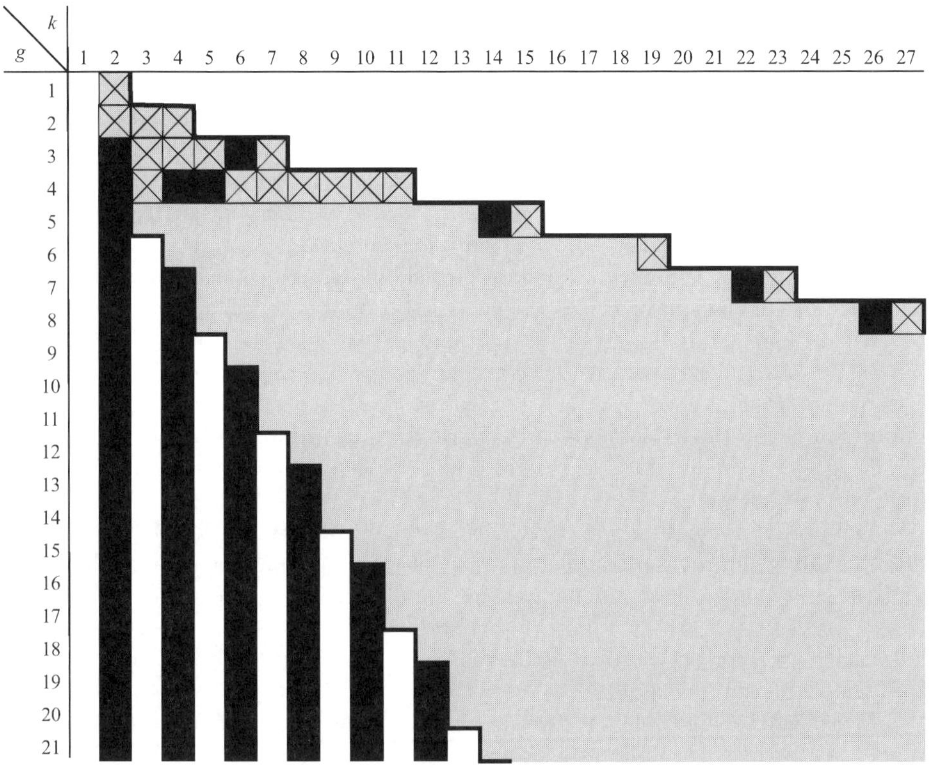

FIGURE 1. A summary of what is known about the cohomology of $\mathrm{Mod}(S_g)$; filled boxes denote nontrivial cohomology groups, the shaded region denotes the unstable region, and in this region X's denote trivial cohomology groups

A more modest question is to determine which $H^k(\mathrm{Mod}(S_g))$ are nonzero.

One of the reasons the cohomology of the mapping class group is important is because it plays the same role for surface bundles that the classical characteristic classes play for vector bundles.

In this vein, we may think of an element of $H^k(\mathrm{Mod}(S_g))$ as follows. It is a function α that assigns to each S_g-bundle over a space B an element of $H^k(B)$; the function α has to satisfy several naturality properties, the most important being that it behaves naturally under pullbacks: if E is a bundle over B and $f : A \to B$ then $\alpha(f^\star(E)) = f^\star(\alpha(E))$. By the naturality of pullback, an element of $H^k(\mathrm{Mod}(S_g))$ is completely determined by its behavior on S_g-bundles over k-manifolds, in which case $H^k(B)$ is a number.

Background. We begin by describing what is known about the cohomology of $\mathrm{Mod}(S_g)$. Throughout, we refer to Figure 1.

In the 1980s Harer proved that the virtual cohomological dimension of $\mathrm{Mod}(S_g)$ is $4g - 5$ for $g \geq 2$, and so $H^k(\mathrm{Mod}(S_g))$ is trivial for $k > 4g - 5$. The latter fact was also proved by Culler–Vogtmann. This explains the empty region at the top right of Figure 1: all of these cohomology groups are zero.

Harer proved that the mapping class groups satisfy homological stability [**76**]. This means that for fixed k, and g large enough, the groups $H^k(\mathrm{Mod}(S_g))$ do not depend on g. Specifically, this has been shown to hold in the "stable range" $k \leq 2\lfloor \frac{g}{3} \rfloor$, by the work of Harer [**75**], Ivanov [**79**], Wahl [**155**], Randal-Williams [**140**], and Boldsen [**21**].

Mumford conjectured that in the stable range the cohomology is generated by the Morita–Mumford–Miller classes $\kappa_1, \kappa_2, \ldots$. Each κ_i is an element of $H^{2i}(\mathrm{Mod}(S_g))$. The κ_i are also called tautological classes.

Here is how κ_1 is defined. As above, κ_1 should assign to an S_g-bundle E over a surface S_h an element of $H^2(S_h)$ (that is, a rational number). The bundle E has a vertical tangent bundle, given by the tangent planes to the fibers. This 2-plane bundle has an Euler class $e \in H^2(E)$. The class κ_1 is the image of e^2 under the Gysin map $H^2(E) \to H^2(S_h)$. In other words, $\kappa_1(E)$ is obtained as the intersection number of three generic sections of the vertical tangent bundle over E. In general, the classes κ_i are defined as the images under the Gysin homomorphism of e^{i+1} (so intersect $i+2$ sections).

Mumford's conjecture was proved in 2006 by Madsen–Weiss [**106**]; see also [**49, 77, 156**]. This gives a complete picture of the cohomology in the stable range: it is the polynomial algebra generated by the κ_i (for fixed g the κ_i are necessarily trivial for $2i > 4g-5$). In Figure 1 this stable range is the region on the bottom left; the reason for the vertical stripes is Harer's stability theorem, plus the fact that the κ_i all have even degree.

In order to solve Problem 9.2 we should thus focus our attention on the unstable range, shaded in gray in Figure 1. A few of the cohomology groups here are known. In fact, we have complete computations of $H^*(\mathrm{Mod}(S_g))$ for $g \leq 4$. The black boxes in the first four rows of Figure 1 represent 1-dimensional cohomology groups and the X's represent trivial groups. The cohomology groups for g equal to 2, 3, and 4 were computed by Igusa in 1960, Looijenga in 1991, and Tommasi in 2005, respectively [**78, 102, 151**].

Dark matter. While very few unstable classes are known explicitly, we do know that many other classes exist. Harer and Zagier [**74**] computed the Euler characteristic of moduli space $\mathcal{M}_g$, thought of as an orbifold. They showed that

$$\chi(\mathcal{M}_g) = \frac{\zeta(1-2g)}{2-2g} \sim (-1)^g \frac{(2g-1)!}{(2-2g)2^{2g-1}\pi^{2g}}.$$

In particular, this number grows super-exponentially in g. This implies that the Betti numbers of torsion-free subgroups of finite index in $\mathrm{Mod}(S_g)$ grow super-exponentially in g. Harer and Zagier promote this to the statement that the Betti numbers of $\mathrm{Mod}(S_g)$ itself grow super-exponentially in g.

The number of stable classes is polynomial in g (since it is a polynomial algebra). This means that there is a super-exponential number of classes in the unstable range waiting to be discovered. What is more, since $\chi(\mathcal{M}_g)$ is negative for g even, we know that there are super-exponentially many classes in odd degree.

Problem 9.2. Find some (or even one!) element of $H^k(\mathrm{Mod}(S_g))$ outside of the stable range.

To be even more greedy, we would like geometrically meaningful interpretations of the unstable elements. The first tautological class κ_1, for example, measures the signatures (and first Pontryagin classes) of surface bundles and relates to the Weil–Petersson volume form on Teichmüller space. Salter [**142**] gives similar characterizations of the other κ_i. We would like to have geometric interpretations for the unstable classes. For instance, what does Looijenga's unstable class in $H^6(\mathrm{Mod}(S_3))$ tell us about S_3-bundles over 6-manifolds?

Farb has called Problem 9.2 the "dark matter problem." (Of course an important difference from the original dark matter problem is that we know the dark matter exists!)

Church–Farb–Putman [**39**] and Morita–Sakasai–Suzuki [**128**] independently proved that $H^{4g-5}(\mathrm{Mod}(S_g))$ is trivial, so there are no unstable classes in the top degree. This might seem like a contradiction at first, since the virtual cohomological dimension of $\mathrm{Mod}(S_g)$ is $4g-5$. But the virtual cohomological dimension being $4g-5$ only requires that $\mathrm{Mod}(S_g)$ has nontrivial cohomology in degree $4g-5$ with *some* (possibly twisted) coefficients. This theorem is indicated by several X's in Figure 1.

Very recently, Chan–Galatius–Payne [**35**] proved that $H^{4g-6}(\mathrm{Mod}(S_g))$ is nontrivial for $g = 5$ and $g \geq 7$. Even more, they show that the dimension grows exponentially in g. This is the first partial answer to Problem 9.2 that fills in infinitely many boxes in Figure 1.

Even if we knew what all of the unstable elements of $H^*(\mathrm{Mod}(S_g))$ were, we would still have the problem of understanding the ring structures for each g. In other words, what are the relations? The Faber conjectures [**52**] describe the structure of the tautological ring, that is, the ring generated by the tautological classes κ_i. The conjectures give explicit proportionality relations between the tautological classes and assert that the tautological ring for $\mathrm{Mod}(S_g)$ "looks like" the algebraic cohomology ring of a smooth projective variety of dimension $g-2$. Much of the conjectures have been proven; see the surveys by Faber [**52**] and Morita [**127**].

Torelli groups. Morita proved that the odd κ_i vanish on the Torelli group $\mathcal{I}(S_g)$. He then asked the following [**126**, Problem 2].

QUESTION 9.3. Are the restrictions of the even κ_i to $\mathcal{I}(S_g)$ trivial or not?

Even for $i = 2$ the answer is unknown. In this case, Morita's question asks whether there is an S_g-bundle over a 4-manifold, with monodromy in $\mathcal{I}(S_g)$, so that if we intersect four generic sections of the vertical tangent bundle we obtain a nonzero number.

10. Pseudo-Anosov theory

The Nielsen–Thurston classification theorem says that mapping classes fall into three types: periodic, reducible, and pseudo-Anosov. To each pseudo-Anosov element there is an associated stretch factor.

Genericity. One of the most famous open problems in this direction is the following.

PROBLEM 10.1. Show that pseudo-Anosov mapping classes are generic.

This problem has been answered in one sense. Maher has shown that a random walk in the mapping class group gives rise to a pseudo-Anosov element with asymptotic probability one [**108**]. Rivin proved the analogous result for the Torelli group [**141**].

Another notion of genericity is that the proportion of pseudo-Anosov elements in the ball of radius r in a Cayley graph for $\mathrm{Mod}(S_g)$ should be asymptotically one. This version is open. Recently Cumplido and Wiest have shown that this proportion is asymptotically positive [**43**].

Fried's question. Another fundamental question is the following.

QUESTION 10.2. Which real numbers are stretch factors of pseudo-Anosov mapping classes?

There are many variants of this question. The most famous version is due to Fried, who posed the following question [**63**, Problem 2]. In the statement, a bi-Perron unit is an algebraic integer α so that all its algebraic conjugates are either equal to $1/\alpha$ or have absolute value in $(1/\alpha, \alpha)$ (the number $1/\alpha$ is not required to be an algebraic conjugate).

QUESTION 10.3. Is is true that every bi-Perron unit has a power that is a stretch factor of a pseudo-Anosov mapping class?

Very little is known about this question. Recently Pankau has shown that every Salem number has a power that is a stretch factor of a pseudo-Anosov mapping class. A Salem number is a real number $\alpha > 1$ so that $1/\alpha$ is an algebraic conjugate and so that all other algebraic conjugates lie on the unit circle.

We remark that the power in Question 10.3 may not be required.

Algebraic degrees. Question 10.2 is still (or more?) interesting if we fix the surface we are considering. Thurston [**150**] proved that the stretch factor of a pseudo-Anosov element of $\mathrm{Mod}(S_g)$ is an algebraic integer with degree between 2 and $6g-6$ (inclusive). Long showed that if the degree is greater than $3g-3$ then it is even. (McMullen has given simple proofs of the Thurston and Long restrictions; see the paper by Shin [**144**].) Strenner then showed that all degrees satisfying the Thurston and Long restrictions do indeed occur [**146**]. Strenner's proof is not completely constructive. This leaves open the following.

PROBLEM 10.4. For each $g \geq 2$ and each even d in between 2 and $6g-6$ and each odd d between 2 and $3g-3$, construct an explicit pseudo-Anosov element of $\mathrm{Mod}(S_g)$ whose stretch factor has algebraic degree d.

Before Strenner's work, Arnoux–Yoccoz [**4**] gave examples of pseudo-Anosov elements of degree g. Also, Shin [**144**] gave examples of pseudo-Anosov elements degree $2g$.

The Arnoux–Yoccoz examples are given in terms of interval exchange transformations; it would be interesting to write them as products of Dehn twists (or other simple mapping classes).[2]

Shin found his examples by doing an extensive computer search. Some algebraic degrees seem to be abundant, and some—such as degree 5 in $\mathrm{Mod}(S_2)$—appear to be rare. We are thus led to the following question.

QUESTION 10.5. Which algebraic degrees are most common? Given g, is there an algebraic degree that is generic?

We also have the following problem inspired by Strenner's result.

[2]This problem was recently solved by Liechti–Strenner [**99**].

PROBLEM 10.6. Which algebraic degrees appear in the Torelli group $\mathcal{I}(S_g)$? What about an arbitrary proper normal subgroup of $\text{Mod}(S_g)$?

Minimal stretch factors. For fixed g the stretch factors in $\text{Mod}(S_g)$ form a discrete subset of the real numbers. In particular, there is for each g a smallest stretch factor λ_g. We thus have the following basic question.

PROBLEM 10.7. Determine the λ_g.

Problem 10.7 has only been solved for g equal to 1 and 2. It is an easy exercise to show that λ_1 is realized by the conjugacy class of the element of $\text{Mod}(S_1)$ corresponding to the matrix

$$\begin{bmatrix} 2 & 1 \\ 1 & 1 \end{bmatrix}.$$

The number λ_2 is the largest real root of the polynomial

$$x^4 - x^3 - x^2 - x + 1.$$

This was proved by Cho–Ham [**36**], who reduced the problem to a finite (but extremely large) check. Lanneau and Thiffeault [**94**] further determined all conjugacy classes in $\text{Mod}(S_2)$ realizing the minimum—there are two—and gave an explicit Dehn twist factorization for each. Minimal stretch factors have also been found for some braid groups [**94**] and some mapping class groups of non-orientable surfaces [**98**].

QUESTION 10.8. What is the smallest stretch factor in $\mathcal{I}(S_g)$? What about $\mathcal{K}(S_g)$? Is the former strictly smaller than the latter?

With Farb and Leininger, the author proved [**57**] that, for g fixed, the smallest stretch factor in the kth term of the Johnson filtration goes to infinity with k. So we may ask more generally if this sequence is monotonically increasing with k.

Normal closures. With Lanier [**93**], the author proved that every pseudo-Anosov mapping class with stretch factor less than $\sqrt{2}$ is a normal generator for $\text{Mod}(S_g)$, that is, its normal closure is the whole group. Said another way, if a pseudo-Anosov mapping class lies in a proper normal subgroup of $\text{Mod}(S_g)$ then its stretch factor is greater than $\sqrt{2}$. On the other hand, the only explicit examples of pseudo-Anosov mapping classes that lie in a proper normal subgroup have stretch factor greater than 62; see [**57**]. So the smallest stretch factor of a pseudo-Anosov mapping class that lies in a proper normal subgroup is between $\sqrt{2}$ and 62.

PROBLEM 10.9. Improve the bounds on the smallest stretch factor of a pseudo-Anosov mapping class lying in a proper normal subgroup of the mapping class group.

With Lanier we also give examples of pseudo-Anosov mapping classes whose normal closure in $\text{Mod}(S_g)$ is equal to any given level m congruence subgroup. We ask the following.

QUESTION 10.10. Can other normal subgroups, such as the Johnson kernel and the other terms of the Johnson filtration, be obtained as the normal closure in $\text{Mod}(S_g)$ of a single pseudo-Anosov mapping class?

We also give examples of pseudo-Anosov mapping classes with the property that arbitrarily large powers are normal generators for the mapping class group. In particular we can take these powers to have arbitrarily large translation lengths on $\mathcal{C}(S_g)$. The measured foliations associated to our examples have nontrivial symmetry groups. Since the generic pseudo-Anosov mapping class (in the sense of random walks) has foliations without symmetries [**111**], we arrive at the following problem.[3]

PROBLEM 10.11. Is the generic element of $\text{Mod}(S_g)$ a normal generator, or not?

The next question is related. In the statement, the extended Torelli group is the preimage of $\{\pm I\}$ under the standard symplectic representation of $\text{Mod}(S_g)$. This group is the normal closure of any hyperelliptic involution.

QUESTION 10.12. If the asymptotic translation length of a pseudo-Anosov mapping class is large, can its normal closure be anything other than the mapping class group or a free group? Can the normal closure be the extended Torelli group?

Small stretch factors. Penner [**133**] proved that there are constants c and C so that

$$\frac{c}{g} \leq \log \lambda_g \leq \frac{C}{g}.$$

In particular, $\lambda_g \to 1$ as $g \to \infty$. In particular, while the set of stretch factors in $\text{Mod}(S_g)$ is discrete for fixed g, the set of all stretch factors is dense in $(1, \infty)$ (powers of stretch factors are also stretch factors).

We can say that a stretch factor λ of an element of $\text{Mod}(S_g)$ is *small* if $g \log \lambda$ is bounded above by Penner's C (or any fixed number greater than or equal to the minimal possible C as above). Let P_C denote the set of pseudo-Anosov mapping classes f with small stretch factor. We emphasize here that the set P_C contains elements of $\text{Mod}(S_g)$ for every g.

McMullen gave an alternate proof of Penner's theorem from the point of view of fibered 3-manifolds [**115**]. Building on work of Fried [**62**] and Thurston [**149**], McMullen [**115**] showed that in a single fibered 3-manifold, there are fibers Σ_g so that the associated monodromies $f_g : \Sigma_g \to \Sigma_g$ have small stretch factor for g sufficiently large.

Inspired by McMullen's work, we consider the set of all mapping tori M_f where $f \in P_C$. For any such f, let M_f° denote the 3-manifold obtained from M_f by removing the suspensions of the singular points of (a representative of) f. With Farb and Leininger [**58**], we proved that the set

$$\Omega_C = \{M_f^\circ \mid f \in P_C\}$$

is finite. The removal of the suspensions of singular points is indeed necessary: Penner [**133**] gave examples of elements of P_C with prongs of arbitrarily large degree; but in a given closed 3-manifold, there is an upper bound to the number of prongs.

QUESTION 10.13. What is the smallest nonzero cardinality of Ω_C? If we vary C in $(1, \infty)$, how does the cardinality grow?

[3]This problem was recently solved by Maher–Tiozzo, who showed that the normal closure of a generic mapping class is free [**109**].

The following related conjecture is due to Farb, Leininger, and the author of this article.

CONJECTURE 10.14 (Symmetry conjecture). *If $f \in \text{Mod}(S_g)$ is a pseudo-Anosov mapping class with stretch factor λ satisfying $g \log \lambda < C$, then f is equal to the product of a periodic element and a reducible element, and moreover the absolute value of the Euler characteristic of the support of the reducible element is bounded above by C, independently of g.*

Leininger has shown that if the stretch factor of f is less than $3/2$ then it is the product of a periodic element and a reducible element. The idea is that in this case we can find a curve c so that $i(c, f(c))$ is at most two, and thus we can find a periodic element taking $f(c)$ back to c. The reducible element constructed here has large complexity. On the other hand, this is nontrivial progress because there are pseudo-Anosov elements of the mapping class group that cannot be written as the product of a periodic element and a reducible element (apply the work of Bestvina–Fujiwara [**11**]).

Pseudo-Anosov surface subgroups. We say that a subgroup of $\text{Mod}(S_g)$ is all pseudo-Anosov if every nontrivial element is pseudo-Anosov. A version of the following question was raised by Farb and Mosher [**60**, Question 1.9].

QUESTION 10.15. Does $\text{Mod}(S_g)$ contain an all pseudo-Anosov surface subgroup?

The motivation for the last question is the study of surface bundles over surfaces. There are no known examples of surface bundles over surfaces that are hyperbolic 4-manifolds. Of course hyperbolic implies negative curvature in the sense of Gromov. The latter is known to imply that the monodromy group is all pseudo-Anosov (if there was a surface bundle over a surface with all pseudo-Anosov monodromy that was not hyperbolic, then this would contradict a conjecture of Gromov, which states that any group G with a finite $K(G,1)$ and no Baumslag–Solitar subgroups is hyperbolic). Therefore, a positive answer to the last question is necessary if there are to exist hyperbolic surface bundles over surfaces.

Leininger–Reid [**96**] produced surface subgroups of $\text{Mod}(S_g)$ where every element outside of a single cyclic subgroup is pseudo-Anosov. The Farb–Mosher paper has inspired a continued interest in the convex cocompact subgroups of the mapping class group.

Bowditch [**22**] showed that there are (at most) finitely many conjugacy classes of all pseudo-Anosov surface subgroups of $\text{Mod}(S_g)$, for each g (this result was also announced at about the same time by Groves and by Sapir). So if such subgroups do exist, they are rare.

Acknowledgments

The author would like to thank Tara Brendle, Lei Chen, Nathan Dunfield, Benson Farb, Autumn Kent, Eiko Kin, Kevin Kordek, Justin Lanier, Johanna Mangahas, Shigeyuki Morita, Andrew Putman, Takuya Sakasai, Nick Salter, Shane Scott, Balázs Strenner, Masaaki Suzuki, and the anonymous referee.

References

[1] Toshiyuki Akita, *Homological infiniteness of Torelli groups*, Topology **40** (2001), no. 2, 213–221, DOI 10.1016/S0040-9383(99)00050-6. MR1808217

[2] Javier Aramayona and Juan Souto, *Automorphisms of the graph of free splittings*, Michigan Math. J. **60** (2011), no. 3, 483–493, DOI 10.1307/mmj/1320763044. MR2861084

[3] V. I. Arnol′d, *A remark on the branching of hyperelliptic integrals as functions of the parameters* (Russian), Funkcional. Anal. i Priložen. **2** (1968), no. 3, 1–3. MR0244266

[4] Pierre Arnoux and Jean-Christophe Yoccoz, *Construction de difféomorphismes pseudo-Anosov* (French, with English summary), C. R. Acad. Sci. Paris Sér. I Math. **292** (1981), no. 1, 75–78. MR610152

[5] Mamoru Asada, *The faithfulness of the monodromy representations associated with certain families of algebraic curves*, J. Pure Appl. Algebra **159** (2001), no. 2-3, 123–147, DOI 10.1016/S0022-4049(00)00056-6. MR1828935

[6] S. Bachmuth, *Braid groups are linear groups*, Adv. Math. **121** (1996), no. 1, 50–61, DOI 10.1006/aima.1996.0046. MR1399602

[7] Gavin Band and Philip Boyland, *The Burau estimate for the entropy of a braid*, Algebr. Geom. Topol. **7** (2007), 1345–1378, DOI 10.2140/agt.2007.7.1345. MR2350285

[8] H. Bass, J. Milnor, and J.-P. Serre, *Solution of the congruence subgroup problem for* $\mathrm{SL}_n\,(n \geq 3)$ *and* $\mathrm{Sp}_{2n}\,(n \geq 2)$, Inst. Hautes Études Sci. Publ. Math. **33** (1967), 59–137. MR0244257

[9] Mladen Bestvina and Martin Bridson. Rigidity of the free factor complex. In preparation.

[10] Mladen Bestvina, Kai-Uwe Bux, and Dan Margalit, *The dimension of the Torelli group*, J. Amer. Math. Soc. **23** (2010), no. 1, 61–105, DOI 10.1090/S0894-0347-09-00643-2. MR2552249

[11] Mladen Bestvina and Koji Fujiwara, *Quasi-homomorphisms on mapping class groups*, Glas. Mat. Ser. III **42(62)** (2007), no. 1, 213–236, DOI 10.3336/gm.42.1.15. MR2332668

[12] Stephen Bigelow, *The Burau representation is not faithful for* $n = 5$, Geom. Topol. **3** (1999), 397–404, DOI 10.2140/gt.1999.3.397. MR1725480

[13] Stephen Bigelow, *Does the Jones polynomial detect the unknot?*, J. Knot Theory Ramifications **11** (2002), no. 4, 493–505, DOI 10.1142/S0218216502001779. Knots 2000 Korea, Vol. 2 (Yongpyong). MR1915491

[14] Stephen J. Bigelow, *Braid groups are linear*, J. Amer. Math. Soc. **14** (2001), no. 2, 471–486, DOI 10.1090/S0894-0347-00-00361-1. MR1815219

[15] Joan S. Birman, *Braids, links, and mapping class groups*, Princeton University Press, Princeton, N.J.; University of Tokyo Press, Tokyo, 1974. Annals of Mathematics Studies, No. 82. MR0375281

[16] Joan S. Birman and Tara E. Brendle, *Braids: a survey*, Handbook of knot theory, Elsevier B. V., Amsterdam, 2005, pp. 19–103, DOI 10.1016/B978-044451452-3/50003-4. MR2179260

[17] Joan S. Birman and Hugh M. Hilden, *On isotopies of homeomorphisms of Riemann surfaces*, Ann. of Math. (2) **97** (1973), 424–439, DOI 10.2307/1970830. MR0325959

[18] M. Boggi. The congruence subgroup property for the hyperelliptic modular group: the closed surface case.

[19] Marco Boggi, *Profinite Teichmüller theory*, Math. Nachr. **279** (2006), no. 9-10, 953–987, DOI 10.1002/mana.200510405. MR2242960

[20] Marco Boggi, *The congruence subgroup property for the hyperelliptic modular group: the open surface case*, Hiroshima Math. J. **39** (2009), no. 3, 351–362. MR2569009

[21] Søren K. Boldsen, *Improved homological stability for the mapping class group with integral or twisted coefficients*, Math. Z. **270** (2012), no. 1-2, 297–329, DOI 10.1007/s00209-010-0798-y. MR2875835

[22] Brian H. Bowditch, *Atoroidal surface bundles over surfaces*, Geom. Funct. Anal. **19** (2009), no. 4, 943–988, DOI 10.1007/s00039-009-0033-3. MR2570310

[23] Tara Brendle and Dan Margalit. Ivanov's metaconjecture: a survey.

[24] Tara Brendle and Dan Margalit, *Normal subgroups of mapping class groups and the metaconjecture of Ivanov*, arXiv:1710.08929 (2017).

[25] Tara Brendle, Dan Margalit, and Andrew Putman, *Generators for the hyperelliptic Torelli group and the kernel of the Burau representation at* $t = -1$, Invent. Math. **200** (2015), no. 1, 263–310, DOI 10.1007/s00222-014-0537-9. MR3323579

[26] Tara E. Brendle and Hessam Hamidi-Tehrani, *On the linearity problem for mapping class groups*, Algebr. Geom. Topol. **1** (2001), 445–468, DOI 10.2140/agt.2001.1.445. MR1852767

[27] Tara E. Brendle and Dan Margalit, *Commensurations of the Johnson kernel*, Geom. Topol. **8** (2004), 1361–1384, DOI 10.2140/gt.2004.8.1361. MR2119299

[28] Tara E. Brendle and Dan Margalit, *Addendum to: "Commensurations of the Johnson kernel" [Geom. Topol. **8** (2004), 1361–1384; MR2119299]*, Geom. Topol. **12** (2008), no. 1, 97–101, DOI 10.2140/gt.2008.12.97. MR2377246

[29] Tara E. Brendle and Dan Margalit, *Point pushing, homology, and the hyperelliptic involution*, Michigan Math. J. **62** (2013), no. 3, 451–473, DOI 10.1307/mmj/1378757883. MR3102525

[30] Tara E. Brendle and Dan Margalit, *Commensurations of the Johnson kernel*, Geom. Topol. **8** (2004), 1361–1384, DOI 10.2140/gt.2004.8.1361. MR2119299

[31] Tara E. Brendle and Dan Margalit, *The level four braid group*, J. Reine Angew. Math. **735** (2018), 249–264, DOI 10.1515/crelle-2015-0032. MR3757477

[32] Martin Bridson, Alexandra Pettet, and Juan Souto. The abstract commensurator of the Johnson kernels, unpublished.

[33] Werner Burau, *Über Zopfgruppen und gleichsinnig verdrillte Verkettungen* (German), Abh. Math. Sem. Univ. Hamburg **11** (1935), no. 1, 179–186, DOI 10.1007/BF02940722. MR3069652

[34] Kai-Uwe Bux, Mikhail V. Ershov, and Andrei S. Rapinchuk, *The congruence subgroup property for* Aut F_2*: a group-theoretic proof of Asada's theorem*, Groups Geom. Dyn. **5** (2011), no. 2, 327–353, DOI 10.4171/GGD/130. MR2782176

[35] Melody Chan, Soren Galatius, and Sam Payne, *Tropical curves, graph homology, and top weight cohomology of* $\mathcal{M}_g$, arXiv:1805.10186 (2018).

[36] Jin-Hwan Cho and Ji-Young Ham, *The minimal dilatation of a genus-two surface*, Experiment. Math. **17** (2008), no. 3, 257–267. MR2455699

[37] Thomas Church, Mikhail Ershov, and Andrew Putman, *On finite generation of the johnson filtrations*, arXiv:1711.04779 (2017).

[38] Thomas Church and Benson Farb, *Infinite generation of the kernels of the Magnus and Burau representations*, Algebr. Geom. Topol. **10** (2010), no. 2, 837–851, DOI 10.2140/agt.2010.10.837. MR2629766

[39] Thomas Church, Benson Farb, and Andrew Putman, *The rational cohomology of the mapping class group vanishes in its virtual cohomological dimension*, Int. Math. Res. Not. IMRN **21** (2012), 5025–5030, DOI 10.1093/imrn/rnr208. MR2993444

[40] Thomas Church and Andrew Putman, *Generating the Johnson filtration*, Geom. Topol. **19** (2015), no. 4, 2217–2255, DOI 10.2140/gt.2015.19.2217. MR3375526

[41] Matt Clay, Johanna Mangahas, and Dan Margalit. Right-angled artin groups as normal subgroups of mapping class groups, in preparation.

[42] James Cooper, *Two* mod-p *Johnson filtrations*, J. Topol. Anal. **7** (2015), no. 2, 309–343, DOI 10.1142/S1793525315500120. MR3326304

[43] María Cumplido and Bert Wiest, *A positive proportion of elements of mapping class groups is pseudo-Anosov*, Bull. Lond. Math. Soc. **50** (2018), no. 3, 390–394, DOI 10.1112/blms.12148. MR3829727

[44] F. Dahmani, V. Guirardel, and D. Osin, *Hyperbolically embedded subgroups and rotating families in groups acting on hyperbolic spaces*, Mem. Amer. Math. Soc. **245** (2017), no. 1156, v+152, DOI 10.1090/memo/1156. MR3589159

[45] Steven Diaz, Ron Donagi, and David Harbater, *Every curve is a Hurwitz space*, Duke Math. J. **59** (1989), no. 3, 737–746, DOI 10.1215/S0012-7094-89-05933-4. MR1046746

[46] Alexandru Dimca, Richard Hain, and Stefan Papadima, *The abelianization of the Johnson kernel*, J. Eur. Math. Soc. (JEMS) **16** (2014), no. 4, 805–822, DOI 10.4171/JEMS/447. MR3191977

[47] Alexandru Dimca and Stefan Papadima, *Arithmetic group symmetry and finiteness properties of Torelli groups*, Ann. of Math. (2) **177** (2013), no. 2, 395–423, DOI 10.4007/annals.2013.177.2.1. MR3010803

[48] Nathan Dunfield. Personal communication. 2018.

[49] Yakov Eliashberg, Søren Galatius, and Nikolai Mishachev, *Madsen-Weiss for geometrically minded topologists*, Geom. Topol. **15** (2011), no. 1, 411–472, DOI 10.2140/gt.2011.15.411. MR2776850

[50] Jordan S. Ellenberg and D. B. McReynolds, *Arithmetic Veech sublattices of* SL(2, **Z**), Duke Math. J. **161** (2012), no. 3, 415–429, DOI 10.1215/00127094-1507412. MR2881227

[51] Mikhail Ershov and Sue He, *On finiteness properties of the Johnson filtrations*, Duke Math. J. **167** (2018), no. 9, 1713–1759, DOI 10.1215/00127094-2018-0005. MR3813595

[52] Carel Faber, *A conjectural description of the tautological ring of the moduli space of curves*, Moduli of curves and abelian varieties, Aspects Math., E33, Friedr. Vieweg, Braunschweig, 1999, pp. 109–129. MR1722541

[53] Benson Farb, *Some problems on mapping class groups and moduli space*, Problems on mapping class groups and related topics, Proc. Sympos. Pure Math., vol. 74, Amer. Math. Soc., Providence, RI, 2006, pp. 11–55, DOI 10.1090/pspum/074/2264130. MR2264130

[54] Benson Farb, *Some problems on mapping class groups and moduli space*, Problems on mapping class groups and related topics, Proc. Sympos. Pure Math., vol. 74, Amer. Math. Soc., Providence, RI, 2006, pp. 11–55, DOI 10.1090/pspum/074/2264130. MR2264130

[55] Benson Farb and Sebastian Hensel, *Moving homology classes in finite covers of graphs*, Israel J. Math. **220** (2017), no. 2, 605–615, DOI 10.1007/s11856-017-1528-5. MR3666439

[56] Benson Farb and Nikolai V. Ivanov, *Torelli buildings and their automorphisms*, arXiv:1410.6223 (2014).

[57] Benson Farb, Christopher J. Leininger, and Dan Margalit, *The lower central series and pseudo-Anosov dilatations*, Amer. J. Math. **130** (2008), no. 3, 799–827, DOI 10.1353/ajm.0.0005. MR2418928

[58] Benson Farb, Christopher J. Leininger, and Dan Margalit, *Small dilatation pseudo-Anosov homeomorphisms and 3-manifolds*, Adv. Math. **228** (2011), no. 3, 1466–1502, DOI 10.1016/j.aim.2011.06.020. MR2824561

[59] Benson Farb and Dan Margalit, *A primer on mapping class groups*, Princeton Mathematical Series, vol. 49, Princeton University Press, Princeton, NJ, 2012. MR2850125

[60] Benson Farb and Lee Mosher, *Convex cocompact subgroups of mapping class groups*, Geom. Topol. **6** (2002), 91–152, DOI 10.2140/gt.2002.6.91. MR1914566

[61] Edward Formanek and Claudio Procesi, *The automorphism group of a free group is not linear*, J. Algebra **149** (1992), no. 2, 494–499, DOI 10.1016/0021-8693(92)90029-L. MR1172442

[62] David Fried, *Flow equivalence, hyperbolic systems and a new zeta function for flows*, Comment. Math. Helv. **57** (1982), no. 2, 237–259, DOI 10.1007/BF02565860. MR684116

[63] David Fried, *Growth rate of surface homeomorphisms and flow equivalence*, Ergodic Theory Dynam. Systems **5** (1985), no. 4, 539–563, DOI 10.1017/S0143385700003151. MR829857

[64] Neil J. Fullarton and Richard Shadrach,. *Observed periodicity related to the four-strand burau representation*, arXiv:1610.04648 (2016). To appear in Experiment. Math. Mathematics.

[65] Louis Funar and Toshitake Kohno, *On Burau's representations at roots of unity*, Geom. Dedicata **169** (2014), 145–163, DOI 10.1007/s10711-013-9847-0. MR3175241

[66] Alexander A. Gaifullin, *On infinitely generated homology of torelli groups*, arXiv:1803.09311 (2018).

[67] Jean-Marc Gambaudo and Étienne Ghys, *Braids and signatures* (English, with English and French summaries), Bull. Soc. Math. France **133** (2005), no. 4, 541–579, DOI 10.24033/bsmf.2496. MR2233695

[68] Betty Jane Gassner, *On braid groups*, Abh. Math. Sem. Univ. Hamburg **25** (1961), 10–22, DOI 10.1007/BF02992772. MR0130309

[69] Geoff Mess. Unit tangent bundle subgroups of the mapping class groups. IHES/M/90/30, 1990.

[70] Fritz Grunewald, Michael Larsen, Alexander Lubotzky, and Justin Malestein, *Arithmetic quotients of the mapping class group*, Geom. Funct. Anal. **25** (2015), no. 5, 1493–1542, DOI 10.1007/s00039-015-0352-5. MR3426060

[71] Fritz Grunewald and Alexander Lubotzky, *Linear representations of the automorphism group of a free group*, Geom. Funct. Anal. **18** (2009), no. 5, 1564–1608, DOI 10.1007/s00039-009-0702-2. MR2481737

[72] Richard Hain, *Finiteness and Torelli spaces*, Problems on mapping class groups and related topics, Proc. Sympos. Pure Math., vol. 74, Amer. Math. Soc., Providence, RI, 2006, pp. 57–70, DOI 10.1090/pspum/074/2264131. MR2264131

[73] Richard M. Hain, *Torelli groups and geometry of moduli spaces of curves*, Current topics in complex algebraic geometry (Berkeley, CA, 1992/93), Math. Sci. Res. Inst. Publ., vol. 28, Cambridge Univ. Press, Cambridge, 1995, pp. 97–143. MR1397061

[74] J. Harer and D. Zagier, *The Euler characteristic of the moduli space of curves*, Invent. Math. **85** (1986), no. 3, 457–485, DOI 10.1007/BF01390325. MR848681

[75] J. L. Harer. Improved homology stability for the homology of the mapping class groups of surfaces (1993) (unpublished).

[76] John L. Harer, *Stability of the homology of the mapping class groups of orientable surfaces*, Ann. of Math. (2) **121** (1985), no. 2, 215–249, DOI 10.2307/1971172. MR786348

[77] Allen Hatcher, *A Short exposition of the Madsen-Weiss theorem*, arXiv:1103.5223 (2011).

[78] Jun-ichi Igusa, *Arithmetic variety of moduli for genus two*, Ann. of Math. (2) **72** (1960), 612–649, DOI 10.2307/1970233. MR0114819

[79] Nikolai V. Ivanov, *On the homology stability for Teichmüller modular groups: closed surfaces and twisted coefficients*, Mapping class groups and moduli spaces of Riemann surfaces (Göttingen, 1991/Seattle, WA, 1991), Contemp. Math., vol. 150, Amer. Math. Soc., Providence, RI, 1993, pp. 149–194, DOI 10.1090/conm/150/01290. MR1234264

[80] Nikolai V. Ivanov, *Automorphism of complexes of curves and of Teichmüller spaces*, Internat. Math. Res. Notices **14** (1997), 651–666, DOI 10.1155/S1073792897000433. MR1460387

[81] Nikolai V. Ivanov, *Fifteen problems about the mapping class groups*, Problems on mapping class groups and related topics, Proc. Sympos. Pure Math., vol. 74, Amer. Math. Soc., Providence, RI, 2006, pp. 71–80, DOI 10.1090/pspum/074/2264532. MR2264532

[82] Dennis Johnson, *The structure of the Torelli group. I. A finite set of generators for $\mathcal{I}$*, Ann. of Math. (2) **118** (1983), no. 3, 423–442, DOI 10.2307/2006977. MR727699

[83] Dennis Johnson, *A survey of the Torelli group*, Low-dimensional topology (San Francisco, Calif., 1981), Contemp. Math., vol. 20, Amer. Math. Soc., Providence, RI, 1983, pp. 165–179, DOI 10.1090/conm/020/718141. MR718141

[84] Dennis Johnson, *The structure of the Torelli group. II. A characterization of the group generated by twists on bounding curves*, Topology **24** (1985), no. 2, 113–126, DOI 10.1016/0040-9383(85)90049-7. MR793178

[85] Dennis Johnson, *The structure of the Torelli group. III. The abelianization of $\mathcal{T}$*, Topology **24** (1985), no. 2, 127–144, DOI 10.1016/0040-9383(85)90050-3. MR793179

[86] Martin Kassabov, *Generating mapping class groups by involutions*, arXiv:math/0311455 (2003).

[87] D. A. Každan, *On the connection of the dual space of a group with the structure of its closed subgroups* (Russian), Funkcional. Anal. i Priložen. **1** (1967), 71–74. MR0209390

[88] Richard Peabody Kent IV, *Congruence kernels around affine curves*, J. Reine Angew. Math. **713** (2016), 1–20, DOI 10.1515/crelle-2014-0023. MR3483623

[89] Rob Kirby, *Problems in low dimensional manifold theory*, Algebraic and geometric topology (Proc. Sympos. Pure Math., Stanford Univ., Stanford, Calif., 1976), Proc. Sympos. Pure Math., XXXII, Amer. Math. Soc., Providence, R.I., 1978, pp. 273–312. MR520548

[90] Mustafa Korkmaz, *Generating the surface mapping class group by two elements*, Trans. Amer. Math. Soc. **357** (2005), no. 8, 3299–3310, DOI 10.1090/S0002-9947-04-03605-0. MR2135748

[91] Daan Krammer, *Braid groups are linear*, Ann. of Math. (2) **155** (2002), no. 1, 131–156, DOI 10.2307/3062152. MR1888796

[92] Justin Lanier, *Generating mapping class groups with elements of fixed finite order*, J. Algebra **511** (2018), 455–470, DOI 10.1016/j.jalgebra.2018.05.037. MR3834783

[93] Justin Lanier and Dan Margalit, *Normal generators for mapping class groups are abundant*, arXiv:1805.03666 (2018).

[94] Erwan Lanneau and Jean-Luc Thiffeault, *On the minimum dilatation of braids on punctured discs*, Geom. Dedicata **152** (2011), 165–182, DOI 10.1007/s10711-010-9551-2. Supplementary material available online. MR2795241

[95] R. J. Lawrence, *Homological representations of the Hecke algebra*, Comm. Math. Phys. **135** (1990), no. 1, 141–191. MR1086755

[96] C. J. Leininger and A. W. Reid, *A combination theorem for Veech subgroups of the mapping class group*, Geom. Funct. Anal. **16** (2006), no. 2, 403–436, DOI 10.1007/s00039-006-0556-9. MR2231468

[97] Christopher J. Leininger and Dan Margalit, *Two-generator subgroups of the pure braid group*, Geom. Dedicata **147** (2010), 107–113, DOI 10.1007/s10711-009-9440-8. MR2660568

[98] Livio Liechti and Balazs Strenner. Small stretch factor pseudo-anosov mapping classes on nonorientable surfaces. Preprint.

[99] Livio Liechti and Balázs Strenner, *The Arnoux-Yoccoz mapping classes via Penner's construction*, arXiv:1805.01248 (2018).

[100] CTQM Problem List. http://ctqm.au.dk/pl/, 03 2006.

[101] D. D. Long and M. Paton, *The Burau representation is not faithful for $n \geq 6$*, Topology **32** (1993), no. 2, 439–447, DOI 10.1016/0040-9383(93)90030-Y. MR1217079

[102] Eduard Looijenga, *Cohomology of $\mathcal{M}_3$ and $\mathcal{M}_3^1$*, Mapping class groups and moduli spaces of Riemann surfaces (Göttingen, 1991/Seattle, WA, 1991), Contemp. Math., vol. 150, Amer. Math. Soc., Providence, RI, 1993, pp. 205–228, DOI 10.1090/conm/150/01292. MR1234266

[103] Eduard Looijenga, *Prym representations of mapping class groups*, Geom. Dedicata **64** (1997), no. 1, 69–83, DOI 10.1023/A:1004909416648. MR1432535

[104] Feng Luo. *Torsion elements in the mapping class group of a surface*, arXiv:math/0004048 (2010).

[105] Colin Maclachlan, *Modulus space is simply-connected*, Proc. Amer. Math. Soc. **29** (1971), 85–86, DOI 10.2307/2037676. MR0286995

[106] Ib Madsen and Michael Weiss, *The stable moduli space of Riemann surfaces: Mumford's conjecture*, Ann. of Math. (2) **165** (2007), no. 3, 843–941, DOI 10.4007/annals.2007.165.843. MR2335797

[107] Wilhelm Magnus, *Beziehungen zwischen Gruppen und Idealen in einem speziellen Ring* (German), Math. Ann. **111** (1935), no. 1, 259–280, DOI 10.1007/BF01472217. MR1512992

[108] Joseph Maher, *Random walks on the mapping class group*, Duke Math. J. **156** (2011), no. 3, 429–468, DOI 10.1215/00127094-2010-216. MR2772067

[109] Joseph Maher and Giulio Tiozzo, *Random walks, wpd actions, and the cremona group*, arXiv:1807.10230 (2018).

[110] Dan Margalit and Rebecca Winarski. The Birman–Hilden theory. Preprint, 2016.

[111] Hidetoshi Masai, *Fibered commensurability and arithmeticity of random mapping tori*, Groups Geom. Dyn. **11** (2017), no. 4, 1253–1279, DOI 10.4171/GGD/428. MR3737282

[112] John McCarthy and Athanase Papadopoulos, *Involutions in surface mapping class groups*, Enseign. Math. (2) **33** (1987), no. 3-4, 275–290. MR925990

[113] John D. McCarthy, *On the first cohomology group of cofinite subgroups in surface mapping class groups*, Topology **40** (2001), no. 2, 401–418, DOI 10.1016/S0040-9383(99)00066-X. MR1808225

[114] John D. McCarthy and Athanase Papadopoulos, *Simplicial actions of mapping class groups*, Handbook of Teichmüller theory. Volume III, IRMA Lect. Math. Theor. Phys., vol. 17, Eur. Math. Soc., Zürich, 2012, pp. 297–423, DOI 10.4171/103-1/6. MR2952767

[115] Curtis T. McMullen, *Polynomial invariants for fibered 3-manifolds and Teichmüller geodesics for foliations* (English, with English and French summaries), Ann. Sci. École Norm. Sup. (4) **33** (2000), no. 4, 519–560, DOI 10.1016/S0012-9593(00)00121-X. MR1832823

[116] R. Taylor McNeill, *A New Filtration of the Magnus Kernel*, ProQuest LLC, Ann Arbor, MI, 2013. Thesis (Ph.D.)–Rice University. MR3211445

[117] D. B. McReynolds, *The congruence subgroup problem for pure braid groups: Thurston's proof*, New York J. Math. **18** (2012), 925–942. MR3007206

[118] Geoffrey Mess, *The Torelli groups for genus 2 and 3 surfaces*, Topology **31** (1992), no. 4, 775–790, DOI 10.1016/0040-9383(92)90008-6. MR1191379

[119] Shinichi Mochizuki, *The profinite Grothendieck conjecture for closed hyperbolic curves over number fields*, J. Math. Sci. Univ. Tokyo **3** (1996), no. 3, 571–627. MR1432110

[120] Shinichi Mochizuki, *The local pro-p anabelian geometry of curves*, Invent. Math. **138** (1999), no. 2, 319–423, DOI 10.1007/s002220050381. MR1720187

[121] Naoyuki Monden, *Generating the mapping class group by torsion elements of small order*, Math. Proc. Cambridge Philos. Soc. **154** (2013), no. 1, 41–62, DOI 10.1017/S0305004112000357. MR3002582

[122] John Atwell Moody, *The Burau representation of the braid group B_n is unfaithful for large n*, Bull. Amer. Math. Soc. (N.S.) **25** (1991), no. 2, 379–384, DOI 10.1090/S0273-0979-1991-16080-5. MR1098347

[123] S. Morita and R. C. Penner, *Torelli groups, extended Johnson homomorphisms, and new cycles on the moduli space of curves*, Math. Proc. Cambridge Philos. Soc. **144** (2008), no. 3, 651–671, DOI 10.1017/S0305004107000990. MR2418709

[124] Shigeyuki Morita, *The extension of Johnson's homomorphism from the Torelli group to the mapping class group*, Invent. Math. **111** (1993), no. 1, 197–224, DOI 10.1007/BF01231286. MR1193604

[125] Shigeyuki Morita, *Structure of the mapping class groups of surfaces: a survey and a prospect*, Proceedings of the Kirbyfest (Berkeley, CA, 1998), Geom. Topol. Monogr., vol. 2, Geom. Topol. Publ., Coventry, 1999, pp. 349–406, DOI 10.2140/gtm.1999.2.349. MR1734418

[126] Shigeyuki Morita, *Cohomological structure of the mapping class group and beyond*, Problems on mapping class groups and related topics, Proc. Sympos. Pure Math., vol. 74, Amer. Math. Soc., Providence, RI, 2006, pp. 329–354, DOI 10.1090/pspum/074/2264550. MR2264550

[127] Shigeyuki Morita. Tautological algebras of moduli spaces : Survey and prospect. *RIMS Kôkyûroku*, 1991:130–139, 2016.

[128] Shigeyuki Morita, Takuya Sakasai, and Masaaki Suzuki, *Abelianizations of derivation Lie algebras of the free associative algebra and the free Lie algebra*, Duke Math. J. **162** (2013), no. 5, 965–1002, DOI 10.1215/00127094-2140560. MR3047471

[129] Shigeyuki Morita, Takuya Sakasai, and Masaaki Suzuki, *Torelli group, johnson kernel and invariants of homology spheres* arXiv:1711.07855 (2017).

[130] Tetsuhiro Moriyama, *The mapping class group action on the homology of the configuration spaces of surfaces*, J. Lond. Math. Soc. (2) **76** (2007), no. 2, 451–466, DOI 10.1112/jlms/jdm077. MR2363426

[131] Suhas Pandit, *The complex of non-separating embedded spheres*, Rocky Mountain J. Math. **44** (2014), no. 6, 2029–2054, DOI 10.1216/RMJ-2014-44-6-2029. MR3310960

[132] Suhas Pandit, *A note on automorphisms of the sphere complex*, Proc. Indian Acad. Sci. Math. Sci. **124** (2014), no. 2, 255–265, DOI 10.1007/s12044-014-0172-9. MR3218895

[133] R. C. Penner, *Bounds on least dilatations*, Proc. Amer. Math. Soc. **113** (1991), no. 2, 443–450, DOI 10.2307/2048530. MR1068128

[134] Thomas W. Müller Peter Abramenko. Are braid group linear? 1999.

[135] Dan Petersen. Which mapping class group representations come from algebraic geometry?, November 2014.

[136] Andrew Putman, *An infinite presentation of the Torelli group*, Geom. Funct. Anal. **19** (2009), no. 2, 591–643, DOI 10.1007/s00039-009-0006-6. MR2545251

[137] Andrew Putman, *A note on the abelianizations of finite-index subgroups of the mapping class group*, Proc. Amer. Math. Soc. **138** (2010), no. 2, 753–758, DOI 10.1090/S0002-9939-09-10124-7. MR2557192

[138] Andrew Putman, *Small generating sets for the Torelli group*, Geom. Topol. **16** (2012), no. 1, 111–125, DOI 10.2140/gt.2012.16.111. MR2872579

[139] Andrew Putman and Ben Wieland, *Abelian quotients of subgroups of the mappings class group and higher Prym representations*, J. Lond. Math. Soc. (2) **88** (2013), no. 1, 79–96, DOI 10.1112/jlms/jdt001. MR3092259

[140] Oscar Randal-Williams, *Resolutions of moduli spaces and homological stability*, J. Eur. Math. Soc. (JEMS) **18** (2016), no. 1, 1–81, DOI 10.4171/JEMS/583. MR3438379

[141] Igor Rivin, *Generic phenomena in groups: some answers and many questions*, Thin groups and superstrong approximation, Math. Sci. Res. Inst. Publ., vol. 61, Cambridge Univ. Press, Cambridge, 2014, pp. 299–324. MR3220895

[142] Nick Salter, *Monodromy and vanishing cycles in toric surfaces*, arXiv:1710.08042 (2017).

[143] Shane Scott. *Combinatorial models for surface and free group symmetries.* PhD thesis, Georgia Institute of Technology, 2018.

[144] Hyunshik Shin, *Algebraic degrees of stretch factors in mapping class groups*, Algebr. Geom. Topol. **16** (2016), no. 3, 1567–1584, DOI 10.2140/agt.2016.16.1567. MR3523049

[145] N. F. Smythe, *The Burau representation of the braid group is pairwise free*, Arch. Math. (Basel) **32** (1979), no. 4, 309–317, DOI 10.1007/BF01238504. MR545151

[146] Balázs Strenner, *Algebraic degrees of pseudo-Anosov stretch factors*, Geom. Funct. Anal. **27** (2017), no. 6, 1497–1539, DOI 10.1007/s00039-017-0429-4. MR3737368

[147] Charalampos Stylianakis, *Congruence subgroups of braid groups*, arXiv:1609.05673 (2016).

[148] Feraydoun Taherkhani, *The Kazhdan property of the mapping class group of closed surfaces and the first cohomology group of its cofinite subgroups*, Experiment. Math. **9** (2000), no. 2, 261–274. MR1780211

[149] William P. Thurston, *A norm for the homology of* 3*-manifolds*, Mem. Amer. Math. Soc. **59** (1986), no. 339, i–vi and 99–130. MR823443

[150] William P. Thurston, *On the geometry and dynamics of diffeomorphisms of surfaces*, Bull. Amer. Math. Soc. (N.S.) **19** (1988), no. 2, 417–431, DOI 10.1090/S0273-0979-1988-15685-6. MR956596

[151] Orsola Tommasi, *Rational cohomology of the moduli space of genus 4 curves*, Compos. Math. **141** (2005), no. 2, 359–384, DOI 10.1112/S0010437X0400123X. MR2134272

[152] Thomas W. Tucker, *Two notes on maps and surface symmetry*, Rigidity and symmetry, Fields Inst. Commun., vol. 70, Springer, New York, 2014, pp. 345–355, DOI 10.1007/978-1-4939-0781-6_17. MR3329282

[153] Vladimir Turaev, *Faithful linear representations of the braid groups*, Astérisque **276** (2002), 389–409. Séminaire Bourbaki, Vol. 1999/2000. MR1886767

[154] T. N. Venkataramana, *Image of the Burau representation at d-th roots of unity*, Ann. of Math. (2) **179** (2014), no. 3, 1041–1083, DOI 10.4007/annals.2014.179.3.4. MR3171758

[155] Nathalie Wahl, *Homological stability for mapping class groups of surfaces*, Handbook of moduli. Vol. III, Adv. Lect. Math. (ALM), vol. 26, Int. Press, Somerville, MA, 2013, pp. 547–583. MR3135444

[156] Nathalie Wahl, *The Mumford conjecture, Madsen-Weiss and homological stability for mapping class groups of surfaces*, Moduli spaces of Riemann surfaces, IAS/Park City Math. Ser., vol. 20, Amer. Math. Soc., Providence, RI, 2013, pp. 109–138. MR3114683

[157] Kazushi Yoshihara. Generating the mapping class group of an orientable surface by elements of order six. In preparation.

School of Mathematics, Georgia Institute of Technology, 686 Cherry St., Atlanta, Georgia 30306

Email address: margalit@math.gatech.edu

Proceedings of Symposia in Pure Mathematics
Volume **102**, 2019
https://doi.org/10.1090/pspum/102/01809

Totally disconnected groups (not) acting on two-manifolds

John Pardon

Abstract. We briefly survey the Hilbert–Smith Conjecture, and we include a proof of it in dimension two (where it is originally due to Montgomery–Zippin).

Hilbert's Fifth Problem [**9, 10**] asks for an investigation into the extent to which assumptions of real analyticity in the theory of Lie group actions on manifolds can be relaxed to mere continuity. This question admits a number of different precise formulations, which currently range from completely solved to wide open. Our brief discussion is far from complete; we refer the reader to Tao [**24**] and Palais [**19**] for more details.

One precise formulation of Hilbert's Fifth Problem is the question of whether every topological manifold G with continuous group operations $m : G \times G \to G$ and $i : G \to G$ has a unique real analytic structure for which the group operations are real analytic (in which case G is called a Lie group). This question was answered positively by Gleason [**6**] and Montgomery–Zippin [**16**]; see also the later work of Hirschfeld [**11**] and Goldbring [**7**]. In fact, the work of Gleason [**5**] and Yamabe [**27, 28**] supplies a stronger result, known as the Gleason–Yamabe theorem, which states that every locally compact group G has an open subgroup $G_0 \subseteq G$ which is an inverse limit of Lie groups (we follow the convention whereby topological groups are always taken to be Hausdorff). A topological group is said to be NSS (or, to have no small subgroups) iff there exists a neighborhood of the identity in G which contains no nontrivial subgroup. A Lie group is obviously NSS (use the exponential map), and it follows immediately from the Gleason–Yamabe theorem that a locally compact group G is Lie iff it is NSS (note, however, that proving the implication NSS $\implies$ Lie is an important step in the proof of the Gleason–Yamabe theorem). This is an amazing rigidity result, characterizing Lie groups among all locally compact groups as those satisfying a simple condition combining the group structure and the topology.

The story for groups acting on manifolds (rather than groups which are themselves manifolds) is less settled. The main open problem is the Hilbert–Smith Conjecture, which asserts that every locally compact group acting faithfully on a

---

2010 *Mathematics Subject Classification.* Primary 57S10, 57M60, 20F36, 57S05, 57N05; Secondary 54H15, 55M35, 57S17.

This research was conducted during the period the author served as a Clay Research Fellow. The author was also partially supported by a Packard Fellowship and by the National Science Foundation under the Alan T. Waterman Award, Grant No. 1747553.

connected manifold M is a Lie group. By a faithful action of G on M, we simply mean a continuous injective homomorphism $G \to \mathrm{Homeo}(M)$, where $\mathrm{Homeo}(M)$ is given the compact-open topology (i.e. the topology of uniform convergence on compact subsets). Connectivity of M (which could equivalently be relaxed to having finitely many connected components) is necessary to avoid examples such as the compact group $\prod_{n\geq 0} \mathbb{Z}/2^n$ acting on $\bigsqcup_{n\geq 0} S^1$ by rotation in each factor. The hypothesis that G be locally compact may seem somewhat artificial, though clearly some hypothesis is needed to exclude $G = \mathrm{Homeo}(M)$, which is certainly not a Lie group. Also, note that even when G is a Lie group, the action of G on M can be rather pathological (as in not locally conjugate to a smooth or even a piecewise linear action). For example, Bing constructed an involution of S^3 whose fixed set is a wildly embedded two-sphere [**1**] as well as a non-manifold X^3 with the property that $\mathbb{R} \times X^3 = \mathbb{R}^4$ [**2**] (thus giving rise to a non-smoothable action of S^1 on the manifold $S^1 \times X^3$).

Since every group acts faithfully on itself by translation, the Hilbert–Smith Conjecture implies the positive answer to the version of Hilbert's Fifth Problem discussed above. In the other direction, the classification of locally compact groups (the Gleason–Yamabe theorem) together with Newman's theorem, to be discussed shortly, allows one to reduce the Hilbert–Smith Conjecture to the special case of $G = A_p = \mathbb{Z}_p := \varprojlim \mathbb{Z}/p^n$, the additive group of p-adic integers (see Tao [**24**] or Lee [**12**]). Note that the group $\mathbb{Z}_p$ is self-similar: the subgroups $p^k\mathbb{Z}_p \subseteq \mathbb{Z}_p$ (which form a neighborhood base at the identity as $k \to \infty$) are all isomorphic to $\mathbb{Z}_p$. In studying the Hilbert–Smith Conjecture, we may therefore, by replacing the given action of $\mathbb{Z}_p$ by the action of $p^k\mathbb{Z}_p$ for some sufficiently large k, assume without loss of generality that the given action of $\mathbb{Z}_p$ is C^0-close to the trivial action. One reason to expect this observation to be helpful is that, by the Gleason–Yamabe theorem, these "small subgroups" are precisely the salient feature of the group $\mathbb{Z}_p$ which distinguishes it from a Lie group. It is also interesting to compare this observation with the result of Newman [**18**] (see also Smith [**23**] and Dress [**4**]) that nontrivial actions of compact Lie groups on manifolds cannot be arbitrarily C^0-close to the trivial action.

THEOREM (Newman [**18**]). *Let M be a connected manifold equipped with a metric d, and let $U \subseteq M$ be a nonempty open subset. There exists $\varepsilon = \varepsilon(U, d|_U) > 0$ such that for every compact Lie group G, the only action of G on M satisfying $d(x, gx) \leq \varepsilon$ for all $x \in U$ and $g \in G$ is the trivial action. In particular, if G acts trivially on U, then it acts trivially on all of M.*

(Note the importance of assuming G is compact: a nontrivial $\mathbb{R}$-action which is arbitrarily C^0-close to the identity can be constructed on any manifold M by integrating any nonzero vector field supported inside a small ball.)

PROOF. It is enough to treat the case $G = \mathbb{Z}/p$ for any prime p. Indeed, the kernel of $G \to \mathrm{Homeo}(M)$ is a closed normal subgroup $H \trianglelefteq G$, so G/H is a Lie group and thus, if nontrivial, contains a subgroup isomorphic to $\mathbb{Z}/p$ for some prime p. Note that the $\varepsilon > 0$ we produce must be independent of p.

We deduce the desired result from the following two claims about $\mathbb{Z}/p$ actions on open subsets of $\mathbb{R}^n$. Suppose $\mathbb{Z}/p$ acts on $V \subseteq \mathbb{R}^n$ and that V contains the unit ball centered at the origin $0 \in \mathbb{R}^n$.

- If $d(x, gx) \leq \frac{1}{2}$ for every $x \in V$ and $g \in \mathbb{Z}/p$, then $\mathbb{Z}/p$ acts trivially on a neighborhood of the origin.
- If $\mathbb{Z}/p$ acts trivially on $V \cap (\mathbb{R}_{\geq 0} \times \mathbb{R}^{n-1})$, then $\mathbb{Z}/p$ acts trivially on a neighborhood of the origin.

To deduce the desired result, note that the first claim produces an $\varepsilon = \varepsilon(U, d|_U) > 0$ such that any G action on M satisfying $d(x, gx) \leq \varepsilon$ for all $x \in U$ acts trivially on some nonempty open subset $U_0 \subseteq U$, and the second claim implies that a G action on M acting trivially on a nonempty open subset is globally trivial.

To prove the first claim, we consider the averaging map

$$A : V \to \mathbb{R}^n,$$
$$x \mapsto \frac{1}{p} \sum_{g \in \mathbb{Z}/p} gx.$$

(I have had less than resounding success in eliciting a laugh by remarking that this proof does not work in characteristic p.) The hypothesis implies that $d(x, A(x)) \leq \frac{1}{2}$. It follows that A is proper over a neighborhood of $0 \in \mathbb{R}^n$, and hence its degree over $0 \in \mathbb{R}^n$ is well-defined. In fact, the inequality $d(x, A(x)) \leq \frac{1}{2}$ implies the degree of A over $0 \in \mathbb{R}^n$ equals 1, using the obvious linear homotopy between A and the identity map. On the other hand, if any point $x \in \mathbb{R}^n$ close to the origin is not fixed by the $\mathbb{Z}/p$ action, then the factorization $V \to V/(\mathbb{Z}/p) \to \mathbb{R}^n$ implies that the degree of A over $x \in \mathbb{R}^n$ is divisible by p (note that $A^{-1}(x) \subseteq V$ is compact and contained in the open locus where $\mathbb{Z}/p$ acts freely). We thus conclude that $\mathbb{Z}/p$ acts trivially on a neighborhood of the origin, as desired.

To prove the second claim, we again consider the degree of the averaging map near the origin. First, we must shrink V to $(\mathbb{Z}/p)(\mathbb{R}_{>-\delta} \times \{x \in \mathbb{R}^{n-1} : |x| < \frac{1}{2}\})$ for sufficiently small $\delta > 0$; this ensures that A is proper over a neighborhood of $0 \in \mathbb{R}^n$, so the degree is well-defined (and locally constant) there. Again, the degree of A near the origin equals 1, this time by considering the degree over nearby points in the interior of the halfspace $\mathbb{R}_{\geq 0} \times \mathbb{R}^{n-1}$. Now just as before, if any point close to the origin were not fixed by the $\mathbb{Z}/p$ action, then the degree of A over that point would be divisible by p, a contradiction. □

It is, of course, tempting to try to adapt the proof of Newman's theorem to treat $\mathbb{Z}_p$ actions as well and thus prove the Hilbert–Smith Conjecture.

There are a number of partial results towards the Hilbert–Smith Conjecture.

- Yang [**29**] showed that for any faithful action of $\mathbb{Z}_p$ on an n-manifold M, the orbit space $M/\mathbb{Z}_p$ must have cohomological dimension $n + 2$.
- For C^2 actions, the Hilbert–Smith Conjecture is due to Bochner–Montgomery [**3**].
- For $C^{0,1}$ actions, the Hilbert–Smith Conjecture is due to Repovš–Ščepin [**22**].
- For $C^{0,\frac{n}{n+2}+\varepsilon}$ actions, the Hilbert–Smith Conjecture is due to Maleshich [**13**].
- For quasiconformal actions, the Hilbert–Smith Conjecture is due to Martin [**14**].
- For uniformly quasisymmetric actions on doubling Ahlfors regular compact metric measure manifolds with Hausdorff dimension in $[1, n+2)$, the Hilbert–Smith Conjecture is due to Mj [**15**].

- For transitive actions, the Hilbert–Smith Conjecture is due to Montgomery–Zippin [**17**, p243] (see also Tao [**24**, §1.6.4]).
- When $\dim M = 1$, the Hilbert–Smith Conjecture is an easy exercise.
- When $\dim M = 2$, the Hilbert–Smith Conjecture is due to Montgomery–Zippin [**17**, p249].
- When $\dim M = 3$, the Hilbert–Smith Conjecture is due to the author [**20**].

By work of Raymond–Williams [**21**], there are faithful actions of $\mathbb{Z}_p$ on n-dimensional compact metric spaces which achieve the cohomological dimension jump of Yang [**29**] for every $n \geq 2$. We also remark that by work of Walsh [**25**, p282 Corollary 5.15.1] there *does* exist a continuous decomposition of any compact PL n-manifold into cantor sets of arbitrarily small diameter if $n \geq 3$ (see also Wilson [**26**, Theorem 3]).

To conclude this short survey, we present a proof of the Hilbert–Smith Conjecture in dimension two. The proof we present is substantially different from the proof given by Montgomery–Zippin [**17**, p249], which is more point set topological. We were led to the proof presented below by trying to adapt the proof from [**20**] to two dimensions.

THEOREM (Hilbert–Smith Conjecture in dimension two). *Any locally compact group admitting a faithful action on a connected two-dimensional manifold is a Lie group.*

PROOF. By the reductions discussed above, it suffices to show that $\mathbb{Z}_p$ does not admit a faithful action on a connected two-dimensional manifold M for any prime p (in fact, our argument below applies to any profinite group in place of $\mathbb{Z}_p$). We suppose the existence of such an action and work to derive a contradiction.

Fix a point $x \in M$ such that no subgroup $p^k\mathbb{Z}_p$ acts trivially on an open neighborhood of x. The existence of such a point x is guaranteed by Newman's theorem. Indeed, suppose that, on the contrary, for every $x \in M$ there exists a $k < \infty$ such that $p^k\mathbb{Z}_p$ acts trivially on an open neighborhood of x. It follows that for every compact subset $K \subseteq M$, there exists a $k < \infty$ such that $p^k\mathbb{Z}_p$ acts trivially on K. Fix some $k_0 < \infty$ such that $p^{k_0}\mathbb{Z}_p$ acts trivially on some nonempty open subset U_0 of M. Now for every pre-compact connected open subset $U \subseteq M$ intersecting U_0, we know that $p^k\mathbb{Z}_p$ acts trivially on U for some $k < \infty$. But now applying Newman's theorem to the action of the finite group $p^{k_0}\mathbb{Z}_p/p^k\mathbb{Z}_p$ on U, we conclude that in fact $p^{k_0}\mathbb{Z}_p$ acts trivially on U. Since U was arbitrary, we conclude that $p^{k_0}\mathbb{Z}_p$ acts trivially on all of M, contradicting faithfulness. This proves the existence of the desired point $x \in M$.

We will derive a contradiction by analyzing the $\mathbb{Z}_p$ action locally near x.

Fix a closed disk $D \subseteq M$ around x, and identify D with the standard unit disk in $\mathbb{R}^2$ with x as its center. By replacing $\mathbb{Z}_p$ with $p^k\mathbb{Z}_p$ for sufficiently large k, we assume that the action is extremely close to the trivial action over D (though note that, of course, $\mathbb{Z}_p$ need not stabilize D).

Fix a point $y \in D$ close to x which is not fixed by $\mathbb{Z}_p$. The orbit $\mathbb{Z}_p y$ is thus either a cantor set or p^k points for some $k \geq 1$ (depending on the stabilizer of y); in particular, $\operatorname{rk} H_1(D \setminus \mathbb{Z}_p y) \geq 2$. Let Q denote the large component of $D \setminus \mathbb{Z}_p B_\varepsilon(y)$ (where $B_\varepsilon(y)$ denotes the closed ε-ball centered at y). By choosing $\varepsilon > 0$ sufficiently small, we can ensure that $\operatorname{rk} H_1(Q) \geq 2$ as well (by a direct limit argument). Now $\mathbb{Z}_p B_\varepsilon(y)$ has finitely many connected components (since $\mathbb{Z}_p$ acts transitively on

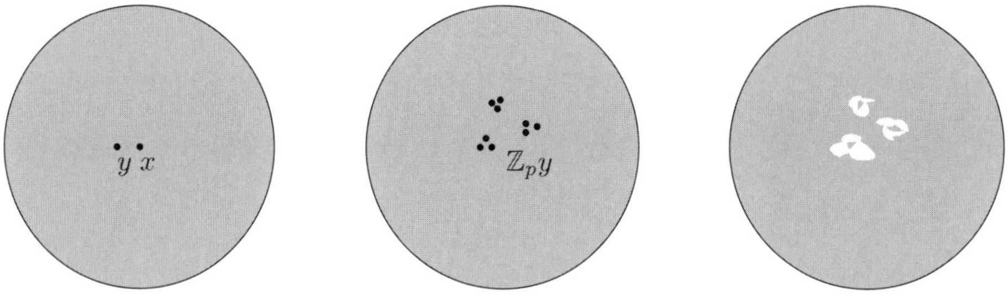

D $\qquad$ D $\qquad$ $D \setminus \mathbb{Z}_p B_\varepsilon(y)$

them and the stabilizer of the component containing y is open), and hence Q is finitely connected. It follows that Q is *homeomorphic to the complement of* $n \geq 2$ *points in the closed disk.*

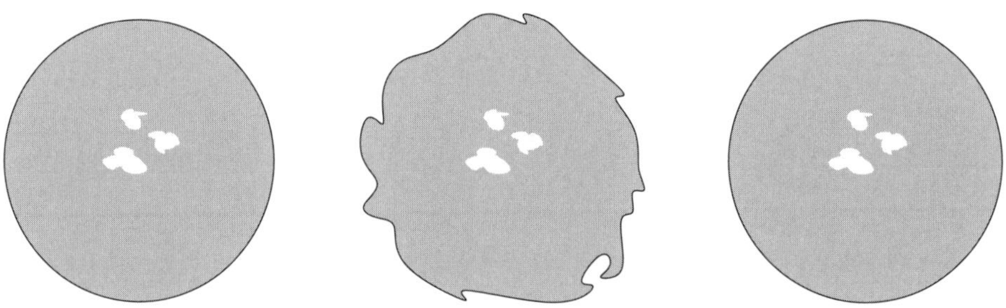

Q $\qquad$ $T_\gamma(Q)$ $\qquad$ $\psi_\gamma(T_\gamma(Q)) = Q$

We now argue that the action of $\mathbb{Z}_p$ on Q gives rise to a homomorphism to the braid group

$$\mathbb{Z}_p \to B_n,$$

where $n \geq 2$ is the number of "holes" in Q. Of course, $\mathbb{Z}_p$ does not truly act on Q, rather it sends the outer boundary ∂D to a nearby circle. Nevertheless, we may still produce a homomorphism to the braid group by observing that, since the action $T_\gamma : Q \to M$ of any $\gamma \in \mathbb{Z}_p$ is very close to the identity, there is a canonical (up to homotopy) map $\psi_\gamma : M \to M$ supported near ∂D such that $\psi_\gamma \circ T_\gamma$ restricts to the identity map on ∂D. Now the map $\mathbb{Z}_p \to B_n$ sends γ to the class $[\psi_\gamma \circ T_\gamma] \in \pi_0 \operatorname{Homeo}(Q, \partial D) = B_n$, and one can check that this map is indeed a group homomorphism.

Now the homomorphism $\mathbb{Z}_p \to B_n$ is continuous, where the target B_n is equipped with the discrete topology, and hence its image is a finite subgroup of B_n. Since $\mathbb{Z}_p$ acts transitively on the finitely many components of $\mathbb{Z}_p B_\varepsilon(y)$, this finite subgroup of B_n must be nontrivial (as it *a fortiori* must have transitive image in the symmetric group S_n and $n \geq 2$). On the other hand, the braid group B_n is well-known to have no nontrivial finite subgroups (see [**8**, §2] for example). This contradiction finishes the proof. □

Acknowledgment

I would like to thank the referee for excellent suggestions for improving the clarity of the proofs.

References

[1] R. H. Bing, *A homeomorphism between the 3-sphere and the sum of two solid horned spheres*, Ann. of Math. (2) **56** (1952), 354–362, DOI 10.2307/1969804. MR0049549

[2] ______, *The cartesian product of a certain nonmanifold and a line is E^4*, Ann. of Math. (2) **70** (1959), 399–412, DOI 10.2307/1970322. MR0107228

[3] S. Bochner and D. Montgomery, *Locally compact groups of differentiable transformations*, Ann. of Math. (2) **47** (1946), 639–653, DOI 10.2307/1969226. MR0018187

[4] A. Dress, *Newman's theorems on transformation groups*, Topology **8** (1969), 203–207, DOI 10.1016/0040-9383(69)90010-X. MR0238353

[5] A. M. Gleason, *The structure of locally compact groups*, Duke Math. J. **18** (1951), 85–104. MR0039730

[6] A. M. Gleason, *Groups without small subgroups*, Ann. of Math. (2) **56** (1952), 193–212, DOI 10.2307/1969795. MR0049203

[7] I. Goldbring, *Hilbert's fifth problem for local groups*, Ann. of Math. (2) **172** (2010), no. 2, 1269–1314, DOI 10.4007/annals.2010.172.1273. MR2680491

[8] J. González-Meneses, *Basic results on braid groups* (English, with English and French summaries), Ann. Math. Blaise Pascal **18** (2011), no. 1, 15–59, DOI 10.5802/ambp.293. MR2830088

[9] D. Hilbert, *Mathematische Probleme. Vortrag, gehalten auf dem internationalen Mathematiker-Congress zu Paris 1900.*, Nachr. Ges. Wiss. Göttingen, Math.-Phys. Kl. **1900** (1900), 253–297 (German).

[10] D. Hilbert, *Mathematical problems*, Bull. Amer. Math. Soc. **8** (1902), no. 10, 437–479, DOI 10.1090/S0002-9904-1902-00923-3. MR1557926

[11] J. Hirschfeld, *The nonstandard treatment of Hilbert's fifth problem*, Trans. Amer. Math. Soc. **321** (1990), no. 1, 379–400, DOI 10.2307/2001608. MR967314

[12] J. S. Lee, *Totally disconnected groups, p-adic groups and the Hilbert-Smith conjecture*, Commun. Korean Math. Soc. **12** (1997), no. 3, 691–699. MR1641899

[13] Ĭ. Maleshich, *The Hilbert-Smith conjecture for Hölder actions* (Russian), Uspekhi Mat. Nauk **52** (1997), no. 2(314), 173–174, DOI 10.1070/RM1997v052n02ABEH001792; English transl., Russian Math. Surveys **52** (1997), no. 2, 407–408. MR1480156

[14] G. J. Martin, *The Hilbert-Smith conjecture for quasiconformal actions*, Electron. Res. Announc. Amer. Math. Soc. **5** (1999), 66–70, DOI 10.1090/S1079-6762-99-00062-1. MR1694197

[15] M. Mj, *Pattern rigidity and the Hilbert-Smith conjecture*, Geom. Topol. **16** (2012), no. 2, 1205–1246, DOI 10.2140/gt.2012.16.1205. MR2946807

[16] D. Montgomery and L. Zippin, *Small subgroups of finite-dimensional groups*, Ann. of Math. (2) **56** (1952), 213–241, DOI 10.2307/1969796. MR0049204

[17] ______, *Topological transformation groups*, Interscience Publishers, New York-London, 1955. MR0073104 (17,383b)

[18] M. H. A. Newman, *A theorem on periodic transformations of spaces*, Quart. J. Math. **os-2** (1931), no. 1, 1–8.

[19] *Andrew M. Gleason 1921–2008*, Notices Amer. Math. Soc. **56** (2009), no. 10, 1236–1267. Ethan D. Bolker, coordinating editor. MR2572754

[20] J. Pardon, *The Hilbert-Smith conjecture for three-manifolds*, J. Amer. Math. Soc. **26** (2013), no. 3, 879–899, DOI 10.1090/S0894-0347-2013-00766-3. MR3037790

[21] F. Raymond and R. F. Williams, *Examples of p-adic transformation groups*, Ann. of Math. (2) **78** (1963), 92–106, DOI 10.2307/1970504. MR0150769

[22] D. Repovš and E. Ščepin, *A proof of the Hilbert-Smith conjecture for actions by Lipschitz maps*, Math. Ann. **308** (1997), no. 2, 361–364, DOI 10.1007/s002080050080. MR1464908

[23] P. A. Smith, *Transformations of finite period. III. Newman's theorem*, Ann. of Math. (2) **42** (1941), 446–458, DOI 10.2307/1968910. MR0004128

[24] T. Tao, *Hilbert's fifth problem and related topics*, Graduate Studies in Mathematics, vol. 153, American Mathematical Society, Providence, RI, 2014. MR3237440

[25] J. J. Walsh, *Light open and open mappings on manifolds. II*, Trans. Amer. Math. Soc. **217** (1976), 271–284, DOI 10.2307/1997570. MR0394674

[26] D. Wilson, *Open mappings on manifolds and a counterexample to the Whyburn conjecture*, Duke Math. J. **40** (1973), 705–716. MR0320989

[27] H. Yamabe, *On the conjecture of Iwasawa and Gleason*, Ann. of Math. (2) **58** (1953), 48–54, DOI 10.2307/1969819. MR0054613

[28] ———, *A generalization of a theorem of Gleason*, Ann. of Math. (2) **58** (1953), 351–365, DOI 10.2307/1969792. MR0058607

[29] C.-T. Yang, *p-adic transformation groups*, Michigan Math. J. **7** (1960), 201–218. MR0120310

DEPARTMENT OF MATHEMATICS, PRINCETON UNIVERSITY, PRINCETON, NEW JERSEY 08544

Proceedings of Symposia in Pure Mathematics
Volume **102**, 2019
https://doi.org/10.1090/pspum/102/01814

Fukaya A_∞-structures associated to Lefschetz fibrations. IV

Paul Seidel

Abstract. We consider Hamiltonian Floer cohomology groups associated to a Lefschetz fibration, and the structure of operations on them. As an application, we will (under an important additional assumption) equip those groups with connections, which differentiate with respect to the Novikov variable.

1. Introduction

This paper continues a discussion [**28**–**31**] of pseudo-holomorphic curve theory as applied to Lefschetz fibrations. Ultimately, our sights are set on Fukaya categories; but here, we remain in the "closed string" context of Hamiltonian Floer cohomology.

Given a symplectic Lefschetz fibration (over the disc), one can single out certain classes of time-dependent Hamiltonians on the total space, which give rise to Floer cohomology groups that are invariants of the fibration. More concretely, the outcome is an infinite family of such groups, indexed by the "rotation number at infinity" $r \in \mathbb{Z}$. All of them vanish if the Lefschetz fibration is trivial (has no critical points). Based on intuition from mirror symmetry, one expects these groups to carry a rich algebraic structure. The initial breakthrough in constructing the desired operations is due to Abouzaid-Ganatra (unpublished), who equipped the $r = 1$ group with a ring structure. Here, we streamline and generalize their insight, by introducing a suitable TQFT framework. Eventually, one wants to compare Hamiltonian Floer cohomology with certain twisted Hochschild cohomology groups of the associated Fukaya category, and thereby relate our operations to ones in homological algebra; however, to carry that out, one needs a specially adapted version of closed-open string maps, which is beyond the scope of the present paper (the literature contains several constructions relating the open and closed string sectors for Lefschetz fibrations, but none of them seems really practical for this purpose).

Our main motivation for pursuing this direction comes from [**31**]. There, symplectic cohomology was shown, under a fundamental additional assumption, to carry a one-parameter family of connections, which differentiate with respect to

2010 *Mathematics Subject Classification.* Primary 53D40; Secondary 14D05, 53D37.

This work was supported by the Simons Foundation, through a Simons Investigator award; by NSF grant DMS-1500954; by the Institute for Advanced Study, through a visiting appointment supported by grants from the Ambrose Monell Foundation and the Simonyi Endowment Fund; and by Columbia University, through an Eilenberg Visiting Professorship.

the Novikov formal parameter. That construction relied on the structure of symplectic cohomology as a BV algebra. Using some of the operations we have made available, a version of the same construction can be carried out for Lefschetz fibrations. The outcome, under a similar assumption as in [**31**], is that each of our Floer cohomology groups carries a connection; for different r, these correspond to different instances of the family of connections on symplectic cohomology. Ultimately, one wants to use these connections, and closed-open string maps, to carry over the enumerative results from [**31**] to Fukaya categories of Lefschetz fibrations; we refer the reader to [**30**, Section 4] for the conjectural outcome and its context.

The structure of this paper may deserve some comment. Floer-theoretic constructions and identities have often been thought of as realizations of abstract TQFT arguments, and that is also true here. Usually, the formal TQFT part is straightforward, and the main work goes into its Floer-theoretic implementation. In our situation, the weight shifts: the TQFT arguments are far from obvious, whereas the Floer theory is quite standard (except possibly for some compactness arguments). Therefore, after outlining our results in Section 2, the bulk of the paper (Sections 3–7) is devoted to TQFT considerations and their geometric prerequisites. After that, we explain the necessary basic Floer theory (Sections 8–9). Putting together the two parts is a routine process, and we will only describe it briefly (Section 10).

2. Main constructions

This section summarizes the structures that will arise from our analysis of Floer-theoretic operations, including connections. This will be done with only the minimal amount of technical details, and without any real attempt at explaining the underlying geometry. At the end, we relate part of our constructions to more familiar ones in symplectic cohomology.

2.1. Background.

Let

$$\pi : E \longrightarrow B \tag{2.1}$$

be a proper symplectic Lefschetz fibration whose base is the open unit disc $B \subset \mathbb{C}$, and whose total space is a $2n$-dimensional symplectic manifold. For simplicity, we assume that

$$c_1(E) = 0, \tag{2.2}$$

and choose a trivialization of the anticanonical bundle (the complex line bundle representing the first Chern class), up to homotopy. The smooth fibres are then closed $(2n-2)$-dimensional symplectic manifolds, again with trivialized anticanonical bundle.

All Floer cohomology groups under discussion will be finite-dimensional $\mathbb{Z}$-graded vector spaces over the Novikov field $\mathbb{K}$ with real coefficients. Elements of $\mathbb{K}$ are of the form

$$f(q) = c_0 q^{d_0} + c_1 q^{d_1} + \cdots, \quad c_i \in \mathbb{R}, \;\; d_i \in \mathbb{R}, \;\; \lim_i d_i = +\infty. \tag{2.3}$$

Concretely, to (2.1) we associate a family of such Floer cohomology groups, denoted by

$$HF^*(E, r), \;\; r \in \mathbb{Z}. \tag{2.4}$$

The $r = 0$ group admits a description in purely topological terms, as the cohomology of E "relative to a fibre at ∞". If we compactify (2.1) to $\bar{\pi} : \bar{E} \to \bar{B}$, then the statement is that

$$HF^*(E,0) \cong H^*(\bar{E}, \bar{E}_w) \quad \text{for some } w \in \partial\bar{B}. \tag{2.5}$$

The other groups (2.4), $r \neq 0$, are not classical topological invariants; they encode symplectic information related to the monodromy around $\partial\bar{B}$. In the simplest case $r = 1$, if μ is that monodromy (a symplectic automorphism of the fibre), with its fixed point Floer cohomology $HF^*(\mu)$, there is a long exact sequence

$$\cdots \to HF^*(\mu) \longrightarrow H^*(E;\mathbb{K}) \longrightarrow HF^*(E,1) \longrightarrow HF^{*+1}(\mu) \to \cdots \tag{2.6}$$

This idea can be generalized to other $r > 0$, leading to a version of the spectral sequence from [**20**], involving powers of μ (see [**29**, Lemma 6.14] for the statement). The situation for $r < 0$ is dual: there are nondegenerate pairings

$$HF^*(E,-r) \otimes HF^{2n-*}(E,r) \longrightarrow \mathbb{K}. \tag{2.7}$$

2.2. Operations. Our main topic is the structure of operations on (2.4). We start with linear maps. Each group $HF^*(E,r)$ has a canonical automorphism Σ, which generates an action of $\mathbb{Z}/r$. For $r = 0$, this automorphism admits a classical interpretation as the relative monodromy map, moving w around $\partial\bar{B}$ in (2.5). For $r \neq 0$, one can think of $HF^*(E,r)$ as the Floer cohomology of the r-th power of a Hamiltonian automorphism, and then Σ is induced by $1/r$ rotation of loops. We also have continuation maps which increase r, and which are invariant under the action of Σ (on either the left or right):

$$C : HF^*(E,r) \longrightarrow HF^*(E,r+1), \tag{2.8}$$

$$D : HF^*(E,r) \longrightarrow HF^{*-1}(E,r+1). \tag{2.9}$$

The structure of multilinear operations is more interesting, and does not seem to admit a single concise description. Two simple consequences of the general theory are the following:

PROPOSITION 2.1. *$HF^*(E,1)$ carries the structure of a Gerstenhaber algebra (a commutative product and a bracket of degree -1, with suitable relations between them).*

PROPOSITION 2.2. *Each $HF^*(E,r)$ is a Gerstenhaber module over $HF^*(E,1)$, in a $\mathbb{Z}/r$-equivariant way (and reducing to the diagonal module for $r = 1$).*

The Gerstenhaber algebra structure includes the Abouzaid-Ganatra product. Propositions 2.1 and 2.2 assign a distingushed role to $HF^*(E,1)$. Operations of this kind are the most important ones for our applications, but they by no means exhaust the structures that are present. Here are two other easily stated results that can be extracted from the general theory.

PROPOSITION 2.3. *There is the structure of a bigraded associative algebra on*

$$\bigoplus_{s\geq 0} HF^*(E,s+1). \tag{2.10}$$

PROPOSITION 2.4. *There is the structure of a bigraded Lie algebra on*

$$\bigoplus_{s\geq 0} HF^{*+1}(E,s+1)^{\mathbb{Z}/(s+1)}. \tag{2.11}$$

The product from Proposition 2.3 contains the algebra structure from Proposition 2.1 (by restricting to $s = 0$), and also the module structure over that algebra from Proposition 2.2, for $r > 0$. It may be worth while spelling out what Proposition 2.4 says, in view of the various degree shifts involved. $HF^i(E,r)^{\mathbb{Z}/r}$ appears in our graded Lie algebra in (cohomological) degree $i-1$, and this determines the signs in the antisymmetry and Jacobi relations. In addition, the bracket is homogeneous with respect to $s = r-1$, hence is given by maps

$$(2.12)\qquad HF^{i_1}(E,r_1)^{\mathbb{Z}/r_1} \otimes HF^{i_2}(E,r_2)^{\mathbb{Z}/r_2} \longrightarrow HF^{i_1+i_2-1}(E, r_1+r_2-1)^{\mathbb{Z}/(r_1+r_2-1)}.$$

The Lie structure from Proposition 2.1 is contained in that of Proposition 2.4, and so is the $\mathbb{Z}/r$-invariant part of the Lie module structure from Proposition 2.2, for $r > 0$.

The product structures from Propositions 2.1–2.3 have a unit

$$(2.13)\qquad e \in HF^0(E,1),$$

which can be thought of as the image of the unit in ordinary cohomology under the map from (2.6). Another important ingredient for us will be the image of the symplectic class, or more precisely of $q^{-1}[\omega_E] \in H^2(E;\mathbb{K})$, under that map. We call this image the Kodaira-Spencer class, and write it as

$$(2.14)\qquad k \in HF^2(E,1).$$

Following the model of [**31**], vanishing of the Kodaira-Spencer class gives rise to another operation, which is not $\mathbb{K}$-linear, but instead differentiates with respect to the Novikov parameter q:

PROPOSITION 2.5. *Suppose that* (2.14) *vanishes. Choose a bounding cochain for its cocycle representative, in the Floer cochain complex underlying $HF^*(E,1)$. That choice determines a connection on each group* (2.4)*, which means a map*

$$(2.15)\qquad \begin{aligned} &\nabla : HF^*(E,r) \longrightarrow HF^*(E,r), \\ &\nabla(fx) = f\nabla x + (\partial_q f)x \quad \textit{for } f \in \mathbb{K}. \end{aligned}$$

Changing the bounding cochain by adding a cocycle representing some class in $HF^*(E,1)$ has the effect of subtracting the Lie action of that class (in the sense of Proposition 2.2*) from* (2.15).

The theory producing all these results is modelled on an abstract notion of TQFT for surfaces equipped with framed $PSL_2(\mathbb{R})$-connections ("framed" means that we consider only connections on the trivial bundle, and do not quotient out by the gauge group). Here, $PSL_2(\mathbb{R})$ is thought of as acting on the base of (2.1) by hyperbolic isometries. Of course, this action cannot be lifted to symplectic automorphisms of the total space, but lifts exist infinitesimally if one only considers the Lefschetz fibration near infinity, and that is sufficient for our purpose. More precisely, the products and Lie brackets mentioned above are all constructed using flat $PSL_2(\mathbb{R})$-connections. More generally, one can allow connections whose curvature belongs to the nonnegative cone inside $\mathfrak{sl}_2(\mathbb{R})$. Nonnegatively curved connections are required to produce the maps (2.8), (2.9), as well as the classes (2.13), (2.14) (and, naturally in view of the latter, play an important role in Proposition 2.5).

At this point, we want to make up for a slight lack of precision in the discussion above. There are many choices involved in setting up $HF^*(E,r)$, but the essential one (other than r, of course) is that of a hyperbolic element $g \in PSL_2(\mathbb{R})$. One uses the action of g on the disc to determine the structure at infinity of a Hamiltonian automorphism of E (and then, r is used to choose an extension of that automorphism over all of E). Different choices of g give rise to isomorphic Floer cohomology groups. However, these isomorphisms are generally unique only up to composition with powers of Σ (it doesn't matter which side the composition is taken on, since the isomorphisms are Σ-equivariant). As an elementary illustration, let's take a second look at the $r = 0$ case. A more precise statement of (2.5) would require that w should lie in the interval in $\partial\bar{B}$ bounded by the repelling fixed point of g (on the left) and its attracting fixed point (on the right). For two choices of g within the same one-parameter semigroup (or equivalently, which have the same attracting and repelling fixed points), the resulting relative cohomology groups $H^*(\bar{E}, \bar{E}_w)$ can be identified canonically; otherwise, one needs to make a choice, of how to move one w point to the other (or more intrinsically, of a path inside the hyperbolic locus connecting the two choices of g). In fact, this is also the situation for general r. We will prove that the algebraic structures from Proposition 2.1 and 2.2 are compatible with those isomorphisms, and a similar argument would apply to Proposition 2.5. In the interest of brevity, we will not address the parallel uniqueness issue for the structures from Propositions 2.3 and 2.4, since those are less central for our intended applications.

REMARK 2.6. The way in which the various pieces fit together (together with the parallel situation in homological algebra [**29**, Section 2f]) suggests the following possible further developments (none of which will be attempted in this paper).

(i) The relation between Proposition 2.3 and the module structures from Proposition 2.2 seems to indicate that the algebra structure (2.10) ought to be extended to $s \in \mathbb{Z}$.

(ii) There might be an extension of (2.11) which adds an $s = -1$ term of the form $HF^*(E, 0)[[u]]$, for u a formal variable of degree 2.

(iii) Take $r > 0$. By using a suitably chosen chain level representative of Σ, one can define the algebraic mapping cone of $id - \Sigma$, which is a graded space $S^*(E, r)$ fitting into a long exact sequence

$$\cdots \to S^*(E,r) \longrightarrow HF^*(E,r) \xrightarrow{id-\Sigma} HF^*(E,r) \to \cdots \tag{2.16}$$

Extend this to $r = 0$ by setting $S^*(E, 0) = HF^*(E, 0)$. Then, the expectation is that

$$\bigoplus_{s \geq -1} S^*(E, s+1) \tag{2.17}$$

carries the structure of a bigraded Batalin-Vilkovisky (BV) algebra. The Lie structure from Proposition 2.4, with its extension from (ii), would be thought of as the "cyclic" counterpart of the bracket on (2.17) (compare [**29**, Remark 2.10]). The maps $S^*(E, r) \to HF^*(E, r)$ would form an algebra homomorphism from (2.17) to (an extended version of) (2.10).

2.3. Relation with symplectic cohomology. Symplectic cohomology is a well-known invariant of noncompact symplectic manifolds, which can be easily adapted to our situation (2.1).

PROPOSITION 2.7. *In the limit $r \to \infty$, the maps* (2.8) *recover the symplectic cohomology of E:*

$$\varinjlim_r HF^*(E,r) \cong SH^*(E). \tag{2.18}$$

We can describe the situation a little more precisely, using versions of symplectic cohomology with "finite slope". These are groups

$$HF^*(E,r), \quad r \in \mathbb{R} \setminus \mathbb{Z}. \tag{2.19}$$

As the notation indicates, they fit in with (2.4), in the sense that there are continuation maps which increase r, and can mix the integer and non-integer cases (we will denote all such maps by C). By definition, symplectic cohomology is the direct limit of $HF^*(E,r)$ using only non-integer values of r. One proves Proposition 2.7 by interleaving this with the integer values. Another familiar property of (2.19) is that it comes with a natural map $H^*(E;\mathbb{K}) \to HF^*(E,r)$ for positive r. If we take $r \in (0,1)$, then composing this map with a continuation map $HF^*(E,r) \to HF^*(E,1)$ recovers the map from (2.6).

REMARK 2.8. It is well-known (see e.g. [**23, 27**]) that symplectic cohomology carries the structure of a BV algebra (this notion has already been mentioned in Remark 2.6). The basic operations are: the (graded commutative) product; and the BV operator Δ, of degree -1. From them, one constructs a bracket of degree -1,

$$[x_1,x_2] = \Delta(x_1 \cdot x_2) - (\Delta x_1)\cdot x_2 - (-1)^{|x_1|} x_1 \cdot (\Delta x_2), \tag{2.20}$$

which yields a Gerstenhaber algebra structure. It is natural to ask how (2.18) relates the operations on (2.4) with those in $SH^*(E)$. For instance, it is easy to show that (2.9) turns into Δ in the limit (see Proposition 7.9(iv) for a sketch of the geometry underlying that argument). While we will not pursue such questions further here, it may be worthwhile summarizing what one can reasonably expect:

(i) The map $HF^*(E,1) \to SH^*(E)$ should be a homomorphism of Gerstenhaber algebras.

(ii) For the Gerstenhaber module structure from Proposition 2.2, the expected behaviour is slightly more complicated. The product is still compatible with that on symplectic cohomology, but the bracket should fit into a commutative diagram

$$\begin{array}{ccc} HF^*(E,1)\otimes HF^*(E,r) & \xrightarrow{[\cdot,\cdot]} & HF^{*-1}(E,r) \\ \downarrow & & \downarrow \\ SH^*(E)\otimes SH^*(E) & \xrightarrow{[\cdot,\cdot]_c} & SH^*(E) \end{array} \tag{2.21}$$

where the bottom $\to$ is the following operation on symplectic cohomology, for $c = r-1$:

$$\begin{aligned} [x_1,x_2]_c &= [x_1,x_2] - c(\Delta x_1)\cdot x_2 \\ &= \Delta(x_1\cdot x_2) - (c+1)(\Delta x_1)\cdot x_2 - (-1)^{|x_1|} x_1 \cdot (\Delta x_2). \end{aligned} \tag{2.22}$$

We will see the same phenomenon in Proposition 2.10 below (which is indeed closely related).

(iii) One expects similar compatibility statements for Propositions 2.3 and (less obviously) 2.4.

For our intended applications, the most important aspect of the relation between (2.4) and (2.19) concerns connections. Let's first summarize the outcome of the construction from [**31**]:

PROPOSITION 2.9. *Fix* $r_1 \in \mathbb{R}^{>0} \setminus \mathbb{Z}$ *and* $r_2 \in \mathbb{R} \setminus \mathbb{Z}$, *such that* $r_1 + r_2 \notin \mathbb{Z}$. *Suppose that the image of* $q^{-1}[\omega_E]$ *in* $HF^2(E, r_1)$ *vanishes, and choose a bounding cochain. This gives rise to a family of "connections" parametrized by* $c \in \mathbb{K}$, *which are maps*

$$(2.23) \qquad \begin{aligned} &\nabla_c : HF^*(E, r_2) \longrightarrow HF^*(E, r_1 + r_2), \\ &\nabla_c(fx) = f\nabla_c x + (\partial_q f)C(x), \end{aligned}$$

related to each other as follows. Associated to the bounding cochain is an $a \in HF^0(E, r_1)$, *and*

$$(2.24) \qquad \nabla_c x = \nabla_1 x + (c-1)\, a \cdot x,$$

where $a \cdot x$ *is the pair-of-pants product* $HF^*(E, r_1) \otimes HF^*(E, r_2) \to HF^*(E, r_1 + r_2)$. *The maps* (2.23) *commute with the continuation maps that increase* r_2. *Hence, in the limit, they induce a family of actual connections on* $SH^*(E)$.

To relate this to Proposition 2.5, one needs to restrict attention to integer values of the parameter c. We will show the following (which is only one of the possible ways of formulating the relationship, and not necessarily the most conceptually satisfying one, but sufficient for our purpose):

PROPOSITION 2.10. *Suppose that the assumption of Proposition* 2.5 *holds; then, so does that of Proposition* 2.9, *for any* $r_1 > 1$. *Choose* $r_1 \in (1, 2)$ *and* $r_2 \notin \mathbb{Z}$, *such that* $r_1 + r_2 \notin \mathbb{Z}$ *and the interval* $(r_2, r_1 + r_2)$ *contains exactly one integer* $r \in \mathbb{Z}$. *Then, for suitably correlated choices of bounding cochains,* (2.23) *for* $c = r - 1$ *factors through* (2.15), *as follows:*

$$(2.25) \qquad HF^*(E, r_2) \xrightarrow{C} HF^*(E, r) \xrightarrow{\nabla} HF^*(E, r) \xrightarrow{C} HF^*(E, r_1 + r_2), \qquad \text{composite} = \nabla_c.$$

3. $\mathfrak{sl}_2(\mathbb{R})$

The Lie algebras which occur naturally in symplectic geometry are those of Hamiltonian vector fields, which are complicated even in two dimensions. Fortunately, to see the structures emerging in our particular application, it is enough consider a finite-dimensional Lie subalgebra. This section contains the necessary elementary material, much of it borrowed from [**15**].

3.1. Nonnegativity and the rotation number. Set $G = PSL_2(\mathbb{R})$ and $\mathfrak{g} = \mathfrak{sl}_2(\mathbb{R})$. Call $\gamma \in \mathfrak{g}$ nonnegative if the associated quadratic form is positive semidefinite:

$$(3.1) \qquad \det(v, \gamma v) \geq 0 \quad \text{for all } v \in \mathbb{R}^2,$$

where $(v, \gamma v)$ is the matrix with columns v and γv; an equivalent notation would be $v \wedge \gamma v \geq 0$. Concretely, nonnegative γ are of the form

$$\gamma = \left(\begin{smallmatrix} \epsilon & \delta-\alpha \\ \delta+\alpha & -\epsilon \end{smallmatrix}\right), \quad \alpha \geq \sqrt{\delta^2+\epsilon^2}. \tag{3.2}$$

If we think of G as acting on $\mathbb{R}P^1$, an element of $\mathfrak{g}$ is nonnegative iff the associated vector field does not point in clockwise direction anywhere. Write $\mathfrak{g}_{\geq 0} \subset \mathfrak{g}$ for the cone of nonnegative elements, which is invariant under the adjoint action. Similarly, we call γ positive if (3.1) is positive definite; those elements constitute the interior $\mathfrak{g}_{>0} \subset \mathfrak{g}_{\geq 0}$. Correspondingly, there is the opposite cone $\mathfrak{g}_{\geq 0} = -\mathfrak{g}_{\leq 0}$ of nonpositive elements, whose interior consists of the negative ones.

There is a related notion for parabolic elements $g \in G$ (the terminology in this paper is that a nontrivial element of G is either elliptic, parabolic, or hyperbolic; the identity $\mathbb{1}$ is considered to be neither). Namely, given such an element, choose the preimage in $SL_2(\mathbb{R})$ with $\mathrm{tr}(g) = 2$. We say that g is a nonnegative parabolic if $\det(v, gv) \geq 0$ for all v; otherwise, it is a nonpositive parabolic (and those are precisely the two parabolic conjugacy classes in G). Equivalently, a parabolic g is nonnegative if it can be written as $g = e^{\gamma}$ for some nonzero nilpotent $\gamma \in \mathfrak{g}_{\geq 0}$, and similarly for the other sign.

A path $g(t) \in G$ is called nonnegative if

$$g'(t)g(t)^{-1} \in \mathfrak{g}_{\geq 0}, \quad \text{or equivalently} \quad g(t)^{-1}g'(t) \in \mathfrak{g}_{\geq 0}. \tag{3.3}$$

Geometrically, a path in G is nonnegative iff along it, points on $\mathbb{R}P^1$ never move clockwise. If $g_1(t)$ and $g_2(t)$ are nonnegative paths, so is their pointwise product, since

$$g_0(t) = g_1(t)g_2(t) \implies g_0'(t)g_0(t)^{-1} = g_1'(t)g_1(t)^{-1} + g_1(t)\big(g_2'(t)g_2(t)^{-1}\big)g_1(t)^{-1}. \tag{3.4}$$

In parallel, we have notions of positive, nonpositive, and negative paths. If $g(t)$ is nonnegative, both $g(t)^{-1}$ and $g(-t)$ are nonpositive, while $g(-t)^{-1}$ is again nonnegative. If $g_1(t)$ is nonnegative, and $g_2(t)$ positive, their product is positive, by (3.4). In particular, any nonnegative path can be perturbed to a positive one, by multiplying it with $e^{t\gamma}$ for some small positive γ.

Let $\tilde{G} \to G$ be the universal cover. A preimage in $\tilde{G}$ of a given $g \in G$ is the same as a lift of the action of g to $\mathbb{R} \to \mathbb{R}/\pi\mathbb{Z} = \mathbb{R}P^1$. Any $\tilde{g} \in \tilde{G}$ has a rotation number, defined in terms of the action on $\mathbb{R}$ by

$$\mathrm{rot}(\tilde{g}) = \lim_{k\to\infty} \frac{\tilde{g}^k(x) - x}{\pi k} \quad \text{for any } x \in \mathbb{R}. \tag{3.5}$$

If one composes $\tilde{g}$ with an element in the central subgroup $ker(\tilde{G} \to G) \cong \mathbb{Z}$, the rotation number changes by adding that same integer. The other general properties of rotation numbers are: they are continuous; conjugation invariant; homogeneous, meaning that

$$\mathrm{rot}(\tilde{g}^k) = k\,\mathrm{rot}(\tilde{g}), \quad k \in \mathbb{Z}; \tag{3.6}$$

they can't decrease along nonnegative paths; and they satisfy the quasimorphism property

$$|\mathrm{rot}(\tilde{g}_1\tilde{g}_2) - \mathrm{rot}(\tilde{g}_1) - \mathrm{rot}(\tilde{g}_2)| \leq 1. \tag{3.7}$$

One can break down the situation into conjugacy types, as follows. Suppose that $\tilde{g}$ is the lift of an elliptic $g \in G$. Then, g is a rotation with angle θ, where $\theta \in$

$(\mathbb{R}/\pi\mathbb{Z}) \setminus \{0\}$ (in our terminology, "rotations" can be with respect to any oriented basis of $\mathbb{R}^2$, hence are not necessarily orthogonal). The choice of $\tilde{g}$ corresponds to a lift $\tilde{\theta} \in \mathbb{R}$, and the rotation number is $\tilde{\theta}/\pi \in \mathbb{R} \setminus \mathbb{Z}$. In the other (hyperbolic or parabolic) cases, the rotation number is an integer, and can be computed as follows. Fix an eigenvector of g. The choice of $\tilde{g}$ determines a homotopy class of paths from the identity to g. The image of the eigenvector along such a path yields a loop in $\mathbb{R}P^1$, whose degree is the rotation number. We would like to state two consequences of this discussion. For the first one, note that the map $\tilde{G} \to G$ factors through $SL_2(\mathbb{R})$. Case-by-case analysis then directly shows that:

LEMMA 3.1. *If $g \in SL_2(\mathbb{R})$ is the image of $\tilde{g} \in \tilde{G}$, then*

$$\begin{cases} \mathrm{tr}(g) > 0 \iff \mathrm{rot}(\tilde{g}) \in 2\mathbb{Z} + (-\frac{1}{2}, \frac{1}{2}), \\ \mathrm{tr}(g) < 0 \iff \mathrm{rot}(\tilde{g}) \in 2\mathbb{Z} + 1 + (-\frac{1}{2}, \frac{1}{2}). \end{cases} \tag{3.8}$$

For our second observation, let's rewrite (3.7) more symmetrically as follows:

$$\tilde{g}_0\tilde{g}_1\tilde{g}_2 = \mathbb{1} \quad \Longrightarrow \quad |\mathrm{rot}(\tilde{g}_0) + \mathrm{rot}(\tilde{g}_1) + \mathrm{rot}(\tilde{g}_2)| \leq 1. \tag{3.9}$$

LEMMA 3.2. (i) *In the situation of* (3.9), *suppose that the image g_k of some $\tilde{g}_k$ is elliptic. Then the inequality is strict.*

(ii) *In the same situation, suppose that g_k is a nonnegative parabolic. Then*

$$\mathrm{rot}(\tilde{g}_0) + \mathrm{rot}(\tilde{g}_1) + \mathrm{rot}(\tilde{g}_2) \leq 0. \tag{3.10}$$

(iii) *Correspondingly, if g_k is a nonpositive parabolic, one has a lower bound of* 0 *for the sum of rotation numbers.*

PROOF. (i) Without loss of generality, one can assume that $k = 0$ and $\mathrm{rot}(\tilde{g}_0) \in (0, 1)$. Then there is a nonnegative path from the identity to g_0, which lifts to a path with endpoint $\tilde{g}_0$. Therefore,

$$\mathrm{rot}(\tilde{g}_0) + \mathrm{rot}(\tilde{g}_1) + \mathrm{rot}(\tilde{g}_2) \leq \mathrm{rot}(\tilde{g}_0) + \mathrm{rot}(\tilde{g}_0\tilde{g}_1) + \mathrm{rot}(\tilde{g}_2) = \mathrm{rot}(\tilde{g}_0) < 1. \tag{3.11}$$

Passing to inverses yields the corresponing lower bound.

(ii) has the same proof, except that this time, a nonnegative path from the identity to $\tilde{g}_0$ exists for $\mathrm{rot}(\tilde{g}_0) = 0$. Again, one passes to inverses to obtain (iii). □

EXAMPLE 3.3. Let $g = \exp(\gamma) \in G$, with $\gamma \in \mathfrak{g}$ sufficiently small. Then, for any integer $r \geq 1$,

$$g(t) = \begin{pmatrix} \cos(r\pi t) & -\sin(r\pi t) \\ \sin(r\pi t) & \cos(r\pi t) \end{pmatrix} \exp(t\gamma) \tag{3.12}$$

is a positive path from the identity to g, because of (3.4). The lift $\tilde{g} \in \tilde{G}$ of g associated to that path differs from the obvious lift of g (close to the identity) by $r \in \mathbb{Z} \subset \tilde{G}$; hence, $\mathrm{rot}(\tilde{g}) \approx r$, with equality unless g is elliptic.

EXAMPLE 3.4. Fix hyperbolic conjugacy classes C_i $(i = 0, 1, 2)$. Up to simultaneous conjugation, there is exactly one triple (g_0, g_1, g_2) with $g_i \in C_i$, $g_0 = g_1g_2$, and such that if we take lifts $\tilde{g}_i$ satisfying $\tilde{g}_0 = \tilde{g}_1\tilde{g}_2$, then

$$\mathrm{rot}(\tilde{g}_0) = \mathrm{rot}(\tilde{g}_1) + \mathrm{rot}(\tilde{g}_2) - 1. \tag{3.13}$$

To see that, let's suppose that our conjugacy classes are determined by $\mathrm{tr}(\tilde{g}_i) = \pm(\lambda_i + \lambda_i^{-1})$ for some $\lambda_i > 1$. By (3.8), if we choose representatives in $SL_2(\mathbb{R})$ such that $\mathrm{tr}(g_1) = \lambda_1 + \lambda_1^{-1}$ and $\mathrm{tr}(g_2) = \lambda_2 + \lambda_2^{-1}$, then $\mathrm{tr}(g_0) = -\lambda_0 - \lambda_0^{-1}$. With that in mind, after common conjugation, the only solutions are

$$(3.14) \qquad g_1 = \begin{pmatrix} \lambda_1 & 0 \\ 0 & \lambda_1^{-1} \end{pmatrix}, \quad g_2 = \begin{pmatrix} \lambda_2 - t & \pm(\lambda_2 - \lambda_2^{-1} - t) \\ \pm t & \lambda_2^{-1} + t \end{pmatrix},$$
$$\text{where } t = \frac{\lambda_0 + \lambda_0^{-1} + \lambda_1\lambda_2 + \lambda_1^{-1}\lambda_2^{-1}}{\lambda_1 - \lambda_1^{-1}}.$$

By looking at the case where $\lambda_1, \lambda_2 \approx 1$ and deforming the matrices g_1, g_2 to the identity, one sees that the solution of (3.14) with the $-$ sign satisfies (3.13) (the solution with the $+$ sign satisfies the version of (3.13) with a $+1$ instead of -1).

EXAMPLE 3.5. Consider elements $\tilde{g}_i$ of elliptic conjugacy classes $\tilde{C}_i \subset \tilde{G}$ ($i = 1, 2$), with

$$(3.15) \qquad \mathrm{rot}(\tilde{g}_1) \in (0,1), \quad \mathrm{rot}(\tilde{g}_2) \in (0,1), \quad \mathrm{rot}(\tilde{g}_1) + \mathrm{rot}(\tilde{g}_2) > 1.$$

For the product $\tilde{g} = \tilde{g}_1\tilde{g}_2$ and its image $g = g_1g_2$ in G, the following is a complete list of possibilities:

$$(3.16) \qquad \begin{cases} g \text{ can be elliptic, with any value } \mathrm{rot}(\tilde{g}) \in (1, \mathrm{rot}(\tilde{g}_1) + \mathrm{rot}(\tilde{g}_2)]; \\ g \text{ can be nonnegative parabolic, and } \mathrm{rot}(\tilde{g}) = 1; \\ g \text{ can lie in any hyperbolic conjugacy class, and } \mathrm{rot}(\tilde{g}) = 1. \end{cases}$$

Moreover, each possibility happens in a way which is unique up to overall conjugation. To see that, we will carry out computations in the associated conjugacy classes in $SL_2(\mathbb{R})$, which correspond to rotations with angles $\theta_i = \pi \mathrm{rot}(\tilde{g}_i) \in (0, \pi)$. The conjugacy class of rotations with angle θ_1 can be identified with the open unit disc in $\mathbb{C}$, by writing

$$(3.17) \qquad g_1 = \begin{pmatrix} \cos(\theta_1) - \sin(\theta_1)\frac{2\,\mathrm{im}(w)}{1-|w|^2} & -\sin(\theta_1)\frac{|1+w|^2}{1-|w|^2} \\ \sin(\theta_1)\frac{|1-w|^2}{1-|w|^2} & \cos(\theta_1) + \sin(\theta_1)\frac{2\,\mathrm{im}(w)}{1-|w|^2} \end{pmatrix}, \quad |w| < 1.$$

Conjugation with standard (Euclidean) rotations acts by rotating w. If g_2 is the standard rotation with angle θ_2, we have

$$(3.18) \qquad \mathrm{tr}(g_1g_2) = 2\cos(\theta_1)\cos(\theta_2) - 2\sin(\theta_1)\sin(\theta_2)\frac{1+|w|^2}{1-|w|^2}.$$

The values of (3.18) are $(-\infty, 2\cos(\theta_1 + \theta_2)]$, each being reached exactly once. The maximum is reached when g_1 is also a standard rotation ($w = 0$), so that g_1g_2 has angle $\theta_1 + \theta_2 > \pi$; from there, as the trace decreases, so does the angle. The case when $\mathrm{tr}(g_1g_2) = -2$ happens as a limit of rotations whose angle approaches π from above. Since g_1g_2 can't be $-\mathbb{1}$, it must therefore be a nonnegative parabolic. We know (from the existence of suitable nonnegative paths) that $\mathrm{rot}(\tilde{g}_i) \leq \mathrm{rot}(\tilde{g}_1\tilde{g}_2) \leq \mathrm{rot}(\tilde{g}_i)+1$. This, together with (3.8), implies the statements about rotation numbers made in (3.16).

REFERENCES. Nonnegative paths and the rotation number quasimorphism generalize to linear symplectic groups (see [**15, 32**] and [**2, 3, 24**], respectively). The original definition of the rotation number as (3.5), due to Poincaré, applies to orientation-preserving homeomorphisms of the circle (for a basic exposition, see [**9**]; and for a proof of the quasimorphism property with the optimal bound (3.7), see [**35**, Proposition 4.2]). Example 3.4 will turn out to be a special case of a general result (Corollary 4.9).

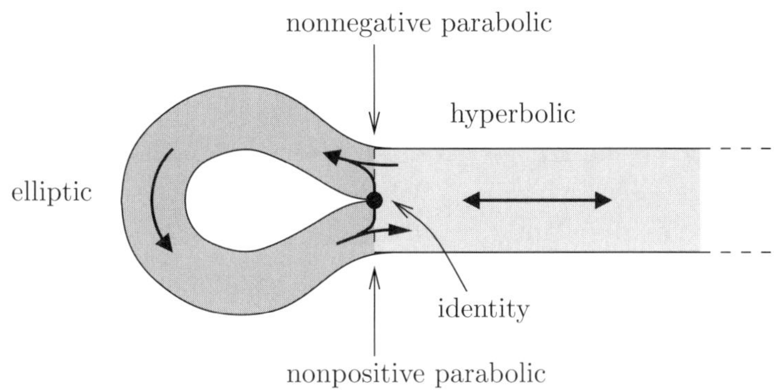

FIGURE 3.1. Traffic flow in nonnegative direction on conjugacy classes (in reality, the space of conjugacy classes is one-dimensional and non-Hausdorff; we've drawn it thickened for the sake of legibility).

3.2. Krein theory. Let's consider the behaviour of nonnegative paths on the level of conjugacy classes. In the elliptic locus, motion can go in one direction only (Krein's Lemma). Namely, for $\gamma \in \mathfrak{g}_{\geq 0}$ written as in (3.2),

$$g = \begin{pmatrix} \cos(\theta) & -\sin(\theta) \\ \sin(\theta) & \cos(\theta) \end{pmatrix} \text{ with } \theta \in (0,\pi) \quad \Longrightarrow \quad \mathrm{tr}(g\gamma) = -2\alpha \sin(\theta) \leq 0, \tag{3.19}$$

so the angle of rotation θ can't decrease along a nonnegative path (and its derivative vanishes exactly when that of the path vanishes). There is no comparable constraint in the hyperbolic locus:

$$g = \begin{pmatrix} \lambda & 0 \\ 0 & \lambda^{-1} \end{pmatrix} \text{ with } \lambda > 1 \quad \Longrightarrow \quad \mathrm{tr}(g\gamma) = (\lambda - \lambda^{-1})\epsilon, \tag{3.20}$$

which can have either sign. The two parabolic conjugacy classes are each permeable in one direction (from elliptic to hyperbolic or vice versa) only. This is again elementary:

$$g = \begin{pmatrix} 1 & 0 \\ \pm 1 & 1 \end{pmatrix} \quad \Longrightarrow \quad \pm\mathrm{tr}(g\gamma) = \delta - \alpha \leq 0 \tag{3.21}$$

(in this case, $\mathrm{tr}(g\gamma) = 0$ iff $\gamma = \begin{pmatrix} 0 & 0 \\ 2\alpha & 0 \end{pmatrix}$; hence, a nonnegative path intersects the parabolic locus either transversally, or tangentially to the one-parameter subgroup at the intersection point). Finally, if a nonnegative path goes through the identity in G, it must leave that element through either nonnegative parabolic elements, or elliptic elements which are small positive rotations (and conversely, it must arrive through nonpositive parabolic elements, or elliptic elements which are small negative rotations). These observations are summarized in Figure 3.1.

REFERENCES. The discussion above is taken from [**15**, **32**] (see more specifically [**32**, Lemma 2.2]), and we will continue to fundamentally follow those references, while making changes and adding material as required for our intended applications. Specifically, Lemma 3.8 below is a special case of [**15**, Proposition 2.4] (our proof fills in some technical details); and Lemmas 3.11–3.13 can be viewed as modified versions of [**32**, Theorem 3.3] (note however that [**32**, Theorem 3.6] is incorrect).

Our next topic is the converse question, of constructing nonnegative paths with prescribed motion on conjugacy classes. We will need a technical refinement of the computations from (3.19) and (3.20). Writing $g = \left(\begin{smallmatrix} a & c \\ b & d \end{smallmatrix}\right) \in SL_2(\mathbb{R})$, the condition $\det(g) = 1$ is equivalent to

$$(a-d)^2 + (c+b)^2 = \mathrm{tr}(g)^2 - 4 + (c-b)^2. \tag{3.22}$$

As a consequence, if we consider a subset of elements with fixed trace, then $T(g) = c - b$ is a proper function on that subset. Note also that $T(g^2) = T(g)\mathrm{tr}(g)$.

LEMMA 3.6. (i) *Let $C_{ell,\theta} \subset SL_2(\mathbb{R})$ be the conjugacy class of rotations with angle $\theta \in (0,\pi)$. There is a function*

$$\begin{cases} \gamma_\theta : C_{ell,\theta} \longrightarrow \mathfrak{g}_{>0}, \\ |T(g\gamma_\theta(g))| \leq A + B|T(g)|, \text{ for some constants } A, B > 0; \\ \mathrm{tr}(g\,\gamma_\theta(g)) \text{ is negative and bounded away from } 0. \end{cases} \tag{3.23}$$

(ii) *Let $C_{hyp,\lambda} \subset SL_2(\mathbb{R})$ be the conjugacy class of hyperbolic elements with eigenvalues $\lambda^{\pm 1}$, for some $\lambda > 1$. There are functions*

$$\begin{cases} \gamma_{\lambda,\pm} : C_{hyp,\lambda} \longrightarrow \mathfrak{g}_{>0}, \\ |T(g\gamma_{\lambda,\pm}(g))| \leq A + B|T(g)|, \text{ for some constants } A, B > 0; \\ \pm\,\mathrm{tr}(g\,\gamma_{\lambda,\pm}(g)) \text{ is positive and bounded away from } 0. \end{cases} \tag{3.24}$$

PROOF. (i) Take

$$\gamma_\theta(g) = \frac{g - \cos(\theta)\,\mathbb{1}}{\sin(\theta)}, \tag{3.25}$$

which is conjugate to $\left(\begin{smallmatrix} 0 & -1 \\ 1 & 0 \end{smallmatrix}\right)$, hence lies in $\mathfrak{g}_{>0}$. Then

$$\mathrm{tr}(g\gamma_\theta(g)) = -2\sin(\theta), \tag{3.26}$$

$$T(g\gamma_\theta(g)) = \frac{\cos(\theta)}{\sin(\theta)} T(g). \tag{3.27}$$

(ii) If $r = r(g)$ is the quantity on either side of (3.22),

$$\gamma_{\lambda,\pm}(g) = \sqrt{r}\left(\begin{smallmatrix} 0 & -1 \\ 1 & 0 \end{smallmatrix}\right) \pm \left(\begin{smallmatrix} a-d & b+c \\ b+c & -(a-d) \end{smallmatrix}\right) \in \mathfrak{g}_{\geq 0} \tag{3.28}$$

satisfies

$$\mathrm{tr}(g\gamma_{\lambda,\pm}(g)) = (c-b)\sqrt{r} \pm r = r\Big(\mathrm{sign}(c-b)\sqrt{1 - \tfrac{\mathrm{tr}(g)^2-4}{r}} \pm 1\Big) \tag{3.29}$$
$$\Longrightarrow \pm\mathrm{tr}(g\gamma_{\lambda,\pm}(g)) \geq \tfrac{1}{2}(\mathrm{tr}(g)^2 - 4),$$

$$T(g\gamma_{\lambda,\pm}(g)) = -\mathrm{tr}(g)\sqrt{r} = -\mathrm{tr}(g)\sqrt{\mathrm{tr}(g)^2 - 4 + T(g)^2}. \tag{3.30}$$

Our choice (3.28) always lies on the boundary of the nonnegative cone. However, one can perturb it into the interior $\mathfrak{g}_{>0}$, and this perturbation can be chosen sufficiently small so as not to disturb its other properties. □

CONVENTION 3.7. When we talk about weak homotopy equivalences for spaces of paths in G (and, later on, for spaces of connections as well), this is understood to refer to smooth maps from finite-dimensional compact manifolds (with boundary or corners) to the relevant space. For instance, saying that some space $\mathcal{P}$ as in (3.34) is weakly contractible means that for every smooth map $P \to \mathcal{P}$ from a compact manifold P, there is a smooth extension $P \times [0,1] \to \mathcal{P}$ which is constant on the other endpoint of the interval. This technical clarification may not be essential, but it slightly simplifies some proofs, and is anyway the natural context for our applications.

LEMMA 3.8. (i) *Fix a nondecreasing function* $\theta(t) \in (0,\pi)$ $(0 \leq t \leq 1)$, *and a* $g_0 \in C_{ell,\theta(0)}$. *Then, the space of nonnegative paths* $g(t)$, *with* $g(0) = g_0$ *and such that* $g(t) \in C_{ell,\theta(t)}$, *is weakly contractible. A corresponding result holds for (strictly) increasing functions and positive paths.*

(ii) *Fix a function* $\lambda(t) > 1$ $(0 \leq t \leq 1)$, *and a* $g_0 \in C_{hyp,\lambda(0)}$. *Then, the space of nonnegative paths* $g(t)$, *with* $g(0) = g_0$ *and such that* $g(t) \in C_{hyp,\lambda(t)}$, *is weakly contractible. The same holds for positive paths.*

PROOF. (i) Let's suppose that we have a smooth family of functions

$$\begin{cases} \gamma_t : C_{ell,\theta(t)} \longrightarrow \mathfrak{g}_{\geq 0}, \\ |T(g\gamma_t(g))| \leq A + B|T(g)|, \text{ for some constants } A, B > 0; \\ \operatorname{tr}(g\,\gamma_t(g)) = -2\sin(\theta(t))\theta'(t). \end{cases} \tag{3.31}$$

The boundedness condition ensures that the ODE

$$g'(t) = g(t)\gamma_t(g(t)) \tag{3.32}$$

can be integrated for any initial value $g(0) \in C_{ell,\theta(0)}$. By construction, each solution is a nonnegative path with $g(t) \in C_{ell,\theta(t)}$.

Functions (3.31) can be constructed from those in Lemma 3.6(i) by multiplying with the bounded nonnegative function

$$g \longmapsto -2\sin(\theta(t))\theta'(t)\operatorname{tr}(g\gamma_{\theta(t)}(g))^{-1}. \tag{3.33}$$

On the other hand, the conditions in (3.31) are convex, allowing us to interpolate between two such functions, and to construct them using partitions of unity. In particular, given a nonnegative path $g(t)$ with the desired behaviour, one can choose γ_t so that this path is a solution of (3.32). More precisely, this kind of argument shows that the space of paths is weakly homotopy equivalent to the space of all (3.31), which implies the desired result. In the positive version of the same result, one takes γ_t which have values in $\mathfrak{g}_{>0}$, and those exist because (3.33) is strictly positive.

The proof of (ii) is parallel, using the corresponding part of Lemma 3.6. □

3.3. Spaces of nonnegative paths. Using Lemma 3.8 as our basic ingredient, we will now discuss the topology of spaces of nonnegative paths with constraints on the endpoints. The general notation will be as follows: if U_0, U_1 are subsets of G,

$$\begin{aligned} &\mathcal{P}(G, U_0, U_1) = \{g : [0,1] \to G \;:\; g(0) \in U_0,\, g(1) \in U_1\}, \\ &\mathcal{P}_{\geq 0}(G, U_0, U_1) = \text{subspace of nonnegative paths}, \\ &\mathcal{P}_{>0}(G, U_0, U_1) = \text{subspace of positive paths}. \end{aligned} \tag{3.34}$$

One can introduce similar notation $\mathcal{P}_{\leq 0}(G, U_0, U_1)$, $\mathcal{P}_{<0}(G, U_0, U_1)$ for nonpositive and negative paths. We will use superscripts $\mathcal{P}(G, U_0, U_1)^r$ to restrict to paths along which the rotation number increases by exactly r; and similarly for inequalities, such as $\mathcal{P}(G, U_0, U_1)^{<r}$.

LEMMA 3.9. *Fix hyperbolic conjugacy classes* C_0 *and* C_1. *Then, evaluation at either endpoint yields a weak homotopy equivalence*

$$\mathcal{P}_{\geq 0}(G, C_0, C_1)^0 \xrightarrow{\simeq} C_i. \tag{3.35}$$

PROOF. By Figure 3.1, any such path remains within the hyperbolic locus, hence falls into the class considered in Lemma 3.8(ii). The rest is straightforward. □

LEMMA 3.10. *Fix hyperbolic conjugacy classes C_i ($i = 0, \dots, m$, for some $m \geq 2$). Consider $g_i \in C_i$ together with a nonnegative path $h(t)$ from $h(0) = g_1 \cdots g_m$ to $h(1) = g_0$, such that if we take lifts of our elements to $\tilde{g}_i \in \tilde{G}$, compatible with the product and path, then*

$$\mathrm{rot}(\tilde{g}_0) = \mathrm{rot}(\tilde{g}_1) + \cdots + \mathrm{rot}(\tilde{g}_m) + 1 - m. \tag{3.36}$$

Any g_i yields a weak homotopy equivalence between the space of all $(g_0, \dots, g_m, h(t))$ and C_i.

PROOF. Because of (3.7) and the fact that rotation numbers don't decrease along nonnegative paths, (3.36) requires that, for any $i \geq 1$,

$$\mathrm{rot}(\tilde{g}_1 \cdots \tilde{g}_i) = \mathrm{rot}(\tilde{g}_1 \cdots \tilde{g}_{i-1}) + \mathrm{rot}(\tilde{g}_i) - 1. \tag{3.37}$$

Suppose that there is a smallest $i \geq 2$ for which the product $g_1 \cdots g_i$ is not hyperbolic. By (3.37) and Lemma 3.2, it would then have to be a nonnegative parabolic. If $i < m$, one would then apply Lemma 3.2 again, to get

$$\mathrm{rot}(\tilde{g}_1 \cdots \tilde{g}_{i+1}) \geq \mathrm{rot}(\tilde{g}_1 \cdots \tilde{g}_i) + \mathrm{rot}(\tilde{g}_{i+1}), \tag{3.38}$$

which is a contradiction to (3.37) for $i + 1$. If $i = m$, the path h would have to go from a nonnegative parabolic to a hyperbolic element without raising the rotation number, which is impossible. We have therefore shown that all products $g_1 \cdots g_i$ are hyperbolic.

Consider the space of all $(g_1, \dots, g_m)$ such that $g_i \in C_i$, the product $g = g_1 \cdots g_m$ is a fixed hyperbolic element, and $\mathrm{rot}(\tilde{g}_1 \cdots \tilde{g}_m) = \mathrm{rot}(\tilde{g}_1) + \cdots + \mathrm{rot}(\tilde{g}_m) + 1 - m$. Then, that space is contractible. One can derive this from Example 3.4 by induction on m, and the hyperbolicity of the intermediate products. We omit the details, since there is a more conceptual alternative, namely to think of it as a special case of Corollary 4.9 below. On the other hand, the space of nonnegative paths $h(t)$ from our fixed hyperbolic g to C_0, along which the rotation number remains constant, is weakly contractible by Lemma 3.9. By taking the product of those spaces, and then letting g vary, one sees that evaluation at $h(0) = g = g_1 \cdots g_m$ gives a weak homotopy equivalence between the space of all $(g_0, g_1, \dots, g_m, h(t))$ and G_{hyp}. Since evaluation at any $h(t)$ lands in the hyperbolic locus, we can evaluate at $h(1) = g_0$ instead. By applying simultaneous conjugation, one extends the result to the other evaluation maps. □

One can interpret Lemma 3.9 as follows. Let $\mathcal{P}(G_{hyp}, C_0, C_1)$ be the space of all paths from C_0 to C_1, but which remain inside the hyperbolic locus. $\mathcal{P}_{\geq 0}(G, C_0, C_1)^0$ is a subspace of $\mathcal{P}(G_{hyp}, C_0, C_1)$, and we have shown that the inclusion yields a weak homotopy equivalence

$$\mathcal{P}_{\geq 0}(G, C_0, C_1)^0 \xrightarrow{\simeq} \mathcal{P}(G_{hyp}, C_0, C_1). \tag{3.39}$$

Alternatively, let's think of the case $C_0 = C_1 = C$. In that case, inclusion of the constant paths yields a weak homotopy equivalence

$$C \xrightarrow{\simeq} \mathcal{P}_{\geq 0}(G, C, C)^0. \tag{3.40}$$

This second kind of interpretation also applies to Lemma 3.10, which turned out to be weakly homotopy equivalent to the subspace for which the path $h(t)$ is constant.

From now on, when deriving results about path spaces, we will often only give the principal steps of the proof, omitting "ϵ–δ level" details (readers interested in seeing such kind of arguments in full may want to look at [**15**]).

LEMMA 3.11. *Fix an elliptic conjugacy class C_0, and a hyperbolic conjugacy class C_1. Then, evaluation at the hyperbolic endpoint yields a weak homotopy equivalence*

$$\mathcal{P}_{\geq 0}(G, C_0, C_1)^{<1} \xrightarrow{\simeq} C_1. \tag{3.41}$$

The same applies to $\mathcal{P}_{\geq 0}(G, C_1, C_0)^{<1}$, still taking the evaluation map at the hyperbolic endpoint.

SKETCH OF PROOF. We begin with a slightly different problem. Let's consider positive paths which start in C_0, end in G_{hyp}, and along which the rotation number increases by less than 1. Such a path necessarily has the following structure:

$$\text{for some } T \in (0,1), \begin{cases} g(t) \text{ is elliptic for } t < T; \\ g(T) \text{ is a nonpositive parabolic;} \\ g(t) \text{ is hyperbolic for } T < t. \end{cases} \tag{3.42}$$

If one fixes T and $g(T)$, then by analyzing $g|[0, T-\epsilon]$ and $g|[T+\epsilon, 1]$ for small $\epsilon > 0$, using Lemma 3.8, one sees that the resulting space of paths is weakly contractible. Allowing T and $g(T)$ to vary yields a path space that is weakly homotopy equivalent to the nonpositive parabolic stratum, hence also to G. Instead of using evaluation at $t = T$, one can use the (homotopic) evaluation at $t = 1$.

We can weaken strict positivity to nonnegativity, without changing the weak homotopy type of the space, since any nonnegative path can be perturbed to a positive one (see (3.4) and the subsequent discussion); for this to work, it's important that the endpoint condition $g(1) \in G_{hyp}$ is an open one. Next, again without changing the weak homotopy type, we can require that the hyperbolic endpoint should lie in a fixed conjugacy class, thanks to Lemma 3.8(ii). This brings us to the desired situation. The other direction (from hyperbolic to elliptic) is proved in the same way. □

LEMMA 3.12. *Fix a hyperbolic conjugacy class C. Evaluation at the endpoint yields a weak homotopy equivalence*

$$\mathcal{P}_{\geq 0}(G, id, C)^1 \xrightarrow{\simeq} C. \tag{3.43}$$

SKETCH OF PROOF. Consider first positive paths which start at the identity, end in G_{hyp}, and along which the rotation number increases by 1. Such a path has the property that $g(t)$ is elliptic for all sufficiently small $t > 0$. By looking at $g|[\epsilon, 1]$ for small $\epsilon > 0$, one reduces their study to Lemma 3.11. The translation back to the original situation also follows that model. □

LEMMA 3.13. *Fix two hyperbolic conjugacy classes C_0 and C_1. Then, evaluation at both endpoints yields a weak homotopy equivalence*

$$\mathcal{P}_{\geq 0}(G, C_0, C_1)^1 \xrightarrow{\simeq} C_0 \times C_1. \tag{3.44}$$

SKETCH OF PROOF. Once more, we start with a slightly different problem. Namely, consider positive paths, such that both endpoints lie in G_{hyp}, and along which the rotation number increases by 1. There is exactly one point T where such a path intersects a fixed elliptic conjugacy class. By breaking up the path into pieces $g|[0,T]$ and $g|[T,1]$, one can think of the whole path space as a fibre product of two spaces of the kind considered in Lemma 3.11. Hence, what we are looking at is the fibre product of two hyperbolic conjugacy classes, formed over a contractible space (the elliptic conjugacy class). As in Lemma 3.11, positivity can be weakened to nonnegativity, and one can use Lemma 3.8(ii) to replace the condition on the endpoints by one of lying in specified hyperbolic conjugacy classes. □

One can interpret these results as follows: the inclusion of nonnegative paths into all paths yields weak homotopy equivalences

$$(3.45)\qquad \begin{aligned} &\left.\begin{aligned}&\mathcal{P}_{\geq 0}(G,C_0,C_1)^{<1} \xrightarrow{\simeq} \mathcal{P}(G,C_0,C_1)^{<1}\\ &\mathcal{P}_{\geq 0}(G,C_1,C_0)^{<1} \xrightarrow{\simeq} \mathcal{P}(G,C_1,C_0)^{<1}\end{aligned}\right\} &&\text{for Lemma 3.11,}\\ &\mathcal{P}_{\geq 0}(G,id,C)^{1} \xrightarrow{\simeq} \mathcal{P}(G,id,C)^{1} &&\text{for Lemma 3.12,}\\ &\mathcal{P}_{\geq 0}(G,C_0,C_1)^{1} \xrightarrow{\simeq} \mathcal{P}(G,C_0,C_1)^{1} &&\text{for Lemma 3.13.}\end{aligned}$$

We will see a few more instances of such behaviour in Section 7. Note that this contrasts with Lemmas 3.9 and 3.10, where including nonnegative paths into all paths (with the same rotation behaviour) would not have yielded a homotopy equivalence.

REMARK 3.14. One can conjecture that statements similar to (3.45) should hold more generally for spaces of nonnegative paths which are "sufficiently long" (meaning that the rotation number increases by a sufficient amount to force transitions between elliptic and hyperbolic elements). This idea is in tune with the results of [**15, 32**].

The next result can be viewed as a weak version of Lemma 3.12 under extra constraints. Instead of determining the topology of the path space under consideration, we will argue by an explicit construction.

LEMMA 3.15. *Fix two hyperbolic conjugacy classes C_1 and C_2. Then there is a nonnegative path $g_1(t)$, starting at the identity and ending in C_1, along which the rotation number increases by 1; together with another path $g_2(t) \in C_2$, such that at all times, $g_1(t)g_2(t) \in C_2$. Moreover, one can write*

$$(3.46)\qquad g_1(t)g_2(t) = k(t)g_2(t)k(t)^{-1}$$

for some path $k(t)$ starting at the identity, whose endpoint $k(1)$ is hyperbolic and (if one lifts the path to $\tilde{G}$ in the obvious way) has rotation number 0.

PROOF. Let's set $g_2 = \mathrm{diag}(\lambda, \lambda^{-1})$ for some $\lambda > 1$. Consider

$$(3.47)\qquad H = \{g \in SL_2(\mathbb{R}) \;:\; \mathrm{tr}(gg_2) = \lambda + \lambda^{-1}\} \xrightarrow{\mathrm{tr}} \mathbb{R}.$$

H is a connected hyperboloid in $\mathbb{R}^3$, and the trace is a Morse function on it, with index 1 critical points at the identity and g_2^{-2}, and corresponding critical levels 2 and $\lambda^2 + \lambda^{-2} > 2$. The open sublevel set $U = \{\mathrm{tr}(g) < 2\} \subset S$ consists of two connected components $U_\pm$. If we write $g = \left(\begin{smallmatrix} a & c \\ b & d\end{smallmatrix}\right)$, those components are

$$(3.48)\qquad U_\pm = \{g \in H \;:\; \pm b > 0, \text{ or equivalently } \pm c < 0\}.$$

We construct a path in H of the following kind. Starting at the identity, the path first moves inside the level set $\{\mathrm{tr}(g) = 2\}$ to $\left(\begin{smallmatrix} 1 & 0 \\ 1 & 1 \end{smallmatrix}\right)$. From there, we strictly decrease the trace, moving into U_+, and then further inside that subset until $\{\mathrm{tr}(g) = -2\}$. At that point, the resulting matrix is necessarily conjugate to $\left(\begin{smallmatrix} -1 & 0 \\ 1 & -1 \end{smallmatrix}\right)$, because it still belongs to U_+. We continue our path downwards into hyperbolic level sets $\{\mathrm{tr}(g) < -2\}$, until we reach the value of the trace (up to sign) prescribed by our choice of C_1. One sees easily that the rotation number increases by 1 along this path.

Denote the path we have just constructed by $g_1(t)$, and take $g_2(t) = g_2$ to be the constant path. Since $g_1(t)g_2$ remains in the conjugacy class of g_2, one can write $g_1(t)g_2 = k(t)g_2k(t)^{-1}$, for some smooth $k(t) = \left(\begin{smallmatrix} p(t) & r(t) \\ q(t) & s(t) \end{smallmatrix}\right)$ with $k(0) = \mathbb{1}$. Then,

$$\begin{aligned} \mathrm{tr}(g_1(t)) = \mathrm{tr}(k(t)g_2k(t)^{-1}g_2^{-1}) &= 2p(t)s(t) - (\lambda + \lambda^{-2})q(t)r(t) \\ &= (1 - p(t)s(t))(\lambda - \lambda^{-1})^2 + 2. \end{aligned} \tag{3.49}$$

By construction $\mathrm{tr}(g_1(1)) < -2$, which with (3.49) yields $p(1)s(1) > 1$, hence

$$\mathrm{tr}(k(1))^2 = (p(1) + s(1))^2 = (p(1) - s(1))^2 + 4p(1)s(1) > 4. \tag{3.50}$$

This shows that $k(1)$ must be hyperbolic (one could derive the same result more elegantly from Theorem 4.7, applied to a punctured torus). Now, bear in mind that $\mathrm{tr}(g_1(t)) \leq 2$ for all t. From this and (3.49), it follows that $p(t)s(t) > 0$ for all t. Hence, both entries $p(t)$ and $s(t)$ remain positive throughout. By tracking how points on the circle move under the action of $k(t)$, one easily concludes that the rotation number of $k(1)$ must vanish.

At this point, it seems that we have been missing an essential condition, namely, that $g_1(t)$ should be a nonnegative path. However, an inspection of each step of the construction (and the use of Lemma 3.8) shows that $g_1(t)$ is conjugate to a nonnegative path, by a time-dependent conjugation. One can apply the same conjugation to $g_2(t)$ and $k(t)$, and that yields the desired result. □

REMARK 3.16. It would be interesting to consider similar questions for nonnegative paths in the larger group $\mathit{Diff}^+(S^1)$ (this idea was already mooted by Abouzaid and Ganatra). Enlarging the group affords one greater freedom, which may simplify the construction of homotopies. On a more fundamental level, the larger group contains many more conjugacy classes: for instance, the pullbacks of hyperbolic elements by finite covers $S^1 \to S^1$, which would be relevant to Fukaya categories with multiple fibres as "stops" [**33**].

4. $\mathfrak{sl}_2(\mathbb{R})$-connections on surfaces

For our TQFT, we will need to consider connections with nonnegative curvature. There is an obvious relation between such connections and nonnegative paths, and (with one exception, Proposition 4.1) all our results will be derived from the elementary considerations in Section 3 in that way. On the other hand, if we restrict attention to flat connections, there is a large body of relevant literature; [**10**] will be particularly important for us.

4.1. Basic notions. Let S be a connected compact oriented surface with nonempty boundary. We consider $\mathfrak{g}$-connections on the trivial bundle over S, meaning operators $\nabla_A = d - A$ for some $A \in \Omega^1(S, \mathfrak{g})$ (we will often refer to A as the connection). Gauge transformations are maps $\Phi : S \to G$, and act on connections (covariantly) by

$$\Phi_* A = \Phi A \Phi^{-1} + (d\Phi)\Phi^{-1}. \tag{4.1}$$

With our sign convention for A, the curvature is

$$F_A = -dA + A \wedge A \in \Omega^2(S, \mathfrak{g}). \tag{4.2}$$

A connection is said to have nonnegative curvature if it satisfies the (gauge invariant) condition

$$F_A(\xi_1, \xi_2) \in \mathfrak{g}_{\geq 0} \quad \text{for any oriented basis } (\xi_1, \xi_2) \in TS. \tag{4.3}$$

Obviously, there is a parallel notion of nonpositive curvature; one can switch from one to the other by changing the orientation of S, or by applying a gauge transformation in $PGL_2^-(\mathbb{R})$. Similarly, one could consider connections with (strictly) positive or negative curvature; but we have no use for those notions.

Given a connection, parallel transport along some $c : [0,1] \to S$ yields a path $g : [0,1] \to G$,

$$g(0) = \mathbb{1}, \quad g'(t)g(t)^{-1} = A_{c(t)}(c'(t)). \tag{4.4}$$

(If c is a closed loop, $g(1)$ is the holonomy of the connection around c.) That path has an obvious lift to $\tilde{G}$, with $\tilde{g}(0)$ the identity; from that, we get a rotation number

$$\mathrm{rot}_c(A) \stackrel{\text{def}}{=} \mathrm{rot}(\tilde{g}(1)) \in \mathbb{R}. \tag{4.5}$$

Since G is homotopy equivalent to a circle, the homotopy class of a gauge transformation can be written as $[\Phi] \in H^1(S)$. The effect of gauge transformations on rotation numbers along closed loops c is

$$\mathrm{rot}_c(\Phi_* A) = \mathrm{rot}_c(A) + \int_c [\Phi]. \tag{4.6}$$

Let's denote the boundary circles by $\partial_i S$, where $i \in I = \pi_0(\partial S)$. By (4.6), the following is a gauge invariant quantity:

$$\mathrm{rot}_{\partial S}(A) = \sum_i \mathrm{rot}_{\partial_i S}(A). \tag{4.7}$$

PROPOSITION 4.1. *Let A be a connection on S, such that the holonomies around the boundary circles are hyperbolic. If the connection is nonnegatively curved,*

$$\mathrm{rot}_{\partial S}(A) \leq -\chi(S). \tag{4.8}$$

PROOF. We consider a slightly different but equivalent situation. Namely, let S be a Riemann surface with tubular ends, carrying a connection A which, in coordinates $(s,t) \in [0,\infty) \times S^1$ on each end, is of the form $a_{i,t}\,dt$, and has hyperbolic holonomy around the S^1 factor. One can define an analogue of (4.7), which we denote by $\mathrm{rot}_{ends}(A) \in \mathbb{Z}$. Consider the $\mathbb{R}$-linear Dolbeault operator associated to the connection,

$$\bar{\partial}_A = \nabla_A^{0,1} : C^\infty_{cpt}(S, \mathbb{C}) \longrightarrow \Omega^{0,1}_{cpt}(S). \tag{4.9}$$

As stated, this applies to compactly supported smooth functions, but in fact one wants to pass to suitable Sobolev completions, let's say from $W^{1,2}$ to L^2. The index theorem (see e.g. [**26**, Section 3.3]) says that the completed operator is Fredholm, with

$$\mathrm{index}(\bar{\partial}_A) = \chi(S) + \mathrm{rot}_{ends}(A). \tag{4.10}$$

The Bochner identity, which applies to any $\xi \in W^{1,2}(S,\mathbb{C})$, is

$$\int_S |\bar\partial_A \xi|^2 = \int_S |\nabla\xi|^2 + \int_S \det(\xi, F_A\xi). \tag{4.11}$$

If we assume that our connection has nonnegative curvature, it follows that any $W^{1,2}$ solution of $\bar\partial_A\xi = 0$ must be zero, hence the index is necessarily ≤ 0. □

By reversing orientation, we get the following:

COROLLARY 4.2. *If A is a flat connection with hyperbolic boundary holonomies,*

$$|\mathrm{rot}_{\partial S}(A)| \leq -\chi(S). \tag{4.12}$$

REFERENCES. Corollary 4.2, proved in [**10**, Theorem 3.3], is a version of the Milnor-Wood inequality [**21**, **34**]. The Milnor-Wood inequality has numerous generalizations (to other Lie groups, and to homeomorphisms of the circle [**34**]). The proof used here applies to linear symplectic groups. One can consider results such as those in [**5**, **16**], proved using pseudo-holomorphic sections of symplectic fibrations over surfaces, as nonlinear analogues.

We will be interested in spaces of connections with fixed boundary behaviour. Concretely, given S, choose $a_i \in \Omega^1(\partial_i S, \mathfrak{g})$ for each $i \in I$, with hyperbolic holonomy. We will use the following associated notation:

$$\begin{array}{ll} g_i & \text{holonomy of } a_i, \\ C_i & \text{conjugacy class of } g_i \text{ in } G, \\ \tilde g_i & \text{preferred lift of } g_i \text{ to } \tilde G, \\ \tilde C_i & \text{conjugacy class of } \tilde g_i \text{ in } \tilde G. \end{array} \tag{4.13}$$

The relevant spaces of connections are

$$\begin{aligned} \mathcal{A}(S,\{a_i\}) &= \{A \in \Omega^1(S,\mathfrak{g}) \;:\; A|\partial_i S = a_i\}, \\ \mathcal{A}_{\geq 0}(S,\{a_i\}) &\text{ subspace of nonnegatively curved connections,} \\ \mathcal{A}_{\mathit{flat}}(S,\{a_i\}) &\text{ subspace of flat connections.} \end{aligned} \tag{4.14}$$

These spaces are acted on by the group

$$\mathcal{G}(S,\{a_i\}) = \{\Phi : S \to G \;:\; (\Phi|\partial_i S)_* a_i = a_i\} \simeq H^1(S,\partial S). \tag{4.15}$$

The weak homotopy equivalence in (4.15) comes from the fact that the stabilizer of each a_i in $C^\infty(\partial_i S, G)$ is a copy of $\mathbb{R}$, hence contractible. In the case of flat connections, the quotient

$$\mathcal{M}_{\mathit{flat}}(S,\{a_i\}) = \mathcal{A}_{\mathit{flat}}(S,\{a_i\})/\mathcal{G}(S,\{a_i\}) \tag{4.16}$$

has a well-known topological interpretation:

$$\begin{aligned} \mathcal{M}_{\mathit{flat}}(S,\{a_i\}) \cong \big\{\sigma : \pi_1(S)^{op} \to G \;:\; &\sigma(\partial_i S) \in C_i, \\ &\text{and } e_{rel}(\sigma) = \textstyle\sum_i \mathrm{rot}(a_i)\big\}/G. \end{aligned} \tag{4.17}$$

Here, $e_{rel}(\sigma)$ is the relative Euler number of the associated $\mathbb{R}P^1$-bundle, with its section over ∂S given by an eigenvector for the holonomy; and G acts by overall conjugation.

REMARK 4.3. In the discussion above, for instance in (4.7), we have always used $\partial_i S$ with its boundary orientation. When looking at specific examples of surfaces, we will sometimes adopt different orientation conventions on some boundary components (and will warn the reader when that happens).

4.2. Spaces of flat connections. For our purpose, what matters is the weak homotopy type of the spaces (4.14) (in the same sense as in Convention 3.7). Flat connections are easier to understand in that respect, because of (4.17).

Let's begin with a case in which the topological aspect is trivial, namely that of the annulus $S = [0,1] \times S^1$. Our convention is that we identify each boundary circle $\partial_i S = \{i\} \times S^1$ with S^1 in the obvious way (which is orientation-reversing for $i = 0$, and orientation-preserving for $i = 1$). In order for $\mathcal{A}_{flat}(S, a_0, a_1)$ to be nonempty, the following condition must be satisfied:

(4.18) a_0, a_1 have conjugate holonomies ($C_0 = C_1$), and the same rotation number $r = \mathrm{rot}(a_0) = \mathrm{rot}(a_1)$.

Equivalently, the conjugacy classes in $\tilde{G}$ are the same ($\tilde{C}_0 = \tilde{C}_1$).

PROPOSITION 4.4. *Take $S = [0,1] \times S^1$, with boundary conditions* (4.18). *Then there is a weak homotopy equivalence*

$$\mathcal{A}_{flat}(S, a_0, a_1) \simeq \mathbb{Z}, \tag{4.19}$$

unique up to adding a constant, and which is compatible with the action of $\mathcal{G}(S, a_0, a_1) \simeq \mathbb{Z}$. To state this more concretely, let $C = C_0 = C_1$. By associating to a connection A its family of holonomies around the loops $\{s\} \times S^1$, one defines a canonical weak homotopy equivalence

$$\mathcal{A}_{flat}(S, a_0, a_1) \longrightarrow \mathcal{P}(C, g_0, g_1). \tag{4.20}$$

PROOF. Clearly, $\mathcal{G}(S, a_0, a_1)$ acts transitively on $\mathcal{A}_{flat}(S, a_0, a_1)$, with stabilizer $\mathbb{R}$ at every point. This implies that each orbit gives a weak homotopy equivalence

$$\mathcal{G}(S, a_0, a_1) \xrightarrow{\simeq} \mathcal{A}_{flat}(S, a_0, a_1). \tag{4.21}$$

We know from (4.15) that $\mathcal{G}(S, a_0, a_1) \simeq \mathbb{Z}$. The map (4.20) is equivariant with respect to the action of ΩG (thought of as gauge transformations in the $[0,1]$ variable on the domain, and as acting by pointwise conjugation on the target space). From that and (4.21), it follows that (4.20) is a weak homotopy equivalence. □

ADDENDUM 4.5. Suppose that $g_0 = g_1 = g$. Take a path

$$c : [0,1] \longrightarrow S, \quad c(0) = (0,0), \quad c(1) = (1,0). \tag{4.22}$$

For any $A \in \mathcal{A}_{flat}(S, a_0, a_1)$, parallel transport along c yields an element of G which commutes with g, hence is either trivial or hyperbolic. The rotation number $\mathrm{rot}_c(A) \in \mathbb{Z}$ then defines an isomorphism $\pi_0(\mathcal{A}_{flat}(S, a_0, a_1)) \cong \mathbb{Z}$. One can take the path to have winding number zero around the S^1 factor, and that gives a preferred choice of (4.19), which is the same one gets from identifying $\pi_0(\mathcal{P}(C, g, g)) \cong \mathbb{Z}$ in (4.20).

ADDENDUM 4.6. In the situation of Proposition 4.4, let $\tau \in \mathit{Diff}(S, \partial S)$ be the (positive, or right-handed) Dehn twist along $\{\frac{1}{2}\} \times S^1$. The action of τ^* on $\mathcal{A}_{flat}(S, a_0, a_1)$ corresponds to subtracting r on the right hand side of (4.19).

We omit the proofs, which are straightforward. The interesting point about Addendum 4.6 is that, even though the topology of the space of flat connections is always the same, the action of the diffeomorphism group depends on the rotation number.

Let's consider more general surfaces S. A classical result concerns the case of extremal rotation numbers, by which we mean

$$\sum_i \mathrm{rot}(a_i) = \pm\chi(S). \tag{4.23}$$

THEOREM 4.7 (Goldman). *Suppose that $\chi(S) < 0$, and that the a_i satisfy (4.23). Then, the representation of $\pi_1(S)$ associated to any $A \in \mathcal{M}_{flat}(S, \{a_i\})$ is (conjugate to) the holonomy representation of a hyperbolic metric on S with geodesic boundaries of length $l_i = \cosh^{-1}\left(\frac{1}{2}|\mathrm{tr}(g_i)|\right)$.*

The holonomy formulation assumes that we have represented our hyperbolic surface S as a quotient of the hyperbolic disc. The sign in (4.23) depends on whether this representation agrees with the given orientation of S or not. Here are two useful implications:

COROLLARY 4.8. *For a flat connection as in Theorem 4.7, the holonomies along non-contractible closed curves is hyperbolic.*

COROLLARY 4.9. *In the situation of Theorem 4.7,*

$$\mathcal{M}_{flat}(S, \{a_i\}) \cong \mathbb{R}^{-3\chi(S)-|\pi_0(\partial S)|}.$$

REFERENCES. Theorem 4.7 is a version of [**10**, Theorem 3.4]. The results from Teichmüller theory needed for Corollary 4.9 can be proved using Fenchel-Nielsen coordinates, see e.g. [**1**] (or for another proof, [**17**]). Even though we will not need that, it is worth while pointing out that Higgs bundle methods can be used to determine the topology of other components of the representation space (as long as the Euler number is nonzero). That idea was pioneered by Hitchin [**11**] for closed surfaces, and has been generalized to surfaces with boundary by Mondello [**22**].

For us, what's important is the following immediate consequence of Corollary 4.9 and (4.15):

PROPOSITION 4.10. *In the situation of Theorem 4.7, there is a weak homotopy equivalence*

$$\mathcal{A}_{flat}(S, \{a_i\}) \simeq H^1(S, \partial S), \tag{4.24}$$

unique up to addition of a constant, and which is compatible with the action of (4.15).

The statement is along the same lines as our previous Proposition 4.4 for the cylinder. In parallel with Addendum 4.5, one can sometimes use rotation numbers along paths which connect different boundary components to reduce the ambiguity in the choice of the map (4.24). This relies on the following.

LEMMA 4.11. *In the situation of Theorem 4.7, choose two distinct boundary components, and assume that, when we identify them in an orientation-reversing way, the boundary conditions agree. More precisely, denoting our boundary components by "left" and "right", what we want is:*

$$\begin{cases} \epsilon_{left} : S^1 \xrightarrow{\cong} \partial_{left} S & \textit{orientation-reversing}, \\ \epsilon_{right} : S^1 \xrightarrow{\cong} \partial_{right} S & \textit{orientation-preserving}. \end{cases} \tag{4.25}$$
$$\textit{such that} \quad \epsilon_{left}^* a_{left} = \epsilon_{right}^* a_{right} \in \Omega^1(S^1, \mathfrak{g}).$$

Then, parallel transport along a path

$$c : [0,1] \longrightarrow S, \quad c(0) = \epsilon_{left}(0), \quad c(1) = \epsilon_{right}(0), \tag{4.26}$$

associates to $A \in \mathcal{A}_{flat}(S, \{a_i\})$ *a hyperbolic element of* G.

PROOF. Glue together $\partial_{left} S$ and $\partial_{right} S$ to obtain a surface of genus one higher than S. Given any $A \in \mathcal{A}_{flat}(S, \{a_i\})$, that surface inherits a flat connection, to which Theorem 4.7 applies. The holonomy of the glued connection around any noncontractible loop must be hyperbolic, by Corollary 4.8; in particular, this holds for the loop obtained by identifying the endpoints of c. □

There is a variation of that idea, which involves the following terminology. For any nonzero integer μ, the μ-fold cover a^μ of some $a \in \Omega^1(S^1, \mathfrak{g})$ is defined to be the pullback by the standard μ-fold self-map of the circle:

$$\text{if } a = a_t\, dt, \quad \text{then } a^\mu = \mu a_{\mu t}\, dt. \tag{4.27}$$

LEMMA 4.12. *Suppose that* S *has at least three boundary components, and that* (4.23) *holds. Choose parametrizations of two boundary components, as before, but now assume that*

$$\begin{cases} \epsilon_{left}^* a_{left} = a^{\mu_{left}}, \\ \epsilon_{right}^* a_{right} = a^{\mu_{right}}, \end{cases} \quad \textit{for some } a \in \Omega^1(S^1, \mathfrak{g}) \textit{ and } \mu_{left}, \mu_{right} > 0. \tag{4.28}$$

Then, parallel transport along any path (4.26) *is again hyperbolic.*

PROOF. Let $\mu = \mu_{left}\mu_{right}$. There is a $\mathbb{Z}/\mu$-cover $\tilde{S} \to S$, such that the preimage of $\partial_{left} S$ has μ_{left} connected components, and similarly for $\partial_{right} S$ (here, we are using the existence of a third boundary component). Choose one component in each preimage, denoting them by $\partial_{left}\tilde{S}$ and $\partial_{right}\tilde{S}$. The pullback of our boundary conditions to those components can be identified with a^μ, in both cases. Hence, if we have $A \in \mathcal{A}_{flat}(S, \{a_i\})$, we can pull it back to $\tilde{S}$, and it will then induce a connection on surface obtained by gluing together $\partial_{left}\tilde{S}$ and $\partial_{right}\tilde{S}$. One can choose the components and gluing map so that a lift of our path c becomes a loop on the glued surface. With this in mind, the same argument as in Lemma 4.11 goes through. □

Returning to our discussion of (4.24), these Lemmas see the following use:

ADDENDUM 4.13. In the situation of Proposition 4.10, suppose that additionally, Lemma 4.11 or Lemma 4.12 applies. Fix a path (4.26). Then, one can choose the map (4.24) so that its composition with $\int_c : H^1(S, \partial S) \to \mathbb{Z}$ equals $\mathrm{rot}_c : \mathcal{A}_{flat}(S, \{a_i\}) \to \mathbb{Z}$.

We now turn to the action of diffeomorphisms $\phi \in \mathit{Diff}(S, \partial S)$. Under (4.24), the pullback action ϕ^* on $\mathcal{A}_{flat}(S, \{a_i\})$ corresponds to an affine automorphism, whose linear part is the usual action of ϕ on $H^1(S, \partial S)$. In particular, if ϕ acts

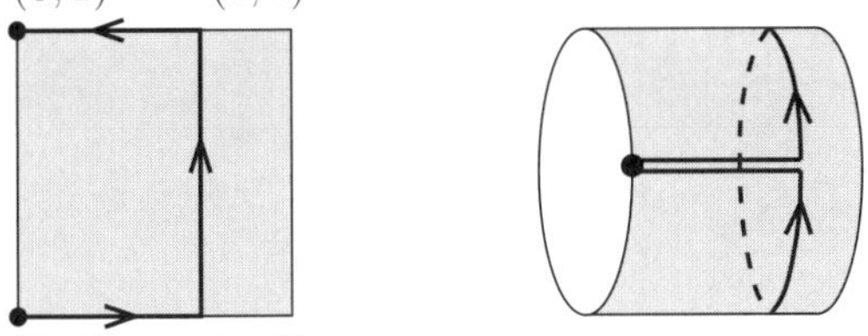

FIGURE 4.1. The paths c_s on the square and the cylinder.

trivially on $H^1(S, \partial S)$, its pullback action on $\mathcal{A}_{flat}(S, \{a_i\})$ can be described up to homotopy as a translation by a certain element

$$T(\phi) \in H^1(S, \partial S). \tag{4.29}$$

We will need the following computation, which can be reduced to Addendum 4.6 by looking at a tubular neighbourhood of the twisting curve.

ADDENDUM 4.14. In the situation of Proposition 4.10, let τ_d be the Dehn twist along a nontrivial simple closed curve d which separates the surface into two parts S_{left} and S_{right}; more precisely, we choose an orientation of d, and take $S_{right} \subset S$ to be the part for which that orientation agrees with the boundary orientation. Then, the associated class (4.29) is

$$\begin{aligned} T(\tau_d) &= \Big(\pm \chi(S_{right}) - \sum_{\partial_i S \subset \partial S_{right}} \mathrm{rot}(a_i) \Big) [d] \\ &= \Big(\mp \chi(S_{left}) + \sum_{\partial_i S \subset \partial S_{left}} \mathrm{rot}(a_i) \Big) [d]. \end{aligned} \tag{4.30}$$

The coefficient in front of $[d]$ in (4.30) is the rotation number of our connections along d, which is prescribed by (4.23) (the sign there determines those in our formula) and the Milnor-Wood inequality for S_{left}, S_{right}.

4.3. Spaces of nonnegatively curved connections. We begin by going back to the basic relationship between curvature and the variation of parallel transport.

LEMMA 4.15. *Given a nonnegatively curved connection on $[0,1]^2$, let $g(s) \in G$ be obtained by parallel transport along the family of paths c_s from Figure 4.1. Then $g(s)$ is a nonpositive path. Conversely, fix a nonpositive path $g(s)$, and a connection $a_0 \in \Omega^1([0,1], \mathfrak{g})$ whose parallel transport along $[0,1]$ yields $g(0)$. Then there is a nonnegatively curved connection A on $[0,1]^2$, with $A|(\{0\} \times [0,1]) = a_0$, such that parallel transport along each c_s recovers $g(s)$. Moreover, the space of all such A is weakly contractible.*

PROOF. We can work up to gauge transformations which are trivial on $\{0\} \times [0,1]$ (the group of such gauge transformations is contractible). Hence, we may restrict our attention to connections which are trivial in s-direction, meaning of the form $A(s,t)dt$; the curvature is then just given by $-\partial_s A$. Let's write $\Pi(s,t) \in G$

for the parallel transport along the straight line from $(s, 0)$ to (s, t). As in (4.4), this is given by

$$\Pi(s, 0) = \mathbb{1}, \quad (\partial_t \Pi)\Pi^{-1} = A. \tag{4.31}$$

Set

$$\Gamma(s, t) = \partial_t(\Pi^{-1}\,\partial_s \Pi) = \Pi^{-1}(\partial_s A)\Pi \in \mathfrak{g}. \tag{4.32}$$

Integrating out yields

$$\Pi^{-1}(\partial_s \Pi) = \int_0^t \Gamma(s, \tau)\,d\tau. \tag{4.33}$$

If the curvature is nonnegative, Γ is nonpositive, which by (4.33) implies that $g(s) = \Pi(s, 1)$ is a nonpositive path.

In the other direction, given a nonpositive path, set $\gamma = g^{-1}\,\partial_s g \in \mathfrak{g}_{\leq 0}$. For A to have nonnegative curvature and the desired parallel transport maps, we need

$$\Gamma(s, t) \in \mathfrak{g}_{\leq 0} \quad \text{and} \quad \int_0^1 \Gamma(s, t)\,dt = \gamma(s). \tag{4.34}$$

The space of Γ satisfying (4.34) is clearly contractible. Given Γ and the parallel transport maps $\Pi(0, t)$ coming from the given a_0, one gets maps $\Pi(s, t)$ by thinking of (4.33) in reverse. This means that we set $\Xi(s, t) = \int_0^t \Gamma(s, \tau)d\tau$, and then define $\Pi(s, t)$ as the solution of the ODE

$$\partial_s \Pi = \Pi\,\Xi. \tag{4.35}$$

Once one has the family $\Pi(s, t)$, there is a unique connection satisfying (4.31), which then automatically restricts to a_0 on $\{0\} \times S^1$. □

LEMMA 4.16. *Fix a nonpositive path $g(s)$, and a connection $a_0 \in \Omega^1(S^1, \mathfrak{g})$ whose holonomy is $g(0)$. Then there is a nonnegatively curved connection A on $S = [0, 1] \times S^1$, with $A|(\{0\} \times S^1) = a_0$, such that the holonomy along each c_s (see again Figure 4.1) is $g(s)$. Moreover, the space of all such A is weakly contractible.*

PROOF. We follow the strategy from Lemma 4.15, but pull everything back to the universal cover $[0, 1] \times \mathbb{R} \to S$, and work $\mathbb{Z}$-periodically in t-direction. Given $\gamma = g^{-1}\partial_s g$, consider $\Gamma : [0, 1] \times \mathbb{R} \to \mathfrak{g}$ which satisfy (4.34) as well as the periodicity condition

$$\Gamma(s, t + 1) = g(s)^{-1}\Gamma(s, t)g(s). \tag{4.36}$$

This means that for a solution of (4.33),

$$\Pi(s, t + 1)^{-1}\partial_s \Pi(s, t + 1) = \gamma(s) + g(s)^{-1}(\Pi(s, t)^{-1}\partial_s \Pi(s, t))g(s) \tag{4.37}$$

$$= (\Pi(s, t)g(s))^{-1}\partial_s(\Pi(s, t)g(s)). \tag{4.38}$$

We start with $a_0 \in \Omega^1(\mathbb{R}, \mathfrak{g})$ which is $\mathbb{Z}$-periodic, and whose parallel transport along $[0, 1]$ is $g(0)$. This means that $\Pi(0, t+1) = \Pi(0, t)g(0)$. Because of (4.37), we then have $\Pi(s, t + 1) = \Pi(s, t)g(s)$ everywhere. As a consequence, the connection determined by $\Pi(s, t)$ through (4.31) is again $\mathbb{Z}$-periodic. □

With that in hand, let us return to the spaces of nonnegatively curved connections from (4.14). We remind the reader of the standing assumption for such spaces, which is that the boundary conditions a_i are connections with hyperbolic holonomy.

PROPOSITION 4.17. *Take $S = S^1 \times [0,1]$. As boundary conditions, fix $a_0, a_1 \in \Omega^1(S^1, \mathfrak{g})$ with the same rotation number $\mathrm{rot}(a_0) = \mathrm{rot}(a_1) = r$. Then there is a weak homotopy equivalence*

$$\mathcal{A}_{\geq 0}(S, a_0, a_1) \simeq \mathbb{Z}, \tag{4.39}$$

unique up to adding a constant on the right, and which is compatible with the action of $\mathcal{G}(a_0, a_1) \simeq \mathbb{Z}$. More explicitly, by associating to a connection A its holonomies around the loops $d_s : S^1 \to S$, $d_s(t) = (s,t)$, one gets a weak homotopy equivalence

$$\mathcal{A}_{\geq 0}(S, a_0, a_1) \longrightarrow \mathcal{P}(G_{hyp}, g_0, g_1). \tag{4.40}$$

PROOF. Recall our general notation (4.13): C_1 is the conjugacy class of the holonomy g_1 of A_1, and $\tilde{C}_1$ its lift to $\tilde{G}$. Consider the space $\mathcal{A}_{\geq 0}(S, a_0, \tilde{C}_1)$ of nonnegatively curved connections which restrict to a_0 on $\{0\} \times S^1$, and whose lifted holonomy around $\{1\} \times S^1$ lies in $\tilde{C}_1$ (or equivalently, the holonomy lies in C_1 and $\mathrm{rot}_{\partial_1 S}(A) = r$). By Lemma 4.16, there is a weak homotopy equivalence between that and a suitable space of nonpositive paths:

$$\mathcal{A}_{\geq 0}(S, a_0, \tilde{C}_1) \simeq \mathcal{P}_{\leq 0}(G, g_0, C_1)^0. \tag{4.41}$$

Lemma 3.9 (applied to the reversed path) shows that the path space in (4.41) is contractible. On the other hand, we have a weak fibration

$$\mathcal{A}_{\geq 0}(S, a_0, a_1) \longrightarrow \mathcal{A}_{\geq 0}(S, a_0, \tilde{C}_1) \longrightarrow C_1. \tag{4.42}$$

The last space in (4.42) should really be the space of connections on S^1 with (lifted) holonomy in $\tilde{C}_1$. By looking at parallel transport, one sees that this is weakly homotopy equivalent to the path space $\mathcal{P}(G, id, C_1)^r$. In turn, evaluation at the endpoint yields a weak homotopy equivalence between that path space and C_1, which is the form in which (4.42) has been stated. It follows that $\mathcal{A}_{\geq 0}(S, a_0, a_1) \simeq \Omega C_1 \simeq \mathbb{Z}$. For any $A \in \mathcal{A}_{\geq 0}(S, a_0, a_1)$, the holonomy around $\{s\} \times S^1$ is hyperbolic. This follows from Lemma 4.16 and the behaviour of nonnegative paths, as described in Lemma 3.9. Hence, (4.40) makes sense. The proof that it is a homotopy equivalence involves looking at gauge transformations in the $[0,1]$-variable, as in Proposition 4.4. More precisely, if PG is the space of all paths in G starting $\mathbb{1}$, the gauge action on a fixed connection yields a commutative diagram extending (4.42),

$$\begin{array}{ccccc} \mathcal{A}_{\geq 0}(S, a_0, a_1) & \longrightarrow & \mathcal{A}_{\geq 0}(S, a_0, \tilde{C}_1) & \longrightarrow & C_1 \\ \uparrow & & \uparrow & & \uparrow \\ \Omega G & \longrightarrow & PG & \longrightarrow & G \end{array} \tag{4.43}$$

Since the middle and right hand $\uparrow$ are weak homotopy equivalences, so is the left hand one, and that implies the desired statement concerning (4.40). □

ADDENDUM 4.18. In the situation from Proposition 4.17, assume additionally that the boundary holonomies belong to the same one-parameter semigroup, meaning that they are of the form

$$g_i = e^{t_i \gamma} \quad \text{with } t_i > 0 \text{ and a common } \gamma. \tag{4.44}$$

Then there is a preferred choice of isomorphism $\pi_0(\mathcal{P}(G_{hyp}, g_0, g_1)) \cong \mathbb{Z}$, by looking at how the eigenvectors rotate; hence, the same holds for (4.39).

ADDENDUM 4.19. In the situation of Proposition 4.17, let $\tau \in \mathit{Diff}(S, \partial S)$ be the Dehn twist along $\{\frac{1}{2}\} \times S^1$. The action of τ^* on $\mathcal{A}_{\geq 0}(S, a_0, a_1)$ corresponds to subtracting r on the right hand side of (4.39).

The first of these Addenda is obvious. The second one follows from the description (4.40) (alternatively, one can deform a_1 so that (4.18) holds, and reduce the computation to Addendum 4.6; this works because the inclusion of the flat connections into the nonnegatively curved ones is a weak homotopy equivalence, by Propositions 4.4 and 4.17).

PROPOSITION 4.20. *Consider the situation from Proposition* 4.17, *except that now,* $\mathrm{rot}(a_0) = \mathrm{rot}(a_1) + 1$. *Then* $\mathcal{A}_{\geq 0}(S, a_0, a_1)$ *is weakly contractible.*

PROOF. The proof is similar to that of Proposition 4.17. By Lemmas 3.13 and 4.16,

$$\mathcal{A}_{\geq 0}(S, a_0, \tilde{C}_1) \simeq \mathcal{P}_{\leq 0}(G, g_0, C_1)^{-1} \simeq C_1. \tag{4.45}$$

One has the same fibration (4.42), and this time it follows that the fibre is weakly contractible. □

PROPOSITION 4.21. *Take S to be a disc, with the boundary parametrized clockwise. As boundary condition, fix $a \in \Omega^1(S^1, \mathfrak{g})$ with* $\mathrm{rot}(a) = 1$. *Then* $\mathcal{A}_{\geq 0}(S, a)$ *is weakly contractible.*

PROOF. Once more, consider first the space $\mathcal{A}_{\geq 0}(S, \tilde{C})$ where we only prescribe the (lifted) conjugacy class of the boundary holonomy. Fix $* \in S \setminus \partial S$, and a family of maps $\phi_t : S \to S$ with $\phi_0 = id$, $\phi_t(*) = *$, $\phi_t|\partial S = id$, $\det(D\phi_t) \geq 0$ everywhere; and such that ϕ_1 contracts a neighbourhood of $*$ to that point. By pulling back via those maps, one sees that $\mathcal{A}_{\geq 0}(S, \tilde{C})$ is weakly homotopy equivalent to the subspace of connections which are trivial near the chosen point. Using this and Lemma 4.16, one shows that the holonomy around concentric loops yields a weak homotopy equivalence

$$\mathcal{A}_{\geq 0}(S, \tilde{C}) \simeq \mathcal{P}_{\leq 0}(G, id, C^{-1})^{-1} = \mathcal{P}_{\geq 0}(G, id, C)^1. \tag{4.46}$$

By combining this with Lemma 3.12, one finds that the holonomy around ∂S yields a weak homotopy equivalence $\mathcal{A}_{\geq 0}(S, \tilde{C}) \simeq C$. Our space $\mathcal{A}_{\geq 0}(S, a)$ is homotopically the fibre of that map. □

PROPOSITION 4.22. *Let S be a genus zero surface with $m+1 \geq 3$ boundary components, labeled by $I = \{0, \dots, m\}$. We choose identifications $\partial_i S \cong S^1$ which are orientation-reversing for $i = 0$, and orientation-preserving for $i > 0$. As boundary conditions, take $a_i \in \Omega^1(S^1, \mathfrak{g})$ which satisfy*

$$\mathrm{rot}(a_0) = \mathrm{rot}(a_1) + \cdots + \mathrm{rot}(a_m) + 1 - m. \tag{4.47}$$

Then there is a weak homotopy equivalence, unique up to adding a constant, and compatible with the action of (4.15),

$$\mathcal{A}_{\geq 0}(S, \{a_i\}) \simeq H^1(S, \partial S). \tag{4.48}$$

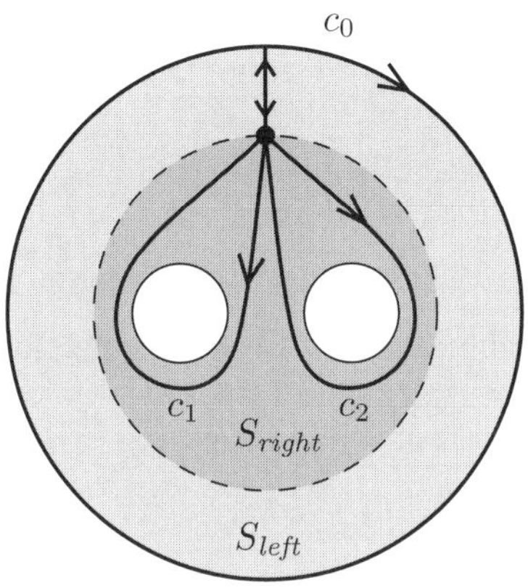

FIGURE 4.2. The loops c_i from the proof of Proposition 4.22, for $m = 2$.

PROOF. Let's start with the space $\mathcal{A}_{\geq 0}(S, \{\tilde{C}_i\})$ where only the conjugacy classes of the (lifted) boundary holonomies are fixed. As in (4.42), this fits into a weak fibration

$$\mathcal{A}_{\geq 0}(S, \{a_i\}) \longrightarrow \mathcal{A}_{\geq 0}(S, \{\tilde{C}_i\}) \longrightarrow \prod_{i=0}^{m} C_i. \tag{4.49}$$

Let's think of S as being glued together, rather trivially, from two parts: S_{left} is a collar neighborhood of $\partial_0 S$, so that the other part $S_{right} \subset S$ is a deformation retract, containing $\partial_i S$ for $i > 0$ (see Figure 4.2). One can find a family of self-maps (ϕ_t) of S, starting at the identity, which are all the identity on ∂S, such that $\det(D\phi_t) \geq 0$ everywhere, and where ϕ_1 retracts S_{right} to a one-dimensional skeleton. From that, it follows (as in the proof of Proposition 4.21) that $\mathcal{A}_{\geq 0}(S, \{\tilde{C}_i\})$ is weakly homotopy equivalent to the subspace of those connections which are flat on S_{right}.

The space of flat connections on S_{right}, with holonomy around $\partial_i S$ ($i > 0$) lying in $\tilde{C}_i$, is weakly homotopy equivalent to $\prod_{i=1}^{m} C_i$, where we can think of the homotopy equivalence as being given by holonomies around loops c_i with a common base point (see again Figure 4.2). In a slight break with our usual notation, we will use g_i to denote those holonomies, and $\tilde{g}_i$ for their lifts to $\tilde{C}_i$. Extending such a connection on S_{right} to a nonnegatively curved one on the whole of S, with the desired behaviour over $\partial_0 S$, amounts (up to homotopy, thanks to Lemma 4.16) to choosing a nonnegative path from $\tilde{g}_1 \cdots \tilde{g}_m$ to some point $\tilde{g}_0 \in \tilde{C}_0$ (the holonomy around c_0). To conclude, we have now shown that $\mathcal{A}_{\geq 0}(S, \{\tilde{C}_i\})$ is weakly homotopy equivalent to the space considered in Lemma 3.10, hence to C_0. By looking at (4.49), one sees that $\mathcal{A}_{\geq 0}(S, \{a_i\}) \simeq \prod_{i>0} \Omega C_i = \mathbb{Z}^m$. Inspection of the argument shows that each gauge orbit yields a weak homotopy equivalence $\mathcal{G}(S, \{a_i\}) \to \mathcal{A}_{\geq 0}(S, \{a_i\})$, and from that we get the statement in its desired form. □

We have arranged the results above roughly in order of complexity of their proofs, which was convenient since they all follow the same strategy. However, this might not be the best way of understanding their meaning. As the reader may have noticed, Proposition 4.17 was a close cousin of the previous discussion concerning flat connections, in Proposition 4.4; similarly, one can compare Propositions 4.10 (specialized to genus zero) and 4.22. The outcome is that we have weak homotopy equivalences

(4.50)

$$\mathcal{A}_{flat}(S, a_0, a_1) \xrightarrow{\simeq} \mathcal{A}_{\geq 0}(S, a_0, a_1) \quad \text{for Proposition 4.4 (as mentioned before),}$$
$$\mathcal{A}_{flat}(S, \{a_i\}) \xrightarrow{\simeq} \mathcal{A}_{\geq 0}(S, \{a_i\}) \quad \text{for Proposition 4.22.}$$

The other two spaces of nonnegatively curved connections that we have analyzed show a different behaviour: they are contractible, which one could interpret as a comparison with the spaces of all connections,

$$(4.51) \quad \begin{aligned} &\mathcal{A}_{\geq 0}(S, a_0, a_1) \xrightarrow{\simeq} \mathcal{A}(S, a_0, a_1) && \text{for Proposition 4.20,} \\ &\mathcal{A}_{\geq 0}(S, a) \xrightarrow{\simeq} \mathcal{A}(S, a) && \text{for Proposition 4.21.} \end{aligned}$$

This reflects our previous distinction between spaces of "short paths" (whose topology could be related to constant paths) and "long paths" (related to spaces of all paths).

REMARK 4.23. It is tempting to think that a comparison result such as (4.50) should hold for higher genus surfaces, as long as the rotation numbers are chosen extremal, meaning that they saturate the bound in Proposition 4.1; and similarly for closed surfaces, with rotation numbers replaced by the Euler class.

5. TQFT considerations

We now introduce our toy model formalism, and explore its implications. In fact, we will go over similar terrain three times: first in a way that only covers TQFT operations in a standard sense; then (in Section 6) with additional features that lead to the construction of connections; and finally, in a mild generalization which allows elliptic holonomies as well (Section 7). Overall, the discussion roughly follows [**31**, Sections 5–6].

5.1. Disc configurations. Our formalism is built as a version of the classical framed little disc operad [**8, 25**], enriched with connections (but not in the gauge theory sense: we keep a fixed trivialization of the bundle).

- Write $D \subset \mathbb{C}$ for the closed unit disc. An m-disc configuration, for $m \geq 0$, is an ordered collection of (closed round) discs $D_1, \dots, D_m$, which are contained in the interior of D, and are pairwise disjoint. To such a configuration, we can associate the (genus zero oriented) surface

$$(5.1) \qquad S = D \setminus \bigcup_{i=1}^{m} (D_i \setminus \partial D_i),$$

  whose boundary circles we label by $\partial_0 S = \partial D$, and $\partial_i S = \partial D_i$ for $i = 1, \dots, m$. We also want to allow one exceptional case, called the identity configuration, which consists of a single disc $D_1 = D$. For that configuration, (5.1) becomes a circle, thought of as an annulus of thickness zero.
- A framing of an m-disc configuration consists of boundary parametrizations

$$(5.2) \qquad \epsilon_i : S^1 \longrightarrow \partial_i S, \quad \epsilon_i(t) = \rho_i e^{-2\pi i t} + \zeta_i, \quad i = 1, \dots, m.$$

Here, ζ_i is the center of D_i, and $|\rho_i| > 0$ is its radius, so only $\arg(\rho_i)$ is a free parameter; for the identity configuration, we specify that the parameter must be $\rho_1 = 1$. To the (5.2) we add a fixed parametrization of the boundary of D,

$$\epsilon_0 : S^1 \longrightarrow \partial_0 S, \quad \epsilon_0(t) = e^{-2\pi i t}. \tag{5.3}$$

The clockwise parametrization convention means that ϵ_0 is inverse to the boundary orientation, while the other ϵ_i are compatible with it.

- Suppose that we have a framed disc configuration, together with $a_i \in \Omega^1(S^1, \mathfrak{g})$ for $i = 0, \dots, m$, which have hyperbolic holonomy (as before, we will write g_i for the holonomies, and C_i for their conjugacy classes). A compatible connection is given by an $A \in \Omega^1(S, \mathfrak{g})$ with nonnegative curvature, and whose pullback by (5.2), (5.3) equals a_i. We also impose the following technical condition: for every boundary circle $\partial_i S$, let r_i be the radial retraction of a neighbourhood onto that circle; then, $A = r_i^*(A|\partial_i S)$ near $\partial_i S$. This means that the behaviour of A close to ∂S is entirely determined by the a_i. Again, we have to mention the exceptional case of the identity configuration, where necessarily $a_0 = a_1$, and there is no additional freedom in choosing A.

A decorated disc configuration is a disc configuration together with a choice of framing and compatible connection. We write $Y = (\epsilon_1, \dots, \epsilon_m, A)$ for the decoration, and usually represent the decorated disc configuration by the pair (S, Y).

There is a gluing process for decorated disc configurations. Start with an m_1-configuration and an m_2-configuration. Choose some $i \in \{1, \dots, m_1\}$, and insert the second configuration into the first one in place of the i-th disc, rescaling and rotating it as prescribed by i-th boundary parametrization. In terms of the associated surfaces, this means that we glue them together along one of their boundary circles, using the given parametrizations. Assuming that the boundary conditions for the connections agree, those will give rise to a connection on the glued surface. The outcome is a decorated configuration consisting of $m = m_1 + m_2 - 1$ discs. The gluing construction is strictly associative. One can formally include the identity configuration in the gluing process, for which (as the name suggests) it is the neutral element.

We will use families of disc configurations, parametrized by smooth manifolds (possibly with boundary or corners) P, and equipped with the same kind of decorations. Aside from one exception, this notion is straightforward:

- The disc configurations vary smoothly depending on the parameter, giving rise to a fibration with fibres (5.1), which we denote by $\underline{S} \to P$.
- Similarly, the framings vary smoothly with P. The connection is given by a one-form in fibre direction on $\underline{S}$. The boundary conditions $(a_0, \dots, a_m)$ are always constant (independent of where we are on P). We usually denote such a decoration by $\underline{Y} = (\underline{\epsilon}_1, \dots, \underline{\epsilon}_m, \underline{A})$.

The exceptional case occurs when we have a family of annuli with the same boundary conditions ($m = 1$ and $a_0 = a_1$). That could include instances of the identity configuration, which means that $\underline{S}$ will be singular in general. This is potentially a problem for the definition of connection on the fibres: to avoid that, we assume that in a neighbourhood of the subset of P where the identity configuration occurs, the framing should be trivial (no rotation), and the connection should be pulled back from radial projection to S^1. The gluing construction extends to families (one starts with a family over P_1 and another over P_2, and ends up with one over $P_1 \times P_2$).

5.2. Spaces of decorations. Fix an m-disc configuration, which is not the identity configuration, and boundary values $(a_0, \dots, a_m)$. Write

$$\begin{aligned} &\mathcal{Y}_{\geq 0}(S, \{a_i\}) \text{ space of all decorations } Y = (\epsilon_1, \dots, \epsilon_m, A) \text{ of } S, \\ &\mathcal{Y}_{\mathit{flat}}(S, \{a_i\}) \text{ subspace where the connection is flat.} \end{aligned} \tag{5.4}$$

With respect to the previously considered spaces $\mathcal{A}_{\geq 0}(S, \{a_i\})$ and $\mathcal{A}_{\mathit{flat}}(S, \{a_i\})$, the only difference (apart from certain conventions concerning boundary orientations) is the additional datum given by the framings. Nevertheless, we find it convenient to explicitly translate the outcome of the discussion from Section 4 into the language of (5.4), since that will be relevant for our applications. The most obvious case is that of the empty disc configuration ($m = 0$), where there is no framing, and Proposition 4.21 says the following:

PROPOSITION 5.1. *Let S be a disc, with* $\mathrm{rot}(a_0) = 1$. *Then $\mathcal{Y}_{\geq 0}(S, a_0)$ is contractible.*

Next, consider the case of the annulus ($m = 1$). Let's introduce the following abelian topological group (depending on $r \in \mathbb{Z}$):

$$\Gamma_r = \begin{cases} \mathbb{Z}/r & r \neq 0, \\ \mathbb{Z} \times S^1 & r = 0. \end{cases} \tag{5.5}$$

PROPOSITION 5.2. (i) *Let S be an annulus, with boundary conditions which satisfy* (4.18). *Then there is a weak homotopy equivalence*

$$\mathcal{Y}_{\mathit{flat}}(S, a_0, a_1) \simeq \Gamma_r \tag{5.6}$$

which is unique up to (homotopy and) adding an element of $\mathbb{Z}/r$. To state this more precisely, choose an identification $[0,1] \times S^1 \cong S$ which restricts to ϵ_i along $\{i\} \times S^1$. Then, the holonomy around the circles $\{s\} \times S^1$ gives a map

$$\mathcal{Y}_{\mathit{flat}}(S, a_0, a_1) \longrightarrow \pi_0(\mathcal{P}(C, g_0, g_1)) \times_{\mathbb{Z}} \mathbb{Z}/r, \tag{5.7}$$

where C is the common conjugacy class of the g_i, and $\mathbb{Z} = \pi_1(C)$ acts by composition of paths with loops. The map (5.7) is canonical (independent of the choice of identification), and yields the discrete part of (5.6); for $r = 0$, $\arg(\rho_1)$ yields the continuous part.

(ii) *Take boundary conditions a_0, a_1 which have the same rotation number r. Then there is a homotopy equivalence*

$$\mathcal{Y}_{\geq 0}(S, a_0, a_1) \simeq \Gamma_r. \tag{5.8}$$

This has the same uniqueness property as (5.6), and can be described as in (5.7), except that the map now takes values in $\pi_0(\mathcal{P}(G_{hyp}, g_0, g_1)) \times_{\mathbb{Z}} \mathbb{Z}/r$. In the more restricted case (4.18), the inclusion $\mathcal{Y}_{\mathit{flat}}(S, a_0, a_1) \subset \mathcal{Y}_{\geq 0}(S, a_0, a_1)$ is a weak homotopy equivalence, and our construction reduces to that in (i).

(iii) *Take boundary conditions a_0, a_1 such that* $\mathrm{rot}(a_0) = \mathrm{rot}(a_1) + 1$. *Then* $\arg(\rho_1)$ *yields a weak homotopy equivalence*

$$\mathcal{Y}_{\geq 0}(S, a_0, a_1) \simeq S^1. \tag{5.9}$$

The proofs are straightforward, given our previous discussion. Each of them is based on considering the fibration

$$\mathcal{A}_{flat}(S, a_0, a_1) \longrightarrow \mathcal{Y}_{flat}(S, a_0, a_1) \xrightarrow{\arg(\rho_1)} S^1, \tag{5.10}$$

or its counterpart for $\mathcal{Y}_{\geq 0}(S, a_0, a_1)$. For (i), Proposition 4.4 tells us that the fibre is homotopy equivalent to $\mathbb{Z}$, and Addendum 4.6 shows that going around the S^1 base corresponds to subtracting r. The same goes for (ii) and (iii), using Proposition 4.17 with Addendum 4.19, respectively Proposition 4.20.

ADDENDUM 5.3. (i) In the situation of Proposition 5.2(i), suppose that $g_0 = g_1 = g$. Then, following Addendum 4.5, there is a preferred choice of (5.6). One can see that from (5.7); or equivalently, it can be described as the map

$$(\epsilon_1, A) \longmapsto \mathrm{rot}_c(A) \in \mathbb{Z}/r, \tag{5.11}$$

where c is any path from $\epsilon_0(0)$ to $\epsilon_1(0)$ (the rotation number mod r is independent of the choice of path).

(ii) In the situation of Proposition 5.2(ii), suppose that g_0 and g_1 lie in the same one-parameter semigroup. Then, there is again a preferred choice of (5.8), see Addendum 4.18 (but it is unclear whether the alternative description (5.11) would still work).

EXAMPLE 5.4. Let $r \neq 0$. Consider the configuration consisting of a disc centered at zero, with parametrization $\epsilon_1(t) = \frac{1}{2}e^{-2\pi i(t+k/r)}$, for some $k \in \mathbb{Z}/r$. Suppose that $a \in \Omega^1(S^1, \mathfrak{g})$ is an r-fold cover, in the sense of (4.27). Radial projection from the disc to the circle equips S with a flat connection A, whose boundary values are $a_0 = a_1 = a$. By looking at (5.11), one sees that this belongs to the k-th component of $\mathcal{Y}_{flat}(S, a, a)$.

PROPOSITION 5.5. (i) *Consider a configuration of $m \geq 2$ discs, with boundary conditions such that* (4.47) *holds, and write $r_i = \mathrm{rot}(a_i)$. Then, there is a weak homotopy equivalence*

$$\mathcal{Y}_{flat}(S, \{a_i\}) \simeq \prod_{i=1}^{m} \Gamma_{r_i}, \tag{5.12}$$

unique up to adding an element of $\prod_{i=1}^m \mathbb{Z}/r_i$.

(ii) *In the same situation, the inclusion $\mathcal{Y}_{flat}(S, \{a_i\}) \subset \mathcal{Y}_{\geq 0}(S, \{a_i\})$ is a weak homotopy equivalence. Hence, $\mathcal{Y}_{\geq 0}(S, \{a_i\})$ admits the same description* (5.12).

To prove this (let's say, for flat connections), one considers the fibration

$$\mathcal{A}_{flat}(S, \{a_i\}) \longrightarrow \mathcal{Y}_{flat}(S, \{a_i\}) \xrightarrow{(\arg(\rho_1), \dots, \arg(\rho_m))} (S^1)^m. \tag{5.13}$$

By Proposition 4.10, the fibre is homotopy equivalent to $H^1(S, \partial S)$, which we identify with $\mathbb{Z}^m$ by taking generators dual to paths c_i $(1 \leq i \leq m)$ from $\partial_0 S$ to $\partial_i S$. The monodromy of (5.13) around the i-th S^1 factor agrees with the pullback action $(\tau_{d_i}^{-1})^*$ of an inverse (or left-handed) Dehn twist along a loop d_i which is a parallel copy of $\partial_i S$. Let's orient this loop opposite to the boundary orientation of $\partial_i S$; it then satisfies $d_i \cdot c_i = -1$, hence its Poincaré dual is minus the i-th unit element in our basis. Addendum 4.14 tells us that the action of $\tau_{d_i}^{-1}$ corresponds to adding the element

$$-T(\tau_{d_i}) = -r_i[d_i] = (0, \dots, r_i, \dots, 0) \in H^1(S, \partial S) \cong \mathbb{Z}^m, \tag{5.14}$$

which implies (i). For (ii), one compares Propositions 4.10 and 4.22, as already mentioned in (4.50).

ADDENDUM 5.6. In the situation of Proposition 5.5(i), suppose that $a_0 = a_m = a$ has rotation number r, and that $\mathrm{rot}(a_i) = 1$ for $i = 1, \dots, m-1$. In analogy with Addendum 5.3(i), an explicit choice of (the discrete part of) (5.12) is then given by

$$(\epsilon_1, \dots, \epsilon_m, A) \longmapsto \mathrm{rot}_c(A) \in \mathbb{Z}/r, \tag{5.15}$$

where c is any path from $\epsilon_0(0)$ to $\epsilon_m(0)$.

Most of the necessary argument is carried over from Addendum 4.13. What remains to be shown is that (5.15) is independent of the choice of path, or equivalently, is invariant under $\mathit{Diff}(S, \partial S)$. To see that, note that $\pi_0(\mathit{Diff}(S, \partial S))$ is generated by Dehn twists along (necessarily separating) curves d. If d does not separate $\partial_0 S$ and $\partial_1 S$, its intersection number with c must be zero, which in terms of Addendum 4.14 means that $T(\tau_d) \cdot [c] = 0$. In the remaining case, $d \cdot c = \pm 1$ and $T(\tau_d)$ is a multiple of $r[d]$, so using Addendum 4.14 again, we get $T(\tau_d) \cdot [c] \in r\mathbb{Z}$. In either case, (5.15) is unchanged under τ_d^*.

One can deepen this discussion by allowing the disc configuration to vary. Let Conf_m (for some $m \geq 2$) be the space of all m-disc configurations (this is homotopy equivalent to the standard ordered configuration space). Over it, we have a universal family of surfaces $\underline{S}$; and on the fibres of that family, we have a space of all possible choices of decorations (let's say, flat ones), denoted by $\mathcal{Y}_{\mathit{flat}}(\underline{S}, \{a_i\})$. This all fits into a diagram, of which the columns and rows are fibrations:

$$\begin{array}{ccccc} \mathbb{Z}^m & \longrightarrow & \mathit{point} & \longrightarrow & (S^1)^m \\ {\scriptstyle \prod_i r_i}\big\downarrow & & \big\downarrow & & \big\downarrow{\scriptstyle \text{inclusion}} \\ \mathbb{Z}^m & \longrightarrow & \mathcal{Y}_{\mathit{flat}}(\underline{S}, \{a_i\}) & \longrightarrow & (S^1)^m \times \mathit{Conf}_m \\ \big\downarrow & & \big\| & & \big\downarrow{\scriptstyle \text{projection}} \\ \prod_{i=1}^m \Gamma_{r_i} & \longrightarrow & \mathcal{Y}_{\mathit{flat}}(\underline{S}, \{a_i\}) & \longrightarrow & \mathit{Conf}_m \end{array} \tag{5.16}$$

Here, $(S^1)^m \times \mathit{Conf}_m$ is the space of framed configurations. The middle row of (5.16) restricts to (5.13) on each configuration, and the description of the map $\mathbb{Z}^m \to \mathbb{Z}^m$ reflects our previous discussion of (5.14). Here is one important additional piece of information concerning the middle row:

ADDENDUM 5.7. In the situation of Proposition 5.5, consider a loop in Conf_m formed by moving the i-th and j-th disc anticlockwise around each other, for some $1 \leq i < j \leq m$. Embed that into $(S^1)^m \times \mathit{Conf}_m$ by taking the $(S^1)^m$ component to be constant. Then, the monodromy of $\mathcal{Y}_{\mathit{flat}}(\underline{S}, \{a_i\}) \to (S^1)^m \times \mathit{Conf}_m$ along that loop is given by adding the element

$$(0, \dots, \overbrace{1 - \mathrm{rot}(a_j)}^{i\text{-th position}}, \dots, \overbrace{1 - \mathrm{rot}(a_i)}^{j\text{-th position}}, \dots, 0) \in \mathbb{Z}^m. \tag{5.17}$$

To see this, take the loop d_{ij} which encloses exactly the i-th and j-th discs. In other words, it divides the surface into pieces S_{left} and S_{right}, of which S_{right} has boundary consisting exactly of d_{ij}, $\partial_i S$ and $\partial_j S$. We orient d_{ij} compatibly with ∂S_{right}, which means that its Poincaré dual in $H^1(S, \partial S)$ is minus the i-th and j-th unit vectors, analogously to (5.14). By Addendum 4.14, the pullback action of $\tau_{d_{ij}}^{-1}$ is given by adding

$$-T(\tau_{d_{ij}}) = (1 - \mathrm{rot}(a_i) - \mathrm{rot}(a_j))[d_{ij}]. \tag{5.18}$$

The monodromy we are looking for (corresponding to moving the discs, but not changing the angles ρ_i or ρ_j) is actually the pullback action of $\tau_{d_{ij}}^{-1}\tau_{d_i}\tau_{d_j}$, where d_i and d_j are as in (5.14). Combining that equation with (5.18) yields the desired result. Of course, by virtue of Proposition 5.5(ii), all this carries over to spaces $\mathcal{Y}_{\geq 0}(\underline{S}, \{a_i\})$.

5.3. The TQFT axioms. Fix a commutative coefficient ring R. We suppose that the following data are given. For every $a \in \Omega^1(S^1, \mathfrak{g})$ with hyperbolic holonomy (our standing assumption), one has a chain complex of R-modules, denoted by $C^*(a)$; its cohomology will be written as $H^*(a)$. For any family of decorated disc configurations whose base is an oriented compact manifold with corners P, and whose boundary values are $a_0, \dots, a_m$, one has a map

$$C^*(a_1) \otimes \cdots \otimes C^*(a_m) \longrightarrow C^{*-\dim(P)}(a_0). \tag{5.19}$$

We will denote (5.19) by ϕ_P, which is concise if somewhat incomplete. The behaviour of these maps is governed by three (groups of) axioms. These are very similar to those in [**31**], hence will be only stated briefly:

- Exchanging the labeling of the m discs permutes the inputs of the associated operation (5.19), with the expected Koszul signs. Reversing the orientation of P flips the sign of (5.19). Taking the disjoint union of two parameter spaces yields the sum of the associated operations. Finally, pulling back a family by a finite covering $\tilde{P} \to P$ of degree μ yields

$$\phi_{\tilde{P}} = \mu \phi_P \tag{5.20}$$

  (the last part, concerning coverings, will be needed only once, in the construction underlying Proposition 5.24).
- The coboundary (defined using the differentials on the chain complexes) of (5.19) equals the sum of the operations associated to the codimension one boundary faces of P. If those boundary faces are written as $\partial_j P$, then the sign convention is that

$$\begin{aligned}(-1)^{\dim(P)} d\phi_P(x_1, \dots, x_m) - \phi_P(dx_1, \dots, x_m) - \cdots \\ \cdots - (-1)^{|x_1|+\cdots+|x_{m-1}|}\phi_P(x_1, \dots, dx_m) + \sum_j \phi_{\partial_j P}(x_1, \dots, x_m) = 0.\end{aligned} \tag{5.21}$$

- Gluing of families corresponds to composition of operations. More precisely, suppose that we have families of m_k-disc configurations, with parameter spaces P_k ($k = 1, 2$), and glue the second one into the i-th place of the first one. The composition axiom says that

$$\begin{aligned}\phi_{P_1 \times P_2}(x_1, \dots, x_{m_1+m_2-1}) = (-1)^{(\dim(P_1)+|x_1|+\cdots+|x_{i-1}|)\dim(P_2)} \\ \phi_{P_1}(x_1, \dots, x_{i-1}, \phi_{P_2}(x_i, \dots, x_{i+m_2-1}), \dots, x_{m_1+m_2-1}).\end{aligned} \tag{5.22}$$

Moreover, the identity configuration induces the identity map $C^*(a) \to C^*(a)$.

We will now explore the consequences of these axioms, remaining on the level of the cohomology groups $H^*(a)$. Take Proposition 5.2(i) in the case $a_0 = a_1 = a$. This implies that $H^*(a)$ carries an action of $H_{-*}(\Gamma_r)$, for $r = \mathrm{rot}(a)$, or in a more concrete formulation:

PROPOSITION 5.8. *$H^*(a)$ carries the following endomorphisms:*

$$\begin{cases} \text{if } r = \mathrm{rot}(a) \neq 0, \text{ an action of } \mathbb{Z}/r; \\ \text{if } r = 0, \text{ a } \mathbb{Z}\text{-action; and additionally, an endomorphism of degree } -1. \end{cases} \tag{5.23}$$

We denote the generator of the $\mathbb{Z}/r$-action by Σ. More generally, Proposition 5.2(ii), together with Addendum 5.3, implies the following:

PROPOSITION 5.9. (i) *Given a_0, a_1 with the same rotation number r, there is an isomorphism*

$$H^*(a_1) \cong H^*(a_0), \tag{5.24}$$

which commutes with the action of (5.23), and is unique up to composition with powers of Σ.

(ii) *If the holonomies of a_0 and a_1 lie in the same one-parameter semigroup, meaning that (4.44) holds, there is a distinguished choice of (5.24), and these choices are compatible with composition.*

The situation for unequal rotation numbers is quite different. Applying Proposition 5.2(iii) yields the following:

PROPOSITION 5.10. *Suppose that $\mathrm{rot}(a_0) = \mathrm{rot}(a_1)+1$. Then, there are canonical maps*

$$C : H^*(a_1) \longrightarrow H^*(a_0), \tag{5.25}$$

$$D : H^*(a_1) \longrightarrow H^{*-1}(a_0). \tag{5.26}$$

The "continuation map" (5.25) is invariant under composition with Σ on the left or right. It is also compatible with (any choice of) the isomorphisms relating different choices of a_0 or a_1, with the same rotation number. The map (5.26) has the same properties.

In a slightly flamboyant formulation, one could say that the continuation maps are more robustly independent of the choice of a than the groups $H^*(a)$ themselves. From Proposition 5.1, we get:

PROPOSITION 5.11. *Suppose that $\mathrm{rot}(a) = 1$. Then there is a distinguished element*

$$e \in H^*(a). \tag{5.27}$$

For different choices of a, these elements are related by (5.24) (which is canonical for this particular rotation number).

We now turn to multilinear operations:

PROPOSITION 5.12. *For $\mathrm{rot}(a) = 1$, $H^*(a)$ carries a canonical structure of a Gerstenhaber algebra, for which (5.27) is the unit.*

PROOF. Take the universal family with boundary conditions $\{a, \dots, a\}$. Consider the $\mathbb{Z}^m$-covering which is the middle row of (5.16). Addendum 5.7 and (5.14) determine this covering, which turns out to be the universal cover of the $(S^1)^m$ factor. This means that the map

$$\mathcal{Y}_{\mathit{flat}}(\underline{S}, a, \dots, a) \longrightarrow (S^1)^m \times \mathit{Conf}_m \tag{5.28}$$

has the following properties: its first component is homotopic to a constant, and its second component is a weak homotopy equivalence. Given any map $P \to \mathit{Conf}_m$, one therefore has a homotopically unique lift to $\mathcal{Y}_{\mathit{flat}}(\underline{S}, a, \dots, a)$. More concretely, a single configuration of two discs ($P = \mathit{point}$) induces the product; two-disc configurations moving anticlockwise around each other ($P = S^1$) induce the bracket; and all the necessary relations between these operations can be implemented by explicit cobordisms, which means that they come from relations in the groups $MSO_d(\mathit{Conf}_m)$ (for $d \leq 2$). We omit the details, since the relation between the little disc operad and Gerstenhaber algebras is classical.

In the construction of the Gerstenhaber structure, we used flat connections, but of course, the same argument goes through with a more general choice of nonnegatively curved ones, thanks to Proposition 5.5(ii). Nonnegatively curved connections appear naturally when discussing unitality, where they arise from the definition of (5.27). The proof of unitality goes as follows: take two copies of the disc which defines (5.27), and glue them into the pair-of-pants which defines the product. After appealing to Proposition 5.1, we conclude that the outcome again defines (5.27). In other words, e is an idempotent. Now take the same pair-of-pants as before, and glue in just one copy of the disc. The outcome is an annulus, whose associated operation, multiplication with e, is necessarily some power of Σ. Because e is idempotent and multiplication is associative, it follows that e is a unit. □

ADDENDUM 5.13. Another application of nonnegatively curved connections is a mild generalization of Proposition 5.12, where one considers "all boundary conditions with rotation number 1 simultaneously". What this means is that for any such a_k, there are maps

$$\cdot : H^*(a_1) \otimes H^*(a_2) \longrightarrow H^*(a_0), \tag{5.29}$$

$$[\cdot, \cdot] : H^*(a_1) \otimes H^*(a_2) \longrightarrow H^{*-1}(a_0). \tag{5.30}$$

These satisfy the Gerstenhaber algebra axioms, for all situations where the compositions make sense. For instance, the associativity equation

$$(x_1 x_2) x_3 = x_1 (x_2 x_3) \in H^*(a_0), \quad \text{where } x_k \in H^*(a_k) \tag{5.31}$$

always holds, where the intermediate terms can be taken to lie in arbitrary groups, $x_1 x_2 \in H^*(a_4)$ and $x_2 x_3 \in H^*(a_5)$. Similarly, inserting a unit element on either the left or right into (5.29) always results in the isomorphism (5.24). From this, it follows that (5.24) is an isomorphism of Gerstenhaber algebras.

Now fix a_1, a_2, with $\mathrm{rot}(a_1) = 1$, and $\mathrm{rot}(a_2) = r$ arbitrary. Consider the universal family of m-disc configurations, and its decorations, where the connections are flat and have boundary values $(a_2, a_1, \dots, a_1, a_2)$. We can use Addendum 5.6 to pick a preferred subspace

$$\mathcal{Y}^0_{\mathit{flat}}(\underline{S}, a_2, a_1, \dots, a_1, a_2) \subset \mathcal{Y}_{\mathit{flat}}(\underline{S}, a_2, a_1, \dots, a_1, a_2), \tag{5.32}$$

namely that on which the map (5.15) vanishes (and in the case where $r = 0$, we require additionally that the angle $\arg(\rho_m)$ should be trivial). In other words, we use (5.15) to provide a trivialization of the fibration which forms the bottom row of (5.16). The resulting homotopy equivalence $\mathcal{Y}^0_{\mathit{flat}}(\underline{S}, a_2, a_1, \dots, a_1, a_2) \rightarrow \mathit{Conf}_m$ gives a distinguished choice of decorations (up to homotopy) for each disc configuration. This choice is compatible with the two gluing processes that will be relevant for us:

- Suppose that we take two configurations, with boundary values $(a_2, a_1, \dots, a_1, a_2)$ and $(a_1, \dots, a_1)$, respectively, and glue the second one into the first one (in one of the possible ways). The rotation number (5.15) of the first configuration remains unchanged under gluing. Hence, this gluing process preserves the subspace $\mathcal{Y}^0_{\mathit{flat}}(\underline{S}, a_2, a_1, \dots, a_1, a_2)$.
- Suppose that we take two decorated disc configurations, both having boundary values $(a_2, a_1, \dots, a_1, a_2)$, and glue them (in the unique possible way). This gluing process adds up the rotation numbers (5.15). Hence, again, if the original configurations lie in $\mathcal{Y}^0_{\mathit{flat}}(\underline{S}, a_2, a_1, \dots, a_1, a_2)$, then so does the glued one.

Let's pass to concrete implications. Using a two-disc configuration, respectively an S^1-family of such configurations, and their lifts to $\mathcal{Y}^0_{\mathit{flat}}(\underline{S}, a_2, a_1, a_2)$, one defines a product and bracket

$$\cdot : H^*(a_1) \otimes H^*(a_2) \longrightarrow H^*(a_2), \tag{5.33}$$

$$[\cdot,\cdot] : H^*(a_1) \otimes H^*(a_2) \longrightarrow H^{*-1}(a_2). \tag{5.34}$$

LEMMA 5.14. *The family of decorated disc configurations underlying* (5.34) *has the following geometry: two discs move around each other; during that rotation, the framing of the first disc rotates* $(1 - r)$ *times, while that of the second one remains constant (up to homotopy).*

PROOF. Consider the middle row of (5.16), specialized to our case:

$$\mathbb{Z}^2 \longrightarrow \mathcal{Y}_{\mathit{flat}}(\underline{S}, a_2, a_1, a_2) \longrightarrow (S^1)^2 \times \mathit{Conf}_2. \tag{5.35}$$

By construction, the holonomy around the two S^1 factors is $(-\mathrm{rot}(a_1), 0)$ and $(0, -\mathrm{rot}(a_2))$, respectively. The operation (5.34) is defined using a loop in $\mathcal{Y}^0_{\mathit{flat}}(\underline{S}, a_2, a_1, a_2)$ which consists of the two discs moving around each other once, and where the framings rotate by some amounts $d_1, d_2 \in \mathbb{Z}$. The image of that loop in $(S^1)^2 \times \mathit{Conf}_2$ must be a loop along which (5.35) has trivial monodromy. On the other hand, by Addendum 5.7, that monodromy is

$$(1 - \mathrm{rot}(a_2), 1 - \mathrm{rot}(a_1)) + d_1(-\mathrm{rot}(a_1), 0) + d_2(0, -\mathrm{rot}(a_2)) = (1 - r - d_1, -rd_2). \tag{5.36}$$

If $r \neq 0$, it follows that

$$d_1 = 1 - r, \quad d_2 = 0. \tag{5.37}$$

If $r = 0$, we know a priori that $d_2 = 0$, by definition of $\mathcal{Y}^0_{\mathit{flat}}(\underline{S}, a_2, a_1, a_2)$, which leads to the same conclusion (5.37). □

PROPOSITION 5.15. *The operations* (5.33) *and* (5.34) *make* $H^*(a_2)$ *into a (unital) Gerstenhaber module over the Gerstenhaber algebra* $H^*(a_1)$. *For* $a_2 = a_1$, *this is the diagonal module.*

PROOF. Let's consider the desired relations (the defining properties of a Gerstenhaber module; see e.g. [**14**, Definition 1.1], which however uses different sign conventions), for $x_1, x_2 \in H^*(a_1)$ and $x_3 \in H^*(a_2)$:

$$(x_1 x_2)x_3 = x_1(x_2 x_3), \tag{5.38}$$

$$[x_1, x_2 x_3] = [x_1, x_2]x_3 + (-1)^{(|x_1|-1)|x_2|} x_2 [x_1, x_3], \tag{5.39}$$

$$[[x_1, x_2], x_3] + (-1)^{|x_1|}[x_1, [x_2, x_3]] + (-1)^{(|x_1|-1)|x_2|}[x_2, [x_1, x_3]] = 0. \tag{5.40}$$

Because the choices of decorations, taken to lie in $\mathcal{Y}^0_{\mathit{flat}}(\underline{S}, a_2, a_1, a_1, a_2)$, are homotopically unique, all we need to show is that the underlying relations hold up to bordism in Conf_3.

The only nontrivial case is (5.40), since there the framings, as analyzed in Lemma 5.14, affect the families obtained from the gluing process. To spell this out in the simplest possible terms, let's temporarily ignore the sizes of discs, and just think of Conf_3 as the ordered three-point configuration space in $\mathbb{C}$. The only thing that depends on r is the first term in (5.40), which (up to homotopy) comes from the family

$$S^1 \times S^1 \longrightarrow \mathit{Conf}_3, \quad (p_1, p_2) \longmapsto (0, \tfrac{1}{2}e^{2\pi i(1-r)p_1 + 2\pi i p_2}, e^{2\pi i p_1}). \tag{5.41}$$

Even there, the r-dependence can be removed by a change of parameters $p_2 \mapsto p_2 + (r-1)p_1$. The remaining argument is therefore the same as in the case $r = 1$, which is standard (and indeed, already entered into Proposition 5.12).

Even though our definition of (5.32) was restricted to flat connections, there is an alternative formulation which works more generally for nonnegatively curved ones. Namely, given a decorated m-disc configuration with boundary values $(a_2, a_1, \dots, a_1, a_2)$, glue in $(m-1)$ discs with boundary value a_1. The outcome is an annulus, with boundary values (a_2, a_2), to which one can associate an element of $\mathbb{Z}/r$ as in Proposition 5.3(ii). The subspace $\mathcal{Y}^0_{\geq 0}(\underline{S}, a_2, a_1, \dots, a_1, a_2)$ is defined by asking that to be zero. One can use this approach to prove unitality (as in Proposition 5.12), and also to prove that the Gerstenhaber module structure is independent of the choice of a_1 and a_2 (as in Addendum 5.13). More precisely, it is compatible with the canonical isomorphisms between different choices of a_1; the isomorphisms between different choices of a_2 are not canonical, but any such choice is a map of Gerstenhaber modules. □

ADDENDUM 5.16. The indirect algebraic argument for unitality (based on first showing $e^2 = e$) used in Propositions 5.12 and 5.15 may strike the reader as a cop-out. We will therefore outline a geometric alternative, which uses the paths constructed in Lemma 3.15. Take the pair-of-pants surface S_1, and equip it with a flat connection whose holonomies around the boundary components, parametrized as in (5.2), are $(g_2(1), g_1(1), g_2(1))$. More precisely, this connection should have boundary values (a_2, a_1, a_2), where $\mathrm{rot}(a_1) = 1$; and additionally, parallel transport along a path from $\epsilon_0(0)$ to $\epsilon_1(0)$ is $k(1)$, whereas parallel transport along a path from $\epsilon_1(0)$ to $\epsilon_2(0)$ is the identity $\mathbb{1}$ (we refer to Figure 5.1, with t set to 1, for the precise picture). The TQFT operation associated to S_2 defines the product $H^*(a_1) \otimes H^*(a_2) \to H^*(a_2)$. Using the nonnegative path $g_1(t)$, one can equip the disc S_2 with a nonnegatively curved connection with boundary value a_1. Gluing together S_1 and S_2 yields the geometric description of the action of (5.27) on $H^*(a_2)$

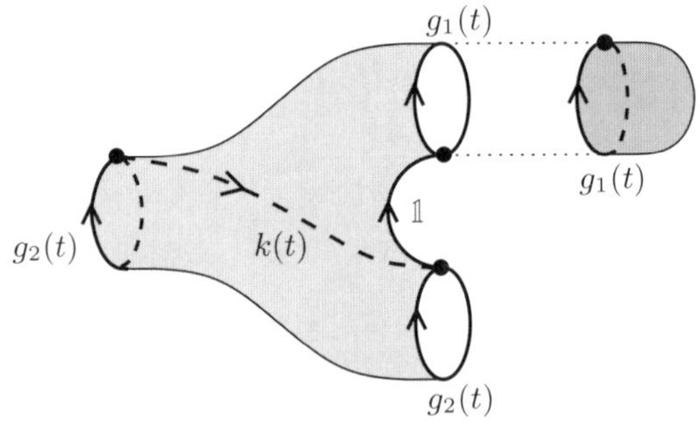

FIGURE 5.1. The geometric version of the proof that e is a unit (Remark 5.16). The flat connection on the pair-of-pants with the given parallel transport maps exists thanks to (3.46).

(to check this, one uses the fact that $k(1)$ has rotation number 0, which is part of Lemma 3.15).

There is a one-parameter family of deformations of this construction, where the boundary holonomies on S_1 are $(g_2(t), g_1(t), g_2(t))$, and the parallel transport maps are $k(t)$ resp. the identity (see again Figure 5.1). Correspondingly, on S_2 one has a connection with boundary holonomy $g_1(t)$. After gluing, one gets a one-parameter family of connections on the cylinder, whose boundary values can still be assumed to be (a_2, a_2) (since the conjugacy class of the holonomies remains the same). For $t = 0$, the outcome is a flat connection on the cylinder, with the additional property that parallel transport along a path joining the two boundary circles is $k(0) = \mathbb{1}$. Hence, the associated operation is the identity map on $H^*(a_2)$.

There are more restrictive notions of Gerstenhaber module than that used above, which are obtained by imposing a relation on $[x_1x_2, x_3]$. The relation that holds in our context depends on the rotation number:

PROPOSITION 5.17. *The Gerstenhaber module structure from Proposition* 5.15 *satisfies*

$$\begin{aligned} &[x_1x_2, x_3] \\ &\quad = (1-r)[x_1, x_2]x_3 + (-1)^{(|x_1|-1)|x_2|}x_2[x_1, x_3] + (-1)^{|x_1|}x_1[x_2, x_3]. \end{aligned} \tag{5.42}$$

PROOF. The idea is the same as for (5.40), with simpler underlying geometry. For $1 \leq i < j \leq 3$, let $\lambda_{ij} \in H_1(\mathit{Conf}_3)$ be the class of a loop where the i-th and j-th point rotate anticlockwise once around each other (while the other point is far away and remains constant). Because of Lemma 5.14, the geometric object underlying $[x_1x_2, x_3]$ is a loop of configurations whose homology class is $(1-r)\lambda_{12} + \lambda_{13} + \lambda_{23}$. The equality (5.42) then amounts to giving each of those three summands a separate geometric representation. □

REMARK 5.18. As an analogue of Proposition 5.17 for $r = 0$, consider the Gerstenhaber algebra $A^* = \Lambda^*(TX)$ of polyvector fields on a manifold X, and the Gerstenhaber module structure of the space $M^* = \Omega^{-*}(X)$ of differential forms.

The first nontrivial instances of this structure (omitting signs) are as follows:
(5.43)

	algebra action on $\theta \in M^*$	Lie action on $\theta \in M^*$
functions $f \in A^0$	$\theta \mapsto f\theta$	$\theta \mapsto df \wedge \theta$
vector fields $Z \in A^1$	$\theta \mapsto i_Z\theta$	$\theta \mapsto L_Z\theta$

Then, (5.42) for $x_1 = f$, $x_2 = Z$, $x_3 = \theta$ (and $r = 0$) becomes

$$L_{fZ}\theta = (Z.f)\theta - i_Z(df \wedge \theta) + f(L_Z\theta), \tag{5.44}$$

which is indeed an easy consequence of the standard Cartan formalism. More generally, suppose that we look at a version of the Gerstenhaber module which consists of differential forms twisted by the $(-r)$-th power of canonical bundle (the bundle of top degree forms). With respect to (5.43), the difference is that the Lie action of $Z \in A^1$ is now written as (with respect to a local volume form η)

$$\theta \otimes \eta^{-r} \longmapsto \Big(L_Z\theta - r\theta\frac{L_Z\eta}{\eta}\Big) \otimes \eta^{-r}. \tag{5.45}$$

This satisfies an instance of (5.42) for the given value of r. If, instead of differential topology, one considers the same formulae in algebraic geometry, this would be (roughly speaking) the B-model mirror of our symplectic geometry constructions.

For the next construction of operations, we switch to a more constrained setup. Fix some a with $\mathrm{rot}(a) = 1$. All our boundary conditions will be covers a^{r_i}, where the degrees $r_i > 0$ satisfy $r_1 + \cdots + r_m - r_0 = m-1$. Moreover, the disc configurations will be such that all discs have centers on the imaginary axis, ordered (say) from bottom to top. We also restrict to the trivial choice of framings, meaning that $\rho_i > 0$. For each $i = 1, \dots, m$, fix a path c_i, going from $\epsilon_0(0) = 1 \in \partial_0 S$ to $\epsilon_i(1)$, and which lies in $S \cap \{\mathrm{re}(z) \geq 0\}$; this assumption fixes the homotopy class of the path (rel endpoints). As explained in Addendum 4.13, we can use such paths to define a map

$$(\mathrm{rot}_{c_1}, \dots, \mathrm{rot}_{c_m}) : \mathcal{A}_{\mathit{flat}}(S, \{a_i\}) \longrightarrow \mathbb{Z}^m \tag{5.46}$$

which, by Proposition 4.10, is a weak homotopy equivalence. For any framed disc configuration in our class, there is a choice of A for which (5.46) vanishes. Such choices are unique up to homotopy, and moreover, they are compatible with gluing discs. This leads to the following:

PROPOSITION 5.19. *Fix some a with* $\mathrm{rot}(a) = 1$. *Then,* $\bigoplus_{s \geq 0} H^*(a^{s+1})$ *has the structure of a bigraded associative algebra, for which* (5.27) *is the unit.*

PROOF. Because of our restrictions, the spaces of choices involved are weakly homotopy equivalent to ordered configuration spaces of points on the real line, which are of course contractible. A single configuration of two discs, with ($r_0 = s_1 + s_2 + 1$, $r_1 = s_1 + 1$, $r_2 = s_2 + 1$), defines the desired product structure. Associativity follows by gluing and considering the resulting three-disc configurations. Since the product $H^*(a) \otimes H^*(a^{s+1}) \rightarrow H^*(a^{s+1})$ agrees with the one previously defined in Proposition 5.15, we already know that e is a left unit; the proof for right unitality is parallel. □

REMARK 5.20. Even though we will not discuss that explicitly, a refined version of our arguments coud be used to construct algebraic structures on the cochain level: for instance, by looking at families parametrized by associahedra, one can get a bigraded A_∞-algebra structure underlying that from Proposition 5.19.

5.4. Multivalued framings. For the final piece of algebraic structure we want to extract from our TQFT, it is convenient to think of disc configurations in a slightly different way.

- When choosing boundary values $a \in \Omega^1(S^1, \mathfrak{g})$, we allow only ones with $r = \mathrm{rot}(a) \geq 1$. Moreover, each such a must be invariant under $1/r$-rotation of the circle. Equivalently, it must be an r-fold cover, in the sense of (4.27) (of some underlying connection, which necessarily has rotation number 1).
- Our disc configurations will come with multivalued framings, which are maps like (5.2), but for which the ρ_i are determined only up to multiplication with r_i-th roots of unity; here, r_i is the rotation number of the associated boundary value a_i.

Given a disc configuration, let $\mathcal{Y}_{multi,\geq 0}(S, \{a_i\})$ be the space of multivalued decorations, and $\mathcal{Y}_{multi,flat}(S, \{a_i\})$ the subspace where the connection is flat. The analogue of Proposition 5.5 (with a similar proof, which we omit) is the following:

PROPOSITION 5.21. *If* (4.47) *holds, both* $\mathcal{Y}_{multi,flat}(S, \{a_i\})$ *and* $\mathcal{Y}_{multi,\geq 0}(S, \{a_i\})$ *are weakly contractible.*

The relevant version of the gluing process goes as follows. Start with two families of configurations with multivalued decorations, parametrized by P_1 and P_2, respectively, and whose associated rotation numbers are $(r_{1,0}, \dots, r_{1,m_1})$ and $(r_{2,0}, \dots, r_{2,m_2})$. Assume additionally that P_2 comes with an (orientation-preserving) action of $\mathbb{Z}/r_{2,0}$, which is such that acting by the generator corresponds to rotating the disc configuration (including multivalued decorations) by $1/r_{2,0}$. Suppose that we have an $1 \leq i \leq m_1$ such that $r_{1,i} = r_{2,0}$, and such that the relevant boundary values agree. Gluing produces a family over a twisted product, which is a fibration

$$P_2 \longrightarrow P_1 \tilde{\times}_i P_2 \longrightarrow P_1. \tag{5.47}$$

More precisely, there is a $\mathbb{Z}/r_{1,i}$-covering $\tilde{P}_{1,i} \to P_1$, which corresponds to choosing lifts of the multivalued framing of the i-th disc to an actual framing. One then takes the quotient

$$\tilde{P}_{1,i} \times P_2 \longrightarrow P_1 \tilde{\times}_i P_2 \stackrel{\text{def}}{=} \tilde{P}_{1,i} \times_{\mathbb{Z}/r_{2,0}} P_2. \tag{5.48}$$

It is straightforward to see that the gluing process, defined a priori on $\tilde{P}_{1,i} \times P_2$, indeed descends to the quotient. Finally, note that if P_1 also carries an action of $\mathbb{Z}/r_{1,0}$ which rotates disc configurations, then the glued family inherits that action.

LEMMA 5.22. *Consider two discs moving around each other, parametrized by* $p \in S^1$. *Using Proposition* 5.21, *we can equip these discs with multivalued decorations, which vary smoothly along our loop, for any rotation numbers* (r_0, r_1, r_2) *such that* $r_0 = r_1 + r_2 - 1$. *The associated multivalued framings are, up to homotopy,*

$$\begin{cases} \epsilon_1(t) = \rho_1 e^{2\pi i(1-r_2)r_1^{-1}p - 2\pi i t} = \rho_1 e^{2\pi i(1-r_0 r_1^{-1})p - 2\pi i t}, \\ \epsilon_2(t) = e^{2\pi i p}\zeta_2 + \rho_2 e^{2\pi i(1-r_1)r_2^{-1}p - 2\pi i t} \rho_2 z = e^{2\pi i p}\zeta_2 + \rho_2 e^{2\pi i(1-r_0 r_2^{-1})p - 2\pi i t}. \end{cases} \tag{5.49}$$

(For the formula to make sense, $|\rho_1|, |\rho_2|$ *need to be sufficiently small, and* $\zeta_2 \neq 0$*). In words, as the discs move, the first one rotates by a fractional amount* $1 - r_0/r_1$*, and the second one by* $1 - r_0/r_2$.

This is the analogue of Lemma 5.14, and is proved in the same way, by appealing to Lemma 5.7; we omit the argument.

ADDENDUM 5.23. We may in fact choose the framings to be exactly given by (5.49). Then, rotating such a configuration by $1/r_0$ yields

$$\begin{cases} e^{2\pi i r_0^{-1}}\epsilon_1(t) = \rho_1 e^{2\pi i r_1^{-1}} e^{2\pi i(1-r_0 r_1^{-1})(p+r_0^{-1})-2\pi i t}, \\ e^{2\pi i r_0^{-1}}\epsilon_2(t) = e^{2\pi t}(p+r_0^{-1})\zeta_2 + \rho_2 e^{2\pi i r_2^{-1}} e^{2\pi i(1-r_0 r_2^{-1})(p+r_0^{-1})-2\pi i t}. \end{cases} \tag{5.50}$$

By definition, the $e^{2\pi i r_k^{-1}}$ factors on the right hand side of (5.50) are irrelevant for multivalued framings. It follows that, after a suitable choice of connections, our family admits a $\mathbb{Z}/r_0$-action of the kind needed in the gluing process, namely $p \mapsto p + r_0^{-1}$.

Let's assume that the coefficient ring of our TQFT satisfies $\mathbb{Q} \subset R$. Consider families of disc configurations with multivalued decorations. Given such a family, with parameter space P, there is a cover $\tilde{P} \to P$ with structure group $\mathbb{Z}/r_1 \times \cdots \times \mathbb{Z}/r_m$, $r_i = \mathrm{rot}(a_i)$, on which the framings become single-valued. We use those covers and our usual TQFT operation (5.19) to define

$$\phi_{multi,P} = r_1^{-1} \cdots r_m^{-1} \phi_{\tilde{P}}. \tag{5.51}$$

The induced cohomology level map

$$H^*(\phi_{multi,P}) : H^*(a_1) \otimes \cdots \otimes H^*(a_m) \longrightarrow H^{*-\dim(P)}(a_0) \tag{5.52}$$

is invariant under the action of $\mathbb{Z}/r_1 \times \cdots \times \mathbb{Z}/r_m$ on its source. If we assume in addition that our family carries an action of $\mathbb{Z}/r_0$ rotating the discs, then (5.52) is invariant under the action of that group on its target. To understand the behaviour of these invariants under gluing, we consider the operation associated to (5.48). If we take the families of decorated disc configurations over $\tilde{P}_1$ and $\tilde{P}_2$ and glue them together, then the outcome descends to $\tilde{P}_1 \times_{\mathbb{Z}/r_{2,0}} \tilde{P}_2$. As a consequence of the axiom (5.20) concerning coverings, and the composition axiom, we get, in abbreviated notation,

$$\phi_{\tilde{P}_1 \times_{\mathbb{Z}/r_{2,0}} \tilde{P}_2} = r_{2,0}^{-1}\phi_{\tilde{P}_1 \times \tilde{P}_2} = r_{2,0}^{-1}\phi_{\tilde{P}_1}\phi_{\tilde{P}_2}, \tag{5.53}$$

and hence

$$\phi_{multi,P_1 \tilde{\times}_i P_2} = \phi_{multi,P_1}\phi_{multi,P_2}. \tag{5.54}$$

Here, "abbreviated notation" means that these composition formulae are shorthand for appropriate analogues of (5.22) (with the same signs). We have been heading for the following application:

PROPOSITION 5.24. *Suppose that the coefficient ring R is a $\mathbb{Q}$-algebra. For each $r > 0$, fix some $a^r \in \Omega^1(S^1, \mathfrak{g})$ with rotation number r, and which is an r-fold cover (not necessarily of the same underlying connection). Then, $\bigoplus_{s \geq 0} H^{*+1}(a^{s+1})^{\mathbb{Z}/(s+1)}$ has the structure of a bigraded Lie algebra.*

PROOF. Consider a loop of two-disc configurations as in Lemma 5.22. One gets a map

$$H^*(\phi_{multi,P}) : H^*(a^{r_1}) \otimes H^*(a^{r_2}) \longrightarrow H^{*-1}(a^{r_0}), \tag{5.55}$$

where $r_0 = r_1 + r_2 - 1$. By definition, this map is $\mathbb{Z}/r_1$-invariant, which means that it remains the same if we compose it with projection (defined by averaging over orbits) to the invariant part of $H^*(a^{r_1})$. In other words, no information is lost if we restrict it to that invariant part. The same applies to $\mathbb{Z}/r_2$ acting on $H^*(a^{r_2})$. Finally, the map takes value in the $\mathbb{Z}/r_0$-invariant part of $H^*(a^{r_0})$. We use it to define our Lie bracket.

Each term in the Jacobi identity corresponds to a family of disc configurations with multivalued decorations, parametrized by T^2. Thanks to Proposition 5.21, we only have to check that these families satisfy the desired relation in the configuration space $Conf_3$. Let's look at the expression $[x_1, [x_2, x_3]]$. Geometrically, this is given by gluing together two copies of the family from Lemma 5.22, with rotation numbers $(r_{1,0} = r_0, r_{1,1} = r_1, r_{1,2} = r_2 + r_3 - 1)$ and $(r_{2,0} = r_2 + r_3 - 1, r_{2,1} = r_2, r_{2,2} = r_3)$ and parameters p_1, p_2. Again using Lemma 5.22, the relevant covering is

$$\tilde{P}_{1,2} = \{(\tilde{p}_1, p_1) \in S^1 \times S^1 \; : \; (r_2 + r_3 - 1)\tilde{p}_1 = (1 - r_1)p_1\}. \tag{5.56}$$

With this at hand, the families over $\tilde{P}_{1,2}$ and P_2 are:

$$\begin{cases} \epsilon_{1,1}(t) = \rho_{1,1} e^{2\pi i(2-r_2+r_3)r_1^{-1}p_1 - 2\pi i t}, \\ \epsilon_{1,2}(t) = e^{2\pi i p_1}\zeta_{1,2} + \rho_{1,2} e^{2\pi i \tilde{p}_1 - 2\pi i t}. \end{cases}$$
$$\begin{cases} \epsilon_{2,1}(t) = \rho_{2,1} e^{2\pi i(1-r_3)r_2^{-1}p_2 - 2\pi i t}, \\ \epsilon_{2,2}(t) = e^{2\pi i p_2}\zeta_{2,2} + \rho_{2,2} e^{2\pi i(1-r_2)r_3^{-1}p_2 - 2\pi i t}. \end{cases} \tag{5.57}$$

The common action of $\mathbb{Z}/(r_2 + r_3 - 1)$ by which we have to divide in (5.47) is generated by $(\tilde{p}_1, p_1, p_2) \mapsto (\tilde{p}_1 + (r_2 + r_3 - 1)^{-1}, p_1, p_2 - (r_2 + r_3 - 1)^{-1})$. Hence, the quotient is parametrized by $(p_1, \tilde{p}_2 = \tilde{p}_1 + p_2)$. The glued family is then of the form

$$\begin{cases} \epsilon_1(t) = \rho_1 e^{2\pi i(2-r_2+r_3)r_1^{-1}p_1 - 2\pi i t}, \\ \epsilon_2(t) = e^{2\pi i p_1}\zeta_2 + \rho_2 e^{2\pi i((1-r_3)\tilde{p}_2 + (1-r_1)p_1)r_2^{-1} - 2\pi i t}, \\ \epsilon_3(t) = e^{2\pi i p_1}\zeta_2 + e^{2\pi i \tilde{p}_2}\zeta_3 + \rho_3 e^{2\pi i((1-r_1)p_2 + (1-r_2)\tilde{p}_2)r_3^{-1} - 2\pi i t}. \end{cases} \tag{5.58}$$

We have not spelled out the relation between the constants appearing here and those in (5.57); the only relevant point is that $|\zeta_3| < |\zeta_2|$, so that the three discs in our configuration (in the given order) form a sun-earth-moon system. In particular, if we ignore framings, and just focus on the position of the celestial bodies, then those are independent of the rotation number. The same holds for the other terms of the Jacobi identity, which are just permuted versions of (5.58); hence, the proof of that identity reduces to a standard argument. □

6. The differentiation axiom

Following the model of [**31**], we now build an additional differentiation operator into the TQFT setup. Because of the intended application, there is also a shift in emphasis, from the cohomology level to the chain level (still remaining fairly ad hoc, rather than giving a description of all the resulting chain level operations).

6.1. The formalism. Assume that our coefficient ring R comes with a derivation $\partial : R \to R$; and that all $C^*(a)$ come with additive graded endomorphisms

$$\begin{aligned} &\nabla\!\!\!\!/ : C^*(a) \longrightarrow C^*(a), \\ &\nabla\!\!\!\!/(fx) = f(\nabla\!\!\!\!/ x) + (\partial f)x \quad \text{for } f \in R,\ x \in C^*(a). \end{aligned} \tag{6.1}$$

Note that we do not ask that $\nabla\!\!\!/$ should commute with the differential. These endomorphisms interact with the geometry as follows:

- The surface (5.1) associated to a decorated disc configuration is allowed to optionally carry a marked point $Z \in S$ (in the exceptional case of the identity configuration, this is a point on the circle.) This additional datum behaves in the obvious way with respect to gluing (where of course only one of the two surfaces involved may carry a marked point). For a family of configurations over P, we similarly allow an optional smooth section $\underline{Z} : P \to \underline{S}$. If our family includes instances of the identity configuration, we ask that in some neighbourhood of the subset of P where that configuration occurs, the section $\underline{Z}$ should take values in the boundary component $\partial_0 \underline{S}$. For the associated TQFT operations, we keep the same axioms as before, treating the additional marked point as formally having degree 2. This means that a family over a compact oriented P with a $\underline{Z}$ section gives rise to an operation

$$\phi_P : C^*(a_1) \otimes \cdots \otimes C^*(a_m) \longrightarrow C^{*-\dim(P)+2}(a_0). \tag{6.2}$$

- Consider the identity configuration, for some a, and turn this into a family over $P = S^1$ by adding the point $Z_p = e^{2\pi i p}$. We denote the associated operation (6.2) by

$$R : C^*(a) \longrightarrow C^{*+1}(a), \tag{6.3}$$

  and require that it should measure the failure of (6.1) to be a chain map:

$$R = \nabla\!\!\!/ d - d\nabla\!\!\!/. \tag{6.4}$$

- Consider any family $\underline{S} \to P$, where none of the fibres is the identity configuration, and which does not come with a marked point (section). We can associate to this a new family, denoted by $\underline{S}_\bullet \to P_\bullet$, as follows. The new parameter space is the total space of the original family, $P_\bullet = \underline{S}$, and $\underline{S}_\bullet$ is the pullback of $\underline{S}$ to that space. This comes with a tautological section $\underline{Z}$. We call this process "adding a marked point in all possible places", and require that the associated operation should be related to the original one by

$$\begin{aligned} &\phi_{P_\bullet}(x_1, \dots, x_m) \\ &\quad = \nabla\!\!\!/ \phi_P(x_1, \dots, x_m) - \phi_P(\nabla\!\!\!/ x_1, \dots, x_m) - \cdots - \phi_P(x_1, \dots, \nabla\!\!\!/ x_m). \end{aligned} \tag{6.5}$$

The second and third part of the requirements above form the "differentiation axiom". We will now explore some immediate consequences. Take the empty configuration ($m = 0$, which means that $S = D$), with the point $Z = 0 \in S$. After equipping this with a nonnegatively curved connection whose boundary value a has rotation number 1 (Proposition 5.1), we get a cocycle

$$k \in C^2(a). \tag{6.6}$$

We call this the Kodaira-Spencer cocycle, and its class in $H^2(a)$ the Kodaira-Spencer class.

LEMMA 6.1. *Fix a_1, a_2 as in Proposition 5.15. Consider a chain level version of the bracket constructed there,*

$$[\cdot,\cdot] : C^*(a_1) \otimes C^*(a_2) \longrightarrow C^{*-1}(a_2). \tag{6.7}$$

Let k be the Kodaira-Spencer cocycle in $C^2(a_1)$. Then there is a chain homotopy

$$\begin{aligned} &\rho : C^*(a_2) \longrightarrow C^*(a_2), \\ &d\rho(x) - \rho(dx) - R(x) + [k, x] = 0. \end{aligned} \tag{6.8}$$

PROOF. Take (6.6) for $a = a_1$, and the family underlying (6.7); glue them together; and then (temporarily) forget the marked point. The outcome is a family, still parametrized by $p \in P = S^1$, in which the surfaces S_p are all annuli, carrying boundary parametrizations and nonnegatively curved connections, with boundary values (a_2, a_2). Proposition 5.2(ii) and Addendum 5.3(ii) say that up to homotopy, our family is classified by a map

$$P \longrightarrow \Gamma_r. \tag{6.9}$$

Assume first that $r \neq 0$, so that Γ_r is discrete. Let's look at a single value of the parameter. Then, the gluing construction precisely describes the module action of (5.27) from Proposition 5.15. In Remark 5.16, we have given a direct geometric proof of the unitality of that action; that argument implies that gluing yields the trivial element of Γ_r, hence that (6.9) is trivial. In the remaining case $r = 0$, what we have just said shows that (6.9) takes values in the trivial connected component of $\Gamma_0 = \mathbb{Z} \times S^1$; one can then use Lemma 5.14 to show that the map $\pi_1(P) \to \pi_1(\Gamma_0)$, which describes the p-dependence of the boundary parametrization of our annuli, is also trivial.

We now know that the glued family can be deformed to a one that is constant in p except for the marked point, which moves once around the annulus. For concreteness, let's give a specific formula for what we can deform our family to (as decorated disc configurations):

$$\begin{cases} D_1 = \frac{1}{4}D, \\ \epsilon_1(t) = \frac{1}{4}e^{-2\pi i t}, \\ A = \text{the pullback of } a_2 \text{ by radial projection}, \\ Z_p = \frac{1}{2}e^{2\pi i p}. \end{cases} \tag{6.10}$$

By deforming the constants $\frac{1}{4}$ and $\frac{1}{2}$ to 1, (6.10) can be connected to the family underlying R. Combining this with the previous deformation yields a family parametrized by an annulus, whose associated TQFT operation ρ precisely satisfies (6.8), as an instance of (5.21). □

In principle, these constructions should be supplemented by suitable uniqueness statements (since we are on the cochain level, this always means uniqueness up to chain homotopy). As a particularly simple instance, (6.6) is homotopically unique, by which we mean the following: if k and $\tilde{k}$ are the cochains arising from two choices of decorations, then

$$\tilde{k} - k = d\kappa; \tag{6.11}$$

and $\kappa \in C^1(a)$, which comes from a family of decorations parametrized by $[0, 1]$, is itself unique up to adding coboundaries of elements in $C^0(a)$ (one could continue this with higher homotopies). This follows immediately from the contractibility statement in Proposition 5.1. Similarly (but with a slightly more complicated argument), ρ is homotopically unique, and there is a relation between the maps ρ obtained from different choices of cochain representative k. Finally, k and ρ are compatible, in a suitable sense, with the quasi-isomorphisms relating the complexes $C^*(a)$ for different choices of a. We omit the details.

6.2. The connection. From now on, our discussion no longer applies to the TQFT formalism in general, but requires a special additional assumption:

ASSUMPTION 6.2. Fix some $a \in \Omega^1(S^1, \mathfrak{g})$ with rotation number 1. Assume that the class of k in $H^2(a)$ is nullhomologous. Given that, we choose a bounding cochain $\theta \in C^1(a)$, $d\theta = k$. Two such cochains will be considered equivalent if their difference is a nullhomologous cocycle.

By adding arbitrary cocycles $t \in C^1(a)$ to θ, the set of equivalence classes becomes an affine space over $H^1(a)$. Note that equivalence classes of θ can be transferred from one representative of k to another: given (6.11) and a choice of θ such that $d\theta = k$, one sets $\tilde{\theta} = \theta + \kappa$. Similarly, one can pass between different a.

PROPOSITION 6.3. *Suppose that Assumption* 6.2 *holds. Then, any group* $H^*(a_2)$ *carries a differentiation operator*

$$\begin{aligned} &\nabla : H^*(a_2) \longrightarrow H^*(a_2), \\ &\nabla(fx) = f\nabla x + (\partial f)x. \end{aligned} \tag{6.12}$$

This operator depends on the equivalence class of the bounding cochain θ*; changing that by some* $t \in H^1(a)$ *has the effect of adding the Lie action* $x \mapsto -[t, x]$ *to* (6.12).

PROOF. The differentiation axiom and Lemma 6.1 show that the Kodaira-Spencer class encodes the essential failure of $\not\nabla$ to be a chain map. Correspondingly, under Assumption 6.2, one can remedy that failure:

$$\begin{aligned} &\nabla : C^*(a_2) \longrightarrow C^*(a_2), \\ &\nabla x = \not\nabla x + \rho(x) - [\theta, x] \end{aligned} \tag{6.13}$$

is a chain map. We define (6.12) to be the induced operation on cohomology. □

There is a uniqueness statement for (6.12), proving the independence of ∇ from all choices made its construction, and its compatibility with the isomorphisms between the groups $H^*(a_2)$ for different choices of a_2. The argument is relatively straighforward, and we omit it. Going further along the lines of [**31**], one could consider the compatibility of ∇ with each of the previously defined algebraic structures; since we have no immediate need or applications for such compatibility results, we prefer not to pursue that line of inquiry here.

7. Elliptic holonomy

We now enlarge our TQFT by additionally allowing, as boundary values, connections with elliptic (which by definition still means nontrivial) holonomy. This allows us to incorporate the parallel theory from [**31**]. There is little new in the strategies of proof; we will therefore focus on a few basic technical statements, and then on the main result, which is the comparison theorem for connections (Proposition 7.13).

7.1. Spaces of nonnegative paths, revisited. We temporarily return to the subject of Section 3. Our first observation concerns the issue of perturbing nonnegative paths to positive ones, while keeping the conjugacy class of the endpoints the same. We have encountered this implicitly before, for instance, in the proof of Lemma 3.11: there, our path ended in the hyperbolic locus, and we used the fact that positive motion on hyperbolic conjugacy classes is unconstrained. Here is an alternative approach, which uses the elliptic locus instead:

LEMMA 7.1. *Let $g(t)$ be a nonconstant nonnegative path, such that $g(1)$ is elliptic. Then there is a deformation $g_s(t)$, with $g_0(t) = g(t)$, such that:*

$$\begin{cases} g_s(0) = g(0), \\ g_s(1) \text{ lies in the same conjugacy class as } g(1), \\ g_s(t) \text{ is a positive path, for all } s > 0. \end{cases} \tag{7.1}$$

The same statement applies to families of paths.

PROOF. This is an adaptation of the argument from Lemma 3.8. Let's first assume, for simplicity, that our path remains entirely inside the elliptic locus, so that $g(t)$ is a rotation with angle $\theta(t) \in (0, \pi)$. Because the path is nonnegative and not constant, we know from (3.19) that $\theta(1) > \theta(0)$. Consider time-dependent functions (3.31). There is such a γ_t which additionally satisfies $g'(t) = g(t)\gamma_t(g(t))$. There is also one more choice $\tilde{\gamma}_t$, obtained from Lemma 3.6(i) by rescaling as in (3.33), which has the additional property that $\tilde{\gamma}_t$ takes values in $\mathfrak{g}_{>0}$ whenever $\theta'(t) > 0$. Take the paths $\tilde{g}_s(t)$ obtained by solving the ODE

$$\tilde{g}_s(0) = g(0), \quad \tilde{g}_s'(t) = \tilde{g}_s(t)\big((1-s)\gamma_t(g_s(t)) + s\tilde{\gamma}_t(g_s(t))\big). \tag{7.2}$$

By construction, these move through the same conjugacy classes as the original path, to which they specialize for $s = 0$. For $s > 0$, the path $\tilde{g}_s$ is positive at all times t where $\theta'(t) > 0$. Now we "redistribute the positivity":

$$\begin{gathered} g_s(t) = \begin{pmatrix} \cos(\alpha_s(t)) & -\sin(\alpha_s(t)) \\ \sin(\alpha_s(t)) & \cos(\alpha_s(t)) \end{pmatrix} \tilde{g}_s(t), \\ \text{where} \quad \begin{cases} \alpha_0(t) = 0, \\ \alpha_s(0) = \alpha_s(1) = 0, \\ \text{for } s > 0,\ \alpha_s'(t) > 0 \text{ at all } t \text{ where } \tilde{g}_s(t) \text{ is not (strictly) positive.} \end{cases} \end{gathered} \tag{7.3}$$

If we choose α_s sufficiently small, (7.3) will be a positive path for all $s > 0$, satisfying (7.1).

In the general case, fix some T such that $g|[T,1]$ is elliptic; one can also arrange that there is a point in $[T,1]$ where the angle of rotation has positive derivative. Now, apply the first part of the argument to $g|[T,1]$, but multiplying $\tilde{\gamma}_t$ by a cutoff function which vanishes on $[0,T]$. The outcome is a family of paths $\tilde{g}_s$ which agree with the original g on $[0,T]$, and which are positive somewhere in $[T,1]$. We then apply the same "redistribution" argument to those paths. The parametrized situation is similar. □

LEMMA 7.2. *Fix elliptic conjugacy classes $C_0, C_1 \subset G$. Let $r_0, r_1 \in \mathbb{R} \setminus \mathbb{Z}$ be rotation numbers of some lifts of those classes to $\tilde{G}$, chosen so that (r_0, r_1) contains exactly one integer, and set $\rho = r_1 - r_0$. Then, $\mathcal{P}_{\geq 0}(G, C_0, C_1)^\rho$ is weakly contractible.*

SKETCH OF PROOF. Lemma 7.1 shows that $\mathcal{P}_{\geq 0}(G, C_0, C_1)^\rho$ is weakly homotopy equivalent to its subspace $\mathcal{P}_{>0}(G, C_0, C_1)^\rho$ of positive paths, so we'll consider that instead. From Figure 3.1, one sees that paths in $\mathcal{P}_{>0}(G, C_0, C_1)^\rho$ can meet the trivial element $\mathbb{1}$ at most once. Let's write

$$\mathcal{P}_{>0}(G, C_0, C_1)^\rho = \mathcal{S}_{open} \cup \mathcal{S}_{id}, \tag{7.4}$$

where $\mathcal{S}_{id}$, the subspace of paths which go through $\mathbb{1}$, is a codimension two submanifold; and $\mathcal{S}_{open}$ is its complement.

Claim (i): *$\mathcal{S}_{id}$ is weakly contractible.*

Except for the time $T \in (0,1)$ where $g(T) = \mathbb{1}$, a path $g(t)$ in $\mathcal{S}_{id}$ must remain in the elliptic locus. To first order near T, the path is determined by $g'(T) \in \mathfrak{g}_{>0}$, which is a contractible choice. On the other hand, Lemma 3.8(i) applies to $g|[0, T-\epsilon]$ and $g|[T+\epsilon, 1]$ for any $\epsilon > 0$, which are therefore determined uniquely up to homotopy by their endpoints $g(T \pm \epsilon)$. Combining these two ideas yields the claimed fact.

Claim (ii): *$\mathcal{S}_{open}$ is weakly homotopy equivalent to a circle.*

Any path $g(t)$ in $\mathcal{S}_{open}$ has the following structure:

$$\text{for some } 0 < T_0 < T_1 < 1, \quad \begin{cases} g(t) \text{ is elliptic for } t < T_0; \\ g(T_0) \text{ is a nonpositive parabolic;} \\ g(t) \text{ is hyperbolic for } T_0 < t < T_1; \\ g(T_1) \text{ is a nonnegative parabolic;} \\ g(t) \text{ is elliptic for } T_1 < t. \end{cases} \tag{7.5}$$

A more precise form of our claim says that, by evaluating such paths at a point in (T_0, T_1) (smoothly depending on the path), we get a weak homotopy equivalence between $\mathcal{S}_{open}$ and the hyperbolic locus G_{hyp}. Indeed, by breaking up the path at such a point, we can reduce our claim to (a fibre product of) two copies of Lemma 3.11, in a version for positive rather than nonnegative paths.

With that settled, we will now show that $\mathcal{P}_{>0}(G, C_0, C_1)^\rho$ is weakly contractible. Take a path in $\mathcal{S}_{id}$, and a two-parameter family of perturbations of that path, which moves it in all directions transverse to $\mathcal{S}_{id}$. As a local model near the identity element, if the original path $g(t)$ is a standard rotation by t, the family could be

$$g_{\delta,\epsilon}(t) = \exp\left(\begin{smallmatrix} \epsilon & \delta - t \\ \delta + t & -\epsilon \end{smallmatrix}\right). \tag{7.6}$$

For (δ, ϵ) nontrivial, $g_{\delta,\epsilon}(0)$ is a hyperbolic element, whose eigenvector for the larger eigenvalue $\exp(\sqrt{\delta^2 + \epsilon^2})$ is

$$\begin{pmatrix} (\sqrt{\delta^2 + \epsilon^2} + \epsilon)^{1/2} \\ \operatorname{sign}(\delta)(\sqrt{\delta^2 + \epsilon^2} - \epsilon)^{1/2} \end{pmatrix}. \tag{7.7}$$

As (δ, ϵ) moves in a small circle around 0, that eigenvector goes once (up to sign, since there's no preferred orientation of the (δ, ϵ) plane) around $\mathbb{R}P^1$. The same holds for general paths and perturbations (by a deformation argument). This shows that the link of the stratum $\mathcal{S}_{id}$ yields a weak equivalence from the circle to $\mathcal{S}_{open}$. Weak contractibility of $\mathcal{P}_{>0}(G, C_0, C_1)^\rho$ follows from that and the previous claims. □

LEMMA 7.3. *Fix conjugacy classes C_0, C_1, C_2 which are, respectively, elliptic, hyperbolic, and elliptic. For the elliptic ones, we require that they correspond to* *rotations with angles* $\theta_0 > \theta_2$ *in* $(0, \pi)$. *Consider* $g_i \in C_i$ *together with a nonnegative path* $h(t)$ *from* $h(0) = g_1 g_2$ *to* $h(1) = g_0$, *such that if we take lifts of our elements to* $\tilde{g}_i \in \tilde{G}$, *compatible with the product and path, the numbers* $\mathrm{rot}(\tilde{g}_0)$ *and* $\mathrm{rot}(\tilde{g}_1) + \mathrm{rot}(\tilde{g}_2)$ *lie in the same component of* $\mathbb{R} \setminus \mathbb{Z}$. *Then, the space of all such* $(g_0, g_1, g_2, h(t))$ *is weakly homotopy equivalent to* C_1, *by projecting to* g_1.

Sketch of proof. We will consider positive paths instead of nonnegative ones. This does not affect the weak homotopy type, by the same argument as in Lemma 3.11, applied to the path $h(t)g_2^{-1}$ (which goes from the hyperbolic element g_1 to $g_0 g_2^{-1}$). Suppose (without loss of generality) that $\mathrm{rot}(\tilde{g}_0) \in (0,1)$, $\mathrm{rot}(\tilde{g}_2) \in (-1, 0)$, which means that $\mathrm{rot}(\tilde{g}_1) = 1$. One can apply the elementary considerations from Example 3.5 to the pair $(\tilde{g}_0, \tilde{g}_2^{-1})$, and that describes all the possible options for $\tilde{g}_0 \tilde{g}_2^{-1}$. In particular, we see that the function $(\tilde{g}_0, \tilde{g}_2) \mapsto \mathrm{tr}(g_0 g_2^{-1})$ is a submersion everywhere except at its maximum value (which is achieved when g_0, g_2 commute). Let's decompose the space $\mathcal{S}$ of all $(g_0, g_1, g_2, h(t))$ into three parts $\mathcal{S}_{ell}$, $\mathcal{S}_{para}$, $\mathcal{S}_{hyp}$, depending on the conjugacy type of $g = g_0 g_2^{-1}$. The two and last subsets are open, and $\mathcal{S}_{para}$ is a hypersurface separating them.

Claim (i): *The projection* $\mathcal{S}_{hyp} \to C_1$ *is a weak homotopy equivalence.*

Given any hyperbolic element g, there is a contractible space of positive paths from C_1 to g along which the rotation number is constant, by (the positive path version of) Lemma 3.9. On the other hand, by Example 3.5, there is an $\mathbb{R}$ worth of choices of (g_0, g_2) such that $g_0 g_2^{-1} = g$. Hence, the map $g_0 g_2^{-1} : \mathcal{S}_{hyp} \to G_{hyp}$ is a weak homotopy equivalence; but it is also homotopic to the map appearing in our claim.

Claim (ii): *The projection* $\mathcal{S}_{para} \to C_1$ *is a weak homotopy equivalence.*

This is similar to the previous case. It requires a statement about positive paths from hyperbolic to nonnegative parabolic elements, but that can be obtained from Lemma 3.9 and a separate consideration of positive paths near the parabolic endpoint, as in Lemma 3.12.

Claim (iii): *The projection* $\mathcal{S}_{ell} \to C_1$ *is a weak homotopy equivalence.*

Let's work with a fixed hyperbolic g_1. Given an elliptic g, the space of positive paths connecting g_1 to g, along which the rotation number increases by less than 1, is weakly contractible, by (the positive path version of) Lemma 3.11. Hence, the homotopy fibre of $\mathcal{S}_{ell} \to C_1$ is weakly homotopy equivalent to the space of (g_0, g_2) such that $g_0 g_2^{-1}$ is elliptic. By the computation in Example 3.5, this space is an open disc.

Putting together the three claims above yields the desired result. □

7.2. Spaces of connections, revisited. As in Section 4, statements about nonnegative paths translate into ones about nonnegatively curved connections. We use the same notation $\mathcal{A}_{\geq 0}(S, \{a_i\})$ as in (4.14), but now allowing boundary values a_i with elliptic as well as hyperbolic holonomy.

Proposition 7.4. *Let* $S = [0,1] \times S^1$. *As boundary conditions, fix* $a_0, a_1 \in \Omega^1(S^1, \mathfrak{g})$ *with elliptic holonomy, such that* $\mathrm{rot}(a_0) \geq \mathrm{rot}(a_1)$, *and where the interval* $(\mathrm{rot}(a_1), \mathrm{rot}(a_0))$ *contains at most one integer. Then* $\mathcal{A}_{\geq 0}(S, a_0, a_1)$ *is contractible.*

PROPOSITION 7.5.

(i) *Let* $S = [0,1] \times S^1$. *As boundary conditions, fix* $a_0 \in \Omega^1(S^1,\mathfrak{g})$ *with hyperbolic holonomy,* $a_1 \in \Omega^1(S^1,\mathfrak{g})$ *with elliptic holonomy, and such that* $\mathrm{rot}(a_1) \in (\mathrm{rot}(a_0)-1, \mathrm{rot}(a_0))$. *Then* $\mathcal{A}_{\geq 0}(S, a_0, a_1)$ *is weakly contractible.*

(ii) *The same is true if* a_0 *has elliptic holonomy,* a_1 *has hyperbolic holonomy, and* $\mathrm{rot}(a_0) \in (\mathrm{rot}(a_1), \mathrm{rot}(a_1)+1)$.

Both statements are proved by the same strategy as Proposition 4.17. The relevant results about path spaces are: for Proposition 7.4, Lemma 3.8(i) or Lemma 7.2 (depending on whether the relevant interval contains an integer or not); and for Proposition 7.5, Lemma 3.11.

PROPOSITION 7.6. *Let* S *be a disc, with the boundary parametrized clockwise. As boundary condition, fix* $a \in \Omega^1(S^1,\mathfrak{g})$ *with elliptic holonomy and* $\mathrm{rot}(a) \in (0,1) \cup (1,2)$. *Then* $\mathcal{A}_{\geq 0}(S,a)$ *is weakly contractible.*

One can reduce Proposition 7.6 to Lemma 3.8(i) or Lemma 7.2 (depending on which interval $\mathrm{rot}(a)$ lies in), just like Proposition 5.1 was derived (via Lemma 3.12) from Lemma 3.11. Our final result follows from Lemma 7.3, in the same way as in Proposition 4.22:

PROPOSITION 7.7. *Let* S *be a pair-of-pants, with boundary parametrizations as in Proposition* 4.22 *(specialized to* $m = 2$*). As boundary conditions, take* $a_0, a_2 \in \Omega^1(S^1,\mathfrak{g})$ *with elliptic holonomy, and* $a_1 \in \Omega^1(S^1,\mathfrak{g})$ *with hyperbolic holonomy, such that for some* $r \in \mathbb{Z}$,

$$\mathrm{rot}(a_0) \in (\mathrm{rot}(a_2)+1, r+2), \quad \mathrm{rot}(a_1) = 1, \quad \mathrm{rot}(a_2) \in (r, r+1). \tag{7.8}$$

Then $\mathcal{A}_{\geq 0}(S, a_0, a_1, a_2)$ *is weakly contractible.*

7.3. The formal setup. We continue in the same axiomatic framework as in Section 5, except that now, the boundary conditions are allowed to have holonomies that are elliptic or hyperbolic. This gives rise to additional cohomology groups $H^*(a)$, whose basic properties can be summarized as follows:

PROPOSITION 7.8.

(i) *Consider connections* $a \in \Omega^1(S^1,\mathfrak{g})$ *with elliptic holonomy. Up to canonical isomorphism,* $H^*(a)$ *depends only on* $\mathrm{rot}(a) \in \mathbb{R} \setminus \mathbb{Z}$.

(ii) *Each such group carries a degree* -1 *endomorphism, the BV operator* Δ.

(iii) *For* a_0, a_1 *as in Proposition* 7.4, *there is a canonical continuation map* $C : H^*(a_1) \to H^*(a_0)$.

(iv) *If* $\mathrm{rot}(a) \in (0,1) \cup (1,2)$, *there is a canonical element* $e \in H^0(a)$.

Parts (i) and (ii) can be constructed using flat connections on an annulus. The third part, obviously, uses Proposition 7.4; and (iv) follows from Proposition 7.6. Our next topic is the comparison between the elliptic and hyperbolic groups $H^*(a)$.

PROPOSITION 7.9.

(i) *Take* a_0, a_1 *as in Proposition* 7.5*(i). Then there is a canonical continuation map* $C : H^*(a_1) \to H^*(a_0)$. *In the case where* $\mathrm{rot}(a_0) = 1$ *and* $\mathrm{rot}(a_1) \in (0,1)$, *this takes the element* $e \in H^0(a_1)$ *from Proposition* 7.8*(iv) to the previously constructed element* (5.27) *of the same name in* $H^0(a_0)$.

(ii) *For a_0, a_1 as in Proposition 7.5(ii), there is a canonical continuation map as before. If* $\mathrm{rot}(a_0) \in (1,2)$ *and* $\mathrm{rot}(a_1) = 1$, *this takes the element* (5.27) *in* $H^0(a_1)$ *to its counterpart from Proposition 7.8(iv).*

(iii) *Composition of the maps from (i) and (ii), in either order, yields the continuation maps from (5.25), respectively from Proposition 7.8(iii). Hence, in the limit we get an isomorphism*

$$\varinjlim_{r\in\mathbb{Z}} H^*(a^r) \cong \varinjlim_{r\in\mathbb{R}\setminus\mathbb{Z}} H^*(a^r). \tag{7.9}$$

Here, on the left hand side of (7.9), we choose for each $r \in \mathbb{Z}$ a connection a^r with hyperbolic holonomy and rotation number r; the direct limit is independent of all choices up to canonical isomorphism, thanks to the properties of continuation maps (Proposition 5.10). On the right hand side, we do the same for elliptic holonomies.

(iv) *The map (5.26) factors through the BV operator. More precisely, given a_0, a_1, a_2 which, respectively, have hyperbolic, elliptic, and hyperbolic holonomy, and such that* $\mathrm{rot}(a_2) < \mathrm{rot}(a_1) < \mathrm{rot}(a_0) = \mathrm{rot}(a_2) + 1$, *we have a commutative diagram*

$$H^*(a_2) \xrightarrow{C} H^*(a_1) \xrightarrow{\Delta} H^{*-1}(a_1) \xrightarrow{C} H^{*-1}(a_0), \qquad D = C\circ\Delta\circ C : H^*(a_2) \to H^{*-1}(a_0). \tag{7.10}$$

Existence and uniqueneness of these "mixed" continuation maps follows from Proposition 7.5. The statements about the unit elements are consequences of Propositions 4.21 and 7.6. In part (iii), the composition of continuation maps is given geometrically by connections on the annulus to which Propositions 4.20 or 7.4 apply. The same applies to (iv), except that in that case, we are dealing with families of connections in which the parametrization of one of the boundary components rotates.

Let's return for a moment to considering just connections with elliptic holonomy. The simplest way to approach the construction of further operations on the associated groups $H^*(a)$ is to significantly constrain the class of connections involved. Concretely, let's consider only those $a = a^r \in \Omega^1(S^1, \mathfrak{g})$ of the form

$$a^r = r\alpha\, dt, \quad r \in \mathbb{R}\setminus\mathbb{Z}, \quad \alpha = \left(\begin{smallmatrix} 0 & -\pi \\ \pi & 0 \end{smallmatrix}\right) \in \mathfrak{g}_{>0}. \tag{7.11}$$

These obviously satisfy $\mathrm{rot}(a^r) = r$. Similarly, on any surface (5.1), we consider

$$A = \beta \otimes \alpha \in \Omega^1(S, \mathfrak{g}), \quad \text{for some } \beta \in \Omega^1(S) \text{ with } d\beta \leq 0. \tag{7.12}$$

Additionally, the restriction of β to each boundary circle must lie in the class (7.11), which means that it has constant coefficients (with respect to the given parametrization) and integral in $\mathbb{R}\setminus\mathbb{Z}$. The inequality in (7.12) expresses the nonnegative curvature condition (see (4.2) for the sign). The advantage of this framework is that (7.12) is a convex condition, hence the choice of β always lies in a contractible space (this is what one usually encounters in discussions of the operations on symplectic cohomology, see e.g. [**27**, Section 8] or [**23**]). Moreover, the choice of boundary parametrizations (5.2) now becomes independent of that of the connection. By applying this idea to the disc, annulus, and pair-of-pants, one gets operations

$$e \in H^0(a^r), \qquad r > 0, \tag{7.13}$$

$$C : H^*(a^{r_1}) \longrightarrow H^*(a^{r_0}), \qquad r_0 \geq r_1, \tag{7.14}$$

$$\cdot : H^*(a^{r_1}) \otimes H^*(a^{r_2}) \longrightarrow H^*(a^{r_0}), \qquad r_0 \geq r_1 + r_2. \tag{7.15}$$

The product (7.15) is associative and commutative in an appropriate sense, and inserting (7.13) into one of its inputs recovers (7.14). Of course, (7.13) and (7.14) agree with the previously defined structures of the same name, in their domains of definition (here, we are using special choices of nonnegatively curved connections, whereas before we allowed any choice and used more sophisticated methods to prove well-definedness). We will be particularly interested in a further family of operations, the twisted brackets

$$[\cdot,\cdot]_c : H^*(a^{r_1}) \otimes H^*(a^{r_2}) \longrightarrow H^{*-1}(a^{r_0}), \quad r_0 \geq r_1 + r_2, \quad c \in \mathbb{Z}. \tag{7.16}$$

These are obtained from a one-parameter family of two-disc configurations, in which: the two discs move once around each other; the boundary parametrization of the first disc rotates $-c$ times, while that of the second disc remains constant. Concretely, one could take the dependence on the parameter $p \in P = S^1$ to be

$$\begin{cases} \epsilon_1(t) = \zeta_1 e^{2\pi i p} + \rho_1 e^{-2\pi i (t+cp)}, \\ \epsilon_2(t) = \rho_2 e^{-2\pi i t}. \end{cases} \tag{7.17}$$

In the case $c = 0$, we write $[\cdot,\cdot]$, without the superscript. The bracket for general c is related to that special case as in (2.22).

7.4. A family of connections. Into the enlarged framework including elliptic holonomy, we now re-introduce the differentiation axiom, as in Section 6. This yields Kodaira-Spencer classes for connections with both hyperbolic and elliptic holonomy, related to each other in the same way as the unit elements in Proposition 7.9.

LEMMA 7.10. *Fix $r_0, r_1, r_2 \in \mathbb{R} \setminus \mathbb{Z}$, such that $r_1 > 0$ and $r_0 \geq r_1 + r_2$, and take a^{r_i} as in (7.11). Consider the following structures on the associated complexes $C^*(a^{r_i})$:*

$$\begin{cases} \textit{the Kodaira-Spencer cocycle } k \in C^2(a^{r_1}), \\ \textit{the continuation map } C : C^*(a^{r_2}) \to C^*(a^{r_0}), \\ \textit{the bracket } [\cdot,\cdot]_c : C^*(a^{r_1}) \otimes C^*(a^{r_2}) \to C^{*-1}(a^{r_0}), \textit{ for some } c \in \mathbb{Z}, \\ \textit{the map } R : C^*(a^{r_2}) \to C^{*+1}(a^{r_2}). \end{cases} \tag{7.18}$$

Then there is a chain homotopy

$$\begin{aligned} &\rho_c : C^*(a^{r_2}) \longrightarrow C^*(a^{r_0}), \\ &d\rho_c(x) - \rho_c(dx) - C(R(x)) + [k,x]_c \\ &\qquad = d\rho_c(x) - \rho_c(dx) - C\nabla\!\!\!/(dx) + dC\nabla\!\!\!/(x) + [k,x]_c = 0. \end{aligned} \tag{7.19}$$

This is the analogue of Lemma 6.1. We omit the proof, which uses the same geometric construction as before; the details are actually simpler this time, because they just amount to choosing suitable one-forms (7.12).

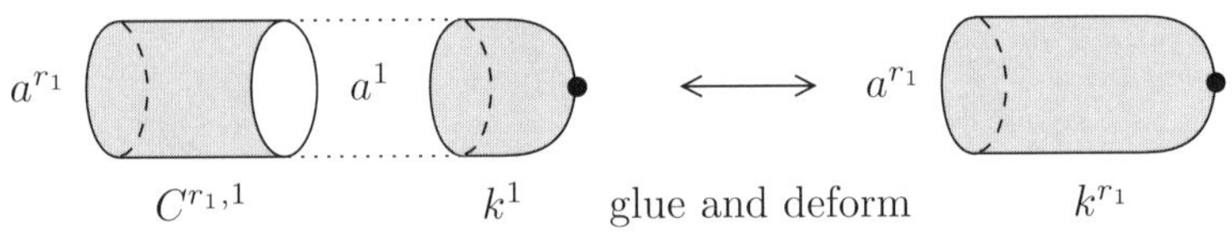

FIGURE 7.1. The family of surfaces underlying (7.25).

ASSUMPTION 7.11. For some $r_1 \in \mathbb{R}^{>0} \setminus \mathbb{N}$, the Kodaira-Spencer cocycle $k \in C^2(a^{r_1})$ is nullhomologous. Given that, we choose a bounding cochain $\theta \in C^1(a^{r_1})$, $d\theta = k$.

PROPOSITION 7.12. *Suppose that Assumption 7.11 holds. Fix $r_0, r_2 \in \mathbb{R} \setminus \mathbb{Z}$ such that $r_0 \geq r_1 + r_2$. Then, there is a family of "connections" with respect to the continuation map $C : H^*(a^{r_2}) \to H^*(a^{r_0})$, indexed by $c \in \mathbb{Z}$. These are maps*

$$(7.20) \qquad \begin{aligned} &\nabla_c : H^*(a^{r_2}) \longrightarrow H^*(a^{r_0}), \\ &\nabla_c(fx) = f\,\nabla_c x + (\partial f)C(x). \end{aligned}$$

This is the counterpart of Proposition 6.3, using the bracket (7.16) and the associated homotopy (7.19): the chain map underlying (7.20) is

$$(7.21) \qquad \nabla_c x = C(\nabla\!\!\!\!/\, x) + \rho_c(x) - [\theta, x]_c.$$

We will not discuss how ∇_c depends on c (but see [**31**, Section 6], which considers the relation between $c = -1$ and $c = 0$; the general case would work in the same way).

PROPOSITION 7.13. *Suppose that Assumption 6.2 holds (and hence, so does Assumption 7.11 for any $r_1 > 1$). Take $r_0, r_2 \in \mathbb{R} \setminus \mathbb{Z}$ such that $r_1 = r_0 - r_2 \in (1, 2)$, and where the interval (r_0, r_2) contains exactly one integer r. Fix some $a \in \Omega^1(S^1, \mathfrak{g})$ whose holonomy is hyperbolic, and which has rotation number r. Then, for appropriately coordinated choices of bounding cochains, (7.20) for $c = r-1$ factors through (6.12), as follows:*

$$(7.22) \qquad H^*(a^{r_2}) \xrightarrow{C} H^*(a) \xrightarrow{\nabla} H^*(a) \xrightarrow{C} H^*(a^{r_0}), \quad \text{with composite } \nabla_c : H^*(a^{r_2}) \to H^*(a^{r_0}).$$

PROOF. Each step in the proof is a fairly straightforward deformation argument, but there are a number of them. Part of the expository challenge is keeping the notation unambiguous, which we will attempt to do by adding rotation numbers as superscripts everywhere. In particular, for the given $r_i \in \mathbb{R} \setminus \mathbb{Z}$ ($i = 0, 1, 2$), we keep the notation a^{r_i} for the connections (with elliptic holonomy) defined in (7.11); but now also write $a^r = a$ for the connection with (hyperbolic holonomy and) rotation number $r \in \mathbb{Z}$, chosen arbitrarily, which appears in (7.22). Additionally, we choose another connection a^1 with (hyperbolic holonomy and) rotation number 1.

As a preliminary step, we need to explain what "appropriately coordinated choices of bounding cochains" means. Two version of the Kodaira-Spencer cocycle are relevant:

$$(7.23) \qquad k^1 \in C^2(a^1) \quad \text{(hyperbolic case)},$$

$$(7.24) \qquad k^{r_1} \in C^2(a^{r_1}) \quad \text{(elliptic case)}.$$

These are related by $C^{r_1,1} : C^*(a^1) \to C^*(a^{r_1})$. More precisely, there is a cochain

$$\begin{aligned} &\kappa^{r_1} \in C^1(a^{r_1}), \\ &k^{r_1} = C^{r_1,1}(k^1) + d\kappa^{r_1}, \end{aligned} \tag{7.25}$$

whose geometric construction is shown schematically in Figure 7.1. It relies on Proposition 7.6, which says that if we glue together the surfaces which define $C^{r_1,1}$ and k^1, the outcome is a surface which (together with its nonnegatively curved connection) can be deformed to that underlying k^{r_1}. We assume that a choice of bounding cochain $\theta^1 \in C^1(a^1)$ for (7.23) is given, and then take that for (7.24) to be

$$\theta^{r_1} = C^{r_1,1}(\theta^1) + \kappa^{r_1} \in C^1(a^{r_1}). \tag{7.26}$$

In our enhanced notation, the definition (7.21) of the "elliptic" connection reads as follows:

$$\nabla_c^{r_0,r_2} x = C^{r_0,r_2}(\not\nabla^{r_2} x) + \rho_c^{r_0,r_2}(x) - [\theta^{r_1}, x]_c^{r_0,r_1,r_2}. \tag{7.27}$$

Our first task is to break up the continuation map in (7.27) into two pieces. As a consequence of Proposition 7.4, the following diagram of continuation maps is homotopy commutative:

$$C^*(a^{r_2}) \xrightarrow{C^{r,r_2}} C^*(a^r) \xrightarrow{C^{r_0,r}} C^*(a^{r_0}), \qquad C^*(a^{r_2}) \xrightarrow{C^{r_0,r_2}} C^*(a^{r_0}). \tag{7.28}$$

Using Proposition 7.4 in the same way, one can modify $\rho_c^{r_0,r_2}$ so that it satisfies a version of (7.19) where the continuation map involved is $C^{r_0,r}C^{r,r_2}$ instead of C^{r_0,r_2}. Denote this new version by $\rho_{broken,c}^{r_0,r_2}$. On the cohomology level, (7.27) can be rewritten as

$$[\nabla_c^{r_0,r_2} x] = \Big[C^{r_0,r}C^{r,r_2}\not\nabla^{r_2} x + \rho_{broken,c}^{r_0,r_2}(x) - [\theta^{r_1}, x]_c^{r_0,r_1,r_2}\Big] \in H^*(a^{r_0}). \tag{7.29}$$

More precisely, let γ^{r_0,r_2} be the chain homotopy which appears in (7.28). The relation between $\rho_c^{r_0,r_2}$ and $\rho_{broken,c}^{r_0,r_2}$ can be encoded schematically as follows:

(7.30)

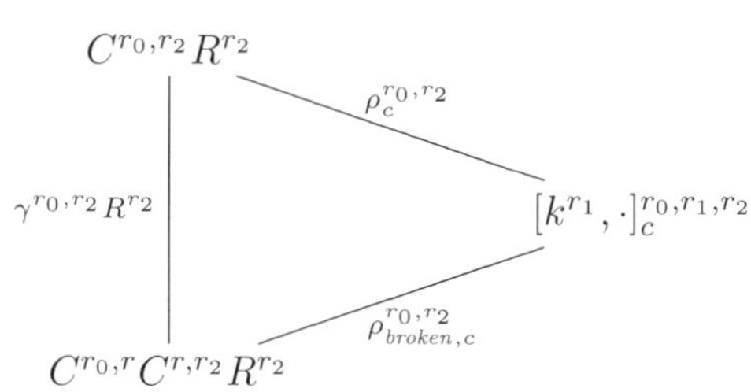

Each side of (7.30) is a homotopy, induced by a family of decorated disc configuration whose parameter space is an annulus $[0,1] \times S^1$ (and over $\{0,1\} \times S^1$, the families can be identified in pairs). There is a cobordism between the union of these three annuli and a family over T^2 (essentially, just rounding off the corners where we have identified the boundary circles of the annuli). One then uses Proposition 7.4 to show that that family can be extended over a solid torus, which provides the (degree -1) operation that "fills in the interior of the triangle" in (7.30). The cochain relating the two sides of (7.21) is given by that operation together with $\gamma^{r_0,r_2}\not\nabla^{r_2}$. We denote the (chain level) right hand side of (7.29) by $\nabla_{broken,c}^{r_0,r_2} x$.

Take the decorated disc configuration underlying C^{r,r_2}, and "add a marked point in all possible places". The result defines an operation $C_{\bullet}^{r,r_2}$, and using (6.5), we can rewrite (7.29) as

$$(7.31)\quad \nabla^{r_0,r_2}_{broken,c} x = C^{r_0,r} \nabla\!\!\!/^{r} C^{r,r_2}(x) + C^{r_0,r} C_{\bullet}^{r,r_2}(x) + \rho^{r_0,r_2}_{broken,c}(x) - [\theta^{r_1}, x]_c^{r_0,r_1,r_2}.$$

Additionally using (7.26), we get

$$(7.32)\quad \begin{aligned} \nabla^{r_0,r_2}_{broken,c} x = {} & C^{r_0,r} \nabla\!\!\!/^{r} C^{r,r_2}(x) + C^{r_0,r} C_{\bullet}^{r,r_2}(x) + \rho^{r_0,r_2}_{broken,c}(x) \\ & - [\kappa^{r_1}, x]_c^{r_0,r_1,r_2} - [C^{r_1,1}(\theta^1), x]_c^{r_0,r_1,r_2}. \end{aligned}$$

We now want to apply a further chain homotopy, which relates two different versions of the Lie bracket:

$$(7.33)\quad \begin{aligned} & \beta_c^{r_0,1,r_2} : C^*(a^1) \otimes C^*(a^{r_2}) \longrightarrow C^{*-2}(a^{r_0}), \\ & [C^{r_1,1}(x_1), x_2]_c^{r_0,r_1,r_2} - C^{r_0,r}[x_1, C^{r_2,r}(x_2)]^{r,1,r} \\ & \quad = d\beta_c^{r_0,1,r_2}(x_1,x_2) + \beta_c^{r_0,1,r_2}(dx_1,x_2) + (-1)^{|x_1|}\beta_c^{r_0,1,r_2}(x_1,dx_2). \end{aligned}$$

This follows by comparing Lemma 5.14 with the definition of (7.16), where the equality $c = r-1$ is key, and then using Proposition 7.7 to construct the family of connections that interpolates between the two sides. From (7.32) we can therefore pass to (the chain homotopic expression)

$$(7.34)\quad \begin{aligned} \nabla^{broken,c}_{r_0,r_2} x \simeq {} & C^{r_0,r} \nabla\!\!\!/^{r} C^{r,r_2}(x) + C^{r_0,r} C_{\bullet}^{r,r_2}(x) + \rho^{r_0,r_2}_{broken,c}(x) \\ & - [\kappa^{r_1}, x]_c^{r_0,r_1,r_2} - C^{r_0,r}[\theta^1, C^{r,r_2}(x)]^{r,1,r} - \beta_c^{r_0,1,r_2}(k^1, x). \end{aligned}$$

Comparing (7.34) with (6.13) shows that the desired statement concerning (7.22) reduces to the following nullhomotopy of maps $C^*(a^{r_2}) \to C^*(a^{r_0})$:

(7.35)

$$C^{r_0,r} C_{\bullet}^{r,r_2}(x) + \rho^{r_0,r_2}_{broken,c}(x) - C^{r_0,r}\rho^{r,r}C^{r,r_2}(x) - [\kappa^{r_1}, x]_c^{r_0,r_1,r_2} - \beta_c^{r_0,1,r_2}(k^1, x) \simeq 0.$$

Importantly, this no longer contains $\nabla\!\!\!/$ or the bounding cochain θ^1: it is purely a relation between TQFT operations. Each term in (7.35) comes from a family parametrized by an annulus, and for the nullhomotopy, the same reasoning as in (7.29) applies. The analogue of (7.30) is this:

(7.36)

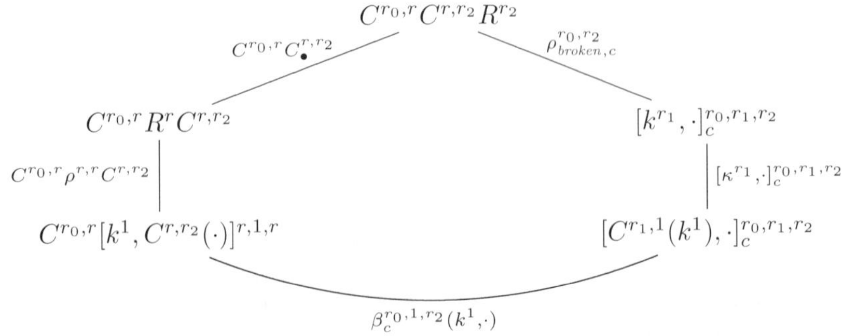

In order to better visualize the situation, we also include a version of (7.36) where the operations are described geometrically, in Figure 7.2. □

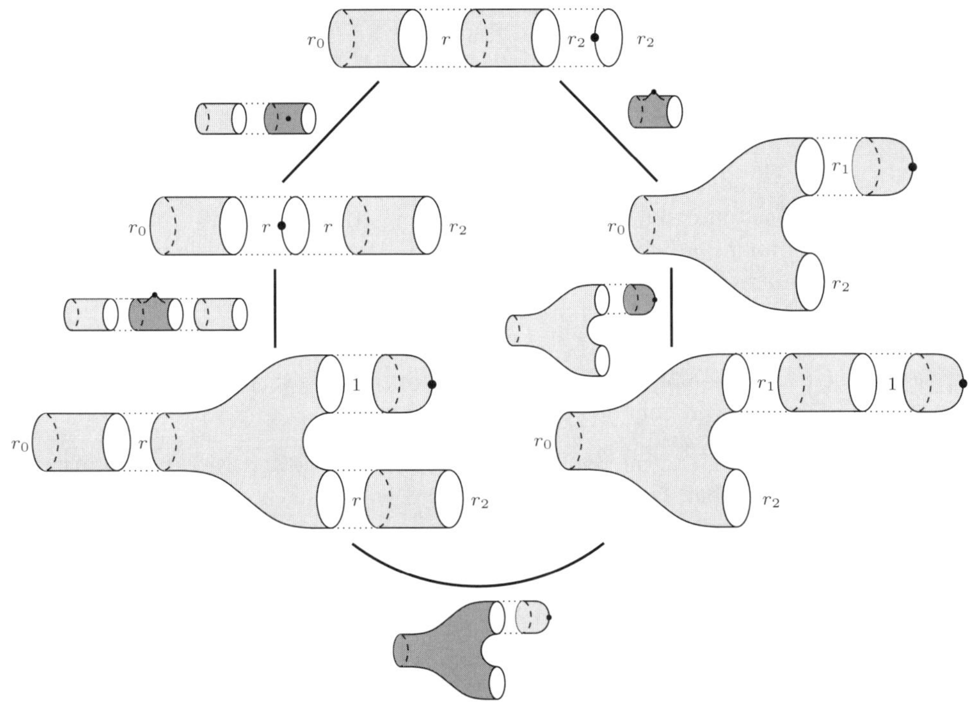

FIGURE 7.2. The families of surfaces appearing in (7.36). The darker shaded pieces are those one which change along the sides of the diagram.

8. Maps to the disc

As a toy model for the subsequent construction of Hamiltonian Floer cohomology, we look at maps from Riemann surfaces to the hyperbolic disc, which satisfy an inhomogeneous Cauchy-Riemann equation. The inhomogeneous term comes from a family of infinitesimal automorphisms, hence is not compactly supported on the disc. We are mostly interested in what happens when the maps approach the boundary circle (at infinity).

8.1. The hyperbolic disc. Let $B \subset \mathbb{C}$ be the open unit disc. The holomorphic automorphism group of B, and its Lie algebra of holomorphic vector fields, are given by:

(8.1)

$$G = PU(1,1) = \left\{ g = \begin{pmatrix} a & b \\ \bar{b} & \bar{a} \end{pmatrix} : |a|^2 - |b|^2 = 1 \right\} / \pm \mathbb{1}, \quad g \mapsto \rho_g(w) = \frac{aw+b}{\bar{b}w + \bar{a}},$$

$$\mathfrak{g} = \mathfrak{u}(1,1) = \left\{ \gamma = \begin{pmatrix} i\alpha & \beta \\ \bar{\beta} & -i\alpha \end{pmatrix} : \alpha \in \mathbb{R},\ \beta \in \mathbb{C} \right\}, \qquad \gamma \mapsto X_\gamma = (\bar{\beta} + 2i\alpha w - \beta w^2)\, \partial_w.$$

The action preserves the hyperbolic metric (in our convention, normalized to have curvature -4)

$$g_{hyp} = \frac{1}{(1-|w|^2)^2}\, dw\, d\bar{w}, \tag{8.2}$$

hence also its symplectic form ω_{hyp}, and distance function dist_{hyp}. There is a unique choice of Hamiltonian function for each vector field X_γ which is linear in γ and compatible with Poisson brackets:

$$H_\gamma = \frac{1}{1-|w|^2}\left(\tfrac{1}{2}(1+|w|^2)\alpha - \mathrm{im}(\beta w)\right). \tag{8.3}$$

The group action extends smoothly to the closed disc $\bar{B}$. Call γ nonnegative if X_γ points in nonnegative direction along $\partial\bar{B}$. In terms of (8.1), the nonnegative cone defined in this way is

$$\mathfrak{g}_{\geq 0} = \{\alpha \geq |\beta|\}. \tag{8.4}$$

In terms of (8.3), γ is nonnegative iff H_γ is nonnegative everywhere (moreover, if $\gamma \notin \mathfrak{g}_{\geq 0}$, H_γ is not even bounded below).

CONVENTION 8.1. As the reader may have noticed, we are switching notation: for the rest of the paper, G and $\mathfrak{g}$ will be as in (8.1). This is not a significant issue, since the two groups that appear under that name are actually isomorphic:

$$PSL_2(\mathbb{R}) \cong PU(1,1), \tag{8.5}$$

and such an isomorphism is canonical up to inner automorphisms (geometrically, it is induced by a holomorphic automorphism of $\mathbb{C}\cup\{\infty\} = \mathbb{C}P^1$ which carries B to the open upper half plane). (8.5) preserves traces; and the associated isomorphism of Lie algebras relates (8.4) to our previous notion of nonnegativity in $\mathfrak{sl}_2(\mathbb{R})$. Hence, the entire discussion from Sections 3–7 carries over immediately to (8.1).

Let S be an oriented surface, equipped with a connection $A \in \Omega^1(S,\mathfrak{g})$. Via the homomorphisms from $\mathfrak{g}$ to the Lie algebras of vector fields and of functions (with the Poisson bracket) on B, this yields one-forms valued in those Lie algebras. We write them as

$$\begin{aligned} &X_A \in C^\infty(S \times B, T^*S \otimes_{\mathbb{R}} \mathbb{C}),\\ &H_A \in C^\infty(S \times B, T^*S). \end{aligned} \tag{8.6}$$

One can apply the same process to the curvature $F_A \in \Omega^2(S,\mathfrak{g})$, which yields

$$\begin{aligned} &X_{F_A} \in C^\infty(S \times B, \Lambda^2_{\mathbb{R}} T^*S \otimes_{\mathbb{R}} \mathbb{C}),\\ &H_{F_A} \in C^\infty(S \times B, \Lambda^2_{\mathbb{R}} T^*S). \end{aligned} \tag{8.7}$$

What X_A describes is the connection induced by A on the trivial fibre bundle $S \times B \rightarrow S$. Like any connection on a fibre bundle, one can also think of this as given by a complementary subbundle to the fibrewise tangent bundle. The choice of such a subbundle determines a lift of ω_{hyp} to a fibrewise symplectic form $\omega_{A,geom} \in \Omega^2(S \times B)$, and this satisfies

$$d\omega_{A,geom} = -dH_{F_A}. \tag{8.8}$$

On the right hand side of (8.8), we think of H_{F_A} as a two-form on $S \times B$ via the projection $T(S \times B) \rightarrow TS$, and then take its exterior derivative. While $\omega_{A,geom}$ is not itself closed, there is an obvious closed replacement (this is an instance of the general theory of symplectic fibrations, see e.g. [**18**, Chapter 6]):

$$\omega_{A,top} = \omega_{A,geom} + H_{F_A}. \tag{8.9}$$

Similarly, the connection determines an almost complex structure J_A on $S \times B$, such that projection to S is J_A-holomorphic, and whose restriction to each fibre is the standard complex structure on B. This satisfies

$$\omega_{A,geom}(\eta, J_A\eta) \geq 0 \quad \text{for all } \eta \in T(S \times B). \tag{8.10}$$

By construction, the induced connection on $S \times B \to S$ extends smoothly to the compactification $S \times \bar{B}$, and so does J_A.

One can be more explicit by working in a local complex coordinate $z = s + it$ on S. If $A = A_1(s,t)ds + A_2(s,t)dt$ with $A_1(s,t)$, $A_2(s,t) \in \mathfrak{g}$, then by definition

$$X_A = X_{A_1} ds + X_{A_2} dt, \tag{8.11}$$

$$H_A = H_{A_1} ds + H_{A_2} dt, \tag{8.12}$$

$$X_{F_A} = \big(X_{\partial_t A_1} - X_{\partial_s A_2} + X_{[A_1,A_2]}\big)\, ds \wedge dt, \tag{8.13}$$

$$H_{F_A} = \big(H_{\partial_t A_1} - H_{\partial_s A_2} + \omega_{hyp}(X_{A_1}, X_{A_2})\big)\, ds \wedge dt. \tag{8.14}$$

The subbundle of $T(S \times B)$ which defines the connection on $S \times B \to S$ is

$$\mathbb{R}(\partial_s + X_{A_1}\partial_w) \oplus \mathbb{R}(\partial_t + X_{A_2}\partial_w). \tag{8.15}$$

The two-form $\omega_{A,geom}$ is characterized by having (8.15) as its nullspace. This means that

$$\omega_{A,geom} = \omega_{hyp} - dH_{A_1} \wedge ds - dH_{A_2} \wedge dt - \omega_{hyp}(X_{A_1}, X_{A_2}) ds \wedge dt, \tag{8.16}$$

$$\omega_{A,top} = \omega_{hyp} - d(H_{A_1} ds) - d(H_{A_2} dt). \tag{8.17}$$

The associated almost complex structure is similarly given by

$$\begin{cases} J_A(\partial_w) = i\partial_w, \\ J_A(\partial_s + X_{A_1}\partial_w) = \partial_t + X_{A_2}\partial_w. \end{cases} \tag{8.18}$$

Suppose that A is nonnegatively curved. This means that, for any oriented basis (ξ_1, ξ_2) in TS, $H_{F_A}(\xi_1, \xi_2)$ is a nonnegative function; and that along $\partial\bar{B}$, $X_{F_A}(\xi_1, \xi_2)$ points in anticlockwise direction. From (8.9) and (8.10), it then follows that

$$0 \leq \omega_{A,geom}(\eta, J_A\eta) \leq \omega_{A,top}(\eta, J_A\eta). \tag{8.19}$$

LEMMA 8.2. *If A has nonnegative curvature, the hypersurface $S \times \partial\bar{B} \subset S \times \bar{B}$ is Levi convex (meaning that its Levi form with respect to J_A is nonnegative).*

PROOF. Take $\psi(z,w) = \frac{1}{2}|w|^2 : S \times \bar{B} \to \mathbb{R}$. In local coordinates on S as before, one gets

$$\begin{aligned} -d(d\psi \circ J_A)|(S \times \partial\bar{B}) = & -\mathrm{im}(X_{A_1} d\bar{w}) \wedge ds - \mathrm{im}(X_{A_2} d\bar{w}) \wedge dt \\ & + \mathrm{im}(\bar{w} X_{\partial_t A_1}) ds \wedge dt - \mathrm{im}(\bar{w} X_{\partial_s A_2}) ds \wedge dt \\ & - \mathrm{im}(\bar{w}\, dX_{A_1}) \wedge ds - \mathrm{im}(\bar{w}\, dX_{A_2}) \wedge dt. \end{aligned} \tag{8.20}$$

The Levi form L_ψ is obtained by restricting (8.20) to the space $T(S \times \partial\bar{B}) \cap J_A T(S \times \partial\bar{B})$, which is (8.15). From (8.20), one gets

$$L_\psi = \mathrm{im}(\bar{w} X_{F_A}). \tag{8.21}$$

Let's spell out the meaning of this. X_{F_A}, defined in (8.7), is a complex-valued skew bilinear form on TS. In (8.21), we identify that space with (8.15) by pullback. The desired property follows immediately from (8.21). □

8.2. The Cauchy-Riemann equation. Now assume that S is a (connected) Riemann surface. We will consider the inhomogeneous Cauchy-Riemann equation

$$\begin{cases} u : S \longrightarrow B, \\ \big(Du - X_A(u)\big)^{0,1} = 0. \end{cases} \tag{8.22}$$

More precisely, X_A is being evaluated at points $\sigma(z) = (z, u(z))$, and the $(0,1)$ part is taken with respect to the given complex structures on the source and target. Equivalently, this is the equation for σ to be a pseudo-holomorphic section:

$$\begin{cases} \sigma = (id, u) : S \longrightarrow S \times B, \\ D\sigma \circ i = J_A(\sigma) \circ d\sigma. \end{cases} \tag{8.23}$$

In local coordinates on S, (8.22) becomes

$$\partial_s u + i\partial_t u = X_{A_1(s,t)}(u) + iX_{A_2(s,t)}(u). \tag{8.24}$$

As one would expect from the appearance of connections, there is an action of the group of gauge transformations. Namely, if $A^\dagger = \Phi_* A$ and $u^\dagger(z) = \rho_{\Phi(z)}(u(z))$ for some $\Phi \in C^\infty(S, G)$, then

$$Du^\dagger - X_{A^\dagger}(u^\dagger) = D\rho_\Phi(Du - X_A(u)), \tag{8.25}$$

where on the right hand side, $D\rho_\Phi$ is the derivative of $\rho_{\Phi(z)} \in Aut(B)$ at the point $u(z)$. Hence, passing from (u, A) to $(u^\dagger, A^\dagger)$ transforms solutions of one equation (8.22) to the other. Equivalently, the gauge transformation determines a diffeomorphism of $S \times B$ which is fibrewise a hyperbolic isometry, and that diffeomorphism relates the almost complex structures J_A and $J_{A^\dagger}$.

We also want to quickly review the notions of energy for solutions of (8.22). The geometric and topological energies are, respectively,

$$E_{geom}(u) = \tfrac{1}{2}\int_S |du - X_A(u)|^2_{hyp}, \tag{8.26}$$

$$E_{top}(u) = \int_S u^*\omega_{hyp} - d(H_A(u)) \tag{8.27}$$

(for general maps u and noncompact S, the integral (8.27) may not converge, but we'll ignore that issue as it will irrelevant for our applications). The difference between the two notions is given by a curvature term:

$$E_{top}(u) = E_{geom}(u) + \int_S H_{F_A}(u). \tag{8.28}$$

In particular, if A is nonnegatively curved,

$$E_{geom}(u) \leq E_{top}(u). \tag{8.29}$$

In other words, the situation is as follows: the two notions of energy are obtained by integrating $\omega_{A,geom}$ and $\omega_{A,top}$ over σ. Then, (8.9) explains (8.28), and the behaviour for nonnegatively curved connections follows from (8.19).

8.3. Convergence and compactness. If the connection A is flat, one can use a local gauge transformation to reduce (8.22) to $\bar\partial u^\dagger = 0$, hence to elementary complex analysis.

LEMMA 8.3. *Suppose that $S = B_r \subset \mathbb{C}$ is the open disc of radius r, and that the connection A is flat. Then for any solution of* (8.22), *we have*

$$|du(z) - X_A(z, u(z))|_{hyp} \leq \frac{r}{r^2 - |z|^2}. \tag{8.30}$$

After a gauge transformation, this is just the Schwarz Lemma (holomorphic maps between hyperbolic Riemann surfaces can't increase distances).

LEMMA 8.4. *Suppose that $S = (-l/2, l/2) \times S^1$ is the open cylinder of length l, and that the connection A is flat, with hyperbolic holonomy $g \in G$ around $\{0\} \times S^1$. Write $\mathrm{tr}(g) = \pm(\lambda + \lambda^{-1})$, for some $\lambda > 1$. Then, a solution of* (8.22) *can exist only if*

$$l \leq \frac{\pi}{\log(\lambda^2)}. \tag{8.31}$$

PROOF. We again apply a gauge transformation to trivialize the connection, but this time working on the universal cover $(-l/2, l/2) \times \mathbb{R}$. After doing that and then again quotienting out by deck transformations, a solution (8.22) turns into a holomorphic map

$$S \longrightarrow B/\rho_g \cong (0, \pi/\log(\lambda^2)) \times S^1, \tag{8.32}$$

which induces an isomorphism of fundamental groups. Such a map can't exist if the source cylinder is longer than the target one; this follows from the Schwarz Lemma, given that the hyperbolic metric on a cylinder of length l has a "waist" of girth $\pi/2l$; see e.g. [**13**, p. 13] (taking into account our scaling convention for the metric). □

LEMMA 8.5. *Suppose that $S = (-l/2, l/2) \times S^1$ as before, and that the connection A is flat, with elliptic holonomy $g \in G$ around $\{0\} \times S^1$. Write $\mathrm{tr}(g) = \pm 2\cos(\theta)$, for some $\theta \in (0, \pi)$. Let $w \in B$ be the unique fixed point of ρ_g. Then, any solution of* (8.22) *satisfies*

$$\sinh(2\,\mathrm{dist}_{hyp}(u(0,0), w)) \leq \frac{\sinh(\pi/2l)}{\sin\theta}. \tag{8.33}$$

PROOF. After the usual gauge transformation, we get a holomorphic map $u^\dagger : (-l/2, l/2) \times \mathbb{R} \to B$, which satisfies $u^\dagger(0,0) = u(0,0)$ and $u^\dagger(s, t+1) = \rho_g(u^\dagger(s,t))$. In the hyperbolic metric on the domain, the distance between the points $(0,0)$ and $(0,1)$ is $\pi/2l$, as in Lemma 8.4. By hyperbolic trigonometry, the distance between $p = u^\dagger(0,0)$ and $q = u^\dagger(0,1)$ satisfies

$$\sin\theta = \frac{\sinh(\mathrm{dist}_{hyp}(p,q))}{\sinh(2\,\mathrm{dist}_{hyp}(p,w))}. \tag{8.34}$$

The rest is once again Schwarz' Lemma. □

LEMMA 8.6. *Let S be any Riemann surface, equipped with a flat connection A, and (u_k) a sequence of solutions of* (8.22). *Suppose that there is a sequence of points z_k, contained in a compact subset of S, such that $u_k(z_k)$ goes to $\partial\bar{B}$. Then u_k goes to $\partial\bar{B}$, in the sense of uniform convergence on compact subsets.*

PROOF. Since we can always replace S by its universal cover, let's assume that it is simply-connected. After a gauge transformation, we get a sequence of bounded holomorphic functions $u_k^\dagger$. This sequence has a convergent (on compact subsets) subsequence, and hence, so does u_k. Let $u_\infty^\dagger, u_\infty : S \to \bar{B}$ be the limits. $u_\infty^\dagger$ is a holomorphic function which meets $\partial\bar{B}$, hence must be constant at a point of $\partial\bar{B}$, by the open mapping theorem. It follows that u_∞ also takes values in $\partial\bar{B}$. By the same argument, every subsequence of (u_k) has a subsequence with the desired behaviour. From this, the result follows. □

To go beyond the flat case, we need to replace holomorphic function theory by more flexible analytic tools. We work with a fixed S and connection A. Bounds for derivatives will be with respect to some metric on S (the choice is irrelevant, since we will always work on compact subsets), and the Euclidean metric on B.

LEMMA 8.7. *Given a compact subset $K \subset S$ and an $r \geq 0$, there is a constant C such that $\|u|K\|_{C^r} \leq C$ for all solutions u of* (8.22).

SKETCH OF PROOF. Let's work in local coordinates on S. There is a uniform bound on the right hand side of (8.24), and of course on u itself. By elliptic regularity, we get a $W^{1,p}$ bound on u (for any $1 < p < \infty$). From there, one proceeds by bootstrapping. □

One can also view Lemma 8.7 as a consequence of the regularity theory for pseudo-holomorphic maps (see e.g. [**19**, Appendix B]) to (8.23); the noncompactness of $S \times B$ does not matter here, since J_A extends to $S \times \bar{B}$.

LEMMA 8.8. *Suppose that A has nonnegative curvature. Let (u_k) be a sequence of solutions of* (8.22). *Suppose that z_k is a sequence of points, contained in a compact subset of S, such that $u_k(z_k)$ goes to $\partial\bar{B}$. Then u_k goes to $\partial\bar{B}$, in the sense of uniform convergence on compact subsets.*

PROOF. Lemma 8.7 shows that, after passing to a subsequence, the u_k converge to some $u_\infty : S \to \bar{B}$. Consider the J_A-holomorphic map $\sigma_\infty(z) = (z, u_\infty(z))$. We know from Lemma 8.2 that $S \times \partial\bar{B}$ is Levi convex, which by [**4**, Corollary 4.7] implies that $\sigma_\infty^{-1}(S \times \partial\bar{B})$ is open and closed. But by assumption, that subset is also nonempty, which shows that u_∞ takes values in $\partial\bar{B}$. The rest is as in Lemma 8.6. □

REMARK 8.9. The expository paper [**7**] provides helpful additional discussion of the result from [**4**] which we have used. The strategy of proof goes as follows. Let $\bar{N}$ be a domain with smooth Levi convex boundary in an almost complex manifold, and N its interior. Suppose that there is a pseudo-holomorphic map from a connected Riemann surface to $\bar{N}$, which meets the boundary but is not entirely contained in it. After restriction to a suitably chosen part of S, we get a pseudo-holomorphic map defined on the unit disc,

$$\begin{cases} \sigma : \bar{B} \longrightarrow \bar{N}, \\ \sigma^{-1}(\partial\bar{N}) = \{1\}, \text{ and } \sigma \text{ is tangent to } \partial\bar{N} \text{ at that point.} \end{cases} \tag{8.35}$$

Near $\sigma(1)$, there is a smooth local defining function r for $\partial\bar{N}$, such that $-(-r)^\theta$ is (strictly) plurisubharmonic in N; the constant $0 < \theta < 1$ can be chosen arbitrarily [**7**, Theorem 2.4], and we will just need $\theta > 1/2$. After shrinking the domain further if necessary, we may assume that $r \circ \sigma$ is well-defined. Since it has vanishing derivative at $z = 1$, we have $r(\sigma(z)) \gtrsim -|z-1|^2$. Hence, $h = -(-r)^\theta \circ \sigma$ is a negative subharmonic function on B, which extends continuously to $\bar{B}$ and vanishes exactly at $1 \in \partial\bar{B}$, such that

$$h(z) \gtrsim -|z-1|^{2\theta}. \tag{8.36}$$

But that contradicts the Hopf Lemma for such functions.

9. Floer cohomology

This section sets up Hamiltonian Floer cohomology theory for Lefschetz fibrations, and the operations on it. The basic approach is classical [**12**]. The issues that are specific to our context arise from the noncompactness of the symplectic manifold. Compared to previous versions of the same formalism (see e.g. [**29**]), the main difference is that at infinity, our Hamiltonian automorphisms are modelled on isometries of the hyperbolic disc (following Section 8), rather than area-preserving diffeomorphisms of the Euclidean plane. Even though the distinction may fundamentally be a technical one, the hyperbolic model turns out to be very helpful for our constructions.

9.1. Geometry of the target space. Let (E^{2n}, ω_E) be a symplectic manifold, together with a trivialization of its anticanonical bundle (for some compatible almost complex structure), and a proper map (2.1) to the hyperbolic disc. Suppose that $\bar{N} \subset \bar{B}$ is a closed collar neighbourhood of $\partial\bar{B}$. We then call $N = \bar{N} \cap B$ a neighbourhood of infinity (for instance, $\{|w| \geq r\}$ for some $r < 1$ is such a neighbourhood). Our main condition is:

- For some neighbourhood of infinity $N = N_E$, there is a (necessarily unique) ω_E-orthogonal splitting of the tangent space at any point $x \in \pi^{-1}(N)$ into horizontal and vertical parts,

$$\begin{cases} TE_x = TE_x^h \oplus TE_x^v, \\ TE_x^v = ker(D\pi_x), \\ D\pi_x : TE_x^h \longrightarrow TB_{\pi(x)} \text{ is a symplectic isomorphism.} \end{cases} \tag{9.1}$$

In other words: x is a regular point of π; TE_x^v is a symplectic subspace; and the symplectic form on its ω_E-orthogonal complement agrees with the pullback of ω_{hyp}. As a consequence, (2.1) restricts to a symplectic fibre bundle over N with a preferred connection, given by TE^h, which is flat. Using that structure, one can construct a canonical compactification

$$\pi : \bar{E} \longrightarrow \bar{B}. \tag{9.2}$$

The total space $\bar{E}$ is a compact manifold with boundary, whose interior is E, and with $\partial\bar{E} = \pi^{-1}(\partial\bar{B})$. The compactification again has the property that, over $\bar{N}$, it is a symplectic fibre bundle with a preferred flat connection. In other words, the splitting in (9.1) extends to points of $\partial\bar{E}$, and so does the symplectic structure on the vertical part. Of course, ω_E itself does not extend to $\bar{E}$, because of the ω_{hyp} component, but there are slightly different symplectic forms that do:

$$\omega_{\bar{E}} = \omega_E - \pi^*(\psi\omega_{hyp}) + \pi^*(\psi\omega_{\bar{B}}), \tag{9.3}$$

where $\omega_{\bar{B}}$ is any (positive) symplectic form on $\bar{B}$, and $\psi : \bar{B} \to [0,1]$ is a function which vanishes outside N, and equals 1 near $\partial\bar{B}$.

- We denote by $\mathcal{J}(E)$ the space of compatible almost complex structures J on E with the following two additional properties. First, there is some neighbourhood of infinity $N = N_J \subset N_E$, such that π is J-holomorphic at all points of $\pi^{-1}(N)$. Secondly, J extends to $\bar{E}$, and that extension is compatible with (9.3).

The last-mentioned condition (compatibility with $\omega_{\bar{E}}$) is independent of the choices made in (9.3). Even before imposing that condition, we already knew that J preserves the extension of the splitting (9.1) to points $x \in \partial\bar{E}$, and is compatible with the orientation of $T\bar{E}^h_x \cong T\bar{B}_{\pi(x)}$. Hence, the additional requirement is that the extension should be compatible with the symplectic structure on $T\bar{E}^v_x$. We want to make one more observation concerning such almost complex structures. Because the hyperbolic metric scales up to infinity as one approaches $\partial\bar{B}$, there is a constant $C = C_J > 0$ such that

$$\omega_{\bar{E}}(X, JX) \leq C\,\omega_E(X, JX) \quad \text{for all } X \in TE. \tag{9.4}$$

In particular, the ω_E-energy of non-constant J-holomorphic spheres is bounded below by a positive constant (since the same holds for $\omega_{\bar{E}}$, which lives on a compact space).

- For any $\gamma \in \mathfrak{g}$, let $\mathcal{H}(E, \gamma)$ be the space of functions $H \in C^\infty(E, \mathbb{R})$ whose restriction to $\pi^{-1}(N)$, for some neighbourhood of infinity $N = N_H \subset N_E$, equals the pullback of the function H_γ from (8.3). Similarly, given $a = a_t\,dt \in \Omega^1(S^1, \mathfrak{g})$, we write $\mathcal{H}(E, a)$ for the space of $H \in C^\infty(S^1 \times E, \mathbb{R})$ such that $H_t \in \mathcal{H}(E, a_t)$ for all $t \in S^1$.

On $\pi^{-1}(N)$, the Hamiltonian vector field X associated to $H \in \mathcal{H}(E, \gamma)$ is the unique horizontal lift of X_γ. Hence, X extends smoothly to $\bar{E}$, and is tangent to $\partial\bar{E}$.

REMARK 9.1. In our application, we will often encounter almost complex structures or functions which depend on additional parameters (i.e. families of such objects). In that case, it is always understood that the relevant neighbourhoods of infinity can be chosen locally constant with respect to the parameter. Equivalently, if P is the parameter space, we are considering a closed collar neighbourhood of $P \times \partial\bar{B}$ inside $P \times \bar{B}$, and then taking its intersection with $P \times B$.

Given $H = (H_t) \in \mathcal{H}(E, a)$ and its time-dependent vector field $X = (X_t)$, we will consider one-periodic orbits

$$\begin{cases} x : S^1 \longrightarrow E, \\ dx/dt = X_t. \end{cases} \tag{9.5}$$

LEMMA 9.2. *If a has nontrivial holonomy, all solutions of* (9.5) *are contained in a compact subset of E.*

PROOF. Otherwise, there would have to be a sequence of one-periodic orbits in E which converge to $\partial\bar{E}$. By projection, the same would have to be true for the corresponding ODE on the hyperbolic disc,

$$\begin{cases} y : S^1 \longrightarrow B, \\ dy/dt = X_{a_t}. \end{cases} \tag{9.6}$$

But that is impossible, since such y correspond to fixed points of the holonomy of a acting on B (of which there is at most one). □

REMARK 9.3. When describing our results in Section 2, we included a Lefschetz (nondegeneracy of critical points) condition, since that puts them into a familiar context. In fact, that condition is irrelevant for the constructions in this paper, and therefore doesn't appear in the definition we have just given.

9.2. Basic transversality and compactness. In spite of the constraints imposed on the almost complex structures and Hamiltonian functions, the transversality results underlying the traditional construction of Floer theory still hold. Here is a sample of the kind of arguments that are required:

LEMMA 9.4. *For generic $J \in \mathfrak{J}(E)$, all simple (non-multiply-covered) J-holomorphic spheres are regular.*

SKETCH OF PROOF. A given J comes with some $N = N_J$ over which $D\pi$ is J-holomorphic. Genericity is understood to be with respect to perturbations which keep that subset fixed. This means that on $\pi^{-1}(N)$, we may change only the TE^v component of J. Let $u : \mathbb{C}P^1 \to E$ be a simple J-holomorphic sphere, and $v = \pi(u)$. At any point z such that $v(z) \in N$, we have $\bar{\partial} v = 0$. Hence, one of the following applies:

(i) There is a point where $v(z) \notin N$. Since the choice of almost complex structure is free (except for the condition of ω_E-compatibility) near $u(z)$, transversality for such u follows from the standard argument.

(ii) There is a point where $v(z) \in N \setminus \partial N$. Note that the degree of v over any point $w \in B$ is the same, hence zero. It follows that for each $w \in N \setminus \partial N$, $v^{-1}(w)$ is open and closed. Under our assumption, this means that u must lie in a single fibre $F = E_w$. A parametrized transversality argument, using the fact that the almost complex structure in vertical direction can be varied freely, shows generic regularity for such maps. To explain the relevance of parametrized transversality, consider the commutative diagram (with exact columns)

(9.7)

$$
\begin{array}{ccc}
0 & & 0 \\
\downarrow & & \downarrow \\
\{\xi \in C^\infty(\mathbb{C}P^1, u^*TE) \,:\, D\pi(\xi) \text{ is constant}\} & \longrightarrow & C^\infty(\mathbb{C}P^1, \Omega^{0,1} \otimes_{\mathbb{C}} u^*TF) \\
\downarrow & & \downarrow \\
C^\infty(\mathbb{C}P^1, u^*TE) & \xrightarrow{D_{E,u}} & C^\infty(\mathbb{C}P^1, \Omega^{0,1} \otimes_{\mathbb{C}} u^*TE) \\
\downarrow{\scriptstyle D\pi} & & \downarrow{\scriptstyle D\pi} \\
C^\infty(\mathbb{C}P^1, \mathbb{C})/\mathit{constants} & \xrightarrow{\bar{\partial}} & C^\infty(\mathbb{C}P^1, \Omega^{0,1}) \\
\downarrow & & \downarrow \\
0 & & 0.
\end{array}
$$

Here, $D_{E,u}$ is the linearization of u as a pseudo-holomorphic map into E. The top $\to$ is the linearization of u as a pseudo-holomorphic map into a variable fibre (its domain is an extension of $C^\infty(\mathbb{C}P^1, u^*TF)$ by $\mathbb{C} = TB_w$). The parametrized transversality theory shows that the top $\to$ is onto for generic J. But since the bottom $\to$ is invertible, we get the same surjectivity result for $D_{E,u}$.

(iii) $v(z) \in \partial N$ for all z. In that case, v is holomorphic, hence must again be constant. The same strategy as in (ii) applies. □

A similar result holds for one-periodic orbits:

LEMMA 9.5. *Suppose that $a \in \Omega^1(S^1, \mathfrak{g})$ has nontrivial holonomy. Then, for generic choice of $H \in \mathcal{H}(E, a)$, all one-periodic orbits* (9.5) *are nondegenerate.*

PROOF. As before, we need to be a bit more precise concerning the statement itself. Each $H_t \in \mathcal{H}(E, a_t)$ comes with some neighbourhood of infinity N_{H_t}, but we can fix an N which is smaller than all of them (see Remark 9.1, and take into account the compactness of the parameter space $P = S^1$). In the case where the holonomy is elliptic, we assume that (9.6) has no solutions satisfying $y(t) \in N$ for all t (for parabolic or hyperbolic holonomy, this condition would be vacuous, since their holonomy acts fixed point freely). That being given, it is clear that perturbing H only outside $\pi^{-1}(N)$ is sufficient to achieve transversality. □

Let's keep the assumption that a has nontrivial holonomy, and choose $H \in \mathcal{H}(E, a)$, satisfying the nondegeneracy condition from Lemma 9.5. Suppose also that we have a family $J = (J_t)$ of almost complex structures in $\mathcal{J}(E)$. Given one-periodic orbits $x_\pm$ as in (9.5), we consider Floer trajectories

$$\begin{cases} u : \mathbb{R} \times S^1 \longrightarrow E, \\ \partial_s u + J_t(u)(\partial_t u - X_t(u)) = 0, \\ \lim_{s \to \pm\infty} u(s,t) = x_\pm(t). \end{cases} \tag{9.8}$$

The pair $\Phi = (J, H)$ will be called a "Floer datum", since this is the structure required to write down Floer's equation (9.8) (nondegeneracy of one-periodic orbits is part of the definition of Floer datum).

LEMMA 9.6. *For generic choice of J (keeping H fixed), all solutions of* (9.8) *are regular. Moreover, if u is a solution whose linearized operator D_u has index ≤ 2, then $u(s,t)$ never lies on a nontrivial J_t-holomorphic sphere.*

This is essentially the transversality result from [**6, 12**]. Applied to solutions $u(s,t) = x(t)$ (D_u is invertible in that case, because of the nondegeneracy of x), the result includes the statement that for any one-periodic orbit, $x(t)$ should not lie on a nontrivial J_t-holomorphic sphere; this is a version of Lemma 9.4 including transversality of evaluations. For the remaining part of Lemma 9.6, one starts with a fixed $\Phi = (J, H)$, and takes some neighbourhood of infinity N which is smaller than N_{J_t} and N_{H_t} for any t, and which has the same property concerning (9.6) as in the proof of Lemma 9.5. Then, since the orbits $x_\pm$ aren't contained in $\pi^{-1}(N)$, the maps u also can't be contained in that subset, which means that perturbing J outside $\pi^{-1}(N)$ gives one enough freedom.

To show that Floer trajectories can't escape to infinity, we will combine elementary complex analysis (as in Section 8.3) and pseudo-holomorphic curve theory. Some basic ingredients are:

- The notion of energy of a Floer trajectory,

$$\begin{aligned} E(u) &= \int_{\mathbb{R} \times S^1} \|\partial_s u\|^2_{J_t} \\ &= \int_{\mathbb{R} \times S^1} u^* \omega_E - \int_{S^1} H_t(x_+(t))\, dt + \int_{S^1} H_t(x_-(t))\, dt. \end{aligned} \tag{9.9}$$

- The "gauge transformation" which makes Floer's equation homogeneous. Namely, let (ϕ_t) be the Hamiltonian isotopy generated by (X_t). The change of variables $u(s,t) = \phi_t(u^\dagger(s,t))$ turns (9.8) into

$$\begin{cases} u^\dagger : \mathbb{R}^2 \longrightarrow E, \\ u^\dagger(s,t) = \phi_1(u^\dagger(s,t+1)), \\ \partial_s u^\dagger + J_t^\dagger(u^\dagger)\partial_t u^\dagger = 0, \\ \lim_{s\to\pm\infty} u^\dagger(s,t) = x_\pm^\dagger, \end{cases} \tag{9.10}$$

  where $J_t = (\phi_t)_* J_t^\dagger$, and $x_\pm^\dagger = x_\pm(0)$. Outside a compact subset, ϕ_t covers an isotopy in $Aut(B)$. Hence, $J_t^\dagger$ again belongs to $\mathcal{J}(E)$.
- Projection to the base. Wherever u is suffficiently close to $\partial\bar{E}$, the projection $v = \pi(u)$ satisfies a special case of (8.22), where A is the pullback of a to the cylinder (hence flat):

$$\partial_s v + i(\partial_t v - X_{a_t}(v)) = 0. \tag{9.11}$$

LEMMA 9.7. *Suppose that (u_k) is a sequence of solutions of (9.8), with an upper bound on the energy. Assume that there are points $(s_k, t_k) \in \mathbb{R} \times S^1$ such that $|s_k|$ is bounded, and $u_k(s_k,t_k) \to \partial\bar{E}$. Then, the u_k converge to $\partial\bar{E}$ uniformly on compact subsets.*

PROOF. Gromov compactness (on compact subsets of the domain) can be applied to the u_k, seen as maps taking values in $\bar{E}$. This may seem suspicious, but is justified by the following argument. Consider the pseudo-holomorphic maps $u_k^\dagger$ as in (9.10). On each compact subset of $\mathbb{R}^2$, we have an upper bound on their energies with respect to ω_E, hence by (8.28) also for $\omega_{\bar{E}}$. When applying Gromov compactness, one thinks of the latter symplectic form on $\bar{E}$.

With this in mind, the Gromov limit (of a subsequence of the u_k) consists of a principal component $u_\infty : \mathbb{R} \times S^1 \to \bar{E}$, which solves the Cauchy-Riemann equation from (9.8), together with bubble components which are pseudo-holomorphic spheres (unlike the usual compactification of Floer trajectory spaces, the notion of Gromov limit we are working with here ignores any components which might split off over the ends $s \to \pm\infty$). By assumption, one of our components must intersect $\partial\bar{E}$. If the principal component does that, then by looking at its projection to $\bar{B}$ (as in Lemma 8.6), one concludes that $u_\infty(\mathbb{R} \times S^1) \subset \partial\bar{E}$. The same argument applies to bubble components. Because the Gromov limit is connected, a component-by-component induction shows that all of it lies in $\partial\bar{E}$. This implies that a subsequence of the u_k converges to $\partial\bar{E}$ on compact subsets. Going from that to the entire sequence works as in Lemma 8.6. □

LEMMA 9.8. *Assume that the holonomy of a is elliptic or hyperbolic. Suppose that (u_k) is a sequence of solutions of (9.8), with an upper bound on the energy. Then, there is a compact subset of E which contains the images of all the u_k.*

PROOF. Let's suppose that on the contrary, there are (possibly after passing to a subsequence) points (s_k, t_k) such that $u_k(s_k, t_k) \to \partial\bar{E}$. After translation in s-direction, we may assume that $|s_k|$ is bounded. Lemma 9.7 implies that, on each relatively compact subset of $\mathbb{R} \times S^1$, $v_k = \pi(u_k)$, for $k \gg 0$, is a solution of (9.11). In the hyperbolic case, one takes that subset to be a sufficiently long finite cylinder $(-l/2, l/2) \times S^1$, and gets a contradiction to Lemma 8.4. In the slightly more complicated elliptic case, Lemma 8.5 shows that $v_k(0,0)$ must lie in a specific compact subset of B, which leads to a contradiction to Lemma 9.7. $\square$

9.3. General surfaces. We now extend (9.8) to other Riemann surfaces, building a familiar kind of TQFT framework (compare e.g. [**26**]). Let S be a punctured (connected, but not closed) Riemann surface. We arbitrarily divide the set I of punctures into subsets I_{in} and I_{out}. The surface should be decorated with the following additional data:

- Tubular ends surrounding the punctures, with pairwise disjoint images,

$$\begin{cases} \epsilon_i : [0,\infty) \times S^1 \hookrightarrow S, & i \in I_{in}, \\ \epsilon_i : (-\infty, 0] \times S^1 \hookrightarrow S, & i \in I_{out}. \end{cases} \tag{9.12}$$

- A connection $A \in \Omega^1(S, \mathfrak{g})$ which is nonnegatively curved and compatible with the tubular ends, meaning that

$$\epsilon_i^* A = a_i = a_{i,t}\, dt \tag{9.13}$$

  for some $a_i \in \Omega^1(S^1, \mathfrak{g})$. Moreover, each a_i should have nontrivial holonomy.

We write $Y = (\{\epsilon_i\}, A)$, and refer to (S, Y) as a decorated surface (this is closely related to, but not exactly the same as, the notion from Section 5; see Section 10 for further discussion). The connection A gives rise to an equation (8.22) for maps $S \to B$, which has the form (9.11) on each end. The corresponding construction for maps to E requires the following additional choices:

- For each end, we fix a Floer datum $\Phi_i = (J_i, H_i)$ in the class determined by a_i.
- Let $J = (J_z)_{z \in S}$ be a family of almost complex structures, $J_z \in \mathcal{J}(E)$, which on the tubular ends satisfies the following asymptotic condition:

$$J_{\epsilon_i(s,t)} \longrightarrow J_{i,t} \quad \text{exponentially fast as } s \to \pm\infty. \tag{9.14}$$

We need to make the last-mentioned condition more precise, in several respects. The difference between $J_{\epsilon_i(s,t)}$ and $J_{i,t}$ is measured after extension to the compactification $\bar{E}$. "Exponentially fast" means that in the C^r sense for any r, the difference between the structures is bounded by a constant times $e^{-\lambda|s|}$, for some $\lambda > 0$. Finally, for each end there should be a neighbourhood of infinity $N = N_{J,i} \subset N_E$, such that all $J_{\epsilon_i(s,t)}$ make π pseudo-holomorphic on $\pi^{-1}(N)$.

- Take $K \in C^\infty(S \times E, T^*S)$, thought of as a one-form on S with values in $C^\infty(E, \mathbb{R})$. This must be such that $K(\xi) \in \mathcal{H}(E, A(\xi))$ for all $\xi \in TS$. Over the ends, we require the following (this time, a strict equality, rather than an asymptotic condition):

$$\epsilon_i^* K = H_{i,t}\, dt. \tag{9.15}$$

We will refer to the pair $\Pi = (J, K)$ as a "perturbation datum". Let $X_K \in C^\infty(S \times E, T^*S \otimes_{\mathbb{R}} TE)$ be the Hamiltonian-vector-field-valued one-form associated to K. We will consider maps

$$\begin{cases} u : S \longrightarrow E, \\ (Du - X_K(u))^{0,1} = 0, \\ \lim_{s \to \pm\infty} u(\epsilon_i(s,t)) = x_i(t). \end{cases} \tag{9.16}$$

The Cauchy-Riemann equation in (9.16) says that $Du - X_K(u) : TS_z \to TE_{u(z)}$ should be complex-linear with respect to J_z. On each tubular end, setting $u_i(s,t) = u(\epsilon_i(s,t))$, this equation reduces to

$$\partial_s u_i + J_{\epsilon_i(s,t)}(\partial_t u_i - X_{i,t}(u_i)) = 0. \tag{9.17}$$

The limits x_i are one-periodic orbits of H_i. In local coordinates $z = s + it$ on S in which $K = K_1\,ds + K_2\,dt$, the equation is

$$\partial_s u + J_{s,t}\partial_t u = X_{K_1(s,t)} + J_{s,t}X_{K_2(s,t)}. \tag{9.18}$$

LEMMA 9.9. *For generic choice of perturbation datum Π, keeping Φ_i fixed, the spaces of solutions of (9.16) are regular. Moreover, if u is a solution whose linearized operator D_u has index ≤ 1, then $u(z)$ never lies on a nontrivial J_z-holomorphc sphere.*

SKETCH OF PROOF. Let's fix a suitably small neighbourhood of infinity N, so that none of the x_i is contained in $\pi^{-1}(N)$. Over each end, one of the following applies (this is a version of [**6**, Lemma 4.1] with additional s-dependence of the almost complex structure, but has the same proof):

(i) The set of points where $\partial_s u_i$ vanishes is discrete. If there is at least one end with this property, there are points (s,t) where $\partial_s u_i \neq 0$ and $u_i(s,t) \notin \pi^{-1}(N)$. Transversality can then be achieved by perturbing the family J_z near $z = \epsilon_i(s,t)$.

(ii) $\partial_s u_i$ vanishes identically. This means that $u_i(s,t) = x_i(t)$. In particular, there are t such that $z = \epsilon_i(0,t)$ satisfies $u(z) \notin \pi^{-1}(N)$. Since this is an open condition, there are also points $z \in S$ which are disjoint from the tubular ends and such that $u(z) \notin \pi^{-1}(N)$. We can then achieve transversality by varying (J,K) near such a point. □

We will now tackle the issue of compactness. This is based on similar ideas as in the case of Floer trajectories:

- There are two notions of energy for solutions of (9.16), paralleling (8.26) and (8.27) from our toy model discussion:

$$E_{geom}(u) = \tfrac{1}{2}\int_S |Du - X_K(u)|^2, \tag{9.19}$$

$$E_{top}(u) = \int_S u^*\omega_E - d(K(u)). \tag{9.20}$$

  These are related by a curvature term,

$$E_{top}(u) = E_{geom}(u) + \int_S R_K(u). \tag{9.21}$$

  We do not want to spell out the definition of R_K, but on a suitable neighbourhood of infinity, it agrees with the pullback of the corresponding term in (8.28). Since we are assuming that A is nonnegatively curved, it follows that the integrand in (9.21) is bounded below. Moreover, it vanishes identically on the tubular ends. It follows that there is a constant $C = C_K$ such that for all u,

$$E_{geom}(u) \leq E_{top}(u) + C. \tag{9.22}$$

- To any perturbation datum $\Pi = (J, K)$ one can associate an almost complex structure J_Π on the trivial fibre bundle $S \times E \to S$, such that (9.16) turns into the equation for J_Π-holomorphic sections. In local coordinates on S, this is given by the same kind of formula as in (8.18),

$$\begin{cases} J_\Pi(Y) = J_{s,t}Y & \text{for } Y \in TE, \\ J_\Pi(\partial_s + X_{K_1}) = \partial_t + X_{K_2}. \end{cases} \tag{9.23}$$

  Near $\partial\bar{E}$, the vector fields $X_{K_1}, X_{K_2} \in C^\infty(E, TE)$ are the horizontal lifts of the corresponding vector fields $X_{A_1}, X_{A_2} \in C^\infty(B, TB)$. Hence, they extend smoothly to $\bar{E}$, and so does J_Π.
- At any point $z \in S$ such that $u(z)$ is sufficiently close to $\partial\bar{E}$, the projection $v = \pi(u)$ satisfies the equation (8.22) associated to A.

LEMMA 9.10. *Suppose that (u_k) is a sequence of solutions of (9.16), with an upper bound on the topological energy. Suppose that there are points z_k, contained in a compact subset of S, such that $u_k(z_k) \to \partial\bar{E}$. Then, the u_k converge to $\partial\bar{E}$ uniformly on compact subsets.*

PROOF. By (9.22), we have an upper bound on the geometric energy. From there on, the proof follows the same strategy as its Floer-theoretic counterpart, Lemma 9.7. This time, the applicability of Gromov compactness has to be explained in a slightly different way. Think of our maps as J_Π-holomorphic sections $S \to S\times\bar{E}$. From the bound on the geometric energy (and the fact that X_K extends smoothly to $\bar{E}$), we get a $W^{1,2}$-bound on those sections, on each fixed compact subset of S. This already suffices for Gromov compactness, without having to introduce a symplectic structure on $S\times\bar{E}$ (see e.g. [**36**] for a suitable exposition). Alternatively, one can argue by the following familiar trick: since $J_\Pi|\{z\} \times \bar{E}$ is compatible with (9.3) for each z, one can add a large positive two-form on S to create a symplectic form on $S \times \bar{E}$ which at least tames J_Π.

The Gromov limit (of a subsequence, in the sense of convergence on compact subsets) consists of a solution $u_\infty : S \to \bar{E}$ of the same $\bar{\partial}$-equation as in (9.16), together with bubble components which are J_z-holomorphic spheres. By projecting to $\bar{B}$ and using the same convexity argument as in Lemma 8.8, one sees that $u_\infty^{-1}(\partial\bar{E})$ is open and closed. The same applies to the bubble components, as in Lemma 9.7, and we again conclude that under our assumptions on u_k, the entire Gromov limit must lie in $\partial\bar{E}$. □

LEMMA 9.11. *Assume that the a_i have elliptic or hyperbolic holonomies. Suppose that (u_k) is a sequence of solutions of (9.16), with an upper bound on the topological energy. Then, there is a compact subset of E which contains the images of all the u_k.*

PROOF. This is modelled on Lemma 9.8. Let's suppose that, on the contrary (after passing to a subsequence), there are z_k such that $u_k(z_k) \to \partial\bar{E}$.

(i) If (again after passing to a subsequence) the z_k are contained in a compact subset, we can apply Lemma 9.10, which implies that $u_k \to \partial\bar{E}$ on compact subsets. In particular, this convergence holds for the restriction of u_k to any fixed finite cylinder $\epsilon_i((s - l/2, s + l/2) \times S^1) \subset S$, which yields a sequence of solutions of (9.17) that approach $\partial\bar{E}$. By taking l to be sufficiently large, and looking at $v_k = \pi(u_k)$, one obtains a contradiction with Lemma 8.4 or Lemma 8.5.

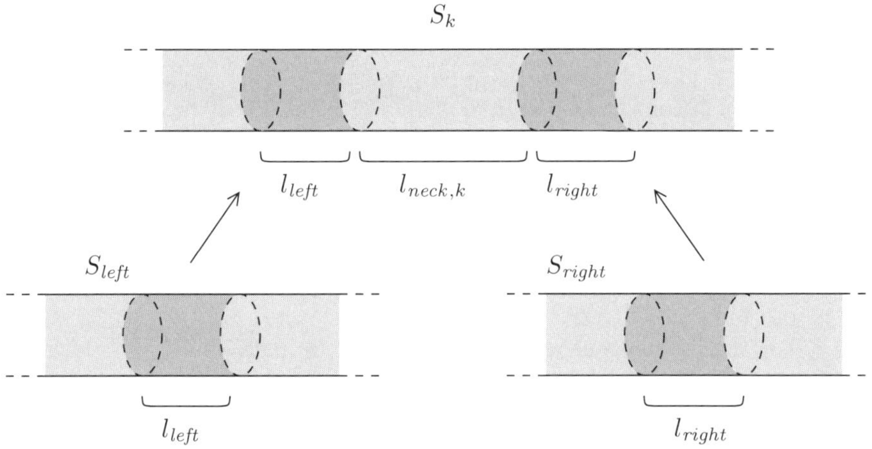

FIGURE 9.1. Notation from the neck-stretching process.

(ii) The other case one has to deal with is where $z_k = \epsilon_i(s_k, t_k)$ for some $|s_k| \to \infty$. Consider the translated versions

$$u_{k,i}(\cdot + s_k, \cdot) : (-|s_k|, |s_k|) \times S^1 \longrightarrow E. \tag{9.24}$$

By definition, the image of $(0, t_k)$ under (9.24) approaches $\partial \bar{E}$. Moreover, these maps satisfy a sequence of Cauchy-Riemann equations which, as $k \to \infty$, converge to an equation (9.8). A version of the argument from Lemma 9.10 can be applied to this sequence; after that, one proceeds as in case (i). □

A frequently encountered class of generalizations is that where, instead of maps defined on a single Riemann surface, one has a family of surfaces. If the parameter space for the family is compact, the previous arguments carry over straightforwardly. A slightly more interesting situation is "neck-stretching", where the Riemann surfaces split into pieces. Let's consider the simplest example of such behaviour, which is one infinite cylinder splitting into two (the general case differs mainly in being even more cumbersome to formulate). The basic situation is:

- Fix $a_{left}, a_{neck}, a_{right} \in \Omega^1(S^1, \mathfrak{g})$, all of which have nontrivial holonomy.
- Fix constants $l_{left}, l_{right} > 0$, and a sequence $l_{neck,k} > 0$, $l_{neck,k} \to \infty$. These govern the length of various pieces of our cylinder, see Figure 9.1.
- Let's write S_k ($k = 1, 2, \dots$), S_{left} and S_{right} for copies of $\mathbb{R} \times S^1$. We will think of these as mapped to each other as follows:

$$\begin{cases} \sigma_{left} : S_{left} \longrightarrow S_k, & \sigma_{left}(s,t) = (s - l_{left}/2 - l_{neck,k}/2, t), \\ \sigma_{right} : S_{right} \longrightarrow S_k, & \sigma_{right}(s,t) = (s + l_{right}/2 + l_{neck,k}/2, t). \end{cases} \tag{9.25}$$

  We give S_{left} tubular ends, whose images are $(-\infty, -l_{left}/2] \times S^1$ and $[l_{left}/2, \infty) \times S^1$; and correspondingly for S_{right}. The ends of S_k have image $(-\infty, -l_{left} - l_{neck,k}/2] \times S^1$ and $[l_{right} + l_{neck,k}/2, \infty) \times S^1$.
- Each S_k carries a nonnegatively curved connection A_k which restricts to a_{left}, a_{right} on its tubular ends, and to a_{neck} on $[-l_{neck,k}/2, l_{neck,k}/2] \times S^1$. Similarly, S_{left} carries a nonnegatively curved connection A_{left} which restricts to a_{left}, a_{neck} on the tubular ends. Moreover, the pullbacks of

A_k by (9.25) converge to A_{left} on compact subsets. On S_{right}, we have a connection A_{right} with corresponding properties.

When setting up the pseudo-holomorphic curve theory, the perturbation data have to be similarly correlated:

- We suppose that Floer data $\Phi_{left} = (J_{left,t}, H_{left,t})$, $\Phi_{neck} = (J_{neck,t}, H_{neck,t})$, and $\Phi_{right} = (J_{right,t}, H_{right,t})$ have been chosen. The Hamiltonians should lie in the classes specified by the previous a's. We also choose perturbation data $\Pi_k = (J_k, K_k)$, $\Pi_{left} = (J_{left}, K_{left})$, and $\Phi_{right} = (J_{right}, K_{right})$, whose behaviour over the ends is governed by the appropriate Floer data.
- For the inhomogeneous terms K_k which are part of Π_k, we additionally prescribe the behaviour on the neck:

$$K_{k,s,t} = H_{neck,t}\,dt \quad \text{for } -l_{neck,k}/2 \le s \le l_{neck,k}/2. \tag{9.26}$$

  The pullback of K_k by (9.25) should converge on compact subsets to K_{left}, K_{right}.
- The requirements for the families of almost complex J_k are the most complicated part of the definition, because of the asymptotic condition (9.14). Pulling back J_k by σ_{left}, and then restricting to $(-\infty, -l_{left}/2] \times S^1$, should yield a family of almost complex structures which converges to J_{left}; and that convergence should be in any C^r topology with sufficiently small exponential weights over the end $s \to -\infty$. Take the same pullback, and restrict to it $[-l_{left}/2, l_{left}/2 + l_{neck,k}/2 + C] \times S^1$ for some constant C, which yields a family of almost complex structures parametrized by increasingly large parts of S_{left}. As $k \to \infty$, that family should converge to J_{left}; and that convergence should be in any C^r topology with sufficiently small exponential weights (where the weight concerns the direction $s \to +\infty$). Parallel requirements apply to the pullback by σ_{right}.

This notion of neck-stretching is designed to allow suitable gluing arguments to go through, while preserving our previous transversality and compactness statements. More precisely, the appropriate version of Lemma 9.9 says that, keeping the Φ's fixed, a generic perturbation of the entire sequence $\Pi_k = (J_k, K_k)$ (within the allowed class) makes the associated moduli spaces regular for all k. Similarly, one gets a version of Lemma 9.11, assuming that the holonomies involved (of $a_{left}, a_{neck}, a_{right}$) are elliptic or hyperbolic. The strategy of proof remains the same, with the breakdown into cases understood in an appropriate way. Namely, take maps $u_k : S_k \to E$ and points $z_k \in S_k$ such that $u_k(z_k) \to \partial\bar{E}$. (i) If (possibly after passing to a subsequence) the z_k have preimages under (9.25) which lie in a compact subset of S_{left}, then one considers a Gromov limit which is a solution of an equation (9.16) involving Π_{left}; and similarly for S_{right}. (ii) In all other cases, one gets a limit which is a Floer trajectory for one of our chosen Floer data.

9.4. Formal aspects of Floer cohomology. The rest of the construction of Floer cohomology is standard, and we will keep the discussion brief. From now on, we will only allow $a \in \Omega^1(S^1, \mathfrak{g})$ with elliptic or hyperbolic holonomy. We also restrict our notion of decorated surface, by requiring the same property for all a_i.

Fix an a. Choose an associated Floer datum $\Phi = (H, J)$, which satisfies the transversality statement from Lemma 9.6. We define a $\mathbb{Z}$-graded cochain complex over the Novikov field (2.3):

$$CF^*(E,H) = \bigoplus_x \mathbb{K}_x, \tag{9.27}$$

where the sum is over the (finitely many) one-periodic orbits (9.5). Each $\mathbb{K}_x$ is an orientation space, which means that it is identified with $\mathbb{K}$ in a way which is canonical up to sign. To simplify the notation, we make an arbitrary choice of such an identification, and denote the resulting basis element in $CF^*(E,H)$ simply by x. The grading is given by Conley-Zehnder indices (which are well-defined since we have chosen a trivialization of the anticanonical bundle of E). The Floer differential is

$$d(x_+) = \sum_u \pm q^{E(u)} x_-, \tag{9.28}$$

where the sum is over all solutions of u of (9.8) which are isolated up to translation. Lemma 9.8, together with more standard compactness arguments, shows that there are only finitely many such u whose energy is lower than any given bound, hence that (9.28) makes sense. The sign with which each u contributes is determined by gluing of determinant lines. One defines $HF^*(E,H)$ to be the cohomology of $(CF^*(E,H),d)$.

Suppose that we have a decorated surface (S,Y), and for each end, a Floer datum Φ_i whose Floer complex is well-defined. Choose a perturbation datum Π satisfying Lemma 9.9. Then, counting solutions of (9.16) yields a chain map of even degree,

$$\begin{aligned} &\phi : \bigotimes_{i\in I_{in}} CF^*(E,H_i) \longrightarrow \Big(\bigotimes_{i\in I_{out}} CF^*(E,H_i)\Big)[n(\chi(S)+|I_{in}|-|I_{out}|)], \\ &\phi(\otimes_{i\in I_{in}} x_i) = \sum_u \pm q^{E_{top}(u)} (\otimes_{i\in I_{out}} x_i). \end{aligned} \tag{9.29}$$

Here, the sum is over solutions with limits $\{x_i\}$ which are isolated; the topological energy is used, since that is locally constant in families of solutions; and there is a sign, exactly as in (9.28). The same holds more generally if we have a family of such surfaces parametrized by a closed oriented manifold P (the Floer data are parameter-independent, but the perturbation data obviously change). The only difference is that the degree of the associated map ϕ_P is $\dim(P)$ less than in (9.29).

The next simplest case is where the parameter space is a compact oriented manifold with corners. In that case, the associated operation is no longer a chain map, but instead a nullhomotopy for the operations associated to the boundary faces. Let's suppose for simplicity that exactly one end is an output, say $I_{out} = \{0\}$ and $I_{in} = \{1,\dots,m\}$; this will be the case in all our applications. Then, the precise expression is exactly as in (5.21).

It remains to take a brief look at gluing. A simple case would be where we have a parameter space P with $\partial P = P_1 \times P_2$. The associated family of surfaces is defined only over $P \setminus \partial P$, and as we approach the boundary, it stretches along a neck, with the limit being two families over P_1 and P_2, respectively. The behaviour of the perturbation data must be as discussed at the end of Section 9.3. One can then still define an operation associated to P, and that satisfies (under the same one-output assumption as before)

$$(9.30)\quad (-1)^{\dim(P)} d\phi_P(x_1,\dots,x_m) - \phi_P(dx_1,\dots,x_m) - \cdots \\ \cdots - (-1)^{|x_1|+\cdots+|x_{m-1}|}\phi_P(x_1,\dots,dx_m) \\ + (-1)^{(\dim(P_1)+|x_1|+\cdots+|x_i|)\dim(P_2)}\phi_{P_1}(x_1,\dots,\phi_{P_2}(x_{i+1},\dots,x_{i+m_2}),\dots,x_m) = 0,$$

where $m = m_1 + m_2 - 1$; note that $(-1)^{\dim(P)} = (-1)^{|\phi_P|}$. As a final example, consider a mixture of the two previously encountered situations, where $\partial P = P_0 \sqcup -(P_1 \times P_2)$ (the minus sign indicates orientation-reversal). The associated family of surfaces should extend over P_0, but on the other boundary component, we have neck-stretching as before. Then, one gets

$$(9.31)\\ (-1)^{\dim(P)} d\phi_P(x_1,\dots,x_m) - \phi_P(dx_1,\dots,x_m) - \cdots \\ \cdots - (-1)^{|x_1|+\cdots+|x_{m-1}|}\phi_P(x_1,\dots,dx_m) + \phi_{P_0}(x_1,\dots,x_m) \\ - (-1)^{(\dim(P_1)+|x_1|+\cdots+|x_i|)\dim(P_2)}\phi_{P_1}(x_1,\dots,\phi_{P_2}(x_{i+1},\dots,x_{i+m_2}),\dots,x_m)=0.$$

In the simplest case where all the P_k are closed and $m_k = 1$, this means that $(-1)^{\dim(P)}\phi_P$ yields a chain homotopy

$$(9.32)\qquad \phi_{P_0} \simeq (-1)^{|\phi_1|\cdot|\phi_2|}\phi_{P_1}\phi_{P_2}.$$

Comparing this with our previous formal TQFT discussion, (9.30) and (9.31) are a combination of (5.21) and (5.22) (the difference reflects the fact that "gluing" is an asymptotic process in the present context). Of course, much more complicated degenerations, where several ends are being stretched, are routinely used in Floer theory (and we will do the same). The discussion above was intended merely as a reminder, including sign conventions.

10. Conclusion

While Floer cohomology is not strictly speaking an instance of the TQFT formalism from Sections 5–7, translating the formal arguments into Floer-theoretic ones is mostly a straightforward task (compare [**31**], which was structured in the same way). We will give a brief sketch of how this leads to the operations from Propositions 2.1–2.4. After that, we will discuss connections, and how the material from Section 6 gives rise to the proofs of Propositions 2.5 and 2.10.

10.1. Floer cohomology groups associated to Lefschetz fibrations. In Section 9, we explained how, given a connection $a \in \Omega^1(S^1, \mathfrak{g})$ with elliptic or hyperbolic holonomy, one chooses a compatible Floer datum $\Phi = (H, J)$ and obtains a Floer cochain complex $CF^*(E, H)$. We want to make a few additional comments concerning that construction:

- Suppose that a has constant coefficients, which are sufficiently small. Then, one can choose H with the same properties, and a version of the argument from [**12**] yields an isomorphism between the Morse and Floer cochain complexes,

$$(10.1)\qquad CM^*(E, H) \cong CF^*(E, H).$$

- The time variable $t \in S^1$ in the definition of Floer cohomology can be shifted. Given $a = a_t dt$, consider ${}^{\tau}a = a_{t+\tau} dt$, where $\tau \in S^1$ is an arbitrary constant. If $\Phi = (H_t, J_t)$ is a Floer datum for a, ${}^{\tau}\Phi = (H_{t+\tau}, J_{t+\tau})$

is a Floer datum for ${}^{\tau}a$. One-periodic orbits and Floer trajectories can be identified accordingly, leading to an isomorphism of chain complexes

$$CF^*(E, {}^{\tau}H) \cong CF^*(E, H). \tag{10.2}$$

- In a similar spirit, let's consider reversing the orientation of S^1, which in the notation of (4.27) means passing from a to a^{-1}. Given a Floer datum $\Phi = (H_t, J_t)$ for a, consider $\Phi^{-1} = (-H_{-t}, J_{-t})$. The flow of the new Hamiltonian is inverse to the old one, and one can similarly relate Floer trajectories by setting $u^{-1}(s,t) = u(-s,-t)$. Taking signs and grading into account, the outcome is an isomorphism of chain complexes

$$CF^*(E, H^{-1}) \cong CF^{2n-*}(E, H)^{\vee}. \tag{10.3}$$

Let's move on to the maps (9.29) between Floer complexes. Following a standard strategy, one uses them first of all to construct continuation maps which show that, up to canonical isomorphism, the choice of Floer datum doesn't affect Floer cohomology. That involves taking a cylinder $S = \mathbb{R} \times S^1$, with A the pullback of a, and a perturbation datum which agrees with one choice of Floer datum on each end. We refer to [**24**] for the original exposition of this process. Once it has been carried out, one can write the Floer cohomology groups as $HF^*(E, a)$. It is elementary to show that (10.2) and (10.3) are compatible with continuation maps. Therefore, one has canonical isomorphisms

$$HF^*(E, {}^{\tau}a) \cong HF^*(E, a) \quad \text{for any } \tau \in S^1, \tag{10.4}$$

$$HF^*(E, a^{-1}) \cong HF^{2n-*}(E, a)^{\vee}. \tag{10.5}$$

EXAMPLE 10.1. The isomorphism (10.4) is particularly interesting for a connection which is a μ-fold pullback, in the sense of (4.27), for $\mu > 1$. In that case, taking $\tau \in (\mu^{-1}\mathbb{Z})/\mathbb{Z}$ yields ${}^{\tau}a^{\mu} = a^{\mu}$, resulting in a $\mathbb{Z}/\mu$-action on $HF^*(E, a^{\mu})$. Note that even though we are working with a connection which has a $\mathbb{Z}/\mu$-symmetry, the associated Floer datum does not have to obey the same restriction (thereby avoiding equivariant transversality problems). Concretely, the chain maps underlying the action are

$$CF^*(E, H) \xrightarrow[\cong]{(10.2)} CF^*(E, {}^{\tau}H) \xrightarrow[\simeq]{\text{continuation}} CF^*(E, H), \tag{10.6}$$

where ${}^{\tau}H \neq H$ in general; this is not a (chain level) isomorphism, but only a homotopy equivalence. Similarly, the group composition law is satisfied only up to homotopy.

Given $a_0, a_1 \in \Omega^1(S^1, \mathfrak{g})$ with the same rotation number, one can find a non-negatively curved connection $A \in \Omega^1(S, \mathfrak{g})$ which agrees with the pullbacks of a_0 and a_1 on the ends (in the hyperbolic case, this is Proposition 4.17; in the elliptic case, the holonomies are conjugate, and one can choose A to be flat). Via (9.29), that gives rise to a map

$$HF^*(E, a_1) \longrightarrow HF^*(E, a_0). \tag{10.7}$$

To understand these maps, it is convenient to allow a little more freedom in the choice of A. Namely, let's require that

$$A_{s,t} = \begin{cases} a_0 = a_{0,t}\, dt & s \ll 0, \\ {}^{\tau}a_1 = a_{1,t+\tau}\, dt & s \gg 0, \text{ for some } \tau \in S^1, \end{cases} \tag{10.8}$$

and also choose the perturbation datum to have the correspondingly rotated asymptotic behaviour as $s \to +\infty$. In view of (10.4), this still gives rise to maps (10.7). An equivalent formulation would be to say that we still consider S as a surface with tubular ends, but where the parametrization of one end has been modified; with respect to that, it still carries a connection that reduces to a_0 and a_1 over the ends. The implications depend on the kind of connections involved:

- For connections with hyperbolic holonomy and rotation number $r \in \mathbb{Z} \setminus \{0\}$, there is a $\mathbb{Z}/r$ worth of choices of (10.7), by Proposition 5.2. A gluing argument (with geometrically nontrivial content, since the curvature condition means that one can't just reverse orientation on S to construct the connection underlying the inverse map) shows that all such maps are isomorphisms. The outcome is just as in Propositions 5.8 and 5.9. Hence, we may write $HF^*(E, r)$ for the associated Floer cohomology group, which is well-defined up to isomorphism (up to canonical isomorphism if one is willing to additionally fix a choice of one-parameter subgroup, to which the holonomy of all connections will belong), and carries a canonical $\mathbb{Z}/r$-action. If we choose our connection to be an r-fold covering, the $\mathbb{Z}/r$-action is that from Example 10.1 (compare Example 5.4).
- What we've said above also applies to the $r = 0$ case, but there is an additional quirk. The abstract theory (Proposition 5.8) predicts that $HF^*(E, 0)$ has an endomorphism of degree -1, obtained by treating τ as a variable in (10.8), and looking at the corresponding parametrized problem (this is the BV operator in Hamiltonian Floer cohomology). However, by looking at the specific choice of connection and Floer datum giving rise to (10.1), one sees that the BV operator vanishes.
- For connections with elliptic holonomy, each group $HF^*(E, r)$ depends only on $r \in \mathbb{R} \setminus \mathbb{Z}$, up to canonical isomorphism, and carries a BV operator; compare Proposition 7.8(i) and (ii).

REMARK 10.2. The vanishing of the BV operator on $HF^*(E, 0)$, and the duality (10.5), can fail in closely related geometric situations. Namely, suppose that we considered a map $\pi : E \to B$ with noncompact fibres, and which itself has a noncompact critical locus (say, of complex Morse-Bott type). We could use a version of Floer cohomology which "wraps" in fibre direction. In this context, neither property would hold any more, even though the formal TQFT framework remains relevant. Therefore, the failure of the abstract theory to predict these behaviours was to be expected.

Along the same lines, one can consider continuation maps $HF^*(E, r_1) \to HF^*(E, r_0)$ which increase the rotation number, $r_0 > r_1$. After adapting the corresponding TQFT results, we find that such maps are canonically defined in the following instances:

- $r_1 \in \mathbb{Z}$, and $r_0 = r_1 + 1$ (hyperbolic to hyperbolic, meaning the holonomies of the connections concerned); compare Proposition 5.10.
- $r_0, r_1 \in \mathbb{R} \setminus \mathbb{Z}$, and (r_1, r_0) contains no integer (elliptic to elliptic); see Proposition 7.8(iii). In this case, the continuation map turns out to be an isomorphism (as in Remark 10.2, this is not a formal TQFT property, but standard from a geometric viewpoint).

- $r_0, r_1 \in \mathbb{R}\setminus\mathbb{Z}$, and (r_1, r_0) contains exactly one integer (elliptic to elliptic). The corresponding TQFT result is again Proposition 7.8(iii)
- $r_0 \in \mathbb{Z}$, and $r_1 \in (r_0 - 1, r_0)$ (elliptic to hyperbolic), see Proposition 7.9(i).
- $r_1 \in \mathbb{Z}$, and $r_0 \in (r_1, r_1+1)$ (hyperbolic to elliptic), see Proposition 7.9(ii).

Moreover, as long as one stays within this list of cases, the composition of continuation maps is again a continuation map. In particular, since $SH^*(E)$ is defined as a direct limit over $HF^*(E, r)$ for $r \notin \mathbb{Z}$, Proposition 2.7 follows immediately, as in Proposition 7.9(iii). Along the same lines, we get maps (2.9), which factor through the BV operator as in Proposition 7.9(iv).

10.2. The algebraic structure of Floer cohomology. Consider operations induced by surfaces other than the cylinder. The translation from the TQFT framework happens in the same way as before: we replace compact surfaces with boundary (5.1) by their analogues with tubular ends (extending the connections in the obvious way), and choose perturbation data. The naive TQFT gluing operation is replaced by neck-stretching. Then,

- Proposition 5.12 carries over to give $HF^*(E, 1)$ the structure of a Gerstenhaber algebra, proving Proposition 2.1. One can show as in Addendum 5.13 that this is well-defined (compatible with the canonical isomorphisms between different choices made in constructing the Floer cohomology group).
- Proposition 5.15 similarly yields a Gerstenhaber module structure on any $HF^*(E, r)$, which is Proposition 2.2 (and moreover, this module satisfies the additional relation from Proposition 5.17). As before, this structure is compatible with the canonical isomorphisms that relate different constructions of $HF^*(E, 1)$, and also with the corresponding non-canonical isomorphisms for $HF^*(E, r)$.
- We can define all the $HF^*(E, r)$, $r \in \{1, 2, \dots\}$, using covers a^r of the same underlying connection $a^1 = a$ with rotation number 1. Then, by proceeding as in Proposition 5.19, we get the algebra structure from Proposition 2.3. The same observation as in Example 10.1 applies: using a^r is useful because it makes it easier to characterize which connections on surfaces to use (via Addendum 4.13), but the Floer data do not have to be $\mathbb{Z}/r$-invariant (or related to each other for different r).

The remaining item is Proposition 2.4, which deserves a little more discussion. The main point can be summarized as follows: the original TQFT construction, in Proposition 5.24, used (5.20), which was formulated as a strict chain level identity. This is potentially suspicious in a Floer-theoretic context, where everything depends on additional choices; however, we will ultimately only need the cohomology level relation, which is unproblematic. Fix positive integers r_i $(i = 0, 1, 2)$, with $r_0 = r_1 + r_2 - 1$. Choose connections a^{r_i} with rotation number r_i and which are r_i-fold covers (not necessarily of the same connection). Also, choose corresponding Floer data Φ_i. To define

$$(10.9) \qquad [\cdot,\cdot] : HF^*(E, r_1)^{\mathbb{Z}/r_1} \otimes HF^*(E, r_2)^{\mathbb{Z}/r_2} \longrightarrow HF^{*-1}(E, r_0)^{\mathbb{Z}/r_0},$$

we start with a family of surfaces parametrized by $p \in P = S^1$. Each such surface S_p is a pair-of-pants, equipped with tubular ends

$$\begin{cases} \epsilon_{0,p} : (-\infty, 0] \times S^1 \longrightarrow S_p, \\ \epsilon_{1,p}, \epsilon_{2,p} : [0, \infty) \times S^1 \longrightarrow S_p, \end{cases} \tag{10.10}$$

of which those for $i = 1, 2$ are determined only up to rotations $\epsilon_{i,p}(s,t) \mapsto \epsilon_{i,p}(s, t + k_i r_i^{-1})$, $k_i \in \mathbb{Z}/r_i$. The p-dependence follows the model from (5.49). One can arrange that there are isomorphisms

$$\begin{cases} \phi_{0,p} : S_p \longrightarrow S_{p+r_0^{-1}}, \\ \phi_{0,p}\epsilon_{0,p}(s,t) = \epsilon_{0,p+r_0^{-1}}(s, t - r_0^{-1}), \\ \phi_{0,p}\epsilon_{i,p} = \epsilon_{i,p+r_0^{-1}} \quad \text{for } i = 1, 2, \end{cases} \tag{10.11}$$

which generate a $\mathbb{Z}/r_0$-action on the family; compare (5.50). We equip our surfaces with connections A_p satisfying $\epsilon_{i,p}^* A_p = a_i$, and which are compatible with (10.11). Again following (5.49), we consider the $\mathbb{Z}/r_1 \times \mathbb{Z}/r_2$-covering $\tilde{P} \to P$ associated to the element

$$(1 - r_2, 1 - r_1) = (-r_0, -r_0) \in H^1(P; \mathbb{Z}/r_1 \times \mathbb{Z}/r_2) = \mathbb{Z}/r_1 \times \mathbb{Z}/r_2. \tag{10.12}$$

The pullback family of surfaces $S_{\tilde{p}}$ comes with well-defined tubular ends $\epsilon_{i,\tilde{p}}$ for $i = 0, 1, 2$, removing the previous ambiguity. The covering group acts on that family, fractionally rotating the ends. Moreover, the $\mathbb{Z}/r_0$-action from (10.11) also lifts to $\tilde{P}$, resulting in a total action of $\mathbb{Z}/r_0 \times \mathbb{Z}/r_1 \times \mathbb{Z}/r_2$. Because that action rotates the tubular ends, it is not compatible with our given Floer data, which means that there is no meaningful notion of equivariant perturbation datum. Instead, we make an arbitrary choice of perturbation datum, and then divide the resulting operation by $r_1 r_2$; the outcome is invariant under the action of $\mathbb{Z}/r_0 \times \mathbb{Z}/r_1 \times \mathbb{Z}/r_2$ on the source and target Floer cohomology groups, and we restrict to the invariant parts to get (10.9). This follows (5.51), except that we remain on the cohomology level (since the Floer cochain complexes do not carry group actions; see Example 10.1).

Now consider the composition of two such brackets, say

$$\begin{aligned} &HF^*(E, a^{r_1})^{\mathbb{Z}/r_1} \otimes HF^*(E, a^{r_2})^{\mathbb{Z}/r_2} \otimes HF^*(E, a^{r_3})^{\mathbb{Z}/r_3} \\ &\xrightarrow{[\cdot,\cdot]\otimes id} HF^{*-1}(E, a^{r_1+r_2-1})^{\mathbb{Z}/r_1+r_2-1} \otimes HF^*(E, a^{r_3})^{\mathbb{Z}/r_3} \\ &\xrightarrow{[\cdot,\cdot]} HF^{*-2}(E, a^{r_1+r_2+r_3-2})^{\mathbb{Z}/r_1+r_2+r_3-2}. \end{aligned} \tag{10.13}$$

Let's write $\tilde{P}_1$ and $\tilde{P}_2$ for the parameter spaces that occur in the two steps of (10.13). Both spaces carry actions of $\mathbb{Z}/r_0$, where $r_0 = r_1 + r_2 - 1$. To prove the Jacobi identity following the strategy from Proposition 5.24, one has to relate (10.13) to the operation derived from the quotient space $\tilde{P}_1 \times_{\mathbb{Z}/r_0} \tilde{P}_2$. For that, one argues as follows: gluing writes (10.13) as an operation arising from a family of surfaces over $\tilde{P}_1 \times \tilde{P}_2$ (for an arbitrary choice of perturbation data). On that family, the $\mathbb{Z}/r_0$-action no longer affects the tubular ends, and hence, we can choose the perturbation data to be equivariant. In other words, our family descends to $\tilde{P}_1 \times_{\mathbb{Z}/r_0} \tilde{P}_2$, including tubular ends; we can make a generic choice of perturbation data on that quotient space, and then pull that choice back to $\tilde{P}_1 \times \tilde{P}_2$. For those specific choices, we have an analogue of (5.20) on the Floer cochain level. Hence, the corresponding Floer cohomology level identity holds.

10.3. The differentiation axiom revisited. We will now implement differentiation with respect to the Novikov parameter q in Floer theory. This closely follows [**31**, Section 9]. As a preliminary step, it is convenient to modify slightly the definition of the Floer differential and of the operations (9.29); the new version is isomorphic to the original one (admittedly, the isomorphism is canonical only up to powers of q, but that is hopefully a venial sin; in any case, all previous constructions also go through with the new version, without any other modifications). Let's fix an $\mathbb{R}$-cycle of codimension two, which means a formal finite linear combination

$$\Omega = \sum_j m_j \Omega_j, \tag{10.14}$$

where the $\Omega_j \subset E$ are properly embedded codimension two submanifolds, and the multiplicities are $m_j \in \mathbb{R}$. This should be such that the Poincaré dual to Ω is the symplectic class $[\omega_E]$. When carrying out any Floer-theoretic construction, we assume that the one-periodic orbits of the relevant Hamiltonians are disjoint from Ω; and when counting maps $u : S \to E$ satisfying a Cauchy-Riemann equation with suitable asymptotics, we replace the expression $q^{\int_u \omega_E}$ with the intersection number

$$q^{u \cdot \Omega} = q^{\sum_j m_j (u \cdot \Omega_j)}. \tag{10.15}$$

The naive differentiation operator on any Floer cochain group is

$$\begin{aligned} &\nabla\!\!\!/ : CF^*(E,H) \longrightarrow CF^*(E,H), \\ &\nabla\!\!\!/(\textstyle\sum_x c_x x) = \textstyle\sum_x (\partial_q c_x) x, \end{aligned} \tag{10.16}$$

where the coefficients are $c_x \in \mathbb{K}$. Let's assume that our Floer datum $\Phi = (H, J)$ has been chosen in such a way that the isolated solutions u of Floer's equation intersect Ω transversally, and no J_t-holomorphic sphere intersects Ω (these conditions are satisfied generically, thanks to transversality of evaluations). Then, one can write

$$\begin{aligned} &R = \nabla\!\!\!/ d - d \nabla\!\!\!/ : CF^*(E,H) \longrightarrow CF^{*+1}(E,H), \\ &R(x_+) = \sum_{x_-} \Big(\sum_u \pm (u \cdot \Omega) q^{u \cdot \Omega - 1} \Big) x_- \end{aligned} \tag{10.17}$$

as an operation associated to a certain parametrized moduli space. Namely, for each Ω_j, we consider pairs (p, u) consisting of $p \in P = S^1$ and a Floer trajectory u, such that at the point $\zeta_p = (0, -p)$, we have $u(\zeta_p) \in \Omega_j$. We count those solutions with multiplicity $q^{-1} m_j$ times the usual power (10.15) of q, and add up over j; see [**31**, Section 9a]. In future, whenever such a weighted count is used, we will refer to it by the shorthand notation "counting those u such that $u(\zeta_p) \in q^{-1}\Omega$".

For expository reasons, let's briefly consider the Floer-theoretic analogue of (5.27), namely the unit class in $HF^0(E, 1)$. This is constructed as follows. Choose some $a_1 \in \Omega^1(S^1, \mathfrak{g})$ with $\mathrm{rot}(a_1) = 1$. Take $S = \mathbb{C}$, considered as a Riemann surface with an output end. We equip that with a nonnegatively curved connection which reduces to a_1 on the end. Choose a Floer datum $\Phi_1 = (H_1, J_1)$ associated to a_1, and a corresponding perturbation datum. Counting solutions of the resulting equation (9.16) then yields a cocycle representative for the unit in $CF^0(E, H_1)$. Now let's additionally fix a point $\zeta \in S$, and impose the condition

$$u(\zeta) \in q^{-1}\Omega \tag{10.18}$$

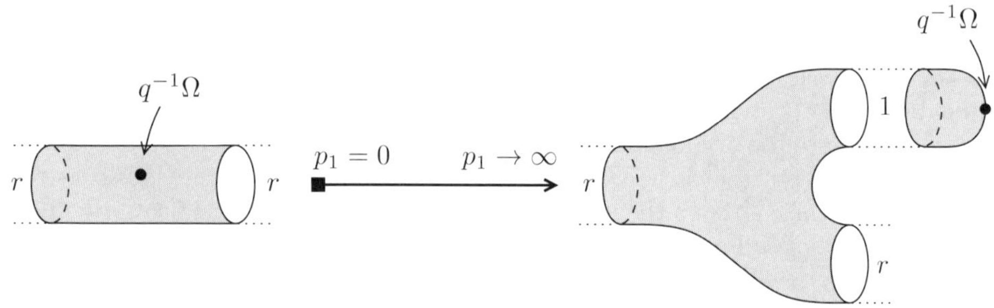

FIGURE 10.1. The geometry behind (10.20), with the rotation numbers of the connections associated to the ends. Note that there is another parameter $p_2 \in S^1$, not shown here.

on the solutions of our equation (9.16). The outcome is the analogue of the Kodaira-Spencer class in $HF^2(E,1)$, represented by a Floer cocycle

$$k \in CF^2(E, H_1). \tag{10.19}$$

In parallel with Lemma 6.1, one has a chain homotopy

$$\begin{aligned} &\rho : CF^*(E, H_2) \longrightarrow CF^*(E, H_2), \\ &d\rho(x) - \rho(dx) - R(x) + [k, x] = 0. \end{aligned} \tag{10.20}$$

The context for this is that one takes a connection a_2 with hyperbolic holonomy and arbitrary rotation number r, as well as an associated Floer datum (H_2, J_2), chosen so that the operator R on $CF^*(E, H_2)$ admits the enumerative interpretation described above. The definition of (10.20) involves a family of Riemann surfaces with marked point (S_p, ζ_p), parametrized by $p = (p_1, p_2) \in P = [0, \infty) \times S^1$, which for $p_1 = 0$ reduce to those defining (10.17). At the other end $p_1 \to \infty$, we "pull out" the marked point so that the surface degenerates into two pieces, one defining the Lie bracket and the other (10.19). The process is shown schematically in Figure 10.1. Of course, our surfaces need to carry connections, and those are constructed exactly as in Lemma 7.10. In particular, this construction constrains how the ends of the limiting pair-of-pants ($p_1 = \infty$) change depending on the remaining parameter $p_2 \in S^1$ (see Lemma 5.14).

The corresponding construction for connections with elliptic holonomy, which is the subject of [**31**], is slightly more complicated. In that case, one can't express R itself as a Lie bracket, but has to combine it with a continuation map, so as to obtain the analogue of (7.19):

$$\begin{aligned} &\rho_c : CF^*(E, H_2) \longrightarrow CF^*(E, H_0), \\ &d\rho_c(x) - \rho_c(dx) - C(R(x)) + [k, x]_c = 0. \end{aligned} \tag{10.21}$$

The context for this equation is as follows. Throughout, we work with connections of the form (7.11) and (7.12). We fix numbers r_i as in Lemma 7.10, and Floer data (H_i, J_i) that are compatible with the connections $r_i \alpha\, dt$. This means that on a neighbourhood of infinity, the Hamiltonian vector field of H_i is (time-independent, and) the horizontal lift of the infinitesimal rotation of B with speed r_i (this is a version of the standard setup for symplectic cohomology). The family of surfaces which defines (10.21) is parametrized by $p = (p_1, p_2) \in P = \mathbb{R} \times S^1$, and splits as

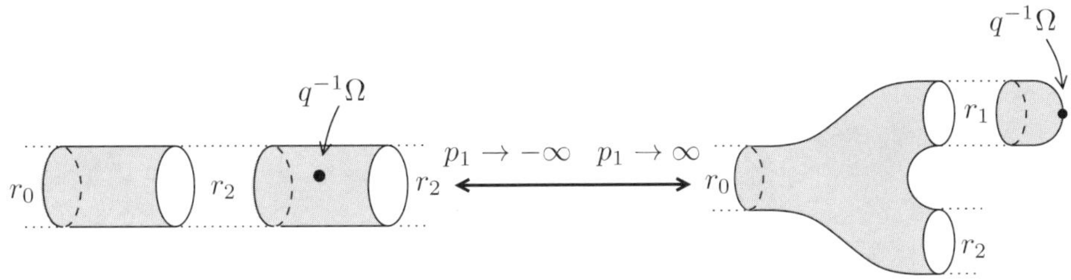

FIGURE 10.2. The geometric construction of (10.21).

we approach either end $p_1 \to \pm\infty$ (see Figure 10.2). The parameter $c \in \mathbb{Z}$ affects only the geometry of the limit $p_1 \to +\infty$. In that limit, the cylinder splits into a pair-of-pants and a copy of $\mathbb{C}$. The copy of $\mathbb{C}$ always carries the data defining a version of the Kodaira-Spencer cocycle, lying in $CF^2(E, H_1)$, which is independent of the remaining parameter p_2. In contrast, the pair-of-pants does depend on p_2 and on c (taking the tubular ends into account). It is convenient to write it as

$$S_{p_2} = (\mathbb{R} \times S^1) \setminus \{(0, -p_2)\}, \tag{10.22}$$

with coordinates (s, t). Then, the two ends asymptotic to $s = \pm\infty$ can be taken to be standard, hence independent of p_2; while the parametrization of the end asymptotic to $(0, p_2)$ rotates $(c + 1)$ times as p_2 goes around the circle.

In the proof of Proposition 2.10 (modelled on that of Proposition 7.13), a variant $\rho_{broken,c}$ of (10.21) occurs, in which the continuation map is replaced by a composition of two such maps. As far as the Floer-theoretic construction is concerned, the only difference is that in the limit $p_1 \to -\infty$, we have a "double stretching" of two necks as the same time (the relative speed at which this happens is irrelevant, as long as both necks stretch unboundedly). The comparison between the two constructions is given by a higher homotopy, which is the counterpart of (7.30). The three-dimensional parameter space involved is drawn schematically in Figure 10.3. Topologically, this space can be thought of as a triangle, with a vertex and the opposite side removed, times S^1. There are two codimension 1 boundary faces (along which the surfaces do not degenerate), plus one more such face at infinity, which would appear when compactifying it (and along which a standard stretching process happens). We want to pay particular attention to what happens at the vertices of our triangle:

- Near the vertex on the right (as drawn in Figure 10.3), the parameter space has local coordinates $(p_1, p_2, p_3) \in \mathbb{R} \times [0, 1] \times S^1$, where $p_1 \gg 0$. We can choose the Riemann surfaces and all their auxiliary data to be independent of p_2, while preserving regularity of the parametrized moduli space. In the region where that holds, no isolated points of that moduli space can occur; and any one-dimensional components that occur are of the form $\{p_1\} \times [0, 1] \times \{p_3\}$.
- Near the top left vertex, we have local coordinates $(p_1, p_2, p_3) \in \mathbb{R} \times (-\infty, 0] \times S^1$, where $p_1 \ll 0$ and p_2 is small. Here, $p_2 = 0$ is the adjacent boundary face, and $p_1 = -\infty$ the "face at infinity". We again choose all data to be p_2-independent, with similar consequences as before.

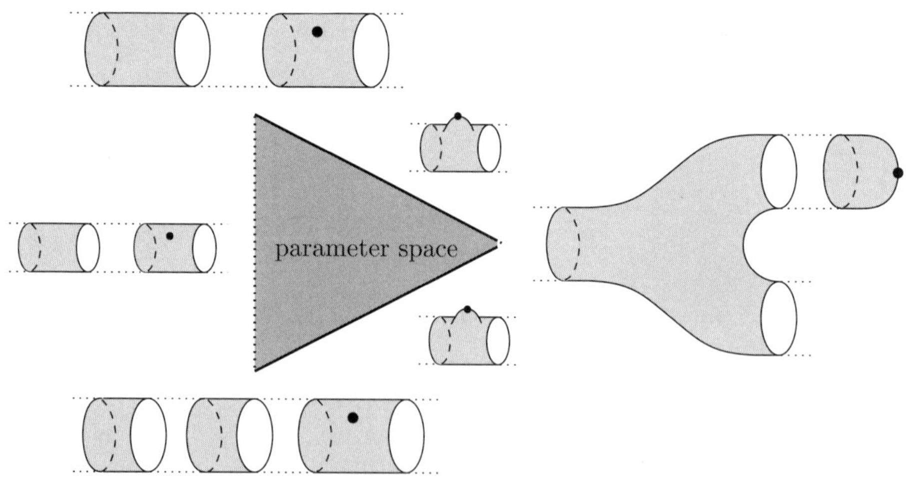

FIGURE 10.3. Relating ρ_c and $\rho_{broken,c}$. As in Figure 10.1, there is an additional S^1 degree of freedom, which we have not represented.

- Near the bottom left vertex, we have local coordinates $(p_1, p_2, p_3) \in \mathbb{R} \times [0, \infty) \times S^1$, where $p_1 \ll 0$ and p_2 is small. Here, we have two "gluing parameters" or stretching lengths. Along the "face at infinity" $p_1 = -\infty$, the p_2 parameter governs the gluing of the left two of the three cylinders (the last cylinder remains the same). Along the boundary face $p_2 = 0$, we carry out both gluing processes simultaneously (with some relation between the gluing parameters). This is, roughly speaking, half of the more standard process near a codimension two corner (since we do not consider a region where only the right two cylinders are glued together).

For the other parameter spaces appearing in the proof of Proposition 2.10 (in particular that from Figure 7.2), similar considerations apply when translating them into Floer-theoretic moduli spaces.

Acknowledgments

I would like to thank Mohammed Abouzaid and Sheel Ganatra for generously sharing their seminal insights; Nick Sheridan for illuminating conversations on further developments; and Alexander Sukhov for patiently answering my questions concerning pseudoconvexity (see Remark 8.9).

References

[1] William Abikoff, *The real analytic theory of Teichmüller space*, Lecture Notes in Mathematics, vol. 820, Springer, Berlin, 1980. MR590044

[2] J. Barge and É. Ghys, *Cocycles d'Euler et de Maslov* (French), Math. Ann. **294** (1992), no. 2, 235–265, DOI 10.1007/BF01934324. MR1183404

[3] Gabi Ben Simon and Dietmar A. Salamon, *Homogeneous quasimorphisms on the symplectic linear group*, Israel J. Math. **175** (2010), 221–224, DOI 10.1007/s11856-010-0010-4. MR2607545

[4] Klas Diederich and Alexandre Sukhov, *Plurisubharmonic exhaustion functions and almost complex Stein structures*, Michigan Math. J. **56** (2008), no. 2, 331–355, DOI 10.1307/mmj/1224783517. MR2492398

[5] Michael Entov, *Commutator length of symplectomorphisms*, Comment. Math. Helv. **79** (2004), no. 1, 58–104, DOI 10.1007/s00014-001-0799-0. MR2031300
[6] Andreas Floer, Helmut Hofer, and Dietmar Salamon, *Transversality in elliptic Morse theory for the symplectic action*, Duke Math. J. **80** (1995), no. 1, 251–292, DOI 10.1215/S0012-7094-95-08010-7. MR1360618
[7] Hervé Gaussier and Alexandre Sukhov, *Levi-flat filling of real two-spheres in symplectic manifolds (I)* (English, with English and French summaries), Ann. Fac. Sci. Toulouse Math. (6) **20** (2011), no. 3, 515–539. MR2894837
[8] E. Getzler, *Batalin-Vilkovisky algebras and two-dimensional topological field theories*, Comm. Math. Phys. **159** (1994), no. 2, 265–285. MR1256989
[9] Étienne Ghys, *Groups acting on the circle*, Enseign. Math. (2) **47** (2001), no. 3-4, 329–407. MR1876932
[10] William M. Goldman, *Topological components of spaces of representations*, Invent. Math. **93** (1988), no. 3, 557–607, DOI 10.1007/BF01410200. MR952283
[11] N. J. Hitchin, *The self-duality equations on a Riemann surface*, Proc. London Math. Soc. (3) **55** (1987), no. 1, 59–126, DOI 10.1112/plms/s3-55.1.59. MR887284
[12] H. Hofer and D. A. Salamon, *Floer homology and Novikov rings*, The Floer memorial volume, Progr. Math., vol. 133, Birkhäuser, Basel, 1995, pp. 483–524. MR1362838
[13] Shoshichi Kobayashi, *Hyperbolic manifolds and holomorphic mappings*, 2nd ed., World Scientific Publishing Co. Pte. Ltd., Hackensack, NJ, 2005. An introduction. MR2194466
[14] Niels Kowalzig and Ulrich Krähmer, *Batalin-Vilkovisky structures on Ext and Tor*, J. Reine Angew. Math. **697** (2014), 159–219, DOI 10.1515/crelle-2012-0086. MR3281654
[15] François Lalonde and Dusa McDuff, *Positive paths in the linear symplectic group*, The Arnold-Gelfand mathematical seminars, Birkhäuser Boston, Boston, MA, 1997, pp. 361–387. MR1429901
[16] François Lalonde and Andrei Teleman, *The g-areas and the commutator length*, Internat. J. Math. **24** (2013), no. 7, 1350057, 13, DOI 10.1142/S0129167X13500572. MR3084738
[17] Feng Luo, *On Teichmüller spaces of surfaces with boundary*, Duke Math. J. **139** (2007), no. 3, 463–482, DOI 10.1215/S0012-7094-07-13932-2. MR2350850
[18] Dusa McDuff and Dietmar Salamon, *Introduction to symplectic topology*, 3rd ed., Oxford Graduate Texts in Mathematics, Oxford University Press, Oxford, 2017. MR3674984
[19] Dusa McDuff and Dietmar Salamon, *J-holomorphic curves and symplectic topology*, American Mathematical Society Colloquium Publications, vol. 52, American Mathematical Society, Providence, RI, 2004. MR2045629
[20] Mark McLean, *Symplectic homology of Lefschetz fibrations and Floer homology of the monodromy map*, Selecta Math. (N.S.) **18** (2012), no. 3, 473–512, DOI 10.1007/s00029-011-0079-6. MR2960024
[21] John Milnor, *On the existence of a connection with curvature zero*, Comment. Math. Helv. **32** (1958), 215–223, DOI 10.1007/BF02564579. MR0095518
[22] Gabriele Mondello, *Topology of representation spaces of surface groups in* $\mathrm{PSL}_2(\mathbb{R})$ *with assigned boundary monodromy and nonzero Euler number*, Pure Appl. Math. Q. **12** (2016), no. 3, 399–462, DOI 10.4310/PAMQ.2016.v12.n3.a3. MR3767231
[23] Alexander F. Ritter, *Topological quantum field theory structure on symplectic cohomology*, J. Topol. **6** (2013), no. 2, 391–489, DOI 10.1112/jtopol/jts038. MR3065181
[24] Dietmar Salamon and Eduard Zehnder, *Morse theory for periodic solutions of Hamiltonian systems and the Maslov index*, Comm. Pure Appl. Math. **45** (1992), no. 10, 1303–1360, DOI 10.1002/cpa.3160451004. MR1181727
[25] Paolo Salvatore and Nathalie Wahl, *Framed discs operads and Batalin-Vilkovisky algebras*, Q. J. Math. **54** (2003), no. 2, 213–231, DOI 10.1093/qjmath/54.2.213. MR1989873
[26] M. Schwarz. *Cohomology operations from* S^1*-cobordisms in Floer homology.* PhD thesis, ETH Zürich, 1995.
[27] Paul Seidel, *A biased view of symplectic cohomology*, Current developments in mathematics, 2006, Int. Press, Somerville, MA, 2008, pp. 211–253. MR2459307
[28] Paul Seidel, *Fukaya* A_∞*-structures associated to Lefschetz fibrations. I*, J. Symplectic Geom. **10** (2012), no. 3, 325–388. MR2983434
[29] Paul Seidel, *Fukaya* A_∞*-structures associated to Lefschetz fibrations. II*, Algebra, geometry, and physics in the 21st century, Progr. Math., vol. 324, Birkhäuser/Springer, Cham, 2017, pp. 295–364. MR3727564

[30] Paul Seidel, *Fukaya A_∞-structures associated to Lefschetz fibrations. II 1/2*, Adv. Theor. Math. Phys. **20** (2016), no. 4, 883–944, DOI 10.4310/ATMP.2016.v20.n4.a5. MR3573028

[31] Paul Seidel, *Fukaya A_∞-structures associated to Lefschetz fibrations. II*, Algebra, geometry, and physics in the 21st century, Progr. Math., vol. 324, Birkhäuser/Springer, Cham, 2017, pp. 295–364. MR3727564

[32] Jennifer Slimowitz, *The positive fundamental group of* Sp(2) *and* Sp(4), Topology Appl. **109** (2001), no. 2, 211–235, DOI 10.1016/S0166-8641(99)00162-5. MR1806336

[33] Zachary Aaron Sylvan, *On partially wrapped Fukaya categories*, ProQuest LLC, Ann Arbor, MI, 2015. Thesis (Ph.D.)–University of California, Berkeley. MR3427304

[34] John W. Wood, *Bundles with totally disconnected structure group*, Comment. Math. Helv. **46** (1971), 257–273, DOI 10.1007/BF02566843. MR0293655

[35] Dongping Zhuang, *Irrational stable commutator length in finitely presented groups*, J. Mod. Dyn. **2** (2008), no. 3, 499–507, DOI 10.3934/jmd.2008.2.499. MR2417483

[36] A. Zinger. Notes on J-holomorphic maps. Preprint ArXiv:1706.00331, 2017.

Massachusetts Institute of Technology, MIT 2-276, 77 Massachusetts Ave, Cambridge, Massachusetts 02139

Email address: pseidel@mit.edu

Proceedings of Symposia in Pure Mathematics
Volume **102**, 2019
https://doi.org/10.1090/pspum/102/01824

2017 GITC Problem Sessions

This is a list of problems asked during the two problem sessions held at the 2017 Georgia International Topology Conference. The name in the parenthesis belongs to the person that brought up the question during the session, not necessarily the individual who first formulated the question. We do not have a record of who posed some of the questions. The sessions were moderated by John Etnyre and Rob Kirby and the list is compiled from notes by Mike Usher, Jean Gutt and Jeremy Van Horn-Morris and slightly edited by David Gay and Gordana Matic. We thank the question authors that replied to our request for proofreading and clarification or comments. The questions are presented in the order in which they were brought up.

QUESTION 1. *(Gay) Does there exist a $(g,0)$-trisection, for some $g < 22$, of a four-manifold that is not the connected sum of $\mathbb{C}P^2, \overline{\mathbb{C}P}^2, S^\times S^2$? Due to a classification result of Meier and Zupan, we know that the answer is no for $g < 3$. Recently Lambert-Cole and Meier found a $(22,0)$ trisection of the $K3$ surface. Thus the interesting range is $3 \leq g \leq 21$.*

Dave Auckly prefaces the following question with this discussion: One may associate a 3-manifold invariant to any suitable quantum group. Associated to $U_q(\mathfrak{sl}_N)$ one has the invariants $Z_{Y^3}(k,N)$ where N is the rank and k is the level. Is it possible to encode these invariants in an easier way?

The so-called gauge-string duality from mathematical physics provides some hope that this may be possible. In particular it relates the quantum invariants of the 3-sphere to the Gromov-Witten invariants of the resolved conifold. The Gromov-Witten invariants are a set of rational numbers with one associated to each possible degree and genus. The Gromov-Witten invariants, in turn are related to the Donaldson-Thomas invariants which are a set of integers. Finally, the Donaldson-Thomas invariants may be described via the so-called BPS states. Given that there is only one example, it is too much to call this a conjecture, but:

QUESTION 2. *(Auckly) Wild Speculation: There is an invariant $Z_{Y^3}(q,a)$ with properties similar to the following:*

- *Evaluating at $q = e^{2\pi i/(N+k)}$, $a = q^N$ yields Witten-Reshetikhin-Turaev invariant of rank N and level k;*
- *It can be expressed as $1 + M(q)\sum_d \sum_{n=0}^{\infty} D_{n,d} q^n a^d$ where $D_{n,d} \in \mathbb{Z}$;*

- *With $q = e^{iu}$, as $u \to 0$ the invariant takes the form*

$$1 + \sum_d \sum_g N_{g,d} u^{2g-2} q^d$$
$$= \sum_{g=0}^{\infty} \sum_{k=1}^{\infty} \sum_d n_{g,d} k^{-1} (-1)^{g-1} \left((-q)^k - 2 + (-q)^{-k} \right)^{g-1}$$

where $n_{g,d} \in \mathbb{Z}$ and $n_{g,d} = 0$.

The $N_{g,d}$ correspond to the Gromov-Witten invariants and the $D_{n,d}$ correspond to the Donaldson-Thomas invariants. The $n_{g,d}$ are the BPS states.

Remark: Formulae of the above form relating these three types of invariants are known to hold in the context of Gromov-Witten theory. The heuristic connection between Chern-Simons theory and Gromov-Witten theory leads one to ask how far the similar structure will go on the Chern-Simons side. Auckly-Koshkin [AK] contains exposition about the work on these ideas in the case of S^3. For S^3 this speculation is essentially true with $n_{0,1} = 1$ and the rest of the BPS invariants equal to zero.

Rob Kirby drew particular attention to two old problems from his problem list:

QUESTION 3. *(Kirby, due to Fox, problem 1.33 in Kirby's problem list.) Is every slice knot a ribbon knot?*

QUESTION 4. *(Kirby, problem 4.97 in Kirby's problem list.) Let X be an almost complex smooth four-manifold. Then gauge theory can be used to show that X isn't a connected sum. Does every irreducible, smooth, closed, simply connected 4–manifold other than S^4 have an almost complex structure?*

QUESTION 5. *(Eliashberg) What do canonical symplectic constructions remember about the smooth topology of the manifold?*

Remarks: To a smooth manifold M, one can associate a canonical symplectic manifold, the cotangent bundle of M, $(T^\star M, d\lambda)$, and a contact manifold, the unit cotangent bundle, $(ST^\star M, \ker(\lambda|_{ST^\star M}))$. What does the contact topology of $ST^* M$ know about the smooth topology of M?

Can we see nontrivial four-dimensional gauge-theoretic invariants via $T^* M$ or $ST^* M$?

Take a four-manifold X, maybe a potential counterexample to the smooth Poincare conjecture in dimension 4. Remove a ball, bounded by a standard sphere to obtain X_0. The cotangent bundle of the ball is flexible, but the boundary of the ball gives a Legendrian submanifold which might be a nontrivial knot. From studying Legendrians on S^3 viewed inside ST_0^X, can one see exotic structures on X?

QUESTION 6. *(Eliashberg)*

Inverting the usual fillability relationship questions, what can be said about symplectic manifolds as boundaries of contact manifolds? The right condition is that the boundary be convex, so it has a dividing set dividing the boundary into two symplectic manifolds. We can abstractly consider pairs of Liouville/Weinstein domain with same contact type boundary. When is such a pair fillable by a tight contact manifold and what are the tight fillings?

Remark: Note that one can always get an overtwisted filling.

QUESTION 7. *(Rasmussen)*
L-spaces are manifolds with simplest-possible Heegaard Floer homology. The L-space conjecture [BGW] states that for a prime oriented manifold M^3 the following are equivalent:
1. M is an L-space
2. $\pi_1(M)$ is not left-orderable
3. M has no co-oriented taut foliation.

Remark: First of all note that this conjecture is only interesting for rational homology spheres, and that for an L-space Y, $dim\widehat{HF}(Y) = |H_1(Y,\mathbb{Z})|$. The conjecture connects different structures arising in low-dimensional topology in a somewhat surprising way, and there is really no apparent reason why it should be true, but it has resisted much effort to disprove it. The only implication known so far is $1 \Rightarrow 3$. i.e. that L-space carries no co-oriented taut foliation. There has been much work by many authors on proving the conjecture for the case of graph manifolds.

A possible route for disproving the conjecture is: Take a three-manifold with torus boundary, which has more than one L-space filling. Then one knows exactly which fillings give L-spaces and should check them for a counterexample. The Snappea census takes many of these fillings and hasn't turned up any counterexamples, but the work so far hasn't gone through them systematically. In particular one could look at the boundary of the interval of slopes giving rise to an L-space.

QUESTION 8. *Given an irreducible closed rational homology sphere with an incompressible torus, is there a taut foliation on it? It would be nice to use classical methods.*

QUESTION 9. *(Lidman) If there exists a degree-one map $Y \to Z$ between rational homology spheres, is* $\dim \widehat{HF}(Y) \geq \dim \widehat{HF}(Z)$*? This is related to the L-space conjecture because non-left-orderability is appropriately preserved under degree one maps between prime three-manifolds. For example this would imply, without geometrization or gauge theory, that homotopy spheres are L-spaces and property P.*

For some partial results along these lines, see [HRW]. For similar results in the case of maps of higher degree, see [LT] and [LM]. It is open if non-zero degree maps more generally preserve the property of being an L-space.

QUESTION 10. *(Mann) Take a closed surface (other than S^2) and form a 3-manifold $\Sigma \times S^1$. Is the space of foliations transverse to S^1 connected? This is a good question in both $C^{0,1}$ and C^∞ categories.*

Remark: The only known case seems to be $C^{0,1}$ on $T^2 \times S^1$, but maybe there are high genus counterexamples.

QUESTION 11. *(Coskunuzer) By using Heegard splittings, or Heegard-Floer theory in general, can we determine a given manifold admits a hyperbolic structure? In other words, can we detect hyperbolic 3-manifolds by using these tools? For a related question, see Hossein Namazi's thesis [Nam].*

Especially after the proof of Virtual Fibering Conjecture, we know that every closed hyperbolic 3-manifold has a finite cover which fibers over the circle. In particular, these manifolds are mapping tori for pseudo-Anosov maps on closed surfaces of genus¿1. With this new input, is it possible to use Heegaard-Floer techniques or Heegaard splittings in general to detect hyperbolic 3-manifolds?

QUESTION 12. *(Affieri) How do the subgroups of the concordance group generated by the standard kinds of examples like L-space knots, algebraic knots, torus knots, intersect with the other kinds of examples like alternating (or quasi-alternating) knots. Is the subgroup generated by quasi-alternating knots modulo alternating knots infinitely-generated?*

QUESTION 13. *(Coskunuzer) Rephrasing a very-well known old question in contact topology: Is there any hyperbolic 3-manifold with no tight contact structure? We know the existence of hyperbolic 3-manifolds without taut foliations. In these manifolds, Tolga Etgu constructed tight contact structures [E].*

Again, with the help of the Virtual Fibering Theorem, is it possible to show every hyperbolic 3-manifold admits a tight contact structure? If not, is it possible to construct a counterexample?

Remark: John Etnyre mentioned that this problem was also discussed in the 2001 Georgia Topology Conference Problem Session.

QUESTION 14. *(Casals) Can we build a Liouville structure on a smooth four-dimensional three-handle, or more generally on a 2n-dimensional $(n+1)$-handle? This Liouville structure should have a concave and a convex end, and endow the smooth cobordism given by the three-handle attachment with the structure of an oriented Liouville cobordism, from the initial contact manifold to the resulting surgered manifold.*

Remark: McDuff gave an example of a Liouville four-manifold whose topology requires three-handles [McD], so one could try isolating the three-handle with a Morse function.

QUESTION 15. *Is the Torelli group finitely presented?*

Remark: Johnson proved that it is finitely generated for $g \geq 3$ and McCullough and Miller proved it is not finitely generated for g=2. It remains open whether the group is finitely presented for $g \geq 3$.

QUESTION 16. *(Gay) Is there a three-component framed link in S^3, with surgery giving S^3, which is not handleslide equivalent to a split link?*

Remark: An affirmative answer to the question about existence of a $(3,0)$-trisection of a four-manifold that is not the connected sum of $\mathbb{C}P^2, \overline{\mathbb{C}P}^2, S^\times S^2$ (first question in this problem list), would give an affirmative answer to this question.

QUESTION 17. *(This is a well known conjecture, brought up by Hutchings) Does every star-shaped region in $\mathbb{R}^{2n}$ have at least n Reeb orbits on its boundary. (This is known for $n = 2$.) More generally, what can be said about the number of periodic orbits.*

QUESTION 18. *(Hutchings) What can we say about the weak version of Viterbo's conjecture asserting that for every compact convex domain $\Omega \subset R^{2n}$ with smooth boundary, there is a Reeb orbit γ on the boundary whose symplectic action $\mathcal{A}(\gamma)$ satisfies $\mathcal{A}(\gamma)^n \leq n vol(\Omega)$. (This is false for star-shaped domains.)*

QUESTION 19. *(Chaidez) Is there an analogue of the Atiyah-Floer conjecture for Kronheimer-Mrowka's instanton homology for webs $J^{\#}$ [KM]? Could this help establish nonvanishing? What would replace the Heegaard splittings in the usual Atiyah-Floer conjecture?*

QUESTION 20. *Does π_1 of every three-manifold other than S^3 have a nontrivial $SU(2)$-representation?*

Remark: By the Kronheimer-Mrowka proof of Property P this is true for surgeries on knots.

QUESTION 21. *(Hall, attributed to Agol)*

(i) *"Virtual four-coloring for surfaces": If G is a graph embedded in a surface Σ, is there a finite-cover of Σ in which the lifted graph is four-colorable? It can be shown that this holds for $\Sigma = T^2$.*

(ii) *Given a triangulated graph in an oriented surface Σ and a four-coloring, use this to construct a map to the oriented four-colored tetrahedral graph on S^2, thus assigning degree ± 1 to each triangle in Σ. The sum of the ± 1's around each vertex will be zero mod 3. If $\Sigma = S^2$ this is enough to reconstruct the original four-coloring, but that is not true for other Σ.*

Conjecture: ± 1's can be assigned for a triangulated graph on any surface in such a way that the sum around any vertex is 0 mod 3, which would yield a 24-fold covering in which the graph is four-colorable (since $24 = |S_4|$). If this is true, it would imply (i).

QUESTION 22. *(Casals) Is there a way of characterizing three-manifolds with arboreal singularities [Na] such that $H_*(X; \mathbb{Z}_3) \neq 0$? These are singular complexes whose singularities are allowed to look like (trivalent vertex) times $\mathbb{R}^2$ or (cone over tetrahedron) times $\mathbb{R}$. The Legendrian surfaces coming from planar graphs [CM, TZ] have Lagrangian fillings with these types of singularities, and the number of $\mathbb{Z}_3$-augmentations coincides with the number of 4-colorings of the planar graph. In particular, if one can build fillings which have a non-vanishing $\mathbb{Z}_3$-homology class, this would imply the existence of a $\mathbb{Z}_3$ -augmentation and thus the four color theorem.*

QUESTION 23. *(Usher) Are $S^1 \times S^1$ and the Chekanov torus the only monotone Lagrangian tori in $\mathbb{R}^4$ up to Hamiltonian isotopy and scaling? In $\mathbb{C}P^2$ Vianna gives infinitely many examples, and Auroux gives infinitely many on $\mathbb{R}^6$*

QUESTION 24. *General question: understand the structure of the three-dimensional homology cobordism group $\Theta^3_{\mathbb{Z}}$ (three-dimensional integral homology three-spheres modulo homology cobordism, with connected sum as group operation).*

(i) *Is there any torsion?*

(ii) *Perhaps easier: does Rohlin invariant one imply infinite order (Manolescu proved it implies order greater than two which disproved the triangulation conjecture for high dimensional manifolds). If the answer is yes then the obstruction to triangulating higher-dimensional manifolds lives in $H^5(-; \mathbb{Z})$.*

(iii) *Does $\Theta^3_{\mathbb{Z}}$ contain $\mathbb{Z} \oplus \mathbb{Z}$, or perhaps even $\mathbb{Z}^\infty$, summands. It's known that it has a $\mathbb{Z}$ summand by Froyshov.*

QUESTION 25. *(Casals) Find an explicit contact open book for T^{2n+1} $(n \geq 2)$ whose page has the homotopy type of an n-dimensional complex. By using S. Donaldon's asymptotically holomorphic techniques, E. Giroux proves ([Gi]) that such an open book exists. The monodromy must be a compactly supported symplectomorphism of a Weinstein domain which cannot be written as the product of Dehn twists, of which we have not very many concrete examples. In general, it is an interesting question to find a compactly supported symplectomorphism of a Weinstein domain which is not symplectically isotopic to a Dehn twist, a fibered Dehn twist or a product of them.*

References

[AK] Dave Auckly and Sergiy Koshkin, *Introduction to the Gopakumar-Vafa large N duality*, The interaction of finite-type and Gromov-Witten invariants (BIRS 2003), Geom. Topol. Monogr., vol. 8, Geom. Topol. Publ., Coventry, 2006, pp. 195–456. MR2404850

[CM] Roger Casals and Emmy Murphy, *Differential algebra of cubic planar graphs*, Adv. Math. **338** (2018), 401–446, DOI 10.1016/j.aim.2018.09.002. MR3861709

[BGW] Steven Boyer, Cameron McA. Gordon, and Liam Watson, *On L-spaces and left-orderable fundamental groups*, Math. Ann. **356** (2013), no. 4, 1213–1245, DOI 10.1007/s00208-012-0852-7. MR3072799

[E] Tolga Etgü, *Tight contact structures on laminar free hyperbolic three-manifolds*, Int. Math. Res. Not. IMRN **20** (2012), 4775–4784, DOI 10.1093/imrn/rnr198. MR2989619

[Gi] Emmanuel Giroux, *Géométrie de contact: de la dimension trois vers les dimensions supérieures* (French, with French summary), Proceedings of the International Congress of Mathematicians, Vol. II (Beijing, 2002), Higher Ed. Press, Beijing, 2002, pp. 405–414. MR1957051

[HRW] Hanselman, Jonathan, and Rasmussen, Jacob, and Watson, Liam. "Bordered Floer homology for manifolds with torus boundary via immersed curves." arXiv:1604.03466 [math.GT]

[KM] P. B. Kronheimer and T. S. Mrowka, *Tait colorings, and an instanton homology for webs and foams*, J. Eur. Math. Soc. (JEMS) **21** (2019), no. 1, 55–119, DOI 10.4171/JEMS/831. MR3880205

[LM] Tye Lidman and Ciprian Manolescu, *Floer homology and covering spaces*, Geom. Topol. **22** (2018), no. 5, 2817–2838, DOI 10.2140/gt.2018.22.2817. MR3811772

[LT] Robert Lipshitz and David Treumann, *Noncommutative Hodge-to-de Rham spectral sequence and the Heegaard Floer homology of double covers*, J. Eur. Math. Soc. (JEMS) **18** (2016), no. 2, 281–325, DOI 10.4171/JEMS/590. MR3459952

[McD] Dusa McDuff, *Symplectic manifolds with contact type boundaries*, Invent. Math. **103** (1991), no. 3, 651–671, DOI 10.1007/BF01239530. MR1091622

[Na] David Nadler, *Arboreal singularities*, Geom. Topol. **21** (2017), no. 2, 1231–1274, DOI 10.2140/gt.2017.21.1231. MR3626601

[Nam] Hossein Namazi, *Heegaard splittings and hyperbolic geometry*, ProQuest LLC, Ann Arbor, MI, 2005. Thesis (Ph.D.)–State University of New York at Stony Brook. MR2707466

[TZ] Treumann, David and Zaslow, Eric. "Cubic Planar Graphs and Legendrian Surface Theory." arXiv:1609.04892 [math.SG].

Published Titles in This Series

ISBN: 978-1-4704-4249-1

9 781470 442491